CALIFORNIA CONSTRUCTION LAW

CALIFORNIA CONSTRUCTION LAW
SIXTEENTH EDITION

KENNETH C. GIBBS
GORDON HUNT

PUBLISHERS

111 Eighth Avenue, New York, NY 10011
www.aspenpublishers.com

This publication is designed to provide accurate and authoritative information in regard to the subject matter covered. It is sold with the understanding that the publisher is not engaged in rendering legal, accounting, or other professional services. If legal advice or other professional assistance is required, the services of a competent professional person should be sought.

—From a *Declaration of Principles* jointly adopted by a Committee of the American Bar Association and a Committee of Publishers and Associations

© 2000 by Aspen Publishers, Inc.
a Wolters Kluwer business
www.aspenpublishers.com

All rights reserved. No part of this publication may be reproduced or transmitted in any form or by any means, electronic or mechanical, including photocopy, recording, or any information storage and retrieval system, without permission in writing from the publisher. Requests for permission to reproduce content should be directed to the Aspen Publishers website at *www.aspenpublishers.com*, or a letter of intent should be faxed to the permissions department at 212-771-0803.

Aspen Publishers, Inc.
Attn: Permissions Department
111 Eighth Avenue, 7th Floor
New York, NY 10011-5201

ISBN 0-7355-0545-4

6 7 8 9 0

Printed in the United States of America

About Aspen Publishers

Aspen Publishers, headquartered in New York City, is a leading information provider for attorneys, business professionals, and law students. Written by preeminent authorities, our products consist of analytical and practical information covering both U.S. and international topics. We publish in the full range of formats, including updated manuals, books, periodicals, CDs, and online products.

Our proprietary content is complemented by 2,500 legal databases, containing over 11 million documents, available through our Loislaw division. Aspen Publishers also offers a wide range of topical legal and business databases linked to Loislaw's primary material. Our mission is to provide accurate, timely, and authoritative content in easily accessible formats, supported by unmatched customer care.

To order any Aspen Publishers title, go to *www.aspenpublishers.com* or call 1-800-638-8437.

To reinstate your manual update service, call 1-800-638-8437.

For more information on Loislaw products, go to *www.loislaw.com* or call 1-800-364-2512.

For Customer Care issues, e-mail CustomerCare@aspenpublishers.com; call 1-800-234-1660; or fax 1-800-901-9075.

Aspen Publishers
a Wolters Kluwer business

SUBSCRIPTION NOTICE

This Aspen Publishers product is updated on a periodic basis with supplements to reflect important changes in the subject matter. If you purchased this product directly from Aspen Publishers, we have already recorded your subscription for the update service.

If, however, you purchased this product from a bookstore and wish to receive future updates and revised or related volumes billed separately with a 30-day examination review, please contact our Customer Service Department at 1-800-234-1660 or send your name, company name (if applicable), address, and the title of the product to:

**ASPEN PUBLISHERS
7201 McKinney Circle
Frederick, MD 21704**

ABOUT THE AUTHORS

Kenneth C. Gibbs is a senior partner in the Los Angeles law firm of Gibbs, Giden, Locher & Turner LLP, specializing in construction law and practicing exclusively in the representation of owners, contractors, subcontractors, material suppliers, and design professionals in construction industry matters and disputes. He obtained his undergraduate degree from the University of California at Berkeley and his J.D. from the University of California at Los Angeles. Mr. Gibbs has written many books and articles on construction industry–related topics and has been a guest lecturer at the University of California, Berkeley on construction contracts. He has also conducted many seminars for construction trade groups and professional associations nationwide and has been a principal speaker on several occasions at the Construction Law Superconference. Mr. Gibbs has also lectured for many years for the California Continuing Education of the Bar on construction law–related topics. Mr. Gibbs is a member of the Public Contract Law Section and the Forum on the Construction Industry of the American Bar Association. He is also a member of the Large and Complex Case Panel of the American Arbitration Association and is an arbitrator on the State of California Public Works Arbitration Panel. Mr. Gibbs serves often as a mediator of construction industry disputes in both the public and the private sectors.

Gordon Hunt is a member of the Pasadena, California, law firm of Hunt, Ortmann, Blasco, Palffy & Rossell, Inc. He has been a leading authority on construction law and litigation in California since the inception of his practice, representing owners, contractors, subcontractors, material suppliers, architects, banks, and surety companies. He obtained his undergraduate degree from the University of California at Los Angeles and his J.D. from the University of Southern California Law School. He is a member of the California, American, and Los Angeles Bar Associations. He has served as chairman of the Real Property Section of the Los Angeles County Bar Association and chairman of the Legal Advisory Committee of the Associated General Contractors of California. Mr. Hunt has lectured for the University of California Continuing Education of the Bar, trade associations and numerous bar associations, and many other groups in the construction industry. He has authored many works on construction law and many articles in bar association journals, trade publications, and elsewhere. In addition to being the coauthor of *California Construction Law,* he is also the author of *Construction Surety and Bonding Handbook* and the author of the California Continuing Education of the Bar Action Guide entitled *Handling Mechanic's Liens and Related Remedies* (*Private Works*) and served as the chief consultant for the California Continuing Education of the Bar publication *California Mechanic's Liens and Related Construction Remedies,* Third Edition, published in 1998. Mr. Hunt also serves as an arbitrator and mediator for the American Arbitration Association and as an arbitrator for the California Public Works arbitration panel. He has conducted seminars for construction trade groups and professional associations throughout the state of California.

SUMMARY OF CONTENTS

Contents		xiii
Preface		xxv
Chapter 1	Licensing	1
Chapter 2	Bidding Private and Public Work	69
Chapter 3	Construction Contracts	99
Chapter 4	Breach of Contract by Owner	135
Chapter 5	Breach of Contract by Contractor	179
Chapter 6	Construction Claims	203
Chapter 7	Scheduling and Proof of Delay Claims	255
Chapter 8	Expanding Liability in the Construction Industry	273
Chapter 9	Mechanics' Liens, Stop Notices, and Bonds on Private Works of Improvement	285
Chapter 10	Stop Notices and Bonds on Public Works	417
Chapter 11	Miller Act	439
Chapter 12	Bankruptcy in the Construction Industry	457
Chapter 13	Home Improvement Contracts	469
Chapter 14	Alternative Dispute Resolution	481
Appendixes		519
Table of Cases		561
Index		589

CONTENTS

Preface xxv

Chapter 1	LICENSING	1
§ 1.01	License Requirement	3
§ 1.02	License Classifications	6
§ 1.03	Contractors' Services	10
§ 1.04	Public Body May Determine Appropriate License for Public Works Construction Projects	14
§ 1.05	Penalties	15
§ 1.06	Exemptions from the License Law	22
	[A] In General	22
	[B] Case Law Exemptions from the License Law	22
	[1] Material Supplier Exemption	23
	[2] Doctrine of Substantial Compliance	25
	[3] Owner/Builder Exemption	31
	[C] License on Federal Projects	31
§ 1.07	Notice Required by License Law	32
§ 1.08	Notice Required by Contract	34
§ 1.09	Grounds for Disciplinary Action	35
	[A] Abandonment of Construction Project	37
	[B] Diversion or Misapplication of Funds	37
	[C] Failure to Pay	37
	[D] Willful Departure from Plans and Specifications	44
	[E] Avoidance or Settlement of Obligations for Less Than Full Amount	45
	[F] Evading the License Law	45
	[G] Willful or Fraudulent Act Injuring Another	48
	[H] Acting under Unlicensed Name or Personnel	48
	[I] Entering into a Contract with an Unlicensed Contractor	49
	[J] Willful Failure to Pay for Materials or Service	49
	[K] Filing False Worker's Compensation Statement	50
	[L] *Terminix:* The Contractor's Defense	50
§ 1.10	Arbitration	51

CALIFORNIA CONSTRUCTION LAW

§ 1.11	Citation Procedure	52
	[A] In General	52
	[B] Board Rules Relating to Citations	53
§ 1.12	Responsible Managing Employee and Responsible Managing Officer	58
§ 1.13	Public Access to Proceedings	59
§ 1.14	License Bonds	60
§ 1.15	License Revocation for Failure to Produce Records	63
§ 1.16	Contracts for Construction of Single-Family Dwelling Must Be in Writing	64
§ 1.17	Unlicensed Contractor May Not Advertise	65
§ 1.18	Time for Filing of Disciplinary Proceedings	65
§ 1.19	Continuing Responsibility for Citations of Member, Officer, Director, or Associate	66
§ 1.20	Contractor's Business Licenses	66
§ 1.21	Contractors Who Prepare Home Inspection Reports Must Not Perform Repairs to the Structure	66
§ 1.22	Kickback by Contractor to Insurance Adjuster Is Crime	67
§ 1.23	Contractors Are Required to Disclose Disciplinary Action	67
§ 1.24	Sureties May Be Exempt from Contractor Licensing Requirement	68
Chapter 2	**BIDDING PRIVATE AND PUBLIC WORK**	**69**
§ 2.01	Rules for Private Work	71
	[A] Bidding Documents	71
	[B] Acceptance	72
	[C] Subcontractor Bids and the Doctrine of Promissory Estoppel	72
	[D] Material Supplier Bids and the Commercial Code	74
§ 2.02	Rules for Public Work	75
	[A] Statutory Framework and Purpose	75
	[B] Applicable Rules and Exceptions	76
	[1] Competitive Bids and Negotiated Proposals	76
	[2] Exceptions	77
	[3] Design-Build	78
	[C] Solicitation of Bids	80
	[1] Or Equal Clause	80
	[2] Prequalification	81
	[D] Submission of Bids	82
	[1] Security	82
	[2] Subcontractor Listing Law	82
	[3] Non-collusion Affidavit	84
	[4] Affirmative Action Forms	85
	[E] Bid Withdrawal and Mistake	85

CONTENTS

	[F]	Evaluation of Bids	86
		[1] Bid Responsiveness	86
		[2] Bidder Responsibility	88
		[3] Alternate Bids	89
		[4] Preferences	90
		[a] Local Firms and Labor	90
		[b] Small and Minority-Owned Businesses	91
	[G]	Bid Protests	94
		[1] Procedures	94
		[2] Remedies	95
		[3] Standard of Review	96

Chapter 3   CONSTRUCTION CONTRACTS 99

§ 3.01	Basic Considerations		101
§ 3.02	Analysis of Owner-Contractor Relationship		101
	[A] Basic Owner-Contractor Agreement		102
	[1] Identity of the Parties		102
	[2] Description of the Work		102
	[3] Time for Performance		102
	[4] Contract Price		103
	[5] Payment Terms		103
	[B] General Conditions		104
	[1] Integration		104
	[2] Interpretation of Contract Documents		104
	[3] Contractor's Review of Documents and Field Conditions		105
	[4] Warranties, Guarantees, and Correction of Work		106
	[5] Subcontractor's Adoption of Contract Documents		106
	[6] Design Delegation		106
	[7] Indemnity		107
	[8] Claims, Damages, and Dispute Resolution		110
	[9] Mediation and Arbitration		111
	[10] Delays and Time Extensions		112
	[11] No Damages for Delay		113
	[12] Concealed or Unknown Conditions		113
	[13] Insurance		114
	[14] Notice Requirements		117
	[15] Termination or Suspension of the Contract		117
§ 3.03	Analysis of Contractor-Subcontractor Relationship		118
	[A] Adoption of Contract Documents and Flow-Down Provisions		118
	[B] Payment Terms and Pay-When-Paid Clause		118

	[C]	Indemnity Obligations	120
	[D]	Bonding Requirements	123
	[E]	Obligation to Continue Work	123
	[F]	Contractor's Right to Terminate Subcontractor	124
§ 3.04		Material Supplier Contracts	126
§ 3.05		Analysis of Owner-Architect Relationship	130
	[A]	Limitation of Liability	131
§ 3.06		Analysis of Design-Build Relationship	132

Chapter 4 BREACH OF CONTRACT BY OWNER 135

§ 4.01	Types of Owner Breach	137
§ 4.02	Failure to Make Progress Payment	137
§ 4.03	Contractor's Right to Stop Work in the Event of Nonpayment: Private Works of Improvement	142
§ 4.04	Prompt Payment Legislation in California	145
§ 4.05	Delay	147
§ 4.06	Defective Plans and Specifications	147
§ 4.07	How Courts Interpret Plans and Specifications	151
[A]	Basic Rule: Implied Warranty of Accuracy of Plans and Specifications	151
[B]	Contractor Not Responsible for Defective Plans and Specifications	153
[C]	Enforceability of Exculpatory Clauses	154
[D]	Rule of Contra Proferentem: Ambiguities in Plans and Specifications Will Be Construed Against the Drafter	155
[E]	Obligation of Contractor to Proceed with the Work	157
§ 4.08	Delay in Start of Performance	158
§ 4.09	Material Change of Contract	159
§ 4.10	Acceleration	161
§ 4.11	Recovery of Consequential Damages	161
§ 4.12	Recovery of Damages for Reduction of Bonding Capacity	163
§ 4.13	Contractor's Claim for Right to Finish Early	167
§ 4.14	Delay Damages on Multi-Prime Projects	172
§ 4.15	Recoverability of Interest for Changed Work	176

Chapter 5 BREACH OF CONTRACT BY CONTRACTOR 179

§ 5.01	Contractor's Performance	181
[A]	Failure to Perform Work per Plans and Specifications	181
[B]	Failure to Complete the Work (Abandonment)	183
§ 5.02	Liquidated Damages for Delay	185
§ 5.03	Loss of Rents for Delay	186
§ 5.04	Loss of Profits for Delay	187
§ 5.05	Fraud	190

§ 5.06	Offsetting Backcharges	191
§ 5.07	Latent and Patent Construction Defects	192
	[A] In General	192
	[B] Latent Defects	193
	[C] Patent Defects	197
	[D] Indemnity Actions Involving Latent Defects	198
§ 5.08	Recovery of Damages for Violation of the Building Code	198
§ 5.09	Termination of Contractor	201
§ 5.10	Impact of Construction Contract Terms	201
§ 5.11	Conclusion	202
Chapter 6	**CONSTRUCTION CLAIMS**	**203**
§ 6.01	Introduction: The Importance of Early Claims Recognition	205
§ 6.02	Overview of Owner's Claims	205
	[A] Owner's Delay Claims	206
	[B] Liquidated Damages Clauses	206
	[C] Contractor's Failure to Perform per Plans and Specifications	208
§ 6.03	Public Works Contractor's Liabilities for False Claims	210
§ 6.04	Terminating the Contractor	216
	[A] Damages Recoverable from the Terminated Contractor	218
	[B] Rescinding the Contract	219
§ 6.05	Overview of Contractor's Claims	219
§ 6.06	Contractor's Time-Related Claims	221
§ 6.07	Contractor's Delay Claims	221
	[A] Delay and Time Extension Claims	222
	[B] Categorizing Types of Delay	222
	[C] Excusable versus Inexcusable Delays	222
	[D] Compensable versus Noncompensable Delays	224
	[E] Concurrent Delays	225
	[F] Causes and Effects of Delay	226
	[G] Extended Home Office Overhead: The *Eichleay* Formula	227
	[H] No Damages for Delay Clauses	230
§ 6.08	Contractor's Disruption Claims	231
	[A] Disruption versus Delay	231
	[B] Examples of Disruption	232
	[C] Proving Disruption Claims	234
§ 6.09	Contractor's Acceleration Claims	234
	[A] Actual Acceleration	235
	[B] Constructive Acceleration	235
§ 6.10	Overview of Leading Causes of Claims	236
	[A] Changes (Actual, Constructive, and Cardinal)	237

	[B]	Actual versus Constructive Changes	238
	[C]	Cardinal Changes	239
§ 6.11		Differing Site Conditions	240
§ 6.12		Suspension of Work	242
§ 6.13		Claim Investigation	243
	[A]	Claim Documentation Checklist	244
	[B]	Claim Outline and Narrative	245
§ 6.14		Quantifying and Proving Damages	246
	[A]	Pros and Cons of the "Total Cost" Method	247
	[B]	Applications of the Jury Verdict Method	249

Chapter 7 SCHEDULING AND PROOF OF DELAY CLAIMS ... 255

§ 7.01	Introduction: Who Has the Burden of Proof and How Is It Met?	257
§ 7.02	Use of CPM Scheduling Techniques Preferred	257
§ 7.03	What Is CPM Scheduling?	262
§ 7.04	How CPM Scheduling Techniques Are Used to Prove Delay	266
§ 7.05	Type of Delay Must Be Determined	269
§ 7.06	Conclusion	272

Chapter 8 EXPANDING LIABILITY IN THE CONSTRUCTION INDUSTRY ... 273

§ 8.01	Introduction	275
§ 8.02	Strict Liability	275
§ 8.03	Fraud	276
§ 8.04	Breach of Implied Warranty	277
§ 8.05	Third-Party Beneficiary	277
§ 8.06	Interference with Prospective Economic Advantage	277
§ 8.07	Owner's Liability to Subcontractor for Defective Plans and Specifications	278
§ 8.08	Failure to Exercise Reasonable Care	280
§ 8.09	Current Trends in California	280
§ 8.10	Certificate of Merit Required by CCP § 411.35	281
§ 8.11	Liability of Consultants	282
§ 8.12	Liability for Defective Product That Has Not Yet Caused Damage: The Economic Loss Rule	283
§ 8.13	"Stigma Damages" Are Not Recoverable	284

Chapter 9 MECHANICS' LIENS, STOP NOTICES, AND BONDS ON PRIVATE WORKS OF IMPROVEMENT ... 285

§ 9.01		Remedies Available	289
	[A]	Mechanic's Lien	289

CONTENTS

		[1]	Statutory Implementation	291
		[2]	Other Persons Entitled to File Liens	292
		[3]	Design Professionals Lien	293
		[4]	Persons Not Entitled to File Liens	294
		[5]	Property to Which Mechanics' Liens Attach	295
		[6]	Notice of Nonresponsibility	297
	[B]	Stop Notices		300
		[1]	Contents of Stop Notices	301
		[2]	Case Law Interpretation	302
			[a] *Familian Corp. v. Imperial Bank*	303
			[b] Negligence in Disbursing Construction Loan Funds	304
			[c] Equitable Liens	306
			[d] General Contractor's Stop Notice Rights	307
			[e] Defending Stop Notice Claims	307
			[f] Effect of Payment Bond on Stop Notices	308
		[3]	Attorney's Fees on Bonded Stop Notice	308
	[C]	Bonds		309
		[1]	Perfecting Payment Bond Claims	310
		[2]	Off-Site Bonds	311
§ 9.02	Procedure for Filing Mechanics' Liens and Stop Notices			311
	[A]	Twenty-Day Notice		311
		[1]	In General	311
			[a] When Given	311
			[b] By Whom Given	312
			[c] To Whom Given	312
			[d] Where Given	312
			[e] How Given	312
			[f] Other Requirements	313
		[2]	Contents of the Notice	314
		[3]	Disciplinary Action	314
		[4]	Filing Preliminary Notices with County Recorder	315
	[B]	Cases Interpreting Preliminary Notice		315
		[1]	*Romak v. Prudential*	315
		[2]	*Brown Co. v. Superior Court*	318
		[3]	*IGA v. Manufacturers Bank*	319
		[4]	*James v. Five Points Ranch*	319
		[5]	*Industrial Asphalt v. Garrett Corp.*	320
		[6]	*Kodiak Industries v. Ellis*	321
		[7]	Additional Case Examples	324
§ 9.03	Recording the Lien			328
	[A]	Recording Requirements		328
		[1]	Notice of Completion	331

	[B]		Multiple Contracts	336
	[C]		Contents of Mechanics' Liens	336
		[1]	Description of Site Sufficient for Identification	337
		[2]	Name of Owner or Reputed Owner	337
		[3]	Name of Person by Whom Claimant Was Employed or to Whom Claimant Furnished Its Materials	338
		[4]	Signature of Claimant	339
	[D]		Time for Filing Suit to Foreclose	339
		[1]	Petition to Remove Lien Not Foreclosed	340
		[2]	Lien Foreclosure Actions Stayed Pending Arbitration	342
	[E]		What Constitutes Completion	344
	[F]		Other Mechanic's Lien Issues	346
	[G]		Stop Notice	347
		[1]	Construction Funds Are Trust Funds	349
		[2]	Service of Stop Notice	350
		[3]	Contents of Stop Notice	350
		[4]	Time for Filing Suit	350
		[5]	Statute of Limitations on Filing Suit to Enforce Stop Notice Is Subject to Doctrine of Equitable Tolling	352
	[H]		Payment (Labor and Materials) Bonds	354
		[1]	Preliminary Notice	355
		[2]	Time for Filing Suit	356
		[3]	Cases Interpreting Bond Claims	356

§ 9.04 Special Problems with Mechanics' Liens 360

	[A]		Priority	360
	[B]		Separate Works of Improvement	366
	[C]		Separate Residential Units	367
	[D]		Segregation of Lien	367
	[E]		Condominiums	369
	[F]		Core Work versus Tenant Work	370
	[G]		Proof	370

§ 9.05 Defenses to Mechanics' Liens and Stop Notices 371

	[A]			Application of Payments	371
	[B]			Releases	372
		[1]		The Historical Context	376
		[2]		The *Halbert's Lumber* Decision	377
		[3]		*Halbert's Lumber* Developments	378
		[4]		Analysis of the New Release Forms	379
			[a]	Definitions	379
			[b]	Exceptions to the Release	380

CONTENTS

		[c]	Inherent Ambiguity of the Unconditional Release Form	381
		[5]	Recommendations and Conclusions	382
	[C]	Effect of Assignment for Benefit of Creditors		383
	[D]	Restrictive Endorsements		384
	[E]	Economic Duress		385
	[F]	Joint Checks		386
	[G]	Unclean Hands: Fraud by the Lien Claimant		387
§ 9.06	Miscellaneous Points			388
	[A]	Removal of Materials: Right to Lien		388
	[B]	Voidable Preferences		389
	[C]	Are Wall-to-Wall Carpets Lienable?		395
		[1]	In General	395
		[2]	Conclusion	397
	[D]	Unreasonable Delay in Payment/Pay-If-Paid Clauses Unenforceable		398
	[E]	Diverting Funds		398
	[F]	Lien Law Liberally Construed		399
	[G]	Oral Agreements for Extras		401
	[H]	Amount of Lien		402
	[I]	Ninety-Day Foreclosure Limit		407
	[J]	Interest		407
	[K]	Jurisdiction		408
	[L]	Forged Endorsement of Joint Check		410
	[M]	Lien Forfeiture: Willful Overstatement		410
	[N]	Slander of Title		410
	[O]	Lien Contents		410
		[1]	Description of Property	410
		[2]	Amount of Liens	410
	[P]	Implied Warranty of Design		411
	[Q]	Personal Liability of Owner		411
	[R]	Parties		411
	[S]	Release Bond		412
	[T]	Contractual Waivers of Mechanic's Lien Remedies		413
Chapter 10	**STOP NOTICES AND BONDS ON PUBLIC WORKS**			**417**
§ 10.01	No Rights to Mechanics' Liens			419
§ 10.02	Stop Notice			419
	[A]	Preliminary 20-Day Notice		419
		[1]	Who Gives Notice?	419
		[2]	When Is Notice Given?	419
		[3]	To Whom Is Notice Given?	420

		[4]	How Is Notice Given?	420
		[5]	Where Is Notice Given?	420
		[6]	When Is Service Complete?	420
		[7]	Contents of Notice	420
		[8]	Disciplinary Action	420
	[B]	Stop Notice Requirements		421
		[1]	Contents of Stop Notice	421
		[2]	Who May Serve Stop Notice?	421
		[3]	How Is Stop Notice Served?	421
		[4]	Upon Whom Is Stop Notice Served?	421
		[5]	Time for Service	422
	[C]	Release of Stop Notice		422
	[D]	Suit on Stop Notice		424
§ 10.03	Labor and Material Bond			424
	[A]	Preliminary 20-Day Notice		424
		[1]	Who Gives 20-Day Notice?	424
		[2]	When Is 20-Day Notice Given?	424
	[B]	Time for Filing Suit on Bond		425
	[C]	Other Requirements		427
§ 10.04	Case Examples			428
	[A]	Decisions Regarding Stop Notice and Bond Remedies		428
	[B]	Attorney's Fees on Payment Bond on Public Works		433
	[C]	Surety's Right to Raise Its Principal's Defenses		436

Chapter 11	**MILLER ACT**		**439**
§ 11.01	Labor and Material Bond		441
§ 11.02	Ninety-Day Notice		441
§ 11.03	Suit on the Bond		442
§ 11.04	Persons Entitled to Recover on the Bond		444
§ 11.05	Proof Required		445
§ 11.06	Practice Notes		450
§ 11.07	Illustrative Cases		450
	[A]	Claim Against Person Certifying Sufficiency of the Surety	450
	[B]	Pay-If-Paid Clause Did Not Waive Miller Act Bond Rights	451
	[C]	General Contractor Not Indispensable Party in an Action on the Miller Act Payment Bond	453
	[D]	Recovery of Money Due under Savings Clause Allowed	454
	[E]	Equitable Lien Not Allowed	455

Chapter 12	**BANKRUPTCY IN THE CONSTRUCTION INDUSTRY**	**457**
§ 12.01	Introduction	459

CONTENTS

§ 12.02	Bankruptcy Process	459
§ 12.03	Performance Issues	459
§ 12.04	Assumption or Rejection of Executory Contract	460
§ 12.05	Time	460
§ 12.06	Cure of Default and Adequate Assurance	461
§ 12.07	Setoff	463
§ 12.08	Mechanic's Lien Claims	463
§ 12.09	Evaluating Potential Claims	465
§ 12.10	Pay Only If Protected	466
§ 12.11	Is the Debtor Bonded?	467
§ 12.12	Completing the Work	468

Chapter 13	**HOME IMPROVEMENT CONTRACTS**	**469**
§ 13.01	Statutory Requirements	471
§ 13.02	Home Improvement Salespersons	471
§ 13.03	Home Improvement Contracts	472
§ 13.04	Swimming Pools	475
§ 13.05	Rescission of Home Solicitation Contracts	476
§ 13.06	Arbitration Provisions in Home Improvement Contracts	478

Chapter 14	**ALTERNATIVE DISPUTE RESOLUTION**		**481**
§ 14.01	Overview of Alternative Dispute Resolution		483
§ 14.02	Arbitration of Construction Disputes		483
	[A]	Advantages of Arbitration	483
	[B]	Disadvantages of Arbitration	485
	[C]	Compelling Arbitration	487
	[D]	Consolidation of Arbitration Proceedings	491
	[E]	Waiver of Right to Arbitrate	493
	[F]	Resisting Arbitration	495
	[G]	Binding Effect of Arbitration Awards	497
	[H]	Disclosure Requirements for Arbitrators	498
§ 14.03	Mediation		499
	[A]	The Mediation Process	499
	[B]	Advantages of Mediation	501
	[C]	Disadvantages of Mediation	502
	[D]	Enforcement of Mediation Agreements	503
§ 14.04	Other Available Types of ADR		506
	[A]	Contractually Mandated ADR	507
	[B]	Court-Ordered and Statutorily Required ADR	509
		[1] Court-Ordered ADR	509
		[2] Statutorily Required ADR	509
		[a] Judicial Arbitration under Civil Procedure Code §§ 1141:10 *et seq.*	509

		[b]	Judicial Mediation under Civil Procedure Code §§ 1775 *et seq.*	510
		[c]	Arbitration Provisions in the Public Contract Code	511
	[C]	"Private Trials" and "Mini-Trials"		513
	[D]	Arbitration Clauses in Residential Construction Contracts		515
§ 14.05	Enforcement of ADR Settlement Agreements			516

Appendixes 519

Appendix 1	Conditional Waiver and Release Upon Progress Payment	521
Appendix 2	Unconditional Waiver and Release Upon Progress Payment	525
Appendix 3	Conditional Waiver and Release Upon Final Payment	529
Appendix 4	Unconditional Waiver and Release Upon Final Payment	533
Appendix 5	California Preliminary Notice, Together With Proof of Service	537
Appendix 6	Mechanics' Lien	541
Appendix 7	Stop Notice	545
Appendix 8	Miller Act Form	549
Appendix 9	Notice to Principal and Surety on Payment Bond on Private Work	553
Appendix 10	Notice to Principal and Surety on Payment Bond on Public Work	557

Table of Cases 561

Index 589

PREFACE

The 16th edition of *California Construction Law* contains substantial revisions from the previous edition. Not only has the book been updated by incorporating new legislation enacted and cases decided through mid-1999, but also various chapters of the book have been completely rewritten and/or reorganized by the authors.

California Construction Law is designed to be used by both construction professionals and attorneys. We have provided practical information that can be used by construction industry participants, as well as detailed legal analyses of California construction law, both as codified in the statutes and as expressed by California courts. The book's layout walks the reader through the legal aspects of a construction project much as one would build a project. First, it deals with preconstruction activities—licensing, bidding, and the formation of the construction contract. Then it discusses what happens when things go wrong—breach of contract by owner and contractor. An in-depth analysis is provided with regard to claims involving delay, disruption, and acceleration, and a new chapter has been added regarding proving and defending delay claims and the use of schedules. Several chapters are devoted to statutory remedies—mechanics' liens, stop notices, and bonds both on public and on private works. Finally, chapters are provided dealing with various other issues and subjects involving the construction industry, including expanding liability, bankruptcy, home improvement contracts, and alternative dispute resolution.

Over the almost 20 years this book has been published, we have attempted to make it as up-to-date and accurate as possible. Both of the authors sincerely appreciate the efforts of attorneys and staff personnel in our respective offices in the researching and preparation of various chapters of the book, and Mr. Gibbs particularly wishes to acknowledge the contributions of his partners, Barbara R. Gadbois and Jeriel C. Smith. We are additionally grateful to the editorial staff of Aspen Law & Business, whose efforts have been instrumental in bringing this 16th edition to fruition.

October 1999

KENNETH C. GIBBS
Los Angeles, California
GORDON HUNT
Pasadena, California

CALIFORNIA CONSTRUCTION LAW

Chapter 1
LICENSING

§ 1.01	License Requirement
§ 1.02	License Classifications
§ 1.03	Contractors' Services
§ 1.04	Public Body May Determine Appropriate License for Public Works Construction Projects
§ 1.05	Penalties
§ 1.06	Exemptions from the License Law
	[A] In General
	[B] Case Law Exemptions from the License Law
	[1] Material Supplier Exemption
	[2] Doctrine of Substantial Compliance
	[3] Owner/Builder Exemption
	[C] License on Federal Projects
§ 1.07	Notice Required by License Law
§ 1.08	Notice Required by Contract
§ 1.09	Grounds for Disciplinary Action
	[A] Abandonment of Construction Project
	[B] Diversion or Misapplication of Funds
	[C] Failure to Pay
	[D] Willful Departure from Plans and Specifications
	[E] Avoidance or Settlement of Obligations for Less Than Full Amount
	[F] Evading the License Law
	[G] Willful or Fraudulent Act Injuring Another
	[H] Acting under Unlicensed Name or Personnel
	[I] Entering into a Contract with an Unlicensed Contractor
	[J] Willful Failure to Pay for Materials or Service
	[K] Filing False Worker's Compensation Statement
	[L] *Terminix:* The Contractor's Defense
§ 1.10	Arbitration
§ 1.11	Citation Procedure
	[A] In General
	[B] Board Rules Relating to Citations

§ 1.12	Responsible Managing Employee and Responsible Managing Officer
§ 1.13	Public Access to Proceedings
§ 1.14	License Bonds
§ 1.15	License Revocation for Failure to Produce Records
§ 1.16	Contract for Construction of Single-Family Dwelling Must Be in Writing
§ 1.17	Unlicensed Contractor May Not Advertise
§ 1.18	Time for Filing of Disciplinary Proceedings
§ 1.19	Continuing Responsibility for Citations of Member, Officer, Director, or Associate
§ 1.20	Contractor's Business Licenses
§ 1.21	Contractors Who Prepare Home Inspection Reports Must Not Perform Repairs to the Structure
§ 1.22	Kickback by Contractor to Insurance Adjuster Is Crime
§ 1.23	Contractors Are Required to Disclose Disciplinary Action
§ 1.24	Sureties May Be Exempt from Contractor Licensing Requirement

§ 1.01 LICENSE REQUIREMENT

In order for a contractor to recover for labor and material furnished on a work of improvement, the contractor must be licensed.[1] *Contractor* is defined in the Contractors License Law as follows:

> The term contractor for the purposes of this chapter is synonymous with the term builder and, within the meaning of this chapter, a contractor is any person who undertakes to or offers to undertake to or purports to have the capacity to undertake or submits a bid to, or does himself or by or through others, construct, alter, repair, add to, subtract from, improve, move, wreck or demolish any building, highway, road, parking facility, railroad, excavation thereof, including the erection of scaffolding or other structures or works in connection therewith, or the cleaning of grounds or structures in connection therewith, and whether or not the performance of work herein described involves the addition to or fabrication into any structure, project, development or improvement herein described of any material or article of merchandise. The term contractor includes subcontractor and specialty contractor.[2]

To seek recovery in the courts under California Business and Professions Code § 7031, all contractors must be able to allege and prove that they were licensed at all times during construction of the project. The leading case of *Hydrotech Systems Ltd. v. Oasis Water Park*[3] illustrates the attitude of the California courts with regard to enforcing the license law. Hydrotech manufactured and installed patented equipment designed to simulate ocean waves. Oasis owned and operated a water park in Palm Springs. Wessman Construction Company was the general contractor on the project. Hydrotech contracted with Wessman to design and construct a 29,000-square-foot "surfing pool" using Hydrotech's wave equipment. Hydrotech sued Wessman and Oasis for $110,000 still due on the contract. Hydrotech included in its suit a claim for breach of contract and for fraud. In the fraud count, Hydrotech alleged (1) because it was concerned about licensing, it wished only to sell and deliver its equipment and to avoid being involved in design or construction of the pool; (2) Oasis insisted that Hydrotech's unique expertise in design and construction was essential; (3) Oasis and Wessman promised that Wessman would arrange for a California contractor to "work with" Hydrotech on any construction activities that required a California contractor's license; (4) Oasis and Wessman promised to pay in full for Hydrotech's equipment and associated equipment and services; (5) Hydrotech relied on these promises and furnished the equipment and services; (6) Oasis and Wessman never intended to honor their promises; and (7) had Hydrotech known that Oasis and Wessman never intended to pay Hydrotech, it would not have performed the contract.

The trial court sustained the demurrers of Oasis and Wessman and dismissed Oasis from the suit on the basis that Hydrotech had not alleged it held a

---

[1] Cal. Bus. & Prof. Code § 7031.
[2] *Id.* § 7026.
[3] 52 Cal. 3d 988 (1991).

contractor's license as required by Business and Professions Code § 7031. The California Supreme Court carefully analyzed the license law issue. The court noted that Business and Professions Code § 7031 states that no person acting in the capacity of a contractor may bring or maintain an action to recover compensation for the performance of any act or contract for which a contractor's license is required without alleging and proving that such person was duly licensed at all times during performance of the act or contract. The purpose of the licensing law is to protect the public from incompetence and dishonesty in those who provide building and construction services. The licensing requirements provide minimal assurance that all persons offering construction services in California have the requisite skill and character, understand applicable laws and cases, and know the rudiments of administering a contracting business. Section 7031 advances those purposes by withholding judicial aid from those who seek compensation for unlicensed work. Section 7031 represents a legislative determination that the importance of deterring unlicensed persons from engaging in the contracting business outweighs any harshness between the parties and that such deterrence can only be realized by denying violators (i.e., unlicensed contractors) the right to maintain any action for compensation in the courts.

The California Supreme Court held that section 7031 barred Hydrotech's right to recover for breach of contract. With regard to the fraud claim, the court said that "regardless of the equities," section 7031 bars "all" actions, however they are categorized, which seek compensation for illegal unlicensed work. The court stated that an unlicensed contractor cannot recover either for the agreed contract price or the reasonable value of the labor and materials and that section 7031 operates even where the person for whom the work was done knew the contractor was unlicensed. Thus, the court held that an unlicensed contractor cannot circumvent section 7031 and its purpose by alleging that when the illegal contract was made the other party had no intention to perform. Section 7031 places the risk of bad faith on the unlicensed contractor. Thus, even fraud in the inducement did not get Hydrotech around the bar of section 7031. It is therefore clear that lack of license is a complete defense to any claim by an unlicensed contractor to recover compensation whether based on contract, unjust enrichment, or even fraud.

An attempt to get around the license requirement and the *Hydrotech* case was made in *Vallejo Development Co. v. Beck Development Co.*[4] by a developer who contended that it was the "master developer" or "administrator" of a project. Vallejo Development Company (VDC) sold certain property to third parties, who were referred to as "merchant builders." The purchase agreements divided the purchase price into two parts: the cost of the land and the improvement costs. The purchase agreements contained provisions pursuant to which VDC promised that, after the close of escrow, it would improve the property in accordance with the city-approved plans and specifications to a finished lot condition. This required VDC to install grading for building pads. The agreements further provided that VDC would be solely responsible for completion of all off-site, on-site, and

---

[4] 24 Cal. App. 4th 929 (1994).

infrastructure improvements, including grading, storm drainage, sanitary systems, curbs, gutters, utilities, streets, street lighting, traffic signals, sidewalks, and landscaping. Pursuant to those agreements, after close of escrow, VDC performed some of that work. VDC did not hold any type of California contractor's license.

The purchasers of the property filed a suit against VDC, seeking rescission and damages for breach of contract, alleging that VDC had failed to complete the agreed-upon infrastructure improvements. VDC, in turn, recorded mechanics' liens for more than $17 million for labor, service, equipment, and materials it claimed to have furnished for grading, storm drains, sewers, water lines, trenches, paving, and other on-site and off-site improvements, and then filed an action to foreclose those mechanics' liens. The purchasers of the property sought dismissal of VDC's complaints on the ground that VDC had failed to allege that it was a duly licensed contractor.

VDC claimed that it acted as an administrator rather than a contractor and further contended that the work was furnished by third-party contractors. The trial court, relying on Business and Professions Code § 7031 and the California Supreme Court's decision in *Hydrotech Systems, Ltd. v. Oasis Water Park*,[5] ruled that because VDC was not a licensed contractor, it could not state a cause of action.

The court of appeal affirmed and held that the actions by VDC were barred by Business and Professions Code § 7031(a), notwithstanding VDC's assertion that it was merely a master developer or administrator, in that VDC was seeking compensation for the performance of acts for which a license was required, and that VDC was in fact acting in the capacity of a contractor.

The court further noted that, even if VDC as the master developer furnished the labor and materials through a licensed general contractor, it was still acting in the capacity of a contractor and had to be licensed. The court also held that VDC did not come within an exception to the licensing requirement for situations where the construction of infrastructure improvements is "incidental" to the parties' overall business relationship. This case, along with the *Hydrotech* case cited above and in § 1.06[B][2], make it clear that both the legislature and the courts will strictly construe the licensing requirements under Business and Professions Code § 7031.

In *Contractors Labor Pool, Inc. v. Westway Contractors, Inc.*,[6] Contractors Labor Pool, Inc. (CLP) entered into an agreement with a subcontractor to provide workers for use on a public work of improvement. When CLP was not paid, it brought an action on the payment bond (the bond is required on public works of improvement in California). CLP recovered at the trial court level, and the bonding company appealed, contending that CLP was not entitled to recover because it had furnished labor on a work of improvement and therefore became a contractor that was required to have a contractor's license under Business and Professions Code § 7031.

---

[5] 52 Cal. 3d 988 (1991).
[6] 53 Cal. App. 4th 152 (1997).

CLP's contract with the subcontractor essentially provided that the workers would remain employees of CLP, but that CLP would provide the workers to the subcontractor, who would have direct supervision and control over them. The court of appeal held that, under that factual scenario, CLP was not a contractor and therefore need not be licensed under section 7031. The essence of the decision is that CLP was merely the employer of workers that had been sent to a licensed subcontractor and therefore CLP did not itself act in the capacity of a contractor. The court of appeal noted that the purpose of section 7031 is to ensure that persons who perform construction work have the necessary skill, character, and understanding of applicable law and codes to provide appropriate construction services. As the law applies only to those who actually perform or supervise the performance of construction services, it is not intended to apply to those who only supply material to be used by others or who only supply laborers who will be supervised by others. Thus, a person or company in the business of supplying equipment or hiring out laborers to be supervised by others is not deemed to be acting in the capacity of a contractor and therefore is not required to have a license.[7]

§ 1.02 LICENSE CLASSIFICATIONS

Subcontractors and specialty contractors must also be licensed. The Contractors State License Board Rules and Regulations have classified contractors as follows:

830. Classification Policy

(a) All contractors to whom licenses are issued shall be classified by the registrar as a specialty contractor, as defined in this article; a general engineering contractor (Class A), as defined in Section 7056 of the Code; or a general building contractor (Class B) as defined in Section 7057 of the Code.

(b) Contractors licensed in one classification shall be prohibited from contracting in the field of any other classification unless they are also licensed in that classification or are permitted to do so by Section 831.

831. Incidental and Supplemental Defined

For purposes of Section 7059, work in other classifications is "incidental and supplemental" to the work for which a specialty contractor is licensed if that work is essential to accomplish the work in which the contractor is classified. A specialty contractor may use subcontractors to complete the incidental and supplemental work, or he may use his own employees to do so.

832. Specialty Contractors Classified

Specialty contractors shall perform their trade using the art, experience, science and skill necessary to satisfactorily organize, administer, construct and complete projects under their classification, in accordance with the standards of their trade.

---

[7] *See* Contractors Dump Truck Serv. Inc. v. Gregg Constr. Co., 237 Cal. App. 2d 1 (1965); Rodoni v. Harbor Eng'rs, 191 Cal. App. 2d 560 (1961).

They are classified into the following subclassifications:

Boiler, Hot Water Heating and Steam Fitting	C-4
Building Moving and Demolition	C-21
Carpentry, Cabinet and Mill Work	C-5
Concrete	C-8
Drywall	C-9
Earthwork and Paving	C-12
Electrical (General)	C-10
Electrical Sign	C-45
Elevator Installation	C-11
Fencing	C-13
Fire Protection Engineering	C-16
Flooring and Floor Covering	C-15
Glazing	C-17
Insulation and Acoustical	C-2
Landscaping	C-27
Limited Specialty	C-61
Lock and Security Equipment	C-28
Low Voltage Systems	C-7
Masonry	C-29
Metal Roofing	C-14
Ornamental Metal	C-23
Painting and Decorating	C-33
Parking and Highway Improvement	C-32
Pipeline	C-34
Plastering and Lathing	C-35
Plumbing	C-36
Refrigeration	C-38
Roofing	C-39
Sanitation System	C-42
Sheet Metal	C-43
Solar	C-46
Steel, Reinforcing	C-50
Steel, Structural	C-51
Swimming Pool	C-53
Tile (Ceramic and Mosaic)	C-54
Warm-Air Heating, Ventilating and Air Conditioning	C-20
Water Conditioning	C-55
Welding	C-60
Well Drilling (Water)	C-57

NOTE: Authority cited: Sections 7008 and 7059, Business and Professions Code. Reference: Sections 7058 and 7059, Business and Professions Code.

The license classification limitations established by the Contractors State License Board (CSLB) were thoroughly reviewed and analyzed in *R.E. Hazard, Jr. Enterprises, Inc. v. Insurance Co. of the West*.[8] R.E. Hazard, Jr. Enterprises,

---

[8] 52 Cal. App. 4th 1088 (1997).

Inc. (Hazard), licensed in 1946 as a general building contractor, began building large commercial projects, such as supermarkets, shopping centers, and hotels, in the late 1950s. In conjunction with construction of those types of projects, Hazard performed more than $100 million worth of site work, including driveways, sidewalks, and grading. Hazard's contracts usually included both site preparation and building construction, but occasionally only site work was involved.

In 1989, Valley Parkway employed Hazard to ready an Escondido site for a commercial subdivision project. The contract, which included no building construction, was confined to site preparation, including surveying, cutting, filling, soils testing, rough and fine grading, relocating underground utilities, and installing canyon drains, retaining walls, fencing, barricades, landscaping, irrigation, street lights, parking stripes, signs, curbs, gutters, and sidewalks. When Valley Parkway refused to pay approximately $600,000 in change order claims, Hazard brought an action to recover on its claims. Valley Parkway denied that Hazard was properly licensed and filed a motion, which was granted, to have the court try the licensing issue first. Valley Parkway contended that, under California Business and Professions Code § 7031, Hazard could not maintain its lawsuit because the site work required a general engineering contractor's (Class A) license. Hazard presented evidence that general building contractors customarily perform the type of site work at issue in the case. The trial court granted Valley Parkway's motion for judgment and awarded attorney's fees to Valley Parkway.

The court of appeal, in reviewing the judgment of the trial court, analyzed various sections of the Contractors License Law, beginning with Business and Professions Code § 7026, which provides that the term *contractor* is synonymous with the term *builder* and includes any person who undertakes to construct, alter, repair, or add to any building, highway, road, parking facility, railroad, excavation, or other structure, project, development, or improvement, or to do any part thereof, including the erection of scaffolding or other structures or works in connection therewith.

Business and Professions Code § 7055 provides for three license classifications: Class A (general engineering contractor); Class B (general building contractor); and Class C (specialty contractors). Under section 7057, a *general building contractor* is one whose principal contracting business is in connection with any structure built, being built, or to be built for the support, shelter, and enclosure of persons, animals, chattels, or movable property of any kind, requiring in its construction the use of more than two unrelated building trades or crafts, or to do or superintend the whole or any part thereof. Section 7056 defines *general engineering contractor* as one whose principal contracting business is in connection with fixed works requiring specialized engineering knowledge and skill, and includes a long list of examples, including highways, streets, and roads.

Section 7059 provides that the CSLB may adopt reasonably necessary rules and regulations to effect the classification of contractors in a manner consistent with the established usage and procedure of the construction business, and may limit the field and scope of operations of a licensed contractor to those in which he or she is classified and qualified to engage (as defined by sections 7055, 7056, 7057, and 7058). Finally, California Code of Regulations title 16, § 830, classifies each contractor as either a specialty contractor, a general engineering contractor, or a general building contractor and states, in subdivision (b), that contractors licensed in one classification shall be prohibited from contracting in the field of any other classification unless they are also licensed in that classification.

On appeal, Valley Parkway relied upon its successful argument to the trial court: that Hazard may do site work under its Class B license only if its contract also includes the construction of a structure. When the contract is limited to grading and other types of work listed in section 7056, Valley Parkway argued, a general engineering contractor's Class A license is required. After review and analysis of the sections previously set forth, the court of appeal rejected Valley Parkway's argument, noting that the plain words of section 7057 show that a general building contractor's operations are not limited solely to construction of structures. Rather, a *general building contractor* is defined as one whose principal contracting business is in connection with any structure built, being built, "or to be built." The court of appeal decided that the phrase "in connection with any structure . . . to be built" would be meaningless unless general building contractors are authorized to perform or superintend work limited to site preparation for future construction or alteration of a structure, or for construction of, for instance, sidewalks, landscaping, pools, fencing, and other exterior components of a residential or commercial development with respect to which the construction of buildings is yet to occur.

The court then rejected Valley Parkway's position that Rule 830(b), by prohibiting contractors licensed in one classification from contracting in the field of any other classification unless they are also licensed in that classification, thereby prohibits general building contractors from performing work that general engineering contractors are also authorized to perform. The court held that CSLB Regulations may only limit a general building contractor's scope of operations to what is defined by section 7057. Because section 7057 specifically authorizes general building contractors to perform work in connection with structures "to be built," Hazard's contract for site work only did not involve "contracting in the field of any other classification" outside Hazard's own field. Quoting Hazard's position, the court of appeal stated, "[W]e think Hazard puts it well":

> "[T]he purpose of the licensing scheme is to insure honesty and competency. No rational end is served by a scheme which permits the [general building] contractor who builds a building to perform the site work therefor, but outlaws the [contract of a general building] contractor who performs the same site work but does not build the building. The relevant concern is not whether the contractor will build a building, but whether the contractor knows enough about construction generally to perform competent site work. If a general building contractor is properly licensed and qualified to perform fixed works in connection with a structure to be built by the contractor, the same contractor is not somehow disqualified because he is not to build the structure."[9]

---

[9] 52 Cal. App. 4th at 1097–1098.

Based on this decision, a general building contractor with a Class B license may take a separate contract to do site work that will ultimately support buildings to be constructed by another, without a general engineering contractor's Class A license.

§ 1.03 CONTRACTORS' SERVICES

Contractors are often in doubt as to what they can or cannot do under their license.

California Business and Professions Code § 7056 defines what a general engineering contractor can do:

> Section 7056. General Engineering Contractor
>
> A general engineering contractor is a contractor whose principal contracting business is in connection with fixed works requiring specialized engineering knowledge and skill, including the following divisions or subjects: irrigation, drainage, water power, water supply, flood control, inland waterways, harbors, docks and wharves, shipyards and ports, dams and hydroelectric projects, levees, river control and reclamation works, railroads, highways, streets and roads, tunnels, airports and airways, sewers and sewage disposal plants and systems, waste reduction plants, bridges, overpasses, underpasses and other similar works, pipelines and other systems for the transmission of petroleum and other liquid or gaseous substances, parks, playgrounds and other recreational works, refineries, chemical plants and similar industrial plants requiring specialized engineering knowledge and skill, powerhouses, power plants and other utility plants and installations, mines and metallurgical plants, land leveling and earth-moving projects, excavating, grading, trenching, paving and surfacing work and cement and concrete works in connection with the above mentioned fixed works.

In *Ron Yates Construction Co. v. Superior Court*,[10] a general engineering contractor contracted with a homeowner to build a seawall, a septic system, and foundation caissons for a beachfront residence in Malibu. The contractor held a Class A general engineering license. The trial judge held that under California Business and Professions Code § 7056, a general engineering contractor (Class A license) could not, as a matter of law, construct a foundation for a residence. When Yates applied for the license with the Contractors State License Board, he flew to Sacramento and was assured by the Board's licensing officer that a company engaged in constructing seawalls and house foundations on the beach should hold a Class A license. When Yates was not paid, he brought a suit against the owners to collect what was due him. The owners had received a letter from the Contractors State License Board, stating that a contractor holding a Class A license could not construct foundations. The owners contended that Yates was not properly licensed, and the trial court ruled in favor of the owners.

---

[10] 186 Cal. App. 3d 337 (1986).

On appeal, the Contractors State License Board, as amicus curiae for the owner, filed a letter brief with the court supporting the owner's position, claiming that a general engineering contractor could not construct foundations. The appellate court reversed the trial court and held that if the construction of foundations required "specialized engineering knowledge and skill,"[11] then the contractor would be properly licensed to construct the foundation.

It is interesting to note in this particular case that Robert Berrigan, a licensing deputy for the board, testified that there was not "a great deal of clarity" on the question as to whether Class A licensees could contract to construct a residence foundation. The court also indicated on appeal that it was not bound by any interpretation of the license law by the Contractors State License Board.

It is evident that there is often confusion as to exactly what licensees may or may not do under their licenses. It can also be seen that licensees may not rely upon the Contractors State License Board's interpretation of the law. When in doubt, licensees should always apply for an additional license for any work they suspect may fall outside their basic license category.

California Business and Professions Code § 7057 defines what a general building contractor can do:

> (a) Except as provided in this section, a general building contractor is a contractor whose principal contracting business is in connection with any structure built, being built, or to be built, for the support, shelter, and enclosure of persons, animals, chattels, or movable property of any kind, requiring in its construction the use of at least two unrelated building trades or crafts, or to do or superintend the whole or any part thereof.
>
> This does not include anyone who merely furnished materials or supplies under Section 7045 without fabricating them into, or consuming them in the performance of the work of the general building contractor.
>
> (b) A general building contractor may take a prime contract or a subcontract for a framing or carpentry project. However, a general building contractor shall not take a prime contract for any project involving trades other than framing or carpentry unless the prime contract requires at least two unrelated building trades or crafts other than framing or carpentry, or unless the general building contractor holds the appropriate specialty license or subcontracts with an appropriately licensed specialty contractor to perform the work. A general building contractor shall not take a subcontract involving trades other than framing or carpentry, unless the subcontract requires at least two unrelated trades or crafts other than framing or carpentry, or unless the general building contractor holds the required special license. The general building contractor may not count framing or carpentry in calculating the two unrelated trades necessary in order for the general building contractor to be able to take a prime contract or subcontract for a project involving other trades.
>
> (c) No general building contractor shall contract for any project that includes the "C-16" Fire Protection classification as provided for in Section 7026.12 or the "C-57" Well Drilling classification as provided for in Section 13750.5

---

[11] As that term is used in Cal. Bus. & Prof. Code § 7056.

of the Water Code, unless the general building contractor holds the specialty license, or subcontracts with the appropriately licensed specialty contractor.

Contractors often believe that they can take a contract to perform work for which they are not licensed if they subcontract out that work to a party licensed to perform the work. This is simply not so. See the *R.E. Hazard, Jr.* case discussed in section 1.02 above.

In the case of *Currie v. Stolowitz*,[12] Currie, in July 1950, contracted with an owner to furnish all the material and labor to install all heating, ventilating, air-conditioning, and plumbing in a hospital. Currie entered into a subcontract with Stolowitz to do the plumbing work. Stolowitz ceased performance under its plumbing subcontract in January 1951. Currie completed the work and filed a notice of completion in April 1952. In September 1955, Currie sued Stolowitz for breach of contract. Currie was licensed as a warm air heating, ventilating, and air-conditioning contractor under the license category C-20. During the progress of the job, Currie, in 1951, obtained an additional specialty license in category C-43 for sheet metal and in category C-4 for steam heating. At no time did Currie hold a C-36 specialty license as a plumbing contractor. Stolowitz was, in fact, a licensed plumbing contractor with a C-36 specialty license. The trial court found that the plumbing work, which Currie had subcontracted to Stolowitz, was not "incidental and supplemental" to the heating, ventilating, and air-conditioning work and concluded therefore that Currie was not properly licensed and could not sue Stolowitz. The appellate court stated that it was admitted that Currie did not at any time have a C-36 plumbing license, and therefore, under Business and Professions Code § 7031, Currie was barred from bringing an action against Stolowitz for breach of contract. The fact that Currie had completed the work and had been paid by the owner was immaterial. The court stated that Business and Professions Code § 7031 bars a suit for breach of contract as well as a suit for collection of compensation. The appellate court opined that the plumbing work for which Currie did not have a license was not "incidental or supplemental," and there was substantial evidence to support the trial court's findings to that effect.

It is clear from this and other cases that a contractor must have a license for all work it undertakes to perform unless the work is "incidental and supplemental." In addition to the *Ron Yates Construction Co.* case, cited above, another case that illustrates this principle is *Roy Bros. Drilling Co. v. Jones*.[13] In that case, the contractor entered into an oral agreement with the owner (an attorney) to drill holes in connection with the foundation work for a residence. The oral agreement provided that the contractor was to drill caisson holes to excavate for the installation of the basement and drill holes for the installation of retaining walls. The oral contract did not in any way relate to the installation of a sanitation system that was being installed on the property. The owner made a motion for summary judgment, contending that the contractor was not properly licensed. The owner

---

[12] 169 Cal. App. 2d 810 (1959).
[13] 123 Cal. App. 3d 175 (1981).

attached, as an exhibit to his declaration, the contractor's bill in the sum of $26,139.65, which bore a license number. The owner also attached a certified statement from the Contractors State License Board showing that that particular license number was issued to the contractor in category C-42, sanitation system. The contractor filed a declaration in opposition which stated that he had received a letter in 1957 from the Contractors State License Board that stated, in effect, that a C-42 license would cover any type of drilling except water well drilling. The owner had filed a cross-complaint against the contractor for breach of contract and fraud. The motion for summary judgment as to the contractor's complaint was granted, and the contractor appealed.

In its opinion, the court of appeal first stated that the documents in support of the owner's motion for summary judgment were adequate and that there was adequate foundation for the documents, which consisted of the bill for services to the owner from the contractor, the letter of inquiry by the owner to the Contractors State License Board, and the reply the owner received from the Contractors State License Board. The court then analyzed the Contractors License Law, as contained in the Business and Professions Code, and the rules and regulations issued by the Contractors State License Board, as contained in the administrative code, and held that the contractor was licensed only in the C-42 category for sanitation systems. Therefore, it could not recover for work within the classification of C-61, drilling. The court stated that there were 36 specialty classifications and that Business and Professions Code § 7059 provided that a specialty contractor could take and execute a contract involving the use of two or more crafts if the work in the crafts or trades for which the contractor was not licensed was "incidental and supplemental" to performance of the craft for which the contractor was licensed. The court found that in this case, the only contract the plaintiff had was to do drilling; the work had nothing to do with sanitation systems and therefore was not "incidental and supplemental" to performance. The court held that the plaintiff was not licensed in the correct category and could not recover by reason of Business and Professions Code § 7031. Furthermore, the court stated that the contractor could not rely upon the letter from the License Board and, with respect thereto, said: "It is a hearsay statement of opinion with respect to a matter of law." Thus, the question of whether a contractor is properly licensed is a question of law for the court.

In *Construction Financial LLC v. Perlite Plastering Co.*,[14] the court of appeal held that a contractor holding a Class B general contractor's license could legally perform specialty work without holding the corresponding Class C licenses. The contractor was, however, found to be in violation of other requirements of the licensing law and could not meet the standards for substantial compliance under Business and Professions Code § 7031(d) then in effect. The *Construction Financial* case is fully discussed in § 1.06[B][2].

---

[14] 53 Cal. App. 4th 170 (1997).

§ 1.04 PUBLIC BODY MAY DETERMINE APPROPRIATE LICENSE FOR PUBLIC WORKS CONSTRUCTION PROJECTS

California Public Contract Code § 3300 provides that any public entity shall specify the classification of the contractor's license that a contractor shall possess at the time a contract is awarded. It further provides that this specification shall be included in any plans prepared for a public project and any notice inviting bids for a public project. Business and Professions Code § 7059(b) provides that in public works contracts, the awarding authority shall determine the license classification necessary to bid and perform the work and that in no case shall the awarding authority award a prime contract to a specialty contractor whose classification constitutes less than a majority of the project.

In *M&B Construction v. Yuba County Water Agency*,[15] the Yuba County Water Agency (the Agency) prepared bid specifications that required that the contractor constructing the project have a Class A general engineering contractor's license. The Agency obtained nine bids for the project, and all bidders had a Class A license except the plaintiff, who held only a Class B general contractor's license and two Class C specialty licenses, to wit, a C-8 for concrete and a C-12 for earthwork and paving. The plaintiff's bid was the lowest bid, but it was rejected by the public body because the plaintiff did not have a Class A license. The Agency's engineer required a Class A license because the project was the construction of a canal and pipeline intended to deliver surface river water to a water company and required the construction of 3,400 feet of earth-lined canal, 903 feet of six-foot diameter concrete pipeline, and related structures. The Agency's engineer concluded that that particular project, including its complexity and the type of equipment involved and the fact that Class A contractors typically performed the type of work involved in constructing heavy reinforced vertical concrete walls, required a Class A license and not a Class C subcontractor's license.

The low bidder challenged the rejection of its bid, and the appellate court upheld the decision of the Agency based upon Public Contract Code § 3300 and Business and Professions Code § 7059, subdivision (b). The court noted that the Agency's interpretation of § 7059(b) was supported by the legislative history of the subdivision. It was sponsored by the Contractors State License Board to clarify when contracts could be let to specialty contractors where the contract involved work outside the specialty contract license. The bill's intention was to reduce some of the requests to the License Board for classification determinations by allowing the awarding authority to determine the licensing classification necessary for bidding a project. Specifically, the awarding authority would have more flexibility and control in deciding who it would be doing business with. In this case, the Agency made a pre-bid determination that the public would be better served in terms of quality and economy by letting the project only to a licensee with the most appropriate experience, while minimizing the need for subcontractors. Such an administrative decision is subject to reversal only if it is arbitrary, capricious, or entirely lacking in evidentiary support or contrary to

---

[15] 68 Cal. App. 4th 1353 (1999).

LICENSING § 1.05

established public policy or unlawful or procedurally unfair. The court held that the Agency's determination in this case did not violate the above standards.

§ 1.05 PENALTIES

There are penalties for not being properly licensed:

1. The claimant may not sue to recover the money due. California Business and Professions Code § 7031 provides as follows:

   (a) Except as provided in subdivision (d), no person engaged in the business or acting in the capacity of a contractor, may bring or maintain any action, or recover in law or equity in any action, in any court of this state for the collection of compensation for the performance of any act or contract for which license is required by this chapter without alleging that he or she was a duly licensed contractor at all times during the performance of that act, or contract, regardless of the merits of the cause of action brought by the person, except that this prohibition shall not apply to contractors who are each individually licensed under this chapter but who fail to comply with Section 7029.

   (b) A security interest taken to secure any payment for the performance of any act or contract for which a license is required by this chapter is unenforceable if the person performing the act or contract was not a duly licensed contractor at all times during the performance of the act or contract.

   (c) If licensure or proper licensure is controverted, then proof of licensure pursuant to this section shall be made by production of verified certificate of licensure from the Contractors' State License Board which establishes that the individual or entity bringing the action was duly licensed in the proper classification or contractors at all times during the performance of any act or contract covered by the action. Nothing herein shall require any person or entity controverting licensure or proper licensure to produce a verified certificate. When licensure or proper licensure is controverted, the burden of proof to establish licensure or proper licensure shall be on the licensee.

   (d) The judicial doctrine of substantial compliance shall not apply under this section where the person who engaged in the business or acted in the capacity of a contractor has never been a duly licensed contractor in this state. However, the court may determine that there has been substantial compliance with licensure requirements under this section if it is shown at an evidentiary hearing that the person who engaged in the business or acted in the capacity of a contractor (1) had been duly licensed as a contractor in this state prior to the performance of the act or contract, (2) acted reasonably and in good faith to maintain proper licensure, and (3) did not know or reasonably should not have known that he or she was not duly licensed. Subdivision (b) of Section 143 does not apply to contractors subject to this subdivision.

(e) The exceptions to the prohibition against the application of the judicial doctrine of substantial compliance found in subdivision (d) shall apply to all contracts entered into on or after January 1, 1992, and to all actions or arbitrations arising therefrom, except that the amendments to subdivisions (d) and (e) enacted during the 1994 portion of the 1993–94 Regular Session of the Legislature shall not apply to either of the following:

(1) Any legal action or arbitration commenced prior to January 1, 1995, regardless of the date on which the parties entered into the contract.

(2) Any legal action or arbitration commenced on or after January 1, 1995, if the legal action or arbitration was commenced prior to January 1, 1995, and was substantially dismissed.

2. It is a misdemeanor to contract without a license.[16]

3. A person who is unlicensed and acts in the capacity of a contractor is subject to a civil penalty of $100 per person per day for each person employed.[17]

4. Fraudulent use of incorrect license number is punishable as follows:

   Any person licensed or unlicensed who willfully and intentionally uses, with intent to defraud, a contractor's license number which does not correspond to the number on a currently valid contractor's license held by that person is punishable by a fine not exceeding ten thousand dollars ($10,000), or by imprisonment in state prison, or in county jail for not more than one year, or by both the fine and imprisonment.[18]

5. It is a misdemeanor to submit a bid to a public agency to act in the capacity of a contractor without having the license therefor, unless the person is exempt from the license law or the bid is on a state project governed by California Public Contract Code § 10164.[19]

6. A bid submitted to a public agency by a contractor who is not licensed shall be considered nonresponsive and shall be rejected by the public agency. The public agency shall, before awarding the bid, verify that the contractor was properly licensed when the contractor submitted the bid.[20]

---

[16] Cal. Bus. & Prof. Code § 7028.
[17] Cal. Lab. Code §§ 1020–1024.
[18] Cal. Bus. & Prof. Code § 7027.3.
[19] Id. § 7028.15(a).
[20] Id. § 7028.15(e).

On occasion, an unlicensed contractor will bid on a private work of improvement. The question arises as to whether other competitive bidders on the same project have any rights against the unlicensed person bidding against them.

In *Settimo Associates v. Environ Systems, Inc.*,[21] Settimo, which held a "B" license, sued Environ, alleging intentional or negligent interference with prospective economic advantage. The complaint alleged that Settimo had twice submitted the lowest bid by any duly licensed contractor for installing a computer support facility. However, Settimo lost each contract to Environ, which held only a specialty contractor's license, which was not sufficient for some of the work involved, but the contracts had been awarded to Environ because it submitted the lowest bid. Settimo claimed that Environ's lack of a general contracting license satisfied the element of unlawful contract interference, so Environ should reimburse Settimo for its lost profits. At the beginning of trial, Environ objected to the introduction of all evidence, alleging that the complaint was fatally defective. The trial court treated that objection as a motion for judgment on the pleadings, granted the motion, and entered judgment for Environ. Settimo appealed, and the appellate court sustained the trial court. The appellate court held that, even though Environ was not properly licensed at the time it was awarded the contracts, that fact did not give rise to an action for unlawful interference with contracts. The court held that there was no statutory authority requiring a private entity to accept bids only from duly licensed contractors. The court stated that the Business and Professions Code, which regulates the licensing of contractors, does not create or deny any civil remedy to bidders who lose projects to unlicensed competitors. The court acknowledged that Environ's conduct amounted to a misdemeanor and foreclosed any possibility of its suing to enforce the contracts, citing Business and Professions Code §§ 7028(a) and 7031. The court concluded, however, that those regulatory sections fall within the responsibility of the Contractors State License Board to take appropriate action in the case of any such misconduct, and that those sections do not create any action for civil damages to a competing bidder.

The bar of Business and Professions Code § 7031 has been applied to a company that contracted with a city to run a "fill" operation on city-owned land. In *K&K Services, Inc. v. City of Irwindale*,[22] K&K Services, Inc., which was not licensed as a contractor, contracted with the city for the right to dump fill material on city-owned property. Under their contract, K&K would provide, place, and compact to 90 percent density clean earth and solid fill material; in addition, K&K was to pave a 300-foot road for incoming and outgoing hauling. The contract provided that the services of K&K were to be at no cost to the city, as K&K was to be "compensated" by virtue of having the right to charge others for any fill material placed on the project. During the progress of the job, the city became concerned that K&K was not placing the solid fill material in the first 15 to 20 feet of the fill to the required 90 percent compaction. A dispute over that issue arose, and K&K refused to correct the work that it had performed. K&K then

---

[21] 14 Cal. App. 4th 842 (1993).
[22] 47 Cal. App. 4th 818 (1996).

brought an action against the city, seeking damages for breach of contract, alleging that the city, by not allowing K&K to continue placing fill on the project, would cause K&K to lose $19 million in profits over the life of the contract. The city raised the defense that the work required a contractor's license, and therefore, K&K's suit was barred by Business and Professions Code § 7031. Based upon K&K's lack of a contractor's license, the city's motion for summary adjudication on its affirmative defense was granted. The court of appeal affirmed, holding that the term "compensation" as used in Business and Professions Code § 7031 included damages allegedly suffered by an unlicensed contractor and that the suit was therefore barred where either (1) the making of the contract or (2) the performance during which the breach occurs comes within the licensing requirement. The court held that, because K&K had agreed to process, grade, compact fill, and pave the project for the city, in exchange for which the city agreed to provide K&K with an exclusive renewable permit for fill rights, these fill rights were plainly "compensation" or an "agreed contract price" for K&K's work. Thus, K&K's lack of a contractor's license precluded recovery, per Business and Professions Code § 7031.

The *K&K Services* case and the case of *Hydrotech Systems, Ltd v. Oasis Water Park*[23] were examined in *Ranchwood Communities Ltd. Partnership v. Jim Beat Construction Co.*[24] In *Ranchwood Communities,* the court of appeal considered the issue of whether a general contractor who worked on a project, but is barred by Business and Professions Code § 7031 from bringing an action for recovery of compensation because the contractor was unlicensed, may nevertheless seek equitable indemnity from subcontractors that it hired to perform work on the project, on the basis that such subcontract work was negligently performed. In the case, an unlicensed contractor sought indemnity from its subcontractors because the unlicensed contractor was at the same time acting in the related capacity of a developer of the overall project and, as the developer, was subject to strict liability for construction defects. (For further discussion of strict liability for construction defects, see Chapter 8).

Ranchwood Park was a 325-unit project developed by Ranchwood Communities (Ranchwood), which acted as developers and general contractors of the project. Ranchwood never held a general contractor's license during the entire time the project was being designed and constructed (1981 to 1988). In 1993, the homeowners' association brought a construction defect action against Ranchwood, alleging that Ranchwood, as the developer/contractor, was liable for damages under strict liability, breach of express and implied warranties, negligence, nuisance, and negligent misrepresentation. Ranchwood answered the complaint and cross-complained against its subcontractors, seeking equitable, implied contractual, express contractual, and total indemnity, as well as contribution and recovery on theories of negligence, breach of contract, breach of warranties, strict liability against certain suppliers, and declaratory relief regarding contractual

---

[23] 52 Cal. 3d 988 (1991).
[24] 49 Cal. App. 4th 1397 (1996).

| LICENSING | § 1.05 |

duties. The case also involved another project, by the name of Ventana, a 168-unit development in LaJolla. Sickles Kellogg Development Company, also not licensed as a general contractor, served as developer/general contractor for the first 85 homes built between 1984 and 1986. Sickles, which had hired a licensed general contractor to supervise construction, contracted directly for trade work with the subcontractors against whom they had filed cross-complaints on the same theories as mentioned above. Sickles's subcontractors filed motions for summary judgment on both cross-complaints, arguing that because both developer/general contractors (Ranchwood and Sickles) were not licensed as general contractors, their entire cross-complaints were barred by Business and Professions Code § 7031.

The subcontractors sought to rely upon *Hydrotech Systems Ltd. v. Oasis Water Park*,[25] arguing that the cross-complaints were actions for compensation for work performed pursuant to illegal contracts entered into by unlicensed persons. Ranchwood and Sickles argued that section 7031 was not a bar to any of the causes of action set forth in their cross-complaints because, as developers and general contractors, they were subject to strict liability and should be allowed to spread their loss among all negligent parties, including the subcontractor cross-defendants. They also contended that they were exempt from the licensing requirements because they were "owner/builders" within the meaning of Business and Professions Code § 7044.

The trial court granted summary judgment in favor of the subcontractors and dismissed both cross-complaints on the grounds that Ranchwood and Sickles were unlicensed and therefore the causes of action were completely barred by Business and Professions Code § 7031. The court of appeal sustained summary judgment with regard to the causes of action based in contract, holding that they could not be maintained without the appropriate contractors' licenses. However, the court reversed the summary judgments as to the indemnity causes of action.

The court first acknowledged that the purpose of the licensing law is to protect the public from incompetence and dishonesty of those who provide building and construction services (citing *Lewis & Queen v. N.M. Ball Sons*[26]). The barring of claims by unlicensed contractors for payment is intended to deter such persons from offering their services or accepting solicitations for their work. A construction contract entered into by an unlicensed entity is illegal. Therefore, regardless of the equities of the case and how the causes of action are characterized, Business and Professions Code § 7031 bars all actions that effectively seek recovery of "compensation" for illegal, unlicensed contract work. As a result, an unlicensed contractor cannot recover either for the agreed contract price or for the reasonable value of the labor and materials furnished.

In a garden-variety dispute over money owed, an unlicensed contractor cannot evade the license law by alleging that the express or implied promise to pay for the contractor's work was fraudulent. The court noted that in *Construction*

---

[25] 52 Cal. 3d 988 (1991).
[26] 48 Cal. 2d 141 (1957).

Financial LLC v. Perlite Plastering Co.,[27] the court of appeal, citing the *Hydrotech* case, again strictly enforced the penalty prescribed in Business and Professions Code § 7031(a), "regardless of the merits of the cause of action" and the harshness of the result to the unlicensed contractor claimant. *Hydrotech* was also cited for the principle that section 7031 does apply even when the person for whom the work was performed knew of the invalidity of the contractor's license. (The *Construction Financial* case is fully discussed in § 1.06[B][2].)

The court further noted that in *Vallejo Development Co. v. Beck Development Co.*[28] (discussed in §1.01) a master developer who agreed to install infrastructure improvements was not allowed to prosecute any of its claims for compensation. The *Ranchwood* court then pointed out that neither the *Hydrotech* case, the *Vallejo* case, nor *K&K Services, Inc.* discussed indemnity issues where a developer of a project, licensed or unlicensed, also acts as its own general contractor and seeks to obtain indemnity from subcontractors who furnished labor, materials, and services to the project. The court therefore looked to the definition of the word "compensation" under the licensing law to see whether indemnity and contribution fit within that definition. The court looked at *K&K Services, Inc.*, which held that "compensation" under Business and Professions Code § 7031 included granting of a city permit for fill rights at a quarry rather than cash and that such compensation was subject to the statutory bar. The court went on to note, however, that the licensing law has been held not to bar certain other forms of recovery by unlicensed contractors. As an example, the court cited two situations in which unlicensed contractors who were also parties to separate contracts were not barred from seeking relief for breach of their separate contracts: *McCarroll v. L.A. County*[29] (separate collective bargaining contract) and *Davis Co. v. Superior Court*[30] (breach of warranty to furnish appropriate materials).

In addition, the court noted that there are cases permitting unlicensed contractors to assert a setoff, based upon a contract for building services, notwithstanding that the contract was otherwise unenforceable due to the absence of a license, citing *Marshal v. Von Zumwalt*,[31] *Steinwinter v. Maxwell*,[32] *Dahl-Beck Electric Co. v. Rogge*,[33] *S&Q Construction Co. v. Palma Seia Development Organization*,[34] and *Culbertson v. Cizek*.[35] The theory of those cases was that, although the unlicensed contractors could not recover for work in their own actions because they would thereby be seeking "compensation" for their services, their lack of licenses did not bar them from offsetting, as a defense to any claims made against them, that which would otherwise have been due under their illegal contracts. The

---

[27] 53 Cal. App. 4th 170 (1997).
[28] 24 Cal. App. 4th 972 (1994).
[29] 49 Cal. 2d 45 (1957).
[30] 1 Cal. App. 3d 156 (1969).
[31] 120 Cal. App. 2d 807 (1953).
[32] 183 Cal. App. 2d 34 (1960).
[33] 275 Cal. App. 2d 893 (1969).
[34] 179 Cal. App. 2d 364 (1960).
[35] 225 Cal. App. 2d 451 (1964).

court also cited *Gaines v. Eastern Pacific*,[36] wherein a contractor who was unlicensed when a subcontract was entered into, but became licensed during its performance, was allowed to recover on a cross-complaint that alleged that the cross-defendant subcontractor had breached the contract by refusing to perform the work properly, so that the contractor was required to pay bills and charges for labor, materials, and mechanics' liens and to incur additional expenses for completion of the project after the subcontractor was ordered to stop work. The court in *Gaines* concluded that those expenses did not represent "compensation for the performance" of the acts called for in the contract, relying upon *American Sheet Metal v. Em-Kay Engineering*,[37] in which it was held that a counterclaim for defective performance that sought damages incurred as a result of the plaintiff's breach was not prohibited by section 7031 because such a cause of action was not within the scope or purpose of the code section and that the contractor was therefore not suing to recover "compensation" under the contract. The court also cited *E.C. Ernst, Inc. v. County of Contra Costa*[38] for the proposition that a contractor with an expired license may be allowed to recover damages for delay in the work and the necessity to perform it out of sequence. From those cases, the court held that the prohibition on actions for "compensation" by unlicensed contractors is subject to numerous exceptions in addition to the express statutory exemptions set forth in the Business and Professions Code.

Turning then to the analysis of the developer/contractors' cross-complaints for indemnity against the subcontractors, the *Ranchwood* court noted that there is a distinction between "contractors" and "developers." Whereas a developer builds improvements for sale to the public after they are completed, a contractor builds improvements on the land of the owner pursuant to a construction contract. A developer may be held liable for defective construction on a strict liability theory, as well as theories of negligence, breach of warranty, nuisance, and fraud or negligent misrepresentation. Noting that indemnity arises from two situations—express contractual language or equitable considerations—the court found that implied contractual indemnity is a form of equitable indemnity.

Thus, the legal theories in cases by developer/contractors against their subcontractors fall into two groups: (1) equitable indemnity, implied contractual and total indemnity, contribution, and recovery in negligence; and (2) express indemnity, breach of contract, breach of warranties, and declaratory relief. As noted above, the developer/contractor's causes of action for express contractual indemnity, breach of contract, breach of warranty, and declaratory relief are based upon illegal and unenforceable contracts and are therefore barred. With regard to the various forms of equitable indemnity, the court took into account the following matters of policy with regard to whether or not an illegal contract may be enforced: (1) the nature of the prescribed penalty; (2) the goal of deterring illegal conduct; and (3) the policy of avoiding disproportionately harsh forfeitures. The

---

[36] 136 Cal. App. 3d 679 (1982).
[37] 478 F. Supp. 809 (E.D. Cal. 1979).
[38] 555 F. Supp. 122 (N.D. Cal. 1982).

court reasoned that the developer/contractors' entitlement to equitable indemnity has a separate genesis from their express contractual indemnity claims. Thus, the court held that—since the primary relief sought by the developer/general contractors was not merely "compensation" for work performed, but rather spreading the cost of the developer/contractors' damages incurred due to defective work performed by others—a developer/contractor being held strictly liable for construction defects, who also acted as an unlicensed general contractor, should not be foreclosed from seeking relief from allegedly negligent subcontractors on theories of equitable indemnity, implied contractual and total indemnity, and contribution. Any ban on a subcontractor's liability for equitable indemnity would result in an overly strict interpretation of the licensing law and would be a windfall and not within the protective purpose of the license law. The court also held that negligence claims, like those for equitable indemnity, are outside the scope of the contractual claims and are thus not barred by the lack of a license. As a result, the *Ranchwood* court reversed the summary judgments and remanded the case for further proceedings in accordance with the principles set forth in that opinion.

§ 1.06 EXEMPTIONS FROM THE LICENSE LAW

[A] In General

A laborer working for wages need not be licensed.[39] In *Blew v. Horner*[40] the plaintiff was an employee of Symons. Symons had been employed by the defendant to perform a roofing subcontract on a project the defendant owned. Symons was unlicensed. The plaintiff was injured on the job and sued the defendant. The court held that because Symons was unlicensed, the plaintiff became an employee of the defendant and, therefore, was entitled only to workers' compensation.

[B] Case Law Exemptions from the License Law

The courts have allowed some deviation from the statutory bar of California Business and Professions Code § 7031 when the unlicensed contractors were not strictly seeking "compensation" for services rendered.

Cases have permitted unlicensed contractors to assert a setoff, based upon a contract for building services, notwithstanding the fact that the contract was otherwise unenforceable due to the absence of a license.[41] The theory of these cases is that although the unlicensed contractors could not recover for work in their own actions because they would thereby be seeking "compensation" for

---

[39] Cal. Bus. & Prof. Code § 7053.
[40] 187 Cal. App. 3d 1380 (1986).
[41] Marshall v. Von Zumwalt, 120 Cal. App. 2d 807 (1953); Steinwinter v. Maxwell, 183 Cal. App. 2d 34 (1960); Dahl-Beck Elec. Co. v. Ragge, 275 Cal. App. 2d 893 (1969); S&Q Constr. Co. v. Palma Seia Dev. Org., 179 Cal. App. 2d 364 (1960); Culbertson v. Cizek, 225 Cal. App. 2d 451 (1964); Ranchwood Communities Partnership v. Jim Beat Constr. Co., 49 Cal. App. 4th 1397 (1996).

their services, their lack of license did not bar them from offsetting, as a defense to any claims made against them, that which would otherwise have been due under their illegal contracts.

In *Gaines v. Eastern Pacific*,[42] a contractor was unlicensed when a subcontract was entered into. The contractor became licensed during performance. In an action against the contractor, the contractor filed a cross-complaint that alleged that the cross-defendant subcontractor had breached the contract by refusing to perform properly so that the contractor was required to pay bills and charge for labor, materials, and mechanics' liens and to incur additional expenses for completion. The court held that those expenses were not "compensation for performance" of the acts called for in the subcontract, and therefore, relying on *American Sheet Metal v. Em-Kay Engineering*,[43] the court held that the cross-complaint was not barred by Business and Professions Code § 7031.[44]

[1] Material Supplier Exemption

A material supplier, who only furnishes materials or supplies without fabricating them into or consuming them in the performance of the work of the contractor, need not be licensed.[45]

The courts in California have had a difficult time determining who falls within the material supplier exemption of California Business and Professions Code § 7052. In the case of *Walke v. Thornsbery*,[46] a manufacturer sold a prefabricated metal restroom to a general contractor. The manufacturer was not licensed. The agreement required the manufacturer to furnish, assemble, and install the prefabricated restroom on a concrete foundation prepared by the general contractor. After the installation of the restroom, the contractor failed to pay the manufacturer the purchase price of the product and claimed that the manufacturer was barred from recovery under Business and Professions Code § 7031. The trial court concluded that the manufacturer was not required to have a contractor's license to place a prefabricated structure on a site prepared by the contractor and therefore was not precluded from recovery.

The appellate court held that the manufacturer's activity was within the purview of the exemption provided by Business and Professions Code § 7045, which exempts from application of the Contractors License Law the sale or installation of any finished products, materials, or articles of merchandise that do not become a fixed part of the structure. The court stated that "[w]hether the goods installed became a fixed part of the structure [is] a question of fact for the trier of fact." Several older decisions held that such things as cold storage plants used in the operation of a hatchery and poultry ranch were within the exemption,

---

[42] 136 Cal. App. 3d 679 (1982).
[43] 478 F. Supp. 809 (E.D. Cal. (1979).
[44] *See also* E.C. Ernst, Inc. v. County of Contra Costa, 555 F. Supp. 122 (N.D. Cal. 1982), which held that a contractor with an expired license could recover damages for delay and working out of sequence.
[45] Cal. Bus. & Prof. Code § 7052.
[46] 97 Cal. App. 3d 842 (1979).

because the equipment installed was prefabricated and did not become a permanent fixture.[47]

In *E.A. Davis & Co. v. Richards*,[48] the court found that the installation of a patented prefabricated kitchen unit (consisting of seven wall cabinets, six base cabinets, a dishwasher, and a sink), which was attached to the floor and walls, came within the exemption to the requirement that the manufacturer be a licensed contractor, despite the fact that the installation required minor plumbing, electrical, and linoleum work incidental to installation of the finished unit. In contrast, if the facts show that substantial installation will be required and the finished product is to become affixed to the underlying realty, the courts will find that the party installing the goods must be licensed.[49] In the *King* case, the court found that the subsurface installation of a prefabricated swimming pool was a substantial installation affixed to the underlying realty, and thereby subjected the contractor to the licensing law. In the *Johnson* case, the court found that installing a sprinkling system, building various signs and setting them in concrete, and excavating and constructing dugouts were more than merely incidental construction activities in connection with the furnishing of finished goods and therefore subjected the contractor to the licensing laws. Thus, the question in each case of whether the party furnishing the work is required to be licensed is a question of fact as to whether the finished products, materials, or articles of merchandise become a fixed part of the structure. If they do not become a fixed part of the structure, then under Business and Professions Code § 7045, the contractor will not be required to be licensed. If they do become a fixed part of the structure, the license law will apply.

In the case of *Steinbrenner v. Waterbury (J.A.) Construction Co.*,[50] the plaintiff entered into an agreement with a general contractor to build cabinets for a particular building. These were not standard "open stock" cabinets. There were custom designed and custom produced. The plaintiff's employees delivered the finished cabinets to the jobsite but did not install them. The court held that the plaintiff was exempt from the license law under Business and Professions Code § 7052.

In *Jackson v. Pancake*,[51] the plaintiff was employed by an owner to do plumbing work. The plaintiff rendered deferred billing, and there was no withholding from the money paid to the plaintiff. There was also evidence that the defendant owner had decided to remodel their theatre on a "force account" basis, hiring carpenters and others who worked under the supervision of the owner and performed the work as directed. The plaintiff billed on an hourly basis. The court held that the plaintiff was an "employee," not a "contractor," and therefore was entitled to recover as an "employee" under Business and Professions Code § 7053. Furthermore, the plaintiff furnished materials incidentally under Business and

---

[47] *See, e.g.,* Costello v. Campbell, 81 Cal. App. 2d 452 (1947).
[48] 120 Cal. App. 2d 237 (1953).
[49] *See* King v. Hinderstein, 122 Cal. App. 3d 430 (1981); Johnson v. Mattox, 257 Cal. App. 2d 714 (1968).
[50] 212 Cal. App. 2d 661 (1963).
[51] 266 Cal. App. 2d 307 (1968).

Professions Code § 7052 and therefore did not act as a contractor in violation of Business and Professions Code §§ 7026 and 7031.

[2] Doctrine of Substantial Compliance

Prior to an amendment to the California Business and Professions Code § 7031, case law had developed what has been commonly referred to as the doctrine of substantial compliance. There are three leading cases in California on the subject.[52] Under this doctrine, a contractor who had not technically complied with the requirements for a license, but had a licensed person in charge of the project, was still protected as if licensed.

Effective January 1, 1990, the Code was amended by adding subdivision (d), which states: "The judicial doctrine of substantial compliance shall not apply to this section." The legislature again amended section 7031 in 1995, significantly qualifying subdivision (d) and adding new subdivision (e). The new language states:

> (d) The judicial doctrine of substantial compliance shall not apply under this section where the person who engaged in the business or acted in the capacity of a contractor has never been a duly licensed contractor in this state. However, the court may determine that there has been substantial compliance with licensure requirements under this section if it is shown at an evidentiary hearing that the person who engaged in the business or acted in the capacity of a contractor (1) had been duly licensed as a contractor in this state prior to the performance of the act or contract, (2) acted reasonably and in good faith to maintain proper licensure, and (3) did not know or reasonably should not have known that he or she was not duly licensed. Subdivision (b) of Section 143 does not apply to contractors subject to this subdivision.
>
> (e) The exceptions to the prohibition against the application of the judicial doctrine of substantial compliance found in subdivision (d) shall apply to all contracts entered into on or after January 1, 1992, and to all actions or arbitrations arising therefrom, except that the amendments to subdivisions (d) and (e) enacted during the 1994 portion of the 1993–94 Regular Session of the Legislature shall not apply to either of the following:
>
> (1) Any legal action or arbitration commenced prior to January 1, 1995, regardless of the date on which the parties entered into the contract.
>
> (2) Any legal action or arbitration commenced on or after January 1, 1995, if the legal action or arbitration was commenced on or after January 1, 1995, and was subsequently dismissed.

It is clear that the limited substantial compliance doctrine enacted in subdivision (d) is much more restricted in its scope than the judicially created substantial compliance doctrine of *Latipac* and its progeny. Under the subdivision

---

[52] Latipac, Inc. v. Superior Court, 64 Cal. 2d 278 (1966); Asdourian v. Araj, 38 Cal. 3d 276 (1985); Knapp Dev. & Design v. Pal-Mal Properties, Ltd., 173 Cal. App. 3d 423 (1985).

(d) exceptions, a court must hold an evidentiary hearing, during which the burden is on the contractor to establish the existence of the elements set forth in subdivision (d).

The first case on the subject after the 1990 amendment was *Hydrotech Systems, Ltd. v. Oasis Water Park.*[53] In this case, the California Supreme Court held that section 7031 bars not only contract actions by unlicensed contractors but tort actions for fraud as well. In its ruling, the court stated:

> Regardless of the equities, Section 7031 bars all actions, however they are characterized, which effectively seek "compensation" for illegal, unlicensed contract work. Thus, an unlicensed contractor cannot recover either for the agreed contract price or for the reasonable value of labor and materials. The statutory prohibition operates even where persons for whom the work was performed knew the contractor was unlicensed.
>
> It follows that an unlicensed contractor may not circumvent the clear provisions and purposes of Section 7031 simply by alleging that when the illegal contract was made, the other party had no intention of performing. Section 7031 places the risk of such bad faith squarely on the unlicenced contractor's shoulders. "Knowing that they will receive no help from the courts and must trust completely to each other's good faith, the parties are less likely to enter an illegal arrangement in the first place."[54]

It is clear that under the newly revised section 7031, as well as under the *Hydrotech* ruling, contractors must make certain that they are properly licensed throughout the course of performance if they have any hope to use the court system to recover amounts they contend are due. In *Pintail Plastering v. Mark Diversified Inc.*,[55] the court, interpreting and applying California law, held that the legislature's nullification of the doctrine of substantial compliance applies to contracts formed prior to the January 1, 1990, effective date of the amendment adding subdivision (d) to Business and Professions Code § 7031.

Pintail, in bankruptcy, filed an adversary proceeding in which it claimed $709,787.65 in damages for delays in its completion of a public works project in Stockton. The defendant general contractor filed a motion to dismiss the adversary proceeding on the ground that Pintail's license had expired on September 30, 1990, and was not renewed until January 4, 1991, and that, during the lapse period, Pintail performed work on the project. The bankruptcy court denied the motion to dismiss, and the general contractor appealed to the district court.

The district court rejected Pintail's argument that because the contract was entered into before the effective date of the statutory amendment, Pintail had a constitutional right to continued application of the judicial doctrine of substantial compliance. In holding that the operative date for application of new section 7031(d) is not the date of execution of the contract, but the time when the contract was actually performed, the district court reasoned that the legislature did

---

[53] 52 Cal. 3d 988 (1991).
[54] *Id.* at 997–998 (citations omitted).
[55] 50 F.3d 15 (9th Cir. 1995).

not change the contractor's licensing statute; it only made clear that a judicial doctrine developed by the courts is inapplicable to that statute. Moreover, the importance of deterring unlicensed persons from engaging in the construction contracting business outweighs any harshness of the result as to these particular parties.

Under the newly revised subdivision (d), referenced *supra,* there are three new requirements for establishing substantial compliance. It now must be shown at an evidentiary hearing that the person acting in the capacity of a contractor "(1) had been duly licensed as a contractor in this state prior to the performance of the act or contract, (2) acted reasonably and in good faith to maintain proper licensure, and (3) did not know or reasonably should not have known that he or she was not duly licensed."

By the new amendment, the legislature has eliminated the restriction to "inadvertent clerical error," as well as the issue of the contractor's negligence, in favor of a standard based on the contractor's reasonableness and good faith. Furthermore, the contractor's prior licensure may have been at an undefined prior time rather than the more restrictive "90 days immediately preceding" the subject contract.

In addition, subdivision (e) has been amended to change the retroactive effectiveness of the subdivision (d) changes, which shall be applied to all contracts entered into on or after January 1, 1992, and to all actions or arbitrations arising therefrom, except that the 1994 amendments to subdivisions (d) and (e) shall not apply to either of the following: (1) any legal action or arbitration commenced prior to January 1, 1995, regardless of the date on which the parties entered into the contract; and (2) any legal action or arbitration commenced on or after January 1, 1995, if the legal action or arbitration was commenced on or after January 1, 1995, and was subsequently dismissed.

Business and Professions Code § 7031(d) was interpreted for the first time in *Construction Financial LLC v. Perlite Plastering Co.*[56] Diversified Gypsum Corporation (Diversified) was formed in 1991 to do subcontract work for Perlite Plastering Company, Inc. (Perlite). One of the principals of Diversified (Johnson) held a valid specialty license (Class C) in drywall (C-9) and painting and decorating (C-33), and owned another company for which he was the responsible managing operator (RMO). On the advice of a representative of Perlite, Diversified did not apply for specialty licenses or designate Johnson as RMO; instead, Diversified applied for and obtained a Class B general contractor's license, through a long-time employee of Perlite, Hector, who went to work for Diversified as its foreman and responsible managing employee (RME). Thus, Diversified held a Class B general contractor's license.

Hector worked for Diversified until June 2, 1992, at which time he was either fired (as Hector understood matters) or it was suggested that he take a long-term leave of absence to deal with certain health and family matters (as Johnson understood matters). Hector then went to work for other companies. Johnson, claiming to believe that Hector had only gone on a long-term leave, did

---

[56] 53 Cal. App. 4th 170 (1997).

not notify the Contractors State License Board (CSLB) or take steps to replace Hector with himself or another as an RME or RMO. Shortly after Hector's departure, Diversified commenced work under a subcontract with Perlite to install drywall in a new library at Pasadena City College.

Moran Construction Company (Moran) was the general contractor, and National Union Fire Insurance Company of Pittsburgh (National Union) was the surety on Moran's payment bond. Diversified's work on the project was scheduled for completion on November 25, 1992; however, as a result of delays, the work was not completed until September 9, 1993. Diversified blamed Perlite for the delays and brought an action against Perlite, claiming damages in excess of $800,000 for breach of contract and for recovery on the Moran payment bond. The case proceeded to a bifurcated trial, pursuant to which the first issue to be tried was whether Diversified held a valid contractor's license at the time it performed the work sued upon; or, if not, whether Diversified was nevertheless in substantial compliance with the licensing requirement. A three-day "evidentiary hearing" was held pursuant to section 7031(d), at the conclusion of which the trial court issued a statement of decision finding that the Class B general contractor's license, which Diversified held at least until June 2, 1992, when Hector left its employ, did not authorize Diversified to enter into a contract with Perlite because the contract did not include at least three trades as required by California Code of Regulations title 16, § 834(b).

The trial court further found that because Diversified's RME, Hector, was no longer employed by Diversified when it commenced work on the project, Diversified did not have a valid contractor's license of any kind at the time it was performing the work. In spite of the fact that Johnson's specialty licenses would have authorized the work, had those licenses been held by Diversified, the trial court concluded that because Diversified did not have a license authorizing its work on the project, section 7031 precluded Diversified from recovering against the bonding company unless it could prove substantial compliance.

The trial court noted that Diversified would have been found in substantial compliance with section 7031 if the case were still governed by the judicial doctrine of substantial compliance, as set forth in *Latipac, Inc. v. Superior Court*[57] and *Asdourian v. Araj*.[58] However, the judicial doctrine of substantial compliance was superseded by statute. (As noted above, section 7031(d) was amended, effective January 1, 1990, to state: "The judicial doctrine of substantial compliance shall not apply to this section." Thus, the present formulation of section 7031(d) came into effect on January 1, 1995.)

The trial court ruled that the Diversified case came under the version of section 7031(d) in effect from January 1, 1991, until the most recent amendment took effect. Under the governing version of the statute at the time the work in this particular case was being performed, in order to prove substantial compliance, Diversified would have to prove (1) that the contractor was a duly licensed contractor during any portion of the 90 days immediately preceding the performance

---

[57] 64 Cal. 2d 278 (1966).
[58] 38 Cal. 3d 276 (1985).

of the contract sued upon; (2) the contractor's category of licensure would have authorized performance of the contract; and (3) noncompliance with the licensure requirement either was the result of inadvertent clerical error or was not the result of negligence of the contractor.

With regard to point 1, the trial court found that Diversified, through Hector, did have a contractor's license within the 90 days preceding commencement of the work. However, point 2 was not satisfied because Hector's Class B license did not authorize the specialty work performed under the subcontract with Perlite. Finally, the trial court found that Diversified's failure to have a license while performing the work was due to its own negligence, in that Johnson had allowed Hector to remain on what Johnson erroneously believed was a long-term leave for 16 months without inquiring whether Hector planned to return. The court noted that, if Johnson had inquired, he would have learned that Hector believed he had been fired and had gone to work for another firm. Thus, Diversified could not prove substantial compliance as defined in the applicable version of section 7031(d).

Diversified appealed, contending that (1) in light of the decision in *Home Depot USA, Inc. v. Contractors State License Board*,[59] issued after entry of judgment in this case, the trial court was incorrect in finding that Diversified's Class B license would not have authorized its work on the project; (2) because the *Home Depot* case invalidated section 834(b), a general contractor can perform specialty contractor work; (3) Diversified's failure to have a valid license was not due to its own negligence; and (4) Diversified had fulfilled the legislative purpose of the law and should therefore be found in substantial compliance.

The court of appeal first noted that the Contractors License Law provides a comprehensive scheme governing contractors doing business in California and reflects a strong public policy in favor of protecting the public against unscrupulous and incompetent contracting work. The supreme court had noted in *Hydrotech Systems Ltd. v. Oasis Water Park*[60] and *Home Depot USA, Inc. v. Contractors State License Board*[61] that the purpose of the licensing law is to protect the public from incompetence and dishonesty of those who provide building and construction services and that the licensing requirements provide minimal assurance that all persons offering such services in California have the requisite skill and character to understand applicable local laws and codes and know the rudiments of administering a contracting business.

The court of appeal noted that if a corporation applies for a contractor's license, the corporation must qualify through either an RMO or an RME who is qualified in the licensing classification for which the corporation is applying.[62] The RMO or RME must be a bona fide officer or employee of the corporation and must be actively engaged in the work covered by the license.[63] The qualifier

---

[59] 41 Cal. App. 4th 1592 (1996).
[60] 52 Cal. 3d 988 (1991).
[61] 41 Cal. 4th 1592 (1996).
[62] *See* Cal. Bus. & Prof. Code § 7068(b)(3).
[63] *Id.* § 7068(d).

must exercise direct supervision over the work for which the license is issued to the extent necessary to secure full compliance with the provisions of the license law.[64] If a corporation's qualifier leaves the corporation's employ, the corporation or the qualifier must notify the Registrar of Contractors in writing and the corporation must replace the qualifier within 90 days (by filing a new application designating a new qualifier). If the corporation's qualifier is not replaced within the 90-day period, the corporation's license is automatically suspended.[65]

The court of appeal next noted the *Home Depot* case, which had ruled California Code of Regulations title 16, § 834(b) invalid, and ruled that Diversified's Class B general contractor's license would have authorized the work. Accordingly, the trial court's finding to the contrary was incorrect. Still, under section 7031(d) in effect when this action was commenced, a contractor that was not duly licensed could maintain an action on the contract only by satisfying all three of the elements outlined earlier. In this particular case, the trial court properly found that Diversified's noncompliance with the license law was caused by its negligence, in that Johnson reasonably could and should have determined that Hector had gone to work for another company, and should have secured a substitute qualifier; Diversified had not presented any argument or evidence to overcome that factual finding. The trial judge had found that Johnson was supervising jobs for two companies, working long hours, dealing with change orders and other immediate problems, and did not think about licensing formalities such as whether he was acting as an RMO or not; however, these factors were simply not sufficient to overcome the conclusion that Johnson (and thus Diversified) had been negligent.

Diversified also argued on appeal that it should not be denied recovery against Perlite because Perlite was at all times aware of Diversified's license status, because Johnson specifically relied on the advice of Perlite's manager in connection with its license, and because the legislature had since provided a more lenient exception to the licensure requirement in the 1994 amendment to section 7031, and therefore the currently enacted standard should be applied. The court of appeal found those arguments unavailing. The court of appeal noted the supreme court's holding in *Hydrotech* that section 7031 does apply even when the person for whom the work was performed knew that the contractor was unlicensed. The supreme court also held (1) that a contractor may not plead reliance upon another person in determining what is required under the Contractors License Law and (2) that unlicensed contractors are held to have knowledge of the law's requirements.

The court of appeal then reviewed the legislative history underlying section 7031(d) and stated that the legislative history supports the view that the legislature intended to create only an extremely narrow exception to the licensure requirement, which would apply only when a contractor was without a license owing to circumstances truly beyond his or her control. Based upon that legislative intent, there was no basis for applying the doctrine of substantial compliance

---

[64] *Id.* § 7068.1.
[65] *Id.* § 7068.2.

under the facts of this case. Courts are not unmindful of the evident harshness of such results; however, the legislature provided (in the 1991 amendment to section 7031) that except as provided in subdivision (d), the bar of subdivision (a) should apply "regardless of the merits of the cause of action" brought by an unlicensed contractor.

It is important to note that the court was interpreting the amendment to section 7031 in effect from 1991 to 1994. The current versions of section 7031(d) and (e), enacted in 1994, are quoted earlier in this section.

[3] Owner/Builder Exemption

Owner/builders are exempt from the Contractors License Law. Specifically, California Business and Professions Code § 7044 states that it does not apply to any of the following:

> (a) An owner of property building improvements thereon who does the work himself through his or her own employees with wages as their sole compensation[,] provided none of the structures are intended or offered for sale;
>
> (b) An owner of property building or improving structures thereon who contracts for such a project with a subcontractor or under the subcontractors' license law. This exemption shall apply to the construction of single family residential structures only if four or fewer of these structures are intended or offered for sale in a calendar year. This limitation shall not apply if the homeowner of the property contracts with a general contractor for the construction.
>
> (c) A homeowner improving his or her principal place of residence or appurtenances thereto, provided that all of the following conditions exist:
>
> (1) The work is performed prior to sale.
>
> (2) The homeowner has actually resided in the residence for the 12 months prior to the completion of the work.
>
> (3) The homeowner has not availed himself or herself of the exemption in this subdivision on more than two structures more than once during any three-year period.
>
> In all actions brought under this chapter, proof of the sale or offering for sale of any such structure by the owner-builder within one year after completion of same constitutes a rebuttable presumption affecting the burden of proof that such structure was undertaken for purposes of sale. Except as otherwise provided in this section proof of the sale or offering for sale of five or more structures by the owner-builder within one year after completion constitutes a conclusive presumption that the structures were undertaken for purposes of sale.

[C] License on Federal Projects

Prior to 1982, Business and Professions Code § 7047 provided that the license law did not apply to any construction, alteration, improvement, or repair carried on within the limits and boundaries of any site or reservation the title of which rests in the federal government. That section was repealed in 1982. The Ninth Circuit Court of Appeals held, however, that enforcement of the California Contractors License Law would interfere with federal contracting and is therefore improper.

In *Gartell Construction, Inc. v. Aubry*,[66] the Department of Industrial Relations attempted to enforce a civil penalty of $100 per employee per day against an unlicensed contractor that was performing work for the Department of the Navy at the Marine Air Corps Station in El Toro. As a federal contractor, Gartell had met the "responsibility" requirements of 48 C.F.R. § 52.236-7 (1990). That particular federal regulation defines *responsible contractor* as one with financial resources, experience, organization, and technical qualifications who has a satisfactory record of performance, integrity, judgment, and skill. The Division of Labor Standards Enforcement cited Gartell for violating California Labor Code § 1021, which provides that unlicensed contractors are subject to civil penalties of $100 per employee for each day of employment. The superior court entered a judgment against Gartell for $57,600. Gartell went to federal court to seek an injunction and declaratory relief to enjoin the Department of Industrial Relations of the State of California from enforcing Labor Code § 1021 against Gartell.

The Ninth Circuit Court of Appeals rendered a judgment in favor of Gartell. The court stated that the United States Supreme Court ruled on this issue 35 years ago, when it held that a state licensing requirement is invalid as applied against a contractor with the federal government because it results in interference with federal government functions and is in conflict with the Federal Procurement Regulations, citing *Leslie Miller, Inc. v. Arkansas*.[67] The court noted that the factors California considers before granting a contractor's license are similar to the factors considered by the federal government in qualifying a contractor to bid for federal work under the Code of Federal Regulations. The court stated that enforcement of the California law would frustrate the federal policy of selecting the lowest responsible bidder and that enforcement of the license law interferes with government contracting. The state argued that Congress had consented to enforcement of the license law because, under standard contract documents, federal contractors are required to obtain necessary licenses and comply with applicable state laws. The court noted, however, following the *Leslie Miller* decision, that state licensing laws cannot be "applicable" or compliance with them "necessary" when such laws are preempted by federal law.

§ 1.07 NOTICE REQUIRED BY LICENSE LAW

Section 7018.5 was added to the California Business and Professions Code effective January 1, 1993, requiring that contractors give a lengthy "notice to owner" before entering into a home improvement contract or swimming pool contract. The notice form was amended to read as follows:

---

[66] 940 F.2d 437 (9th Cir. 1991).
[67] 352 U.S. 187, 77 S. Ct. 257 (1956).

NOTICE TO OWNER

Under the California Mechanics' Lien Law, any contractor, subcontractor, laborer, supplier, or other person or entity who helps to improve your property, but is not paid for his or her work or supplies, has a right to place a lien on your home, land, or property where the work was performed and to sue you in court to obtain payment.

This means that after a court hearing, your home, land, and property could be sold by a court officer and the proceeds of the sale used to satisfy what you owe. This can happen even if you have paid your contractor in full if the contractor's subcontractors, laborers, or suppliers remain unpaid.

To preserve their rights to file a claim or lien against your property, certain claimants such as subcontractors or material suppliers are each required to provide you with a document called a "Preliminary Notice." Contractors and laborers who contract with owners directly do not have to provide such notice since you are aware of their existence as an owner. A preliminary notice is not a lien against your property. Its purpose is to notify you of persons or entities that may have a right to file a lien against your property if they are not paid. In order to perfect their lien rights, a contractor, subcontractor, supplier, or laborer must file a mechanics' lien with the county recorder which then becomes a recorded lien against your property. Generally, the maximum time allowed for filing a mechanics' lien against your property is 90 days after substantial completion of your project.

TO INSURE EXTRA PROTECTION FOR YOURSELF AND YOUR PROPERTY, YOU MAY WISH TO TAKE ONE OR MORE OF THE FOLLOWING STEPS:

(1) Require that your contractor supply you with a payment and performance bond (not a license bond), which provides that the bonding company will either complete the project or pay damages up to the amount of the bond. This payment and performance bond as well as a copy of the construction contract should be filed with the county recorder for your further protection. The payment and performance bond will usually cost from 1 to 5 percent of the contract amount depending on the contractor's bonding ability. If a contractor cannot obtain such bonding, it may indicate his or her financial incapacity.

(2) Require that payments be made directly to subcontractors and material suppliers through a joint control. Funding services may be available, for a fee, in your area which will establish voucher or other means of payment to your contractor. These services may also provide you with lien waivers and other forms of protection. Any joint control agreement should include the addendum approved by the registrar.

(3) Issue joint checks for payment, made out to both your contractor and subcontractors or material suppliers involved in the project. The joint checks should be made payable to the persons or entities which send preliminary notices to you. Those persons or entities have indicated that they may have lien rights on your property, therefore you need to protect yourself. This will help to insure that all persons due payment are actually paid.

(4) Upon making payment on any completed phase of the project, and before making any further payments, require your contractor to provide you with unconditional "Waiver and Release" forms signed by each material supplier, subcontractor, and laborer involved in that portion of the work for which payment was made. The statutory lien releases are set forth in exact language in Section 3262 of the Civil Code. Most stationery stores will sell the "Waiver and Release" forms if your contractor does not have them. The material suppliers, subcontractors, and laborers that you obtain releases from are those persons or entities who have filed preliminary notices with you. If you are not certain of the material suppliers, subcontractors, and laborers working on your project, you may obtain a list from your contractor. On projects involving improvements to a single-family residence or a duplex owned by individuals, the persons signing these releases lose the right to file a mechanics' lien claim against your property. In other types of construction, this protection may still be important, but may not be as complete.

To protect yourself under this option, you must be certain that all material suppliers, subcontractors, and laborers have signed the "Waiver and Release" form. If a mechanics' lien has been filed against your property, it can only be voluntarily released by a recorded "Release of Mechanics' Lien" signed by the person or entity that filed the mechanics' lien against your property unless the lawsuit to enforce the lien was not timely filed. You should not make any final payments until any and all such liens are removed. You should consult an attorney if a lien is filed against your property.

§ 1.08   NOTICE REQUIRED BY CONTRACT

Every licensed contractor taking on prime contracts must include in prominent type in the contract the following statements:

Section 7030. Licensee's Statements on Contracts:

"Contractors are required by law to be licensed and regulated by the Contractors' State License Board which has jurisdiction to investigate complaints against contractors if the complaint regarding a patent act or omission is filed within four years of the date of the alleged violation. A complaint regarding a latent act or omission pertaining to structural defects must be filed within 10 years of the date of the alleged violation. Any questions concerning a contractor may be referred to the Registrar, Contractors' State License Board, P. O. Box 26000, Sacramento, California 95826."

All licensed contractors must also include their license number in contracts, subcontracts, calls for bid, and advertising:

Section 7030.5. Inclusion of License Number in Contracts, Advertising, Etc.

Every person licensed pursuant to this chapter shall include his license number in: (a) all construction contracts; (b) subcontracts and calls for bid; and (c) all forms of advertising, as prescribed by the Registrar of Contractors, used by such a person.

LICENSING

§ 1.09

The board has adopted the following rule as to what constit[utes]

794.1 License Number Required in Advertising

As used in Section 7030.5 of the Code, the term advertisi[ng is] not limited to the following: any card, contract proposal, si[gn let]tering on vehicles, registered in this or any other state, bro[chure,] circular, newspaper, magazine, airwave transmission and a[ny direc]tory under any listing denoting "Contractor" or any word o[f sim]ilar import or meaning requesting any work for which a license is required by the Contractors License Law.

Upon a showing of good cause, the registrar may grant an exemption to a licensee engaged in interstate contracting from the requirement that the licensee's license number be included in any advertising lettering on a vehicle registered in this or any other state. A request for an exemption shall be submitted on a form prescribed by the registrar.

NOTE: Authority cited: Section 7008, Business and Professions Code. Reference: Section 7030.5, Business and Professions Code.

§ 1.09 GROUNDS FOR DISCIPLINARY ACTION

The following list contains items that constitute a violation of California statute regarding licensed contractors. Several items on the list are discussed more fully in the sections noted. If any of these violations occurs, the contractor would be subject to disciplinary action.

1. Abandonment of a construction project[68] (see § 1.09[A])
2. Diversion or misapplication of funds[69] (see § 1.09[B])
3. Contractor's failure to pay the subcontractor a progress payment within 10 days of being paid[70] (see § 1.09[C].)
4. Willful departure from plans and specifications or from accepted trade standards for good and workmanlike construction[71] (see § 1.09[D])
5. Violation of safety provisions in the state Labor Code, resulting in death or serious injury[72]
6. Willful or deliberate disregard and violation of building laws, safety laws, labor laws, or compensation insurance laws[73]

---

[68] Cal. Bus. & Prof. Code § 7107.
[69] *Id.* § 7108.
[70] *Id.* § 7108.5
[71] *Id.* § 7109.
[72] *Id.* § 7109.5.
[73] *Id.* § 7010. enclosing an existing porch with a ceiling lower than the required by a cidy code and allowing a floor raised less than the code's clearance requirements were violations of section 7110. Barry v. Contractors State License Bd., 85 Cal. App. 2d 600 (1948).

7. Contractor's requirement that an unpaid laborer grant a release[74]
8. Labor commissioner's finding of willful or deliberate violation of the Labor Code[75]
9. Failure to make and keep records for inspection[76]
10. Misrepresentation in obtaining a license[77]
11. Material failure to complete the project for the price stated in the contract[78]
12. Avoidance or settlement of obligations for less than the full amount (see § 1.09[E])
13. Aiding, abetting, or conspiring with an unlicensed person to evade the license law (see § 1.09[F])
14. Material failure to comply with the license laws[79]
15. Willful or fraudulent act injuring another[80] (see § 1.09[G])
16. Acting as a contractor under an unlicensed name or unlicensed personnel[81] (see § 1.09[H])
17. Acting as a contractor under an inactive license[82]
18. Entering into a contract with an unlicensed contractor[83] (see § 1.09[I])
19. Willful failure to prosecute work diligently[84]
20. Willful failure to pay money when due for materials or services, or false denial of liability to obtain discount or delay[85] (see § 1.09[J])
21. Conviction of a felony related to qualification, functions, or duties[86]
22. Violation of any rules and regulations of the Contractors State License Board or regulations issued by the registrar.[87]

---

[74] Cal. Bus. & Prof. Code § 7110.1.

[75] *Id.* § 7110.5.

[76] *Id.* § 7111; West Coast Home Improvement Co. v. Contractors State License Bd., 72 Cal, App. 2d 287 (1945).

[77] Cal. Bus. & Prof. Code § 7112. In a corporate application, the managing employee falsely stated that no person listed as personnel had been convicted of a felony. the license was revoked. Falsely stating there were no unpaid bills was also grounds for discipline. Nelson Valley Bldg. Co. v. Morrisey, 135 Cal. App. 2d 738 (1955).

[78] Cal. Bus. & Prof. Code § 7113.

[79] *Id.* § 7115.

[80] *Id.* § 7116.

[81] *Id.* § 7117.

[82] *Id.* § 7117.5.

[83] *Id.* § 7118.

[84] Cal. Bus. & Prof. Code § 7119.

[85] *Id.* § 7120.

[86] *Id.* § 7123.

[87] Rule 860, Contractors State License Board Rules and Regulations.

[A] Abandonment of Construction Project

In *Bailey-Sperber, Inc. v. Yosemite Insurance Co.*,[88] the court held that a subcontractor who never began performance of a subcontract had abandoned it under California Business and Professions Code § 7107 and also had failed in a material respect to complete the contract under section 7113.

In *Viking Pools, Inc. v. Maloney*,[89] a swimming pool contractor who refused to perform work under an express warranty in the contract was held to have been guilty of abandonment of and a material failure to complete the construction project and was therefore in violation of Business and Professions Code §§ 7107 and 7113.

[B] Diversion or Misapplication of Funds

Section 7108 provides that failure to substantially account for the application or use of funds or property on the construction project constitutes a cause for disciplinary action. In order to recover on a contractor's license bond alleging a violation of section 7108, the claimant would be obligated to prove that there was a willful or fraudulent diversion or nonpayment of funds.[90]

[C] Failure to Pay

Failure of the prime contractor or subcontractor to pay any subcontractor a progress payment within 10 days, unless otherwise agreed in writing, of payment of the amount allowed the prime contractor on account of the subcontractor's work would justify disciplinary action. This applies to public and private works of improvement.[91] In the event there is a bona fide dispute, the prime contractor or subcontractor may withhold 150 percent of the amount in dispute. In any action to recover the amount withheld, the prevailing party shall be entitled to attorneys' fees. Wrongful withholding shall subject the licensee to a penalty of 2 percent per month. A subcontractor was allowed recovery of the 2 percent per month penalty in an action on a payment bond on a public work project under California Public Contract Code § 10262.5, which provides a similar penalty.[92]

Many subcontractors think that the prime contractor must pay them within 10 days of each progress payment. This section would seem to support that. The prime contractor who does withhold progress payments will have to justify the withholding. The phrase "unless otherwise agreed in writing" will be used to justify the withholding if the subcontract agreement has wording in it allowing the prime contractor to withhold payment for delay, defective work, cleanup, backcharges, and other such events. If the withholding of the progress payment

---

[88] 64 Cal. App. 3d 725 (1976).
[89] 48 Cal. 3d 602 (1989).
[90] All Bay Mill & Lumber Co. v. Surety Co., 208 Cal. App. 3d 11 (1989).
[91] Cal. Bus. & Prof. Code § 7108.5.
[92] *See* Washington Int'l Ins. Co. v. Superior Court, 62 Cal. App. 4th 981 (1998).

§ 1.09[C] CALIFORNIA CONSTRUCTION LAW

turns out to be improper, the licensee withholding the money shall be subject to interest at 2 percent per month plus attorneys' fees.

See § 4.04 in Chapter 4 for details regarding California's comprehensive legislative scheme governing payment of retention and progress payments on both public and private works of improvement within the state.

Pay-When-Paid Clauses

If there is no dispute as to the subcontractor's work, the subcontractor is entitled to be paid in a reasonable time even if the owner has withheld retention from the prime contractor and even if the subcontract provides that the subcontractor is to be paid when the owner pays the prime contractor. In the case *Yamanishi v. Bleily & Collishaw, Inc.*,[93] the subcontract provided: "Contractor agrees to pay to the subcontractor upon receipt of each payment received from the owner the portion of said payment allowed to contractor on account of subcontractor's work to the extent of subcontractor's interest therein less any percentage retained under said general contract."

The subcontractor's work in that case had been approved by all parties. The prime contractor and owner had a dispute about matters other than the subcontractor's work, and the owner had withheld money from the prime, who, in turn, had withheld from the subcontractor. The subcontractor sued and prevailed. The court stated:

> Yamanishi contended in the superior court, and now contends here, that the subcontracts' paragraph (r) did not create a condition precedent. Instead he argues that this provision simply stated the times at which the subcontractors would ordinarily be entitled to progress payments for their work, without any intent that it might operate to deny payment to the subcontractor if the contractor was denied payment through no fault of the subcontractor. In such latter event, it is argued, no specific time for payment to the subcontractor being provided, payment was due upon performance of the subcontract (*see Johnstone v. E. & J. Mfg Co.*, 45 Cal. App. 2d 586, 588 [114 P.2d 658]), or within a reasonable time thereafter (*see Bank of America v. Engleman*, 101 Cal. App. 2d 390, 394 [225 P.2d 597]).[94]

The court agreed with Yamanishi's interpretation and held:

> Defendant's interpretation of paragraph (r) would postpone payments earned by a subcontractor, itself without fault, until a dispute between the contractor and the owner is resolved, perhaps months or even years later. Indeed, it gives no reasonable assurance that such a dispute would ever be resolved. While the question is unsettled, the contractor continues unobligated to the subcontractor. On the other hand, if the dispute be lost because of the contractor's fault, then surely the contractor must pay his subcontractor creditor from other funds; if won, he must apply all or a substantial part of the money

---

[93] 29 Cal. App. 3d 457 (1972).
[94] *Id.* at 462.

he receives toward his subcontractual obligations. His interest would seem more likely to benefit from avoidance of any settlement with the owner. It is unlikely that such a result was intended by the contracting parties. The rules announced in *Hertzka & Knowles v. Salter, supra,* and *Hawley v. Orange County Flood etc. Dist., supra,* are applicable.[95]

Many other cases have reached the same result as the court did in *Yamanishi.*[96]

The question whether a contractor must pay a subcontractor when the contractor has not yet been paid by the owner depends largely upon the specific language of the payment clause in the subcontract agreement. The *Yamanishi* case, which is cited and discussed above, held that the contractor had to pay the subcontractor even though the contractor had not been paid by the owner by reason of the fact that the subcontract did not make it clear that payment by the owner was a "condition precedent" to the contractor's obligation to pay the subcontractor. Footnote 96 cites many of the cases on both sides of this issue. Essentially, the law across the United States appears to be that, in order for the contractor to insulate itself from liability, the payment clause in its subcontracts must provide that payment by the owner to the contractor is an express "condition precedent" to the contractor's obligation to pay the subcontractor. As discussed below the California Supreme Court has invalidated "condition precedent" clauses.

One question that often arises, even in the case of a well-drafted "condition precedent" subcontract payment clause, is what efforts, if any, the contractor must make to recover payments from the owner. While there is no California case on point, *Urban Masonry Corp. v. N&N Contractors, Inc.*[97] does address this issue in the context of a sub-subcontract.

---

[95] *Id.* at 463.

[96] Seal Tite Corp. v. Ehret, Inc., 589 F. Supp. 701 (D.N.J. 1984); Byler v. Great Am. Ins. Co., 395 F.2d 273 10th Cir. 1968); Trinity Universal Ins. Co. v. Smithwick, 222 F.2d 16 (8th Cir), *cert denied,* 350 U.S. 837 (1955); Thos. J. Dyer Co. v. Bishop Int'l Eng'g Co., 303 F.2d 655 (6th Cir. 1962); Darrell T. Stuart Contractor v. Bridges, 402 P.2d 143 (Ariz. Ct. Appl. 1965); Peacock Constr. Co. v. Modern Air Conditioning, Inc., 353 So. 2d 840 (Fla. 1977); Fishman Constr. Co. v. Hansen 209 A.2d 605 (Md. 1965); A.J. Wolfe Co. v. Baltimore Contractors, Inc., 244 N.E. 2d 717 (Mass. 1969); Schuler-Haas Elec. v. Aetna Cas. & Sur., 371 N.Y.S.2d 207 (App. Div. 1975); Howard-Green Elec. Co. v. Chaney & James Constr. Co., 182 S.E.2d 601 (N.C. Ct. App. 1971); Mignot v. Parkhill, 391 P.2d 755 (Or. 1964); Aesco Steel, Inc. v. J.A. Jones Constr. Co., 621 F. Supp. 1576 (E.D. La. 1985).

For scholarly analysis, see Restatement (Second) of Contracts § 227 cmt. b, illus. 1 (1981); Calamari & Perillo, The Law of Contracts 390–91 (2d ed. 1977); 3A Corbin on Contracts §§ 636, 641 (1960); 5 Williston, Contracts § 799 (3d ed 1961); For a contrary view, see *Hood v. Gordy Homes, Inc.,* 267 F.2d 882 (4th Cir. 1959); *Mascioni v. I.B. Miller, Inc.,* 184 N.E. 473 (N.Y. 1933). When the contract expressly provides that pyament by the owner to the prime contractor is a condition precedent to the prime contractor's obligation to pay the sub, most courts enforce such a clause. Crown Plastering v. Elite Assocs., 560 N.Y.S.2d 694 (App. Div. 1990); Gilbane Bldg. Co. v. Brish Waterproofing Co., 585 A.2d 248 (Md. 1991). For other cases on this subject, see Kirksey & Brown, *The Pay-When-Paid/Pay-If-Paid Dichotomy and the Florida Trilogy—The Bright Line or Murky Fog?,* 11 Construction Law. No. 4, at 8 (Oct. 1991).

[97] 676 A.2d 26 (D.C. Ct. App. 1996).

Blake Construction Company, the prime contractor for construction of an office building in the District of Columbia, subcontracted the masonry work to Urban Masonry Corporation, which subcontracted the installation of precast concrete panels to N&N Contractors, Inc. The subcontract between Urban and N&N stated: "Payments will be made to N&N promptly as they are received. Receipt of payment by Urban shall be a condition precedent to payment being owed to N&N." Thus, the subcontract made payment by Blake to Urban a condition precedent to Urban's obligation to pay N&N.

While performing the job, N&N discovered that its agreement with Urban had significantly understated the number of pieces requiring welding. N&N alleged that this delayed its progress and increased its labor and equipment costs. Prior to resolving N&N's claim, Urban met with Blake to resolve numerous disputes between them. Urban and Blake agreed to a settlement, pursuant to which each agreed to waive any claim against the other arising out of the project. Then, when N&N pursued its claim against Urban, Urban raised the "condition precedent" language in the subcontract as a defense. Urban took the position that, inasmuch as it had not been paid by Blake for N&N's alleged increased costs, Urban had no obligation to pay N&N.

The District of Columbia Court of Appeals ruled that the condition precedent did not apply under these facts, because Urban had intentionally frustrated the fulfillment of the condition. Specifically, the court stated that when Urban entered into its settlement with Blake, Urban knew of N&N's outstanding claim for additional compensation and, of course, the "condition precedent" language contained in its subcontract with N&N. Notwithstanding this knowledge, Urban agreed to a settlement with Blake that failed to secure payments for N&N. As a result, the court ruled that Urban had itself breached the implied condition that imposed upon Urban a duty not to frustrate the fulfillment of the condition precedent. Because Urban could not benefit from its own willful hindrance of the condition precedent, it was liable for breach of its contract with N&N.

The *Urban Masonry* case may be considered persuasive authority, though it is not a binding precedent under California law, for an exception to the enforceability of the "condition precedent" language in subcontract payment provisions.

In a landmark decision, the California Supreme Court affirmed the *Yamanishi* case and went one step further by declaring "condition precedent" payment clauses in subcontracts to be void and unenforceable. In *Wm. R. Clarke Corp. v. Safeco Insurance Co.*,[98] Keller Construction Company Ltd. (Keller), as a general contractor under a contract with an owner to perform rehabilitation work on a commercial building, entered into subcontracts with various subcontractors. Each subcontract contained a "pay if paid" provision. An addendum reiterated that the pay if paid limitation did not waive the subcontractor's mechanic's lien rights, and provided that each subcontractor's mechanic's lien right was to be its "sole remedy" in the event the owner failed to pay Keller.

Keller's prime contract with the owner required Keller to obtain a labor and material payment bond to protect the owner from mechanic's lien claims by

---

[98] 15 Cal. 4th 882 (1997).

subcontractors or material suppliers. Keller obtained from Safeco Insurance Company of America (Safeco) a payment bond as defined in California Civil Code § 3096, which recited that it had been executed to comply with the Mechanic's Lien Law. The bond also stated that Keller, as principal, and Safeco, as surety, were bound to any and all persons who performed labor upon or furnished materials used in the project. The condition of the bond was that if the principal (Keller) paid the persons performing labor or furnishing materials, then the bond would be void; otherwise, it would remain in full force and effect. In other words, if Keller failed to pay the claims of subcontractors and suppliers, Safeco would be obligated to do so.

The paid if paid clause included in Keller's subcontracts was extensive; it contained the following provisions:

1. "Receipt of funds by Contractor from Owner is a condition precedent to the Contractor's obligation to pay Subcontractor under this Agreement, regardless of the reason for Owner's nonpayment, whether attributable to the fault of the Owner, Contractor, Subcontractor or due to any other cause."

2. "Contractor shall have no obligation, legal, equitable or otherwise, to pay Subcontractor for Work performed by Subcontractor unless and until Contractor is paid by the Owner for the Work performed by Subcontractor. Furthermore, in the event Contractor is never paid by Owner for Subcontractor's Work, then Subcontractor shall forever be barred from making, and hereby waives, in perpetuity, any claim against Contractor therefor."

3. "Nothing in this Addendum shall be interpreted as limiting Subcontractor's right to enforce its statutory mechanic's lien rights or remedies, if any, against Project property and Subcontractor expressly agrees that such mechanic's lien rights, if any, shall be its sole remedy and means for payment (regardless of whether the value [of] Project property is sufficient or insufficient, for any reason, to satisfy Subcontractor's claim) on account of Work performed by Subcontractor for which Contractor has not been paid by Owner."

4. "Subcontractor shall not seek payment from Contractor for, and shall forever refrain from instituting any legal or equitable action for collection of, money and/or compensation for Work performed by Subcontractor for which Contractor is not paid [by] Owner."

Obviously, the subcontract had very strong pay if paid provisions, the purpose of which was to shift the risk of nonpayment from Keller to its subcontractors. After substantial work had been completed on the project, as a result of the owner's insolvency, it stopped making payments to Keller. Keller declined to pay those subcontractors who had recorded mechanics' liens and filed actions against Keller and Safeco on the payment bond. When the trial judge ruled in favor of the subcontractors and the court of appeal affirmed, Safeco, as the surety on the payment bond, appealed to the California Supreme Court.

The California Supreme Court began by acknowledging that, in recent years, general contractors in California had begun to insert pay if paid provisions into their subcontractor agreements. A pay if paid clause makes payment by the owner to the general contractor a "condition precedent" to the general contractor's obligation to pay the subcontractor for work performed. A legal "condition precedent" is either an act of a party that must be performed, or an uncertain event that must occur, before a right accrues or a duty arises under a contract. In this case, the pay if paid clause made payment by the owner a condition precedent to the contractor's obligation to pay its subcontractors; therefore, if the payment-by-owner event never occurred, the contractor would have no obligation ever to pay its subcontractors.

The supreme court then took note of many cases across the United States that have addressed this issue. In some of those cases, the condition precedent language was upheld; in others it was not. The highest court of New York concluded that a true pay if paid provision in a subcontract for construction work is void as against public policy; pay if paid provisions were declared void and unenforceable by statute in Illinois, North Carolina, and Wisconsin. The California Supreme Court noted that it had granted review of this case to determine whether a subcontractor may recover on a general contractor's payment bond for work it performed under a contract containing a pay if paid provision when the owner has not paid the general contractor. The supreme court held:

> We conclude that pay if paid provisions like the one at issue here are contrary to the public policy of this state and therefore unenforceable because they effect an impermissible indirect waiver or forfeiture of the subcontractors' constitutionally protected mechanic's lien rights in the event of nonpayment by the owner. Because they are unenforceable, pay if paid provisions in construction subcontracts do not insulate either general contractors or their payment bond sureties from their contractual obligations to pay subcontractors for work performed.
>
> Having concluded that a general contractor's liability to a subcontractor for work performed may not be made contingent on the owner's payment to the general contractor, we conclude that Keller was liable to the subcontractors under their subcontracts for the work they performed and that Safeco, as Keller's surety, was likewise liable on the payment bond.

The court, in so concluding, noted that the California state constitution provides:

> Mechanics, persons furnishing materials, artisans, and laborers of every class, shall have a lien upon the property upon which they have bestowed labor or furnished material for the value of such labor done and material furnished; and the Legislature shall provide, by law, for the speedy and efficient enforcement of such liens. Cal. Const. art. XIV, § 3.

Moreover, the court noted, the lien law should be liberally construed:

> "The mechanic's lien is the only creditors' remedy stemming from constitutional command and our courts "have uniformly classified the mechanics' lien laws as remedial legislation, to be liberally construed for the protection of laborers and materialmen." . . . (*Hutnick v. United States Fidelity & Guaranty Co.* (1988) 47 Cal.3d 456, 462.) "[S]tate policy strongly supports the preservation of laws which give the laborer and materialman security for their claims." (*Connolly Development, Inc. v. Superior Court* (1976) 17 Cal.3d 803, 827.)

The court then held that, by virtue of California Civil Code § 3262, a subcontractor cannot waive its mechanic's lien rights except under certain specified circumstances; specifically, section 3262(d) provides that a waiver and release of mechanic's lien rights shall be null and void and unenforceable unless it substantially follows the forms set forth in the statute. Noting the four statutory forms (i.e., the Conditional and Unconditional Waiver and Release Upon Progress Payment and the Conditional and Unconditional Waiver and Release Upon Final Payment), the supreme court concluded that a subcontractor could waive its mechanic's lien rights only under the provisions of Civil Code § 3262:

> Thus, under our mechanic's lien law, waiver and release of mechanic's lien rights is permitted only in conjunction with payment, or a promise of payment, and a conditional release is effective only if the claimant is actually paid. (See Cal. Mechanics' Liens and Other Remedies (Cont.Ed.Bar 1988) § 4.21, p. 200.[99]

> By closely and carefully circumscribing subcontractors' freedom to waive mechanic's lien rights, and by forbidding waivers not accompanied by payment, or a promise of payment, the Legislature has already determined that there are policy considerations here that override the value of freedom of contract. We merely recognize and enforce that legislative policy determination.[100]

> Having concluded that a general contractor's liability to a subcontractor for work performed may not be contingent on the owner's payment to the general contractor, we conclude that Keller was liable to the subcontractors under their subcontracts for the work they performed and that Safeco, as Keller's surety, was likewise liable on the payment bond.[101]

The majority opinion of the California Supreme Court was concurred in by four justices, but a strong dissent was rendered by the other three justices. The dissenting justices would have held the pay if paid clause to be valid to enforce a risk that the subcontractors agreed to take when they signed their subcontracts with the general contractor. Furthermore, the dissent noted, it has long been the

---

[99] 15 Cal. 4th at 889.
[100] *Id.* at 891–892.
[101] *Id.* at 896–897.

law in California that a surety's liability is no less than, nor greater than, that of its principal. In this case, because Keller was not liable to the subcontractors by reason of the pay if paid clause, the surety could not be held liable to the subcontractors on the payment bond. Specifically, the dissent concluded that

> [T]he contract between the contractor and the subcontractors does not affect the mechanic's lien rights of either against the property except to establish the value of the lien. The parties validly could and did agree among themselves that each would assume equally the risk of owner default while preserving their mutual mechanic's lien remedies against the property. The payment bond answered for the contractor's obligations, not the owner's, and therefore, neither the principal nor the surety is liable in the absence of the contractor's default. This conclusion is reasonable, straightforward, and consistent with the parties' intent, the law of mechanic's liens and surety, and the public policy in favor of enforcing freely negotiated contracts. The subcontractors entered into their agreements voluntarily. They should be bound by those agreements.

The *Clarke* case was followed in the case of *Capitol Steel Fabricators, Inc. v. Mega Construction Co.*,[102] holding "condition precedent" clauses in subcontracts unenforceable on public works projects as well.

[D] Willful Departure from Plans and Specifications

California Business and Professions Code § 7109 was violated when a contractor undertook to put asbestos siding on houses and spray the underpinnings for termites but did not spray the work on a number of jobs or, where the spraying was done, it was inconsequential and valueless.[103] Failure to pour a concrete slab properly was cause for discipline under this section.[104]

In a case interpreting Business and Professions Code §§ 7109 and 7113, the court held that an owner who built a home could not be disciplined under the license law. In *Linda Jones General Builder v. Contractors State License Board*,[105] Linda Jones owned a piece of property and built a home on it with the intention of residing there. When it was finished, she sold it. The following winter, there was a heavy snowfall. After viewing the property, Linda Jones wrote a letter to the new owners telling them that a covered walkway was in a hazardous condition from the load of snow on it. Linda Jones refused to pay for the necessary repairs, contending that the problem was due to the owners' failure to remove the heavy buildup of snow. The owners repaired it themselves at a cost of approximately $20,000.

Because Linda Jones was also a licensed contractor, she was charged by the Contractors State License Board with violating Business and Professions Code

---

[102] 58 Cal. App. 4th 1049 (1997).
[103] West Coast Home Improvement Co. v. Contractors State License Bd., 72 Cal. App. 2d 287 (1945).
[104] Mickelson Concrete Co. v. Contractors State License Bd., 95 Cal. App. 3d 631 (1979).
[105] 194 Cal. App. 3d 1320 (1987).

§§ 7109 and 7113 (failure to complete contract for the agreed price). The Contractors State License Board found numerous violations of the Contractors License Law and revoked all of Linda Jones's general builder's licenses. Linda Jones filed an action in superior court to restrain the Contractors State License Board from revoking her license, but the superior court denied her request for an injunction. The court of appeal reversed. The court held that Linda Jones General Builder could not be disciplined under the license law, because those sections pertain to the conduct of a contractor in performing a contract with an owner. In this case, Linda Jones had no contract with an owner, as she was the owner of the property herself when the work was done.

[E] **Avoidance or Settlement of Obligations for Less Than Full Amount**

Contractors are in violation of statute if they avoid or settle for less than the full amount of their lawful obligation by:

1. Composition, arrangement, or reorganization with creditors under state law
2. Composition, arrangement, or reorganization with creditors under any agreement or understanding
3. Receivership
4. Assignment for the benefit of creditors
5. Trusteeship
6. Dissolution.[106]

This does not apply to a licensee's individual settlement with a creditor that is not a part of the settlement with other creditors. No proceedings can be brought against the licensee for any type of a bankruptcy, composition, arrangement, or reorganization under the federal Bankruptcy Code. Accordingly, a creditor of a bankrupt licensee was not allowed to recover on the license bond.[107]

[F] **Evading the License Law**

Persons in the construction industry often attempt to circumvent the license law by "using," "borrowing," or "paying for the use" of someone else's license. This is improper and can result in disciplinary action and loss of a contractor's right to sue in court.[108] For example, in *Rushing v. Powell*,[109] the plaintiff entered into a contract with the defendant to construct a swimming pool. Rushing held an individual contractor's license as a concrete contractor in Class C-8. Junior

---

[106] Cal. Bus. & Prof. Code § 7113.5.
[107] FAJ, Inc. v. Surety Co. of Pac., 68 Cal. App. 3d Supp. 20 (1977).
[108] Cal. Bus. & Prof. Code § 7114.
[109] 61 Cal. App. 3d 597 (1976).

Ray Anderson, a swimming pool contractor licensed in Class C-53, entered into a verbal arrangement with Rushing in order to obtain a joint license to engage in swimming pool construction under Class C-53 under the name of Star Dust Pools. Star Dust Pools qualified for the issuance of the joint swimming pool contractor's license because Anderson was represented to be a member of Star Dust under the provisions of California Business and Professions Code § 7068, that is, as a responsible managing employee (RME) or responsible managing officer (RMO). See § 1.12.

Because Anderson was already licensed as a swimming pool contractor, the examination and demonstration of fitness were waived. In return for the use of Anderson's name to procure the C-53 license in the name of Star Dust Pools, the plaintiff transferred to Anderson $6,000 worth of equipment. The evidence was that Anderson at no time performed any function whatsoever in the management or operation of the business. He participated in no pool construction, he was not consulted in connection therewith, he shared no management prerogatives, and he incurred no profit opportunity. For a period of time, the plaintiff entered a partnership with Louis G. Galloway, doing business as Star Dust Pools. There was no notification to the License Board that there had been a change.

Subsequently, the plaintiff filed a notice of dissolution of partnership and certificate of doing business under the fictitious name, declaring under penalty of perjury that the business would be conducted by Eric Rushing, under the fictitious name of Star Dust Pools.

The contract in this case was signed by Keith Hutchison, a salesman, on behalf of Star Dust Pools and the owners. It was signed on June 30, 1972, and the construction was completed on August 15, 1972. The joint license that had been previously issued was canceled in November 1972.

When the plaintiff was not paid the $5,000 for the work done on the pool, he brought a lawsuit in his own name. The owner raised the issue of failure to comply with the fictitious business name statute and the Contractors License Law by demurrer, both at the pretrial conference and by motion for nonsuit.

At the presentation of evidence, the court found that the license law had been substantially complied with, that a valid fictitious name certificate was not on file, and that the plaintiff was entitled to a judgment at such time that he proved that a fictitious business name had been filed showing that he and Anderson had done business as Star Dust Pools. When the plaintiff filed a fictitious business name statement showing that Eric Rushing and Junior Ray Anderson formerly did business as Star Dust Pools, the owner appealed. The appellate court reversed the lower court's judgment.

The appellate court held that Anderson could not "lend his license" to Rushing and that he had to be an active participant in the business. The court first stated that the Contractors License Law is intended to protect the public against dishonesty and incompetence in the operation of the contracting business. It further held that California Business and Professions Code § 7031 (disallowing recovery without proper licensing) was a legislative determination that the importance of deterring unlicensed persons from engaging in the contracting business outweighs any harshness between the parties, and that such deterrence can best be realized by denying violators the right to maintain any action for compensation. The court stated that an essential ingredient to the accomplishment of that public purpose is the assurance that those performing particular types of specialty contracting are qualified by training and experience. To hold that a person not so qualified can nevertheless act as a contractor in the particular field of work would be inimical to that purpose and objective and would emasculate certain clear expressions of the statute.

The court stated that the plaintiff in this case had not been required to qualify personally as a swimming pool contractor by taking an examination or demonstrating his competence, but was able to participate in the business of swimming pool construction only by virtue of Anderson's license. The total inactivity of Anderson in the swimming pool construction enterprise permitted the plaintiff in substance to engage in a specialty in which he had not demonstrated qualification by taking and passing the appropriate examination and for which he had not been licensed as an individual, thus in violation of the stated purpose and the spirit of the Contractors License Law.[110]

The court stated that, from California Business and Professions Code § 7068 and the California Code of Regulations, it is clear that the legislature intended that a person qualifying on behalf of the business entity be an active participant in the business.

The court cited the case of *Frank v. Kozlovsky*,[111] which held that when a partner had departed from the licensed partnership, the license was thereby terminated and the continuing party could not sue for work accomplished under the terminated license, emphasizing that to hold otherwise would be to defeat the expressed legislative policy of deterring unlicensed persons from engaging in the contracting business.

The court also cited *Weeks,* noting that the case held that establishing contractor's licensing requirements to protect the public against imposition and ensure that unlicensed persons would not engage in the contracting business would be circumvented if a licensed contractor could contract and then delegate performance to an unlicensed contractor and recover on the contract.

The court stated that the full protection the statute contemplates requires that the work performed be done by or under the supervision of a licensed contractor. In this case, the plaintiff was denied recovery. Thus, the various schemes used in the industry to avoid the license law will not be countenanced by the courts.

In the case of *Buzgheia v. Leasco Sierra Grove*,[112] the court held that where the licensee has qualified through the use of an RME (responsible managing employee), the contractor has the burden of proving not only that the firm is properly licensed but also that the RME requirements have been met. Specifically, the contractor must prove that the RME has exercised supervision and control of the construction operation as required by Business and Professions Code

---

[110] *See* Weeks v. Merritt Bldg. & Constr. Co., 39 Cal. App. 3d 520 (1974).
[111] 13 Cal. App. 3d 120 (1970).
[112] 60 Cal. App. 4th 374 (1997).

§ 7068.1 and that Rule 823 of the Contractors State License Board has been complied with, which rule provides that the RME must be a bona fide employee who is permanently employed, is actively engaged in the operation of the business, and is so engaged at least 32 hours per week that such contracting business is in operation.

[G] Willful or Fraudulent Act Injuring Another

The courts have held that California Business and Professions Code §§ 7109, 7110, and 7116 require willful misconduct as a basis for disciplinary action although other sections do not.[113]

[H] Acting under Unlicensed Name or Personnel

In *West Coast Home Improvement Co. v. Contractors State License Board*,[114] the corporation's salesmen had little or no control exercised over them by the corporation. They would negotiate and sign contracts with no review by the corporation. The court affirmed that this was a violation of California Business and Professions Code §§ 7114, 7117, and 7118. The court also clearly set forth the purpose of disciplinary proceedings:

> The object of establishing the Contractors State License Board and vesting in the Registrar of Contractors disciplinary powers is for the protection of the public. The law is intended primarily to keep the contracting business clean and wholesome, to the end that it may merit the respect and confidence of the public in general and in particular those who have recourse to contractors in the construction or improvement of their properties. Therefore, the purpose of a disciplinary proceeding such as the one with which we are here concerned is to determine the fitness of a licensed contractor to continue in that capacity. It is not intended for the punishment of the individual contractor, but for the protection of the contracting business as well as the public by removing, in proper cases, either permanently or temporarily, from the conduct of a contractor's business a licensee whose method of doing business indicates a lack of integrity upon his part or a tendency to impose upon those who deal with him. A disciplinary proceeding such as the one in the instant case is comparable to such a proceeding under the State Bar Act.[115]

It should also be noted that licensees should always conduct their business in the exact name that appears on their license. Many contractors will change their name without changing their license, which could cause problems of pleading and proof of license status under California Business and Professions Code § 7031. Contracts, letterheads, invoices, statements, business cards, and all other written documents used by a licensee should be identical to the name shown on the license.

---

[113] Bailey-Sperber, Inc. v. Yosemite Ins. Co., 64 Cal. App. 3d 725 (1976).
[114] 72 Cal. App. 2d 287 (1945).
[115] *Id.* at 301–302.

[I] Entering into a Contract with an Unlicensed Contractor

In *Wang v. Division of Labor Standards Enforcement*,[116] Joseph Wang, a prime contractor, entered into a subcontract with a plastering subcontractor for lath and plaster work. The prime contractor did not inquire with the Contractors State License Board as to whether the subcontractor was licensed. When the subcontractor signed the subcontract, he inserted a license number on the subcontract. It ultimately turned out that the license number belonged to another company and that the subcontractor did not, in fact, have a valid license. The Division of Labor Standards Enforcement discovered that the subcontractor was unlicensed and assessed a penalty of $9,700, which represented $100 per day for the 97 working days of the contract.

This penalty was assessed under California Labor Code § 1021.5, which pertains to a person who holds a valid state contractor's license and who "willingly and knowingly" enters into a contract with any person to perform services for which a license is required. If that person is not an independent contractor or does not hold a valid state contractor's license, then the contractor shall be subject to a civil penalty. The *Wang* court held that under the facts of this case, the prime contractor did not have a duty to contact the Contractors State License Board to determine whether the subcontractor was licensed, and because the contractor did not in fact know that the subcontractor was unlicensed, the contractor had not "willingly and knowingly" entered into a contract with an unlicensed person.

In coming to that conclusion, the court commented upon California Business and Professions Code § 7118, which provides that a licensed contractor is subject to discipline by the Contractors State License Board for entering into a contract with an unlicensed contractor. In that respect, the court commented that when section 7118 was originally enacted in 1939, it stated that a contractor was subject to discipline for "knowingly entering into a contract." The court went on to note that the word "knowingly" was deleted from section 7118 in 1975. The court therefore distinguished section 7118 from California Labor Code § 1021.5.

Although the issue was not before the court and was therefore not squarely decided, it is arguable under this opinion that because the word "knowingly" has been deleted by the legislature from section 7118, it would be incumbent upon a contractor to do more than merely rely upon a license number supplied by the subcontractor, and that the contractor might have a duty to make an inquiry of the Contractors State License Board as to whether the subcontractor was licensed.

[J] Willful Failure to Pay for Materials or Service

In *Beach v. Contractors State License Board*,[117] the contractor was paid the full contract price but left certain subcontractors unpaid. The court held this to be a violation of California Business and Professions Code § 7120.

---

[116] 219 Cal. App. 3d 1152 (1990).
[117] 151 Cal. App. 2d 117 (1957).

On the other hand, however, the court in *All Bay Mill & Lumber Co. v. Surety Co.*[118] held that the mere fact that the licensee had received a payment for lumber that it failed to pay to the lumber supplier did not conclusively prove that there was a willful or fraudulent diversion or nonpayment of funds in violation of section 7120.

[K] Filing False Workers' Compensation Statement

Effective January 1, 1997, California Business and Professions Code § 7125 was amended to provide that a contractor who files a false statement, to the effect that he or she does not employ any person so as to become subject to the workers' compensation laws of California, is subject to disciplinary action.

[L] *Terminix:* The Contractor's Defense

One of the leading cases in the area of disciplinary proceedings centers around an interpretation of California Business and Professions Code § 7113. In *Terminix Co. v. Contractors State License Board*,[119] the licensee had been charged with violating sections 7109 (willful departure from plans), 7113 (failure to complete for price stated), 7115 (material failure to comply with license law), 7116 (committing a willful or fraudulent act injuring another), and 7119 (entering into a contract with an unlicenced contractor). The charges arose out of termite control work done by Terminix for six customers. The Board found that Terminix had violated all sections except 7115 and suspended their license for six months. Some of the charges related to the failure of Terminix to perform oral promises made by its salesmen. The contracts with the customers provided that "no representative has any power or authority to make any alterations of this contract or any promises or representations other than contained herein, and this contract contains the entire agreement of the parties." The court held that this clause put the customer on notice that the salesmen were powerless to vary by oral promise the terms of the written contract and, therefore, Terminix could not be disciplined for violation of unenforceable oral promises.

The court then stated that sections 7109, 7113, 7116, and 7119 require material prejudice or substantial injury to the owner. Before the complaint was filed by the Board against Terminix, Terminix had made settlements with the owners, granting them credits substantially greater than the hearing officer found was owed them. The court held, therefore, that because there was no injury or prejudice to the owner by reason of the aforesaid credits that exceeded their actual damages, this also precluded disciplinary action. With regard to one of the owners, no settlement had been made, but Terminix had offered, prior to the initiation of proceedings, to settle in terms more favorable than the owner was entitled to under the findings of the hearing officer. The court stated:

---

[118] 208 Cal. App. 3d 11 (1989).
[119] 84 Cal. App. 2d 167 (1948).

LICENSING § 1.10

> The statute does not penalize a licensee for making overcharges which the owner does not pay, nor for poor work which is done over to the satisfaction of the owner. A contractor cannot be held guilty of a violation of the act so long as he stands ready, able and willing to fulfill his contract.[120]

The court also stated:

> Of course a debtor who has been overcharged by his creditor suffers no detriment unless he pays the overcharge. A contractor who has done inferior work is not a violator of the statute if, before he has made any settlement with the owner, he offers and is able and willing to replace the inferior work with good work at no expense to the owner. In determining whether injury has resulted to the owner, the positions of the owner and contractor must be judged as of the time when their business is concluded. Although during the course of the work there may be mistakes and failures on the part of the contractor to keep his agreement, if, when the time for payment for the work arrives, he makes a fair and satisfactory settlement with the owner, he has been guilty of no breach of the law and the same, of course, is true if he corrects any overcharges whether they were made inadvertently or intentionally.[121]

The importance of *Terminix* cannot be overemphasized. Whenever contractors learn of legitimate claims being made against them, they may be able to avoid disciplinary proceedings by making satisfactory settlements with the complaining witnesses. If the claimants refuse to settle, contractors should always make a written offer to settle on terms that will result in no prejudice to the complaining witnesses. *Terminix* is an important but little known case that can be used by licensees and their attorneys to take the sting out of disciplinary proceedings.

§ 1.10 ARBITRATION

The license laws contain certain provisions for arbitration. Those provisions are contained in California Business and Professions Code §§ 7085 to 7085.8. The arbitration procedure is an attempt to provide a vehicle for resolution of complaints between homeowners and contractors that is short of formal disciplinary proceedings. It often proves to be a better method than lengthy and costly court proceedings and formal disciplinary proceedings taken against the contractor.

After investigating a verified complaint alleging a violation of sections 7107 (abandonment of contract), 7109 (disregard of specifications or accepted trade standards), 7110 (violation of safety provision), 7113 (breach of contract), 7119 (willful failure to prosecute construction project), or 7159 (home improvement contract requirements), and upon finding a possible violation, the registrar of contractors may, with the concurrence of both the licensee and the complainant, refer the alleged violation and any dispute between the licensee and the complainant to arbitration.[122]

---

[120] *Id.* at 174.
[121] *Id.* at 175, 176.
[122] Cal. Bus. & Prof. Code § 7085.

In order for such a dispute to be eligible for arbitration, the registrar must find that:

1. There is evidence that the complaining party has suffered, or is likely to suffer, material damages as a result of a violation of the sections in question.

2. There are reasonable grounds to believe that the public interest would be better served by arbitration than by disciplinary action.

3. The licensee does not have a history of repeated or similar violations.

4. The licensee is in good standing.

5. The licensee does not have any outstanding disciplinary actions filed against him or her.

6. The licensee has not requested, nor have the parties agreed to, private arbitration of the dispute.

7. The damages are not less than $2,500 nor more than $25,000.[123]

Once the registrar of contractors determines that arbitration would be a suitable means of resolving the dispute, the complainant and the licensee are notified and the complainant is informed of the consequences of selecting arbitration over judicial remedies. The registrar provides forms to be filled out by the complainant and the licensee and returned to the registrar within 30 days, authorizing the registrar to proceed with administrative arbitration. Return of the "agreement to arbitrate" by the parties shall authorize the registrar to proceed with administrative arbitration.

Once the complaining witness and the licensee both authorize the registrar to proceed with arbitration, the matter is referred to the Office of Administrative Hearings, which will then conduct a hearing.[124]

The failure of a licensee to comply with an arbitration award constitutes grounds for suspension or revocation of the license.[125]

§ 1.11 CITATION PROCEDURE

[A] In General

In addition to addressing arbitration proceedings, the license laws also contain a citation procedure.[126] Both of these procedures have been enacted by the legislature in order to speed up the resolution of disputes between contractors and homeowners.

---

[123] *Id.*
[124] *Id.* § 7085.4.
[125] *Id.* § 7085.6.
[126] The citation procedures are contained in Cal. Bus. & Prof. Code §§ 7099–7099.9.

LICENSING § 1.11[B]

If, upon investigation, the registrar of contractors has probable cause to believe that a licensee's violation of the license law is grounds for revocation or suspension, the registrar may, in lieu of formal disciplinary proceedings, issue a citation to the licensee. The citation shall be in writing and shall describe with particularity the nature of the violation, including the reference to the particular division of the license law that has been violated. In addition, the citation may contain both an order of correction, fixing a reasonable time for correction of the violation, and an assessment of a civil penalty.[127]

The Contractors State License Board has the duty to promulgate regulations covering the formulation of an *order of correction,* which gives due consideration to both the time required to correct and the practical feasibility of the correction. The Board also promulgates regulations covering the assessment of civil penalties, giving due consideration to the gravity of the violation, the good faith of the licensee, and the history of previous violations. In no event shall the civil penalty be greater than $2,000.[128]

[B] Board Rules Relating to Citations

The rules adopted by the Contractors State License Board to implement the citation procedure are as follows:

Article 8. Citation 880. Order of Correction—Practical Feasibility.

Before including an order of correction in a citation, due consideration shall be given to the practical feasibility of correction in accordance with, but not limited to, the following criteria:

(a) An order of correction is appropriate where it would not result in excessive destruction of or substantial waste of existing acceptable construction.

(b) An order of correction is appropriate where the owner of the construction project is willing to allow the cited licensee to correct.

(c) An order of correction is appropriate where it appears to the registrar that the cited licensee has competence or ability to correct.

NOTE: Authority cited: Section 7008, Business and Professions Code. Reference: Sections 7008 and 7099.1, Business and Professions Code.

881. Order of Correction—Alternative Compliance.

A cited licensee may comply with an order of correction by having and paying for another licensee to do the corrective work. The cited licensee remains responsible, however, for any failure to fully comply with the order of correction.

An order of correction may, but need not, contain the alternative that the cited person may pay a specified sum to the owner of the construction project in lieu of correcting.

---

[127] *Id.* § 7099.
[128] *Id.* §§ 7099.1–7099.2.

NOTE: Authority cited: Sections 7008 and 7099.1, Business and Professions Code. Reference: Sections 7099 and 7099.1, Business and Professions Code.

882. Order of Correction—Time Required to Correct.

Where an order of correction is included in a citation, due consideration shall be given to the time required to correct in accordance with, but not limited to, the following criteria:

(a) Accepted industry practice in that area relating to performance of such work under certain climate or weather conditions.

(b) A reasonable time in which to obtain necessary materials.

(c) The number of working days the construction project will be made accessible by the owner for corrections.

NOTE: Authority cited: Sections 7008 and 7099.1, Business and Professions Code. Reference: Sections 7099 and 7099.1, Business and Professions Code.

883. Order of Correction—Extension of Time to Correct.

If the cited person, after exercising substantial efforts and reasonable diligence, is unable to complete the correction within the time allowed because of conditions beyond his control, he may request an extension of time in which to correct. Such request must be made in writing, and must be made prior to the expiration of the time allowed in the order of correction. An extension may be granted upon showing of good cause which determination is within the discretion of the registrar. If a request for extension of time is not made prior to the expiration of time allowed in the order of correction, failure to correct within the time allowed shall constitute a violation of the order of correction whether or not good cause for an extension of time existed.

NOTE: Authority cited: Sections 7008 and 7099.1, Business and Professions Code. Reference: Sections 7099 and 7099.1, Business and Professions Code.

884. Recommended Assessments of Civil Penalties.

In assessing a civil penalty against a person who has not previously been cited for violation of the same or similar section of the Contractors License Law, the registrar shall give due consideration to the following guidelines:

Section Violation	Recommended Minimum Civil Penalty	Recommended Maximum Civil Penalty
7030	$ 50.	$ 100.
7030.5	50.	100.
7071.13	50.	100.
7075	50.	100.
7083	50.	100.
7018.5	50.	150.
7026.7	50.	150.
7029	50.	150.

7029.5	50.	150.
7029.6	50.	150.
7029.7	50.	150.
7034	50.	150.
7068.2	50.	150.
7110.1	50.	150.
7117	50.	150.
7125	50.	150.
7168	50.	150.
7028	50.	500.
7028.5	50.	500.
7071.11	50.	500.
7111	50.	500.
7113.5	50.	500.
7117.5	50.	500.
7154	50.	500.
7157	50.	500.
7110	50.	1000.
7068.1	50.	1000.
7159	50.	1000.
7167	50.	1000.
7107	100.	1000.
7108	100.	1000.
7108.5	100.	1000.
7114	100.	1000.
7118	100.	1000.
7119	100.	1000.
7120	100.	1000.
7109	50.	1500.
7109.5	50.	1500.
7113	50.	1500.
7116	50.	1500.
7158	50.	1500.
7161	50.	1500.
7123	500.	1500.

The minimum and maximum civil penalties as set forth above are advisory only. Where there is more than one violation, where, in the judgment of the registrar, a person has exhibited bad faith or where, in the judgment of the registrar, the violation is grave, the maximum recommended civil penalty shall be $2,000.00.

Where a cited person has a history of violations of the same or similar sections of the Contractors License Law, the maximum recommended civil penalty shall be $2,000.00.

Where a citation lists more than one violation the amount of assessed civil penalty shall be stated separately for each section violated.

Where a citation lists more than one violation and each of the violations relates to the same construction project, the total penalty assessment in each citation shall not exceed $2,000.00.

NOTE: Authority cited: Sections 7008 and 7099.2, Business and Professions Code. Reference: Sections 7099 and 7099.1, Business and Professions Code.

885. Appeal of Citation.

Any person served with a citation pursuant to Section 7099 of the Business and Professions Code may contest the citation by appealing to the registrar within 15 working days from the receipt of such citation. The 15 day period may be extended upon showing of good cause which determination is within the discretion of the registrar.

The cited person may contest any or all of the following aspects of the citation:

1. The occurrence of a violation of the Contractors License Law;
2. The reasonableness of the order of correction, if an order of correction is included in the citation;
3. The period of time allowed for correction, if an order of correction is included in the citation;
4. The amount of the civil penalty, if a civil penalty is assessed in the citation.

NOTE: Authority cited: Section 7008, Business and Professions Code. Reference: Sections 7099.3, 7099.4 and 7099.5, Business and Professions Code.

886. Service of Citation.

Service of a citation shall be made in accordance with the provisions of Section 11505(c) of the Government Code, and, further, that a copy of the citation be sent by regular mail.

NOTE: Authority cited: Section 7008, Business and Professions Code. Reference: Sections 7099.3, 7099.4 and 7099.5, Business and Professions Code.

887. Criteria to Evaluate the Gravity of a Violation of Business and Professions Code, Section 7028.7.

Before assessing a civil penalty under Section 7028.7 of the Business and Professions Code, the registrar shall give due consideration to the gravity of the violation, including, but not limited to, a consideration of whether the cited person did one or more of the following:

1. Falsely represented that he/she was licensed.
2. Failed to perform work for which money was received.
3. Executed or used any false or misleading documents in order to induce a person to enter into a contract or to pay money.
4. Made false or misleading statements in order to induce a person to enter into a contract or pay money.

5. Failed to apply funds which were received for the purpose of obtaining or paying for services, labor, materials, or equipment.
6. Performed work that was potentially hazardous to the health, safety, or general welfare of the public.
7. Performed work in violation of the building laws, safety laws, labor laws, compensation insurance laws, or unemployment insurance laws.
8. Performed work that did not meet acceptable trade standards for good and workmanlike construction.
9. Was convicted of a crime in connection with the violation.
10. Committed any act which would be cause for disciplinary action against a licensee.
11. Committed numerous or repeated violations.

NOTE: Authority: Sections 7008 and 7028.7, Business and Professions Code.
Reference: Section 7028.7, Business and Professions Code.

A licensee who is served with a citation may appeal to the registrar of contractors within 15 working days from receipt of the citation with respect to the violations alleged by the registrar, the correction periods, the amount of penalties, and the reasonableness of the change required by the registrar to correct the condition.[129]

If the licensee fails to notify the registrar within 15 working days from receipt of the citation of the intention to contest the citation, then the citation shall be deemed to be a final order of the registrar and may not be subject to review by any court or agency. The registrar may extend that 15-day period.[130]

If the licensee notifies the registrar of the intention to contest the citation, then the registrar shall afford the licensee a hearing. After the hearing, the registrar issues a decision, based on findings of fact, affirming, modifying, or vacating the citation or penalty, or directing other appropriate relief.[131]

The failure of a licensee to comply with an order of correction or pay any civil penalty assessed after the order or assessment is final is ground for suspension or revocation of the license.[132] It is also a misdemeanor.[133]

These citation provisions apply to an unlicensed individual acting in the capacity of a contractor.[134]

---

[129] Cal. Bus. & Prof. Code § 7099.3.
[130] Id. § 7099.4.
[131] Id. § 7099.5.
[132] Id. § 7099.6.
[133] Id. § 7099.85(a).
[134] Id. § 7099.8.

§ 1.12 RESPONSIBLE MANAGING EMPLOYEE AND RESPONSIBLE MANAGING OFFICER

An individual applying for a license qualifies by personal appearance or by that of a responsible managing employee (RME).[135] A partnership qualifies by the appearance of a general partner or by an RME.[136] A corporation qualifies by the appearance of a responsible managing officer (RMO) or an RME.[137] An *RME* is a "bona fide employee" of the applicant and is actively engaged in the classification of work for which the RME is the qualifying person on behalf of the applicant.[138] A person applying as a qualifying partner, RME, or RMO shall not hold any other license unless responsible for exercising direct supervision and control of the employer's or principal's construction operation, as is necessary to secure full compliance with the license law, and one of the following exists:

1. There is 20 percent common ownership of the equity of each individual or firm for which the applicant qualifies.
2. The additional firm is a subsidiary or joint venturer with the first.
3. The partnership or corporation has a majority of officers or partners that are the same.[139]

It should be noted that the Contractors State License Board has adopted a Board rule that defines *bona fide employee* as used in California Business and Professions Code § 7068 as follows:

> 823. Definitions: Bona Fide Employee; Direct Supervision and Control.
>
> (a) For purposes of Section 7068 of the Code, "bona fide employee" of the applicant means an employee who is permanently employed by the applicant and is actively engaged in the operation of the applicant's contracting business for at least 32 hours or 80 percent of the total hours per week such business is in operation, whichever is less.
>
> (b) For purposes of Section 7068.1 of the Code, "direct supervision and control" includes any one or any combination of the following activities: supervising construction, managing construction activities by making technical and administrative decisions, checking jobs for proper workmanship, or direct supervision on construction job sites.
>
> **NOTE:** Authority: Section 7008, Business and Professions Code. Reference: Sections 7068 and 7069.1, Business and Professions Code.

These provisions were further fortified by the enactment of Business and Professions Code § 7068.1. That provision states that the person who serves as

---

[135] Cal. Bus. & Prof. Code § 7068(B)(1).
[136] *Id.* § 7068(B)(2).
[137] *Id.* § 7068(B)(3).
[138] *Id.* § 7068(D).
[139] *Id.* §§ 7068, 7068.1(a), (b), (c).

the qualifier for a licensed contractor must exercise such direct supervision and control of construction operations as is necessary to secure full compliance with the contractors license law and the rules and regulations of the Contractors State License Board. It further provides that a person may not act in the capacity of a qualifying person for more than one firm, unless there is common ownership of at least 20 percent of the equity of each firm.

Effective January 1, 1992, a qualifying individual may act as the qualifier for no more than three firms in any one-year period.

As noted in § 1.09[F] supra, in the case of *Buzgheia v. Leasco Sierra Grove*,[140] the court held that where the licensee qualifies through an RME, the licensee must prove compliance with both Business and Professions Code § 7068.1 and Rule 823 of the Contractors State License Board.

§ 1.13 PUBLIC ACCESS TO PROCEEDINGS

The Contractors State License Board has adopted a rule giving the public access to the Board's disciplinary proceedings. Rule 863 reads as follows:

> 863. Public Access to Information.
>
> The registrar shall establish a system whereby members of the public may obtain from board records information regarding complaints made against licensed contractors, their history of legal actions taken by the board, and license status, as hereafter specified. For purposes of this section, "complaint" means a written allegation that a licensee has violated any provision of the Contractors License Law or board regulations.
>
> (a) Information to Be Released Regarding Complaints. The registrar shall maintain records showing the complaints received against licensees and, with respect to such complaints, shall make available to members of the public, upon request, the following information, in writing:
>
> (1) The nature of all complaints on file against a licensee which are currently assigned to a deputy registrar for investigation after a screening process that indicated sufficient merit existed to warrant further investigation. Information regarding complaints currently assigned to a consumer services representative for screening shall not be disclosed.
>
> (2) The nature of all complaints on file against a licensee which were previously assigned to a deputy registrar which have been referred for legal action within the one (1) year period prior to the date of the request.
>
> (3) Such general cautionary statements, if any, as may be considered appropriate regarding the usefulness of complaint information to individual consumers in their selection of a contractor. Any complaints disclosed pursuant to subsection (1) above shall also be accompanied by a cautionary statement that no conclusions or judgments as to the validity of an open complaint may be assumed pending a completed investigation.
>
> (4) Whenever complaint information is requested, the information disclosable under subsections (c) and (d) below shall also be released in writing.

---

[140] 60 Cal. App. 4th 374 (1997).

(b) If a complaint which was previously assigned to a deputy registrar for investigation is subsequently determined by the registrar or the Office of the Attorney General not to have merit, it shall be deleted from the complaint disclosure system.

(c) Information to Be Released Regarding Disciplinary Legal Actions. The registrar shall maintain records showing a history of any legal actions taken by the board against all current license holders and shall make available to members of the public, upon request, all the following information:

(1) Whether any current license holder has ever been disciplined by the registrar and, if so, when and for what offense; and

(2) Whether any current licensee has ever been cited, and, if so, when and for what offense, and, whether such citation is on appeal or has been dismissed or complied with.

(3) Whether any current license holder is named as a respondent in any currently pending disciplinary action.

(d) Information to Be Released Regarding License and Bond Status. The registrar shall maintain records showing certain licensing and bonding information for all current license holders and shall make available to members of the public, upon request, all the following information regarding current license holders:

(1) The name of the licensee as it appears on the board's records; and

(2) The license number; and

(3) The classification(s) held; and

(4) The address of record; and

(5) The personnel of the licensee; and

(6) The date of original licensure; and

(7) Whether a bond or cash deposit is maintained and, if so, its amount; and

(8) If the licensee maintains a bond, the name and address of the bonding company and the bond's identification number, if any.

(e) Limitation of Access to Information. Further, the registrar may set reasonable limits upon the number of requests for information responded to per month from any one requester.

NOTE: Authority: Section 7008, Business and Professions Code. Reference: Sections 7124.5 and 7124.6, Business and Professions Code.

§ 1.14 LICENSE BONDS

Under California Business and Professions Code § 7071.6(a), the Contractor's State License Board requires as a condition precedent to the issuance, reinstatement, reactivation, or renewal of a contractor's license that the applicant have on file a contractor's license bond in the sum of $7,500. The contractor may substitute for the bond a cash deposit in that amount.

Under § 7071.6(b), the amount of the bond for a swimming pool contractor is $10,000. Under this subdivision, the Contractors State License Board shall require, as a condition precedent to the issuance or reinstatement or reactivation of a license, that any applicant having been previously found to have failed or refused to pay a contractor, subcontractor, consumer, material supplier, or employee, based upon an entered and unsatisfied final judgment from a court of law, file or have on file with the Contractors State License Board a judgment bond or other security sufficient to guarantee payment of an amount equal to the unsatisfied final judgment. The contractor has 90 days from the date of notification by the Board to file the bond, or the application becomes void. The Board may not issue, reinstate, or reactivate a license until the judgment bond is filed with the Board. The judgment bond is in addition to the contractor's license bond. The judgment bond is to remain on file for a minimum of one year and after that time may be removed when the contractor submits proof of satisfaction of all debts. The contractor can provide the Board with a notarized copy of any accord reached with the judgment creditor holding an unsatisfied final judgment, to satisfy the debt in lieu of filing the bond. Failure to maintain the bond or failure to abide by the accord shall result in an automatic suspension of the license. The judgment bond remains in full force in the amount posted until the entire debt is satisfied.

In light of these bond requirements, a creditor of a licensed contractor should check to determine whether the appropriate bond is posted and may seek recovery on the bond. Similarly, if a contractor, subcontractor, consumer, material supplier, or employee obtains a judgment against a licensed contractor, that person should notify the Contractors State License Board of that fact so that the Board, in turn, will require the contractor to post the judgment bond.

The judgment bond requirement applies only to an unsatisfied judgment that is "substantially related to the qualifications, functions, or duties on the license being applied for." Section 7071.6(e) requires a licensee to notify the registrar in writing of any entered and unsatisfied judgment within 90 days from the date of the judgment. If the licensee fails to notify the registrar in writing within 90 days, the license shall be automatically suspended retroactively to 90 days from the date of the judgment, and the suspension shall not be removed until proof of satisfaction of the judgment is submitted to the Contractors State License Board. If the licensee does notify the registrar within 90 days of the judgment, and the judgment is entered and unsatisfied, the Board shall require, as a condition to the continued maintenance of the license, that the licensee file or have on file with the Board a judgment bond or other security sufficient to guarantee payment of an amount equal to the unsatisfied judgment. The licensee has 90 days from the date of notification by the Board to file the bond or, at the end of the 90-day period, the license shall be automatically suspended.

Under section 7071.6(f), the Board shall take the action required by that section upon notification by any party having knowledge of the outstanding judgment and a showing of proof of the judgment. The term "judgment" also includes any final arbitration award.

Finally, section 7071.6(h) provides that as a condition precedent to acquiring a license, the Board may require the applicant to post a contractor's bond in twice the amount required under section 7071.6(a) and (b) if the applicant for the license has been convicted of contracting without a license and that violation constitutes a substantial injury to the public.

Business and Professions Code § 7071.5 identifies the persons benefited by the contractor's license bond. The bond inures to the benefit of a homeowner contracting for improvement of the homeowner's personal family residence who has been damaged as a result of violation of the license law.[141] The bond or cash deposit also insures to the benefit of any person damaged as a result of a willful and deliberate violation of the license law, or as a result of fraud by the licensee.[142] It also inures to the benefit of an employee damaged by the failure of the licensee to pay wages.[143] Finally, it inures to the benefit of trust funds.[144]

Section 7071.11 sets forth the procedures for filing an action to enforce a claim on a bond. The persons who may recover on a license bond include an employee making a claim for wages and fringe benefits, which claim has priority.[145] The action on the bond must be brought within two years after the expiration of the license period during which the act or omission occurred giving rise to the claim against the bond.[146]

The courts have interpreted what types of claims may be made on license bonds. For example, in *Nelson Supply Co. v. Surety Co. of Pacific*,[147] the court held that a material supplier, in order to recover under the bond under Business and Professions Code § 7120 (Willful or Deliberate Failure to Pay), had to show the following elements:

1. That there was a willful or deliberate failure to pay;
2. That the moneys due for the materials or labor were rendered in connection with the licensee's operations as a contractor; and
3. That the contractor had the capacity to pay, or that the contractor had received sufficient funds to pay for the material supplies from the project on which the materials were used.

Additionally, in *Surety Co. of Pacific v. Piver*,[148] the court of appeal addressed the priority of competing claims against a single license bond. In that case, a claimant obtained a judgment in Marin County against the surety on the license bond. Subsequently, a second lawsuit by a different claimant was brought against the surety on the same license bond. The court held that the surety could restrain the enforcement of the judgment in the first action and could require interpleader of the various claims on the bond so that the court would be able to

---

[141] Cal. Bus. & Prof. Code § 7071.5(a).
[142] *Id.* § 7071.5(b).
[143] *Id.* § 7071.5(c).
[144] *Id.* § 7071.5(d).
[145] *Id.* § 7071.11(a).
[146] *Id.*
[147] 161 Cal. App. 3d 490 (1984).
[148] 149 Cal. App. 3d Supp. 29 (1983).

apportion the amount of the bond among the various claimants. In other words, the court held that, just because one claimant obtained its judgment before other claims were made on the bond, that did not give the first claimant priority over subsequent competing claims.

Several cases have held that attorneys' fees are recoverable in an action on a license bond. In *Sweeney v. McClaran*,[149] the court held that it could, in its discretion, award reasonable attorneys' fees and costs to the surety company, even where there were multiple competing claims made on the bond exceeding the penal obligation of the bond, based upon the language of the interpleader statute under which the surety brought its action. Similarly, in *Bolivar v. Surety Co.*,[150] an action was brought by a homeowner against the license bond surety. The court held that the homeowner could recover attorneys' fees against the surety because the underlying contract between the homeowner and the contractor had an attorneys' fee clause. The court reasoned that, under California Civil Code § 2808, the surety's liability was commensurate with that of its principal and, under Civil Code § 2809, the obligation of a surety is neither larger nor more burdensome than that of its principal. The court therefore held that because the surety's principal, the contractor, was liable for attorneys' fees, the surety's liability would be commensurate with that of its principal, and therefore the surety was likewise liable for attorneys' fees.

In *Bailey-Sperber, Inc. v. Yosemite Insurance Co.*,[151] the court of appeal held that a prime contractor could sue the surety on the subcontractor's license bond after the subcontractor died and failed to perform the subcontract. The basis for recovery, as noted by the court, was that the subcontractor abandoned the job, in violation of Business and Professions Code § 7101; furthermore, the subcontractor failed to complete the project, in violation of Business and Professions Code § 7113.

In *Carpenters Health & Welfare Trust Fund v. Surety Co. of Pacific*,[152] the court of appeal held that Business and Professions Code § 7105(d), indicating that trust funds can bring an action on license bonds, was preempted by ERISA. Therefore, the union could not re-seek recovery of fringe benefits on the license bond.

§ 1.15 LICENSE REVOCATION FOR FAILURE TO PRODUCE RECORDS

In *Pinney v. Phillips*,[153] the registrar of contractors revoked the license of an electrical contractor for the contractor's failure to produce its business records after a written demand was made upon the contractor under California Business and Professions Code § 7111. Section 7111 provides that the refusal by a licensee to comply with the written request of the registrar to make records available for

---

[149] 58 Cal. App. 3d 824 (1976).
[150] 72 Cal. App. 3d Supp. 22 (1977).
[151] 64 Cal. App. 3d 725 (1976).
[152] 13 Cal. App. 4th 1406 (1993).
[153] 230 Cal. App. 3d 1570 (1991).

inspection constitutes a cause for disciplinary action. The registrar had received information that an electrical contractor might be involved in receiving stolen property. The Contractors State License Board had also received complaints that the contractor was not paying prevailing wages. An investigator for the License Board issued a written demand that the contractor make available all records pertaining to construction activities, including contracts, vouchers, checks, invoices, accounts receivable, and accounts payable, from August 1, 1981, to August 31, 1984.

There were numerous offers, counteroffers, and demands between the License Board and the contractor. Finally, the License Board filed an accusation against the contractor. The administrative law judge found that the contractor had failed to make its records available, so the ALJ revoked the contractor's license and prohibited the contractor from serving as an officer, director, associate, partner, or qualifying individual of any licensee.

The decision of the administrative law judge was adopted by the registrar of contractors, and the contractor filed a petition for writ of administrative mandamus with the court. The trial court granted the petition and ordered the registrar to vacate its decision. The registrar of contractors appealed, and the appellate court affirmed the decision of the trial court. The Fourth Amendment to the United States Constitution prohibits unreasonable searches and seizures. Under the "closely regulated business exception," certain businesses may be inspected without a warrant. The court noted that that exception applies only to businesses that, by their nature, have such a high risk of illegal conduct that unannounced inspections are essential, such as junkyards, mines, gun shops, liquor stores, and massage parlors. The court held that the contracting business is not a closely regulated business for purposes of the Fourth Amendment. The court, therefore, concluded that Business and Professions Code § 7111 cannot constitutionally permit a warrantless inspection or production of records without a subpoena. The court held that section 7111 is constitutional if it is used in connection with a subpoena and a search warrant that provide the licensee with an opportunity for prior judicial review. The court held that because the registrar of contractors cannot constitutionally compel the production of records without a subpoena, the registrar may not revoke a contractor's license for failure to produce the records.

§ 1.16 CONTRACT FOR CONSTRUCTION OF SINGLE-FAMILY DWELLING MUST BE IN WRITING

California Business and Professions Code § 7164 requires that every contract, and any changes in any contract, between an owner and a contractor for the construction of a single-family dwelling, which dwelling is to be retained by the owner for at least one year, shall be evidenced in a writing signed by both parties. Furthermore, section 7164 provides that the writing must contain the following:

1. Name, address, and license number of the contractor
2. Approximate dates when work will begin and be substantially completed

3. Legal description of the location of the work
4. Language of the notice required pursuant to section 7018.5 (see § 1.07)
5. Other matters agreed to by the parties
6. Writing that is legible and that clearly describes any other document that is to be incorporated into the contract
7. A provision, in close proximity to the signature line and printed in at least 10-point bold type or all capital letters, that the owner has the right to require the contractor to obtain a performance and payment bond, and that the expense of the bond may be borne by the owner.

Prior to commencement of the work, the contractor must furnish the owner with a copy of the contract.

§ 1.17 UNLICENSED CONTRACTOR MAY NOT ADVERTISE

It is a misdemeanor for any person to advertise for construction or for a work of improvement unless that person holds a valid contractor's license in the particular classification so advertised.[154] A licensed building or engineering contractor may advertise as a general contractor.[155] The term "advertise" as used in California Business and Professions Code § 7021.1 includes, but is not limited to, the issuance of any card, sign, or device to any person causing, permitting, or allowing any sign or marking on or in any building or structure or in any newspaper, magazine, or by airwave transmission or in any directory under a listing for construction or work of improvement covered by the Contractors License Law.[156] A violation of section 7021.1 is punishable by a fine of not less than $700 or more than $1,000; if the registrar of contractors has probable cause to believe an unlicensed individual is in violation of section 7027.1, the registrar may issue a citation to that person.[157]

§ 1.18 TIME FOR FILING OF DISCIPLINARY PROCEEDINGS

California Business and Professions Code § 7091 provides that a complaint against a licensee alleging the commission of any "patent acts or omissions" that may be grounds for legal action shall be filed in writing with the registrar of contractors within four years after the act or omission alleged as the grounds for disciplinary action. An accusation or citation against a licensee must be filed within four years after the "patent act or omission" alleged as the ground for disciplinary action or within 18 months from the date of the filing of the complaint with the registrar of contractors, whichever is later.

---

[154] Cal. Bus. & Prof. Code § 7021.1(a).
[155] Id. § 7027.1(a).
[156] Id. § 7021.1(b).
[157] Id. § 7021.1(c), (d).

With respect to an accusation alleging a violation of section 7112 (misrepresentation in obtaining license), the accusation may be filed within two years after the discovery by the registrar of contractors of the alleged facts constituting the fraud or misrepresentation prohibited by section 7112.

A complaint against a licensee alleging commission of any "latent acts or omissions" that may be grounds for legal action pursuant to section 7109 (departure from accepted trade standards or departures from or disregard of plans and specifications) regarding structural defects, as defined by the regulations of the Contractors State License Board, shall be filed in writing with the registrar of contractors within 10 years after the act or omission alleged as the ground for disciplinary action. The accusation or citation against the licensee must be filed within 10 years after the latent act or omission or within 18 months from the date of the filing of the complaint with the registrar of contractors, whichever is later, with the same exception for violation of section 7112, as mentioned above. As used in section 7112, *latent act or omission* means an act or omission that is not apparent by reasonable inspection.

§ 1.19 CONTINUING RESPONSIBILITY FOR CITATIONS OF MEMBER, OFFICER, DIRECTOR, OR ASSOCIATE

Under California Business and Professions Code § 7121.1, the disassociation of any member, officer, director, or associate from the license of any partnership, corporation, firm, or association whose license has been cited pursuant to section 7099 shall not relieve that member, officer, director, or associate from responsibility for complying with the citation if that person had knowledge of or participated in any of the prohibited acts for which the citation was issued. Likewise, the disassociation of any qualifying partner, responsible managing officer, or responsible managing employee from the license that has been cited pursuant to section 7099 shall not relieve that qualifying partner, responsible managing officer, or responsible managing employee from any responsibility for complying with the citation.[158]

§ 1.20 CONTRACTOR'S BUSINESS LICENSES

California Business and Professions Code §§ 16000 and 16100 were amended, effective January 1, 1997, to provide that cities and counties, before issuing a business license for an applicant to conduct a contracting business, must verify that the applicant is licensed by the Contractors State License Board.

§ 1.21 CONTRACTORS WHO PREPARE HOME INSPECTION REPORTS MUST NOT PERFORM REPAIRS TO THE STRUCTURE

California Business and Professions Code §§ 7195 *et seq.* were added effective January 1, 1997, dealing with "home inspectors." These statutes define *home*

---

[158] Cal. Bus. & Prof. Code § 7122.1.

inspection as a noninvasive physical examination, performed for a fee, in connection with the transfer of residential property of one to four units. *Home inspection report* is defined as a report of conditions observed or recommendations made; a *home improvement inspector* is any individual who performs a home inspection.

The law, which does not apply to registered engineers, licensed land surveyors, or licensed architects, provides that home inspectors who are not licensed general contractors, structural pest control operators, architects, or registered professional engineers are required to conduct a home inspection with a degree of care that a reasonably prudent home inspector would exercise. Any contractual provision that purports to waive the duty of care or to limit the liability of a home inspector is void. The time for commencement of a legal action against a home inspector is no more than four years from the date of the inspection. It is an unfair business practice for a home inspector to perform or offer to perform repairs to the structures inspected; to inspect a property in which the home inspector has any financial interests; to offer any inducement to the owner of the property for referral of business; or to contract for a fee that is in any way contingent upon the conclusions of the report or the close of escrow. The new law prohibits even home inspectors who are licensed contractors from contracting to perform repairs and thus will have an impact on the many contractors who have added home inspection services to their line of business.

§ 1.22 KICKBACK BY CONTRACTOR TO INSURANCE ADJUSTER IS CRIME

California Penal Code § 551 provides that it is unlawful for a contractor or an agent or employee of a contractor to offer to any insurance agent or adjuster any commission or direct or indirect consideration for referring an insured person to a contractor for repairs to, or replacement of, a structure covered by the insurance policy. It is a crime for a contractor to offer or give to an insured any discount intended to offset a deductible required by an insurance policy covering repairs to, or replacement of, a residential or commercial structure.

§ 1.23 CONTRACTORS ARE REQUIRED TO DISCLOSE DISCIPLINARY ACTION

California Business and Professions Code § 7030.1 was added, effective January 1, 1997, to strengthen the requirement that contractors disclose disciplinary actions. Under the statute, a contractor whose license has been suspended or revoked two or more times within an eight-year period must disclose the suspension or revocation in a document provided to prospective customers before entering into a contract to perform work on residential property with four or fewer units. For a first violation, a penalty of $1,000 is assessed; for a second violation, $2,500; for a third violation, $5,000 plus a one-year suspension of license; for a fourth violation, revocation of the contractor's license.

§ 1.24 SURETIES MAY BE EXEMPT FROM CONTRACTOR LICENSING REQUIREMENT

Under existing California law, a surety undertaking to perform construction work in order to fulfill its obligations under a performance bond is required to have a contractor's license.[159] Effective January 1, 1997, California Business and Professions Code § 7044.2 was added to provide that an admitted surety insurer that engages a contractor to complete construction work for which the surety has issued a performance bond is exempt from the requirements of the Contractors License Law, provided that all actual construction work is performed by duly licensed contractors. The exemption appears to be available only when the surety "engages a contractor" to complete the work; therefore, a surety would not be exempt from the contractor license requirements if it subcontracted everything out or employed a number of different trade contractors to complete the project.

---

[159] *See* General Ins. Co. v. St. Paul Fire & Marine Ins. Co., 38 Cal. App. 3d 760 (1974).

Chapter 2
BIDDING PRIVATE AND PUBLIC WORK

§ 2.01 Rules for Private Work
    [A] Bidding Documents
    [B] Acceptance
    [C] Subcontractor Bids and the Doctrine of Promissory Estoppel
    [D] Material Supplier Bids and the Commercial Code

§ 2.02 Rules for Public Work
    [A] Statutory Framework and Purpose
    [B] Applicable Rules and Exceptions
        [1] Competitive Bids and Negotiated Proposals
        [2] Exceptions
        [3] Design-Build
    [C] Solicitation of Bids
        [1] Or Equal Clause
        [2] Prequalification
    [D] Submission of Bids
        [1] Security
        [2] Subcontractor Listing Law
        [3] Non-collusion Affidavit
        [4] Affirmative Action Forms
    [E] Bid Withdrawal and Mistake
    [F] Evaluation of Bids
        [1] Bid Responsiveness
        [2] Bidder Responsibility
        [3] Alternate Bids
        [4] Preferences
            [a] Local Firms and Labor
            [b] Small and Minority-Owned Businesses
    [G] Bid Protests
        [1] Procedures
        [2] Remedies
        [3] Standard of Review

§ 2.01 RULES FOR PRIVATE WORK

Unlike public entities, private builders are free to adopt whatever procedures they choose in awarding construction contracts, limited only by lender requirements, business practices, and market conditions. Thus, the private owner is free to invite competitive bids or simply to negotiate directly with potential prime contractors regarding all terms and conditions of the construction contract. Private owners who obtain competitive bids from a number of different contractors are primarily concerned with obtaining a responsible bidder to complete the work set forth in the contract documents at the lowest possible price. While some private owners may utilize open bidding procedures—i.e., placing an advertisement in a local newspaper to solicit bids from all qualified contractors—many private owners utilize a "short list" and restrict bidding to those contractors who are identified and invited to bid by the owner.

[A] Bidding Documents

Bidding documents on private works typically consist of one or more of the following: invitation to bid; instructions to bidders; bid forms, e.g., contractor qualification statement and price proposal; and form of owner-contractor agreement. Unlike public works, there is no required format or content for the bidding documents.

Issues that frequently arise with respect to private bidding documents include the degree to which the bidder is entitled to rely on the information contained in the bidding documents in calculating its bid price; what knowledge is imputed to the bidder; the extent of the bidder's duty to investigate the site; and the effect of owner disclaimers in the bid documents. A long line of cases, beginning with *United States v. Spearin*,[1] have held that the owner impliedly warrants that the plans and specifications it issues are free from defects and that the project can be constructed from the plans and specifications. This rule, often referred to as the *Spearin* doctrine, has been adopted by California courts, as discussed in Chapter 4.[2] Given the implied warranty of fitness of the plans and specifications, a general disclaimer by the owner may not be sufficient to transfer the risk of defective plans and specifications to the contractor; however, express disclaimers applicable to specific statements in the bidding documents may overcome the implied warranty.[3]

---

[1] 248 U.S. 132 (1918).

[2] *See* Gogo v. Los Angeles Flood Control Dist., 45 Cal. App. 2d 334 (1941); Souza & McCue Constr. Co. v. Superior Court, 57 Cal. 2d 508 (1962); Jasper Constr., Inc. v. Foothill Junior College Dist., 91 Cal. App. 3d 1 (1979), *reh'g denied;* Tonkin Constr. Co. v. County of Humboldt, 188 Cal. App. 3d 828 (1987).

[3] *See, e.g.,* Wunderlich v. State, 65 Cal. 2d 777 (1967) (contractor's claim for extra compensation denied where express terms of agreement assigned risk of unknown site conditions to contractor).

[B] Acceptance

Even if a private owner solicits multiple bids, there is no obligation to award a contract to the lowest bidder or even to award the contract at all. The advertisement or invitation to bid is not itself an offer, and no rights are created until the owner accepts the bid as the contract price.[4] After having received bids, the private owner may enter into negotiations with any prime contractor, regardless of whether that contractor's bid was low, or whether the contractor even participated in the bidding, unless the owner binds itself to make the award in a particular manner in the bidding documents.

Generally, the owner must communicate or provide notice of its acceptance of the bid to the contractor. An owner's silence in the face of an offer is not an acceptance unless there is a relationship between the parties or a previous course of dealing pursuant to which silence would be understood as acceptance.[5]

Once a bid for private work has been accepted by the owner, there are relatively few statutory rules governing formation of the contract. The Contractors State License Law[6] is the primary California code that governs private works. It provides, among other things, that it is a misdemeanor for a person to act as a contractor without a license, and that unlicensed contractors have no standing to bring suit to recover the cost of work performed.

[C] Subcontractor Bids and the Doctrine of Promissory Estoppel

A frequent bidding dispute on private works concerns whether a subcontractor can withdraw or refuse to honor its bid to the prime contractor. The relationship between subcontractors and prime contractors on private construction projects is not regulated, except by certain prompt payment statutes; rather, it is governed only by the basic principles of common-law and equity. The traditional common-law principle that a bid is not binding until accepted produces a harsh result when the prime contractor relies on the subcontractor's bid in submitting its own bid to the owner. Under these circumstances, the equitable doctrine of promissory estoppel is applied to require a subcontractor to honor its bid, even if it makes a mistake in calculating the bid price.

In the leading case of *Drennan v. Star Paving Co.*,[7] a subcontractor sought to withdraw its bid before the prime contractor had accepted it but after the prime contractor has submitted its bid to the owner. The California Supreme Court held the subcontractor's bid implied a promise not to withdraw if the prime contractor was going to rely upon it, to its detriment, in preparing its bid to the owner. Therefore, unless the prime contractor has some reason to know that a bidding subcontractor has made an error, the subcontractor will be bound to perform in

---

[4] Klose v. Sequoia Union High Sch. Dist., 118 Cal. App. 2d 636 (1953).
[5] Southern California Acoustics Co. v. C.V. Holder, Inc., 71 Cal. 2d 719, 722 (1969).
[6] Cal. Bus. & Prof. Code §§ 7000 *et seq.;* Albaugh v. Moss Constr. Co., 125 Cal. App. 2d 126 (1954).
[7] 52 Cal. 2d 409 (1958).

accordance with its bid. If the subcontractor is incapable of actually performing, the doctrine of promissory estoppel gives rise to a cause of action allowing the prime contractor to be awarded damages against the nonperforming subcontractor. Such damages include the difference between the nonperforming subcontractor's bid and the amount of the subcontract entered into by the prime contractor with a replacement subcontractor. The prime contractor may also be entitled to recover damages for delay or disruption to the project by proving that such damages were proximately caused by the subcontractor's failure to honor its bid. The usual measure of damages under a promissory estoppel theory, however, is the difference between the original subcontractor's bid and the cost of the replacement subcontract.[8]

Many cases have required the courts to decide whether the prime contractor's recovery is barred by actual or constructive knowledge that the subcontractor has made a mistake in its bid. These cases address whether it was reasonable for the prime contractor to rely upon a subcontractor's low bid, or whether the bid was so low in comparison to other bids that any such reliance was inherently unreasonable. For example, in *Norcross v. Winters*,[9] a subcontractor bid $4,800 for a job. The court of appeals held that the disparity between this bid and the other bids for the same job, which ranged from $7,490 to $8,101, was not sufficient to put the prime contractor on constructive notice of the subcontractor's mistake. Thus, courts will generally require a very large price difference for a subcontractor to argue successfully that the prime contractor should have known about the subcontractor's mistake in calculating and submitting the bid.[10]

Other potential defenses to a contractor claim against a subcontractor based upon promissory estoppel include:

1. unreasonable delay by the prime contractor in accepting subcontractor bids;

2. lack of actual reliance, inferable by the fact that the prime contractor used the subcontractor's bid as a starting point in attempting to obtain lower bids, i.e., bid shopping or bid peddling;

3. subcontractor's justified refusal to sign a subcontract containing unfair and/or unreasonable terms and conditions; and

4. prime contractor's ability, upon ascertaining the subcontractor's mistake, to attempt to withdraw its bid to the owner.

Another frequent bidding dispute on private works concerns whether a prime contractor may be required to award to a particular subcontractor when the prime contractor actually used that subcontractor's bid in computing its bid to the owner.

---

[8] *See, e.g.,* C & K Eng'g Contractors v. Amber Steel Co., 23 Cal. 3d 1 (1978) (under promissory estoppel, damages are limited "as justice requires").

[9] 209 Cal. App. 2d 207 (1962).

[10] *See, e.g.,* H.W. Stanfield Constr. Corp. v. Robert McMullan & Sons, 14 Cal. App. 3d 848 (1971); Saliba-Kringlen Corp. v. Allen Eng'g Co., 15 Cal. App. 3d 95 (1971).

This would be the expected result if the rationale underlying the promissory estoppel doctrine were to be applied to prime contractors in the same manner as subcontractors. This does not hold true, however, because a bidding subcontractor does not generally take any action to its detriment in "reliance" on a prime contractor's using its bid. Such reliance arises only in the unusual situation when the prime contractor has made an express promise to award the subcontract to the lowest bidding subcontractor. In such a case, the subcontractor, having relied upon the prime contractor's promise, may enforce it through application of the doctrine of promissory estoppel.[11]

[D] Material Supplier Bids and the Commercial Code

Under the terms of California Commercial Code § 2205, bids by material suppliers are subject to the following rules:

1. A subcontractor's oral or written bid is considered an offer to supply goods to a licensed contractor and is irrevocable.

2. A material supplier's signed written bid that, by its terms, gives assurances that it will be held open is not revocable for lack of consideration, for the time stated in the bid, or if no time is stated, for a reasonable time not to exceed three months.

3. Such bids shall remain irrevocable, notwithstanding lack of consideration, for a period of 10 days following the award of the contract to the prime contractor, but in no event more than 90 days following the date the bid was made by the material supplier, except if the bid is oral and the price of the goods exceeds $2,500, the material supplier is released from the offer unless the contractor confirms the bid in writing within 48 hours.

4. The material supplier may provide in the bid or offer that it will be held open for less than the time set forth in the statute. If the bid is oral, the material supplier must follow up with a written letter stating the period of time during which the bid will remain open.[12]

Thus, whenever oral bids for supplies are submitted, the contractor should routinely confirm the bid in writing if it wants to hold the price for 90 days. Conversely, if the supplier does not intend that its oral bid remain firm for 90 days, it should immediately follow up with a letter to the contractor identifying when the quote expires.

---

[11] *See* Electrical Constr. & Maintenance Co. v. Maeda Pac. Corp., 764 F.2d 619 (9th Cir. 1985) (enforcing oral promise by prime contractor that it would award subcontract to lowest bid).

[12] Cal. Com. Code § 2205. *See generally* Foremost Pro Color, Inc. v. Eastman Kodak Co. 703 F.2d 534 (9th Cir. 1983).

§ 2.02 RULES FOR PUBLIC WORK

[A] Statutory Framework and Purpose

The laws concerning bidding on public projects are very different from those regulating private projects. Not only the contractor, but also the public agency itself, is subject to regulation by statute. The California codes set forth a complicated array of statutes governing public contracts. While the basic rules are consolidated in the California Public Contract Code, additional requirements are found in numerous other codes, state regulations, city charters, and local ordinances.[13] The Public Contract Code is divided into five major sections: (1) Administrative Provisions;[14] (2) Contracting by State Agencies;[15] (3) Contracting by Local Agencies;[16] (4) Arbitration of Public Works Contract Claims;[17] and (5) Withheld Contract Funds.[18]

As evidenced by the foregoing divisions, not all provisions of the Public Contract Code apply to every public agency. In fact, much of the underlying contract preparation and many bidding disputes involve determining which code provisions apply to a particular public entity. The statutory scheme, however, can be understood and applied in a consistent and predictable manner if construction practitioners familiarize themselves with the structure and underlying purpose of public works law.

The goals of competitive bidding are to guard against favoritism, improvidence, extravagance, fraud, and corruption; to prevent the waste of public funds; and to obtain the best economic result for the public.[19] Competitive bidding requirements exist for the benefit of the public and were not established to protect the individual bidder.[20] In consolidating the majority of competitive bidding statutes in the Public Contract Code, the legislature identified the following additional objectives: to clarify the law with respect to competitive bidding requirements, to ensure full compliance with competitive bidding statutes as a means of protecting the public from misuse of public funds, and to provide all qualified bidders with a fair opportunity to enter the bidding process, thereby stimulating competition in a manner conducive to sound fiscal practices.[21]

To implement the foregoing goals and objectives, the legislature mandated specific procedures for solicitation, submission, and evaluation of bids. The resulting bid procedures can be cumbersome and place administrative burdens on

---

[13] *See, e.g.,* Cal. Gov't Code §§ 6250–6260 (Public Records Act); Cal. Lab. Code §§ 1770 *et seq.* (Prevailing Wage Law).

[14] Cal. Pub. Cont. Code §§ 1100 *et seq.*

[15] *Id.* §§ 10100 *et seq.*

[16] *Id.* §§ 20100 *et seq.*

[17] *Id.* §§ 22200 *et seq.*

[18] *Id.* §§ 22300 *et seq.*

[19] Graydon v. Pasadena Redevelopment Agency, 104 Cal. App. 3d 631 (1980), *cert. denied,* 449 U.S. 983, *reh'g denied,* 449 U.S. 1104 (1981).

[20] *See* East Bay Garbage Co. v. Washington Township Sanitation Co., 52 Cal. 2d 708 (1959); Steelguard, Inc. v. Jannesen, 171 Cal. App. 3d 79 (1985).

[21] Cal. Pub. Cont. Code § 100.

bidders and agency contract administrators. However, once mastered, a contractor's ability to accurately document the required components of a bid to a public entity often determines the successful bidder. Moreover, familiarity with competitive bidding laws facilitates prompt resolution of bid protests and award of public contracts.

[B] Applicable Rules and Exceptions

Most state and local agencies must follow the Public Contract Code requirements for force accounts and for competitive bidding for public works projects. Although certain appellate decisions have ruled that local agencies may use their own workforce on projects, regardless of the threshold statutory amount for competitive bidding,[22] in practice, most state and local agencies use their own forces only for work that falls below the competitive bidding monetary threshold. Agencies are prohibited from splitting large projects into smaller pieces to avoid competitive bidding procedures.[23] However, under certain circumstances, several projects may be united and bid as a single project.[24] For projects that fall above the monetary threshold, public agencies must use the formal bidding procedures, including public notice of the projects and selection of the lowest responsive and responsible bidder.[25]

[1] Competitive Bids and Negotiated Proposals

Traditionally, state and local agencies construct buildings and improvements through a three-part, "design-bid-build," process.[26] Under this process, agencies may either use in-house architects and engineers to develop plans, specifications, and cost estimates for construction, or adopt selection procedures, pursuant to California Government Code §§ 4525 *et seq.,* commonly referred to as the "The Little Brooks Act," and issue requests for proposals from qualified designers. The Little Brooks Act generally requires public agencies to select the best offer on the basis of demonstrated competence and qualifications and then, if possible, negotiate a fair and reasonable price with that offerer. A construction management contract, although nominally falling under the Little Brooks Act, at times has been held to require competitive bidding where the construction manager also serves as the contractor or guarantees the cost of construction.[27] After the plans and specifications are complete, most state and local agencies must utilize a second competitive bidding process to select the construction contractor.

---

[22] Construction Indus. Force Account Council v. Delta Wetlands, 2 Cal. App. 4th 1589 (1992).
[23] *See, e.g.,* Cal. Pub. Cont. Code § 20116.
[24] *See, e.g.,* Cal. Pub. Cont. Code § 10127.
[25] *See, e.g., id.* §§ 10108, 20672.
[26] Many public agencies, however, use other solicitation procedures, including competitive negotiation, design-build, and prequalification.
[27] *See, e.g.,* City of Inglewood-L.A. County Civic Ctr. Auth. v. Superior Court, 7 Cal. 3d 861 (1972).

[2] Exceptions

Although state and local agencies are generally required to award construction contracts following preparation of full, complete, and accurate plans and specifications and estimates of costs and to award to the lowest responsive and responsible bidder, exceptions to these competitive bidding rules include, but are not limited to:

1. emergency works;[28]
2. projects such as courts and jails, for which the public authority is acting under existing court orders to expedite construction;[29]
3. projects requiring specialized equipment;[30]
4. instances in which no statute expressly requires competitive bidding;[31]
5. scenarios in which all of the work in a project is to be performed by the public agency's own employees;[32]
6. charter cities, which are not subject to statutory competitive bidding requirements but must comply with requirements set forth in the charter or by local ordinance or code;[33]
7. undesirability or impossibility to advertise an award on the basis of low bid;[34] and
8. infrastructure projects constructed under public/private development agreements.[35]

Although there are numerous exceptions to the use of competitive bidding procedures, the penalties for ignoring or purposefully evading competitive bidding procedures can be severe. In order to guard against corruption and favoritism as well as protect taxpayer funds from unauthorized or improper expenditures, courts have declared contracts void where there was a complete failure to follow competitive bidding requirements and the legal requirements for contracting with the agency were clear. A contractor is charged with knowledge of

---

[28] *See, e.g.,* Cal. Pub. Cont. Code §§ 10340(b)(1), 20134(a), 22050.

[29] *See, e.g., id.* § 20134(b) (West Supp. 1998).

[30] *See, e.g.,* Cal. Pub. Util. Code § 130238.

[31] *See, e.g.,* Associated Builders & Contractors v. Contra Costa Water Dist., 37 Cal. App. 4th 466 (1995).

[32] *See, e.g.,* Construction Indus. Force Account Council v. Delta Wetlands, 2 Cal. App. 4th 1589 (1992).

[33] *See, e.g.,* Stacy & Witbeck, Inc. v. City & County of S.F., 36 Cal. App. 4th 1074 (1995); Committee of Seven Thousand v. Superior Court, 45 Cal. 3d 491 (1988); Piledrivers Local Union No. 2375 v. City of Santa Monica, 151 Cal. App. 3d 509 (1984); Smith v. Riverside, 34 Cal. App. 3d 529 (1973); Pasadena v. Charleville, 215 Cal. 384 (1932).

[34] Los Angeles Dredging Co. v. Long Beach, 210 Cal. 3d 348 (1930); Hiller v. City of L.A., 197 Cal. App. 2d 685 (1961).

[35] Cal. Gov't Code §§ 5956 *et seq.*

the limits of an agency's contracting authority and thus may be precluded from recovering the cost of work performed in violation of competitive bidding rules or ordered by an individual with no authority to bind the agency.[36] However, certain courts have recognized that when the enforceability of contracts or the propriety of a procurement arises after work has been performed, or where the state of the law was not settled at the time of the award, the contract should be nullified only when the illegality is plain.[37]

[3] Design-Build

The benefits of statutorily required competitive bidding procedures are price competition, clear identification of the work purchased, and avoidance of corruption, fraud, favoritism, and politics in the expenditure of public money through open, public, objective selection bidding methods. The three-part procedure, however, can be time consuming and costly. Moreover, if problems arise during construction, the designer and contractor often seek to blame one another for errors and extra costs, creating delays and increasing the number of change orders. In an effort to save time and costs and to avoid the conflicts that often arise between designer and contractor, an increasing number of public and private projects are being delivered utilizing design-build methods.[38]

In general, state agencies[39] are subject to legislative restrictions that prevent the use of design-build in the bidding and award of public contracts. Most local agencies are subject to statutory competitive bidding requirements for construction contracts but may have the ability to use design-build on a low bid basis. The local agency may "prequalify" the design member of the team[40] or utilize professional qualifications as one of the criteria for selection of responsive and responsible bidders. As noted above, charter cities are exempt from Public Contract Code competitive bidding requirements but must satisfy any restrictions contained in their own charters. Many city charters provide for the use of a "competitive negotiation" process, which works well for a design-build procurement. Different rules apply to different agencies; however, in general, express enabling legislation is required for most state and local agencies to use design-build as a project delivery system on a particular project.[41]

Although combining the selection of design and construction work in one package may shorten the schedule for completion and reduce the number of change orders, it creates conflicts concerning well-established principles of law and traditional methods of risk allocation. For example, California Civil Code

---

[36] *See, e.g.,* Reams v. Cooley, 171 Cal. 150 (1915); Miller v. McKinnon, 20 Cal. 2d 83 (1942).

[37] *See, e.g.,* John Reiner & Co. v. United States, 325 F.2d 438 (Ct. Cl. 1963), *cert. denied,* 377 U.S. 931 (1964).

[38] *See* discussion in Chapter 3, § 3.06.

[39] Other than the University of California and California State University—*see* Cal. Pub. Cont. Code §§ 10503, and 10708.

[40] *See* § 2.02[C][2].

[41] *See, e.g.,* Cal. Gov't Code § 14016; Cal. Pub. Cont. Code §§ 10503, 10708.

§ 3247(a) requires every original contractor on a public project to provide a payment bond, however, subsection (c) of the statute indicates that architects and engineers are not required to post payment bonds. Many public agencies also require performance bonds, which are designed to give the project owner added protection that the project is timely completed in accordance with the plans and specifications. Surety companies may be reluctant to write performance bonds that cover both design and construction because surety companies are accustomed to guaranteeing only the construction activities of the contractors, not the performance of design professionals.

Another problematic area for public design-build projects is the Subletting and Subcontracting Fair Practices Act,[42] which requires contractors to name major subcontractors at the time of the bid. This requirement is difficult to meet since the design has not been finalized at the time subcontractors must bid a fixed price for the work. Public agencies that require subcontractors to be listed at the time a design-build proposal is submitted are often faced with numerous substitution requests, as the listed subcontractors refuse to perform for the price quoted to the prime contractor after they review the complete scope of work in the fully developed plans and specifications.

Enacted in 1998, California Government Code §§ 13332.19 and 14661 provide comprehensive design-build procedures that address some of the problematic areas of design-build procurements. The legislature expressly found and declared that it is in the best interest of the state to construct state offices and other facilities in a cost-efficient manner that represents the best overall value to taxpayers, and that the design-build process can be an attractive option to the state by accelerating delivery schedules and achieving cost savings by promoting improved coordination between contractor and architect, shifting management risk from the state to the design-build team and minimizing change orders through early collaboration between design and construction disciplines.

Government Code § 14661 authorizes the state to enter into design-build contracts for the construction of state facilities under specified conditions:

1. The legislature approves the use of the design-build process for a specific project.

2. The department has prepared a program setting forth the scope of the project and to establishing a competitive prequalification and selection process for design-build teams. (The statute limits prequalification criteria to specified areas, including licensing, experience, management plan, bonding, and insurance capacity and claims history.)

3. The process for selection of the design-builder must be one of three identified methods, as deemed to be in the best interest of the state.

4. Design-build contractors must comply with the Subletting and Subcontracting Fair Practices Act during each phase of the project. The state

---

[42] Cal. Pub. Cont. Code §§ 4100 *et seq.* See § 2.02[D][2].

may identify types of subcontractors, by license classification, that must be listed by the design-build entity at the time of the bid within certain limits. All subcontractors that are listed at the time of bid are afforded the protection of the Subletting and Subcontracting Fair Practices Act. Moreover, all subcontracts not listed at the time of the bid shall be competitively bid, with public notice, a fixed date and time for receiving bids, reasonable prequalification criteria, and award made to the lowest responsive and responsible bidder.

5. Design-build contractors must be bonded, and the state agency must work with the surety insurance community to agree upon an acceptable form of bond for the design-build projects.

The foregoing statute will, no doubt, serve as a model for many local agencies and charter cities in developing design-build procurement packages. An examination of basic competitive bidding rules, however, is necessary even with respect to design-build contracting.

[C] Solicitation of Bids

After preparing "full, complete and accurate plans and specifications and estimates of cost,"[43] the agency must publish a notice inviting bids by advertising in news media and public postings within specified time periods.[44] The contents of the invitation to bid vary depending on the agency. Invitations must include information such as the location and deadline for submission of bids; specification of the license classification that a contractor shall possess at the time a contract is awarded;[45] provisions to permit the substitution of securities for any monies withheld by a public agency to insure performance under a contract;[46] and certain contractual indemnity provisions.[47] Specifications provided for bidding the work must be sufficiently detailed, definite, and precise.[48]

[1] Or Equal Clause

California Public Contract Code § 3400 prohibits "sole source" specifications on public work, i.e, specifications referencing a material or product by brand name or manufacturer, unless:

1. at least two trade names are listed followed by the words "or equal" (listing the specified material and an equal product manufactured in California followed by the words "or equal");

---

[43] Cal. Pub. Cont. Code §§ 10120, 10180.
[44] *See, e.g., id.* §§ 10140, 22037.
[45] *Id.* §§ 10141, 22037, 3300; Cal. Bus. & Prof. Code § 7028.15.
[46] Cal. Pub. Cont. Code §§ 10263, 22300.
[47] *Id.* Code § 20103.6.
[48] *See generally* Baldwin-Lima-Hamilton Corp. v. Superior Court, 208 Cal. App. 2d 803 (1962).

2. the product is designated to match others in use;

3. use of a unique or novel product is in the public's interest;

4. only one brand or trade name is known to the public agency; or

5. the governing body, by resolution, makes a finding included in the specifications that the product is designated by trade name in order that a field test or experiment be made to determine suitability for future use.

Section 3400 further requires that the specifications provide a period of time prior to the award of the contract for submission of data substantiating a bidder's request for substitution of an "equal" item.[49]

[2] Prequalification

In an attempt to streamline bidding procedures and to specify in detail the criteria for determining whether a contractor is qualified or responsible to perform the work,[50] many state and local agencies have adopted prequalification procedures. Agencies may prequalify prime contractors for a single project or may seek to establish a pool of qualified contractors or subcontractors who will be invited to bid on a series of similar projects over a designated period of time. Many contractors, however, assert that if a contractor is properly licensed and can provide a surety bond, it should be permitted to bid public work, regardless of the builder's level of past experience or claims history. Despite the complaints of some contractors, the trend in public works is to utilize prequalification procedures in conjunction with award based on low bid, particularly for complex projects that have severe budget and time constraints.

Certain statutes provide that prequalification procedures are discretionary;[51] however, other legislation mandates the use of prequalification in specified circumstances.[52] Under such procedures, potential bidders must qualify in advance of submitting a monetary bid, utilizing standard questionnaires covering experience, quality, and timeliness of past performance, insurance and bonding capacity, and safety, labor, and tax violations.[53] Whenever prequalification is required of any bidder on a contract, it must be required for all prospective bidders to the contract. The agency must adopt a uniform system of rating bidders. If a bidder is determined to be unqualified, it must be given an opportunity to rebut this determination; however, a quasi-judicial or formal hearing is not required.[54] After a pool of prequalified bidders is established, the agency solicits price proposals, and award is made to the lowest responsible bidder.

---

[49] *See* Cal. Pub. Cont. Code § 3400(a).
[50] *See* § 2.02[F][2].
[51] *See e.g.,* Cal. Pub. Cont. Code § 10160.
[52] *See, e.g.,* Cal. Pub. Util. Code § 130051.21.
[53] *See, e.g.,* Cal. Gov't Code § 14661.
[54] City of Inglewood-L.A. County Civic Auth. v. Superior Court, 7 Cal. 3d 861 (1972).

[D] Submission of Bids

Under competitive bidding rules, regardless of whether prequalification is used, the bid or price proposal must be delivered, under seal, on time, to the exact location listed in the bid package. Under most statutory, code, and charter provisions, the public entity cannot receive or consider any late bids.[55]

[1] Security

Most bids for public works contracts must include bid security, either in the form of cash, cashier's check, certified check, or a bid bond issued by an approved surety in the amount of 10 percent of the bid.[56] The bond guarantees that the bidder will sign the contract for the amount stated in its bid. Failure to provide bid security may result in rejection of the bid for nonresponsiveness.

[2] Subcontractor Listing Law

The Subletting and Subcontracting Fair Practices Act (the "Act") requires prime contractors to designate, in their bids, the names and addresses of each subcontractor who will perform work the value of which exceeds one-half of 1 percent of the prime contractor's total bid or, in the cases of construction of streets/highways and bridges, work the value of which exceeds one-half of 1 percent of the total bid or $10,000, whichever is greater. The Act is designed to prevent bid shopping and bid peddling that take place after award of the prime contract. In implementing the Act, a variety of local agencies have required prime contractors to provide additional information at the time of bid, not expressly required by the statute. Such information includes the subcontractor's address, telephone and fax number(s), contact person, license number, license expiration date, etc. The detailed bid information required by the Act, coupled with the common practice of local agencies to require additional detailed information about subcontractors, often creates confusion at the hectic bid deadline. Not surprisingly, the statutory and local agency requirements have fueled countless bid protests over the years.[57]

In response to increasingly complicated local requirements and case law, California Public Contract Code § 4104 was amended in 1998 to provide that a public entity, at its option, may permit any subcontractor information, other than the name and location of the business, to be submitted up to 24 hours after the deadline for bids.[58] If the public entity makes a material change in the bid specifications (i.e., a change with substantial cost impact) less than 72 hours before

---

[55] *See, e.g.,* Cal. Pub. Cont. Code §§ 10167, 10168.

[56] *See, e.g., id.* §§ 10167, 20129.

[57] *See, e.g.,* MCM Constr., Inc. v. City & County of S.F., 66 Cal. App. 4th 359 (1998), *req. for depublication denied;* Valley Crest Landscape, Inc. v. City Council, 41 Cal. App. 4th 1432 (1996); Ghilotti Constr. Co. v. City of Richmond, 45 Cal. App. 4th 897 (1996).

[58] *See* Cal. Pub. Cont. Code § 4104(a)(2)(A).

the bid deadline, the deadline must be extended to no less than 72 hours from the original or current approved deadline for bid submittal.[59] The prime contractor whose bid is accepted by the public entity is ordinarily prohibited from substituting any person or entity as a subcontractor in place of those listed in its original bid. However, there are eight specific instances in which the awarding entity may consent to the substitution of a subcontractor:

(1) when the subcontractor listed in the bid, after having had a reasonable opportunity to do so, fails or refuses to execute a written contract, when that written contract, based upon the general terms, conditions, plans and specifications for the project involved or the terms of that subcontractor's written bid, is presented to the subcontractor by the prime contractor;

(2) when the listed subcontractor becomes bankrupt or insolvent;

(3) when the listed subcontractor fails or refuses to perform his or her subcontract;

(4) when the listed subcontractor fails or refuses to meet the bond requirements of the prime contractor as set forth in section 4108;

(5) when the subcontractor was listed as the result of an inadvertent clerical error;

(6) when the listed subcontractor is not licensed pursuant to the Contractor's License law;

(7) when the awarding authority, or its duly authorized officer, determines that the work performed by the listed subcontractor is substantially unsatisfactory and not in substantial accordance with the plans and specifications, or that the subcontractor is substantially delaying or disrupting the progress of the work; and

(8) when the listed subcontractor is ineligible to work on public works projects pursuant to § 1777.1 or 1777.7 of the Labor Code (dealing with subcontractors who have defrauded their apprentices with respect to wages).[60]

If such a substitution is proposed, the listed subcontractor is entitled to receive notice of the requested substitution and the reasons for it. If the listed subcontractor objects to the substitution, it has five working days to submit written objections. Failure to file written objections shall constitute the listed subcontractor's consent to the substitution. If written objections are filed, the authority must hold a hearing and provide the listed subcontractor written notice at least five working days prior to the hearing.[61] The courts have identified an implied procedure requiring the agency not only to consent to substitution of the listed subcontractor but also to approve the replacement subcontractor.[62]

---

[59] *See id.* § 4104.5.
[60] *Id.* § 4107(a)(1)–(8).
[61] *Id.* § 4107(a).
[62] E. F. Brady Co. v. M. H. Golden Co., 58 Cal. App. 4th 182 (1997).

If a prime contractor asserts a claim of inadvertent clerical error, it must give, within two days after the bid opening, written notice of its claim to the awarding public entity, with copies to the subcontractor involved. Courts have permitted contractors to substantially comply with the notice by alternate means.[63] The subcontractor then has six working days from the opening of the prime bids within which to submit to the awarding public entity any written objection to the prime contractor's claim of clerical error.[64] In *Cal-Air Conditioning, Inc. v. M.P. Allen General Contractors*,[65] the court of appeal applied the doctrine of substantial compliance to a prime contractor's obligation, under California Public Contract Code § 4107.5, to provide written notice to a subcontractor of the prime contractor's claim of inadvertent clerical error in listing that subcontractor in a successful bid on a public works project. A listed subcontractor has a cause of action against the prime contractor for any damages (including lost profits) resulting when the prime contractor unlawfully substitutes a subcontractor.[66] Moreover, an illegally substituted subcontract is itself unenforceable. Thus, a prime contractor who has unlawfully substituted a subcontractor may not thereafter file suit against the substituted subcontractor.[67]

Any failure to specify a subcontractor will be interpreted to indicate the prime contractor's intent to perform the work with its own forces.[68] A prime contractor does not breach the statute and is not required to request permission from the awarding agency to perform work itself when a listed subcontractor refuses to perform the work after having a reasonable opportunity to do so.[69] If a bidding prime contractor fails to comply with the subcontractor listing law, the awarding public entity may either cancel its contract or assess a monetary penalty not exceeding 10 percent of the subcontract price.[70] Moreover, any violation of the subcontractor listing law may be cause for disciplinary action by the Contractors State License Board.[71]

[3] Non-collusion Affidavit

Each public works contract must include a non-collusion affidavit signed and submitted with the bid swearing, among other things, that the bid is not made on behalf of any undisclosed person, that the bid is not collusive or sham, that the bidder has not solicited or conspired with any other bidder to submit a false

---

[63] Cal-Air Conditioning, Inc. v. Auburn Union Sch. Dist., 21 Cal. App. 4th 65 (1993).
[64] Cal. Pub. Cont. Code § 4107.5.
[65] 21 Cal. App. 4th 655 (1993).
[66] *See* Southern Cal. Acoustics Co. v. C.V. Holder, Inc., 71 Cal. 2d 719 (1969); Coast Pump Ass'n v. Stephen Tyler Corp., 62 Cal. App. 3d 421 (1976); C.L. Smith Co. v. Roger Ducharme, Inc., 65 Cal. App. 3d 735 (1977); Interior Sys., Inc. v. Del E. Webb Corp., 121 Cal. App. 3d 312 (1981).
[67] *See* Kiely Corp. v. H.C. Gibson, 231 Cal. App. 2d 39 (1964).
[68] *See* Cal. Pub. Cont. Code § 4106.
[69] W. J. Lewis Corp. v. Harper Constr. Co., 116 Cal. App. 3d 27 (1981).
[70] *See* Cal. Pub. Cont. Code § 4111.
[71] *See id.* § 4107.

bid or refrain from bidding, that the bidder has made no agreements for price fixing or payment of fees in connection with the bid, and that all statements contained in the bid are true.[72]

[4] Affirmative Action Forms

If the project is subject to one of the affirmative action programs discussed in more detail in § 2.02[F][3], bidders will be required to submit certain documentation relating to participation levels of and good-faith outreach efforts to disadvantaged business enterprises or other designated classifications. Failure to submit the required documents will render a bid nonresponsive.

[E] Bid Withdrawal and Mistake

A bidder may withdraw its bid upon written request submitted to the public entity before the deadline for bid submissions. A bidder who has withdrawn a bid may submit a new bid if the bid submission deadline has not already passed.[73] Once the bid is submitted, the bidder has made a binding offer for the contract work, which the public entity can accept upon bid opening.

A bidder who discovers that it has made a mistake in its bid after the bids have been opened may seek relief from its mistake under certain circumstances.[74] Historically, the only remedy available to a contractor seeking to withdraw a bid because of a mistake was to sue for rescission.[75] In 1971 the legislature codified the grounds for obtaining relief for mistakes in bids and eliminated the remedy of rescission for agencies bound by the California Public Contract Code. In order to be relieved from its bid, the bidder must establish:

1. a mistake was made;

2. the contractor gave the public entity written notice of the mistake within five days after the opening bids, specifying in the notice in detail how the mistake occurred;

3. the mistake made the bid materially different than the bidder intended it to be; and

4. the mistake was made in filling out the bid and not due to error in judgment or to carelessness in inspecting the site of the work, or in reading the plans or specifications.[76]

---

[72] Cal. Pub. Cont. Code § 7106.
[73] *See, e.g.,* Cal. Pub. Cont. Code § 10167.
[74] *Id.* § 5101.
[75] M. F. Kemper Constr. Co. v. City of L.A., 37 Cal. 2d 696 (1951); Elsinore Union Sch. Dist. v. Kastorf, 54 Cal. 2d 380 (1960).
[76] Cal. Pub. Cont. Code § 5103.

Failure to comply with the notice requirements constitutes a sufficient ground, standing alone, to deny relief to a bidder.[77]

Although a bidder may not change its bid, if the agency refuses to allow the contractor to withdraw its bid, the contractor must perform the work at the price bid or forfeit its bid security. If the contractor accepts award of the contract in spite of a mistake in its bid, it may not thereafter seek to rescind or reform the contract.[78]

The contractor may file suit against the public agency to recover the amount of any forfeited security, without interest or costs, within 90 days after the opening of the bid. Actual forfeiture is not a prerequisite to filing suit.[79] If the bidder fails to recover judgment, it must pay all costs incurred by the public agency in the suit, including attorneys' fees.[80]

[F] Evaluation of Bids

In order to provide all qualified bidders with a fair opportunity to enter the bidding process, public agencies utilize standard forms, receive sealed bids within a published deadline, and refrain from *ex parte* or private communications with bidders. To maintain fairness during the evaluation process, bids must be rejected if they are not responsive to the invitation and specifications or if a bidder is not responsible to perform the work. An awarding agency has a high degree of discretion in determining the responsiveness of a bid or the responsibility of a bidder. Moreover, the awarding agency always has the discretion to reject all bids and readvertise or to abandon the project.[81] Practical considerations, such as delay to the project and the cost of rebidding, may be among the many factors considered in determining whether to reject all bids and rebid the project.

[1] Bid Responsiveness

To be responsive, a bid must conform to the material terms of the bid package.[82] Usually, whether a bid is responsive can be determined from the face of the bid without outside investigation or information.[83] Properly worded performance specifications, however, permit the agency to exercise discretion in determining whether a bid meets the specifications. An agency may require the bidder to demonstrate proposed systems or submit additional information in order to determine whether the bid meets the specifications.[84] A bid that varies materially

---

[77] *See* A&A Elec., Inc. v. City of King, 54 Cal. App. 3d 457 (1976).

[78] *See* Lemorge Elec. v. County of San Mateo, 46 Cal. 2d 659 (1956).

[79] Ballret Bros. Constr. v. Regents of the Univ. of Cal., 80 Cal. App. 3d 321 (1978).

[80] Cal. Pub. Cont. Code §§ 5101, 5102.

[81] Charles L. Harney, Inc. v. Durkee, 107 Cal. App. 2d 803 (1962).

[82] Menefee v. County of Fresno, 163 Cal. App. 3d 1175 (1985); Southern Check Exch. v. San Diego, 5 Cal. App. 3d 81 (1970).

[83] MCM Constr., Inc. v. City & County of S.F., 66 Cal. App. 4th 359 (1998), *req. for depublication denied.*

[84] *See, e.g.,* National Identification Sys., Inc. v. State Bd. of Control, 11 Cal. App. 4th 1446 (1992).

from the specifications must be rejected.[85] Responsiveness is evaluated on a case-by-case basis. Typically, the material terms of a bid are those that affect price, quantity, quality, or delivery and those terms that the bid package clearly identifies as mandatory.[86] Moreover, an agency cannot reject a bid on the basis of arbitrary or unwritten policies.[87] A bid fails to comply materially with the bid package if it gives the bidder a substantial economic advantage or benefit not enjoyed by other bidders.[88] A substantially conforming bid, although not strictly responsive, may be accepted if the variants cannot have affected the amount of the bid or given a bidder an advantage or benefit not allowed other bidders.[89] However, a local agency is not required to waive insubstantial deviations from bid requirements because the power to waive immaterial bid deviations is discretionary, not mandatory.[90] A bid may be rejected as nonresponsive without a hearing.[91]

The case of *Menefee v. County of Fresno*[92] illustrates the application of the rules concerning responsiveness and waiver of immaterial deviations. In *Menefee*, the low bid was challenged because the bidder failed to sign the appropriate line on the proposal sheet of its bid form, although the form was signed in other places and was accompanied by a signed bid bond. The court upheld the award of the contract to the low bidder. It did not matter where the bid was signed, as long as it was signed by the appropriate parties. The court concluded that because the bidder gained no advantage, the public agency could waive the immaterial defect. Under the facts presented in *Menefee*, the court held that the low bidder could not refuse to enter a contract based upon its failure to sign the bid form, and thus affirmed the trial court's ruling validating the county's decision not to consider the unsigned bid as nonresponsive.

The court in *Valley Crest Landscape, Inc. v. City Council of Davis*[93] applied the analysis set forth in *Menefee* to overturn award of a contract to a bidder that was permitted to recalculate the amount of work to be performed by the contractor's own forces. The court concluded that the low bidder in the *Valley Crest* case had an unfair advantage because it could have withdrawn its bid. Misstating the correct percentage of work to be done by a subcontractor is in the nature of a typographical or arithmetic error. It makes the bid materially different and is a mistake in filling out the bid; therefore, under California Public Contract

---

[85] Stimson v. Hanley, 151 Cal. 379 (1907).
[86] Pozar v. Department of Transp., 145 Cal. App. 3d 269 (1983).
[87] *See, e.g.,* Monterey Mechanical Co. v. Sacramento Reg'l County Sanitation Dist., 44 Cal. App. 4th 1391 (1996).
[88] Menefee v. County of Fresno, 163 Cal. App. 3d 1175 (1985); 471 Op. Cal. Att'y. Gen. 54 (1949).
[89] *See* Universal By-Products v. Modesto, 43 Cal. App. 3d 145 (1974).
[90] MCM Constr., Inc. v. City & County of S.F., 66 Cal. App. 4th 359 (1998), *req. for depublication denied.*
[91] *See, e.g.,* Educational & Recreational Servs., Inc. v. Pasadena Unified Sch. Dist., 65 Cal. App. 3d 775 (1977).
[92] 163 Cal. App. 3d 1175 (1985).
[93] 41 Cal. App. 4th 1432 (1996).

Code § 5103, the low bidder could have sought relief by giving the city notice of the mistake within five days of the bid opening. The fact that the low bidder did not seek such relief was irrelevant. The key point was that such relief was available and the low bidder thus had a benefit not available to the other bidders since it could have backed out of its bid. Therefore, its mistake could not be waived as an immaterial irregularity. *Valley Crest* is also noteworthy because, by the time the case was decided, the contractor who received the award had substantially performed its contract. The court of appeal declined to consider whether the contractor was entitled to the reasonable value of the work performed notwithstanding the fact that the court had declared the contract void.[94]

[2] Bidder Responsibility

In addition to submitting a responsive bid, the contractor must be "responsible." The term "responsible bidder" has been held to refer not only to trustworthiness but also to the quality, fitness, and capacity of the bidder to perform the particular requirements of the proposed work satisfactorily.[95] A bidder is responsible if it can perform the contract as promised. A determination that a bidder is responsible is a complex matter depending, often, on information received outside the bidding process.[96] The factors that determine responsibility depend upon the particular project; however, in general, they are those factors used as prequalification criteria discussed in § 2.02[C][2], i.e., financial resources, surety and insurance capacity, construction experience, personnel, and many others. For example, a state agency may reject a bidder who has been disqualified from bidding on other projects because of a violation of law or safety ordinance.[97]

Although public agencies have discretion to determine which bidders are responsible, they may not, if they determine that more than one bidder is responsible, make the award on the basis of relative superiority.[98] To permit the awarding authority to reject the lowest qualified bid and award the contract to a more qualified bidder would frustrate the purpose of competitive bidding by promoting favoritism.[99] Nevertheless, an awarding authority is given a great deal of discretion in determining whether a bidder is even minimally responsible.[100]

---

[94] A case decided shortly after *Valley Crest* involved strikingly similar circumstances but reached a different result. *See* Ghilotti Constr. Co. v. City of Richmond, 45 Cal. App. 4th 897 (1996), *review denied*.

[95] *See* West v. City of Oakland, 30 Cal. App. 566 (1916); Boydston v. Napa Sanitation Dist., 222 Cal. App. 3d 1362 (1990).

[96] Taylor Bus Serv., Inc. v. San Diego Bd. of Educ., 195 Cal. App. 3d 1331 (1987).

[97] Cal. Pub. Cont. Code § 10162.

[98] City of Inglewood-Los Angeles County Civic Ctr. Auth. v. Superior Court, 7 Cal. 3d 861 (1972); Boydston v. Napa Sanitation Dist., 222 Cal. App. 3d 1362 (1990).

[99] City of Inglewood-Los Angeles County Civic Ctr. Auth. v. Superior Court, 7 Cal. 3d 861 (1972).

[100] *See* Raymond v. Fresno City Unified Sch. Dist., 123 Cal. App. 2d 626 (1954); R. & A. Vending Serv., Inc. v. City of L.A., 172 Cal. App. 3d 1188 (1985).

Before rejecting a low bidder on grounds of non-responsibility, the agency must notify the bidder of the evidence supporting that finding and afford the bidder an opportunity to demonstrate that it is qualified to perform the contract.[101]

[3] Alternate Bids

A common practice of state and local agencies is to require bidders to submit a base bid as well as a number of alternate items of work that may be added or deleted from the base bid to determine the low bidder. California Public Contract Code § 10126 expressly authorizes the use of alternate bids on state projects, provided (1) the alternates are based on estimates that do not exceed 10 percent of the cost of the project, (2) available funds are sufficient to cover the base project, (3) the low bid is determined through adding or deducting alternates in listed numerical sequence, and (4) the contract is awarded to the lowest bidder providing the lowest total costs for the base bid and the alternates accepted. For local projects, there are no statewide statutes addressing the use of alternates, and in light of home rule principles, it is unlikely that a statewide statute could impose uniform rules about alternate bids on charter cities and counties.[102]

The use of alternate bids is primarily budget driven. Alternate bids are used as a means to ensure that a project can be awarded for available funds, through the use of deductive alternates, without the need for revising the plans and specifications and readvertising the project. Typically, the local agency reserves the right to select any combination of alternates with the base bid, to arrive at the lowest cost and best value. Depending upon the number of alternates and the nature of the work, the combination of alternates selected may often result in an award to other than the low bidder on the base contract work. Frustrated bidders have challenged the use of alternate bids as violating competitive bidding rules, alleging that allowing the agency to pick and choose alternate bids after the identity of bidders is known gives the agency the opportunity to unfairly manipulate the bidding process to either ensure that a favored contractor is determined to be the low bidder or to prevent award to a disfavored contractor.

While the argument that the alternate bid process gives the public agency the opportunity to improperly exercise favoritism is a compelling one, it must be balanced against the standard of review discussed in § 2.02[G][3]. A public agency's determination of the lowest responsible bidder will only be set aside upon proof of fraud or a clear and convincing showing of abuse of discretion.[103] Thus, the mere opportunity to exercise favoritism would not seem to be sufficient to set aside an award based on selection of alternates. Certain appellate courts have stated that because of the potential for abuse in letting public contracts, contracts awarded without strict compliance with bidding requirements

---

[101] City of Inglewood-Los Angeles County Civic Ctr. Auth. v. Superior Court, 7 Cal. 3d 861 (1972).

[102] *See, e.g.,* Piledrivers Local Union No. 2375 v. City of Santa Monica, 151 Cal. App. 3d 509 (1984).

[103] West v. City of Oakland, 30 Cal. App. 566 (1916); City Council of Beverly Hills v. Superior Court, 272 Cal. App. 2d 876 (1969).

will be set aside even where it is certain there was in fact no corruption;[104] however no published opinion in California addresses the use of alternate bids in the context of competitive bidding requirements.

In order to avoid protest when alternates are used, some local agencies have implemented procedures aimed at minimizing the opportunity to exercise favoritism, such as awarding on the basis of low bid, regardless of the pricing of the alternates, or implementing "anonymous" procedures so that the identity of the bidders is not known until after the alternates are selected. Even these procedures are problematic given the requirement that bids be opened and publicly announced and that agencies must know the identity of bidders in order to make a determination concerning responsibility. As such, the use of alternates and the means of determining the successful bidder must be carefully considered on a project-by-project basis.

[4] Preferences

[a] Local Firms and Labor

Under section 6107 of the Public Contract Code, California state agencies are required to give California companies a bid preference over out-of-state bidders equal to any preference that the non-California contractor would have been afforded in its home state. The statute defines *California company* to be any contractor (1) whose principal place of business is in California, or (2) whose principal place of business is in another state with no such bid-preference laws, or (3) that has paid at least $5,000 in California sales or use taxes on construction activities in each of the preceding five years. The provisions of the equal protection and privilege and immunity clauses of the U.S. Constitution are generally held to prohibit local preferences when federal funds are utilized; however, states and municipalities may enact such preferences to ameliorate disadvantages suffered by local businesses.[105]

Public agencies and private owners are both using project labor agreements to prevent labor disputes and resulting project delays. Typically, the agency or owner enters into a project labor agreement with local unions, and the selected contractor is required to comply with the project labor agreement. Compliance ordinarily requires the contractor to pay union scale wages and benefits in exchange for the union's agreement not to strike on the project. Nonunion contractors have challenged the validity of such agreements as violative of competitive bidding statutes. The U.S. Supreme Court held that project labor agreements were not invalid under federal law in the *Boston Harbor* case,[106] but left open

---

[104] Konica Bus. Machs. U.S.A., Inc. v. Regents of Univ. of Cal., 206 Cal. App. 3d 449 (1988).

[105] *See, e.g.,* Associated Gen'l Contractors, Inc. v. City & County of S.F., 813 F.2d 922 (9th Cir. 1987).

[106] Building & Constr. Trades Council of the Metro. Dist. v. Associated Builders & Contractors of Mass./R.I., Inc., 507 U.S. 218 (1993).

the question of validity under state law. As of the date of this edition, the California Supreme Court has held that one project labor agreement is constitutional and it has granted review of another case.[107]

[b] Small and Minority-Owned Businesses

Although the genesis of affirmative action programs can be traced to formation of the Small Business Administration in the 1960s, the major legislation that affects California public works was enacted in the 1980s. In 1983, federal legislation required California public works receiving funds from the United States Department of Transportation to adopt programs in compliance with federal regulations. In 1986, California Public Contract Code §§ 2000 *et seq.* were enacted, setting forth the components of a discretionary program for local agencies who choose to implement affirmative action programs. In 1988, Public Contract Code §§ 10115 *et seq.* were enacted, mandating a program for state contracts.

Whether federal, state, or local, all programs set goals for participation by minority-owned business enterprises (MBEs), women-owned business enterprises (WBEs), disabled veteran–owned business enterprises (DVBEs), or small disadvantaged business enterprises (DBEs) in public contracts and require bidders to either meet the goals or demonstrate that good-faith efforts were made to meet the goals. In the decade that followed implementation of affirmative action programs, bidders protested the use of such programs as violating city charters, competitive bidding rules and constitutional rights.

In the landmark U.S. Supreme Court decision *Richmond v. J.A. Croson Co.*,[108] the court found that a municipality's mandatory set-aside for minority-owned subcontractors violated the Equal Protection clause of the United States Constitution because there was no direct evidence of past discrimination. In so holding, the court applied a "strict scrutiny" standard to state and local affirmative action programs.

In *Domar Electric, Inc. v. City of Los Angeles*,[109] the California Supreme Court ultimately held that a city's outreach program did not violate its charter and that the outreach program's objectives were consistent with the goals of competitive bidding where outreach programs were required of all bidders, including MBE and WBE bidders, and directed to all subcontractors, including MBE and WBE subcontractors. On remand, the court of appeal held that the city was bound by the provisions of Public Contract Code § 2000, which sets forth participation goals for MBEs and WBEs and good-faith outreach requirements.[110] The court specifically found that the city's program for good-faith outreach paralleled the ten criteria listed in section 2000. The court in *Domar* further held

---

[107] Associated Builders and Contractors, Inc. v. San Francisco, 21 Cal. 4th 352 (1999); Associated Builders and Contractors, Inc. v. Metropolitan Water Dist. of S. Cal., 59 Cal. App. 4th 1503 (1997), *review granted, depublished,* 72 Cal. Rptr. 2d 215 (1998).

[108] 488 U.S. 469 (1989).

[109] 9 Cal. 4th 161 (1994).

[110] Domar Elec., Inc. v. City of L.A., 41 Cal. App. 4th 810 (1995).

that the legislative intent behind section 2000 was to provide two types of programs for affirmative action: (1) one whose goals and requirements could be met either by achieving the MBE/WBE participation goal or by demonstrating a good-faith, but unsuccessful, effort to encourage such participation; and (2) one whose goals and requirements involved only the making of a good-faith outreach effort, irrespective of the attainment of the stated goal. In order to comply with the *Croson* decision, requiring proof of discrimination, the *Domar* court opined that a public entity that does not have a proper disparity study is limited to the second option.

In *Adarand Constructors, Inc. v. Pena*,[111] the U.S. Supreme Court held that equal protection under the Fifth Amendment, like the Fourteenth Amendment, requires that affirmative action programs conducted pursuant to federal regulations meet the constitutional test of "strict scrutiny." Because race was a factor in awarding the subcontract in *Adarand*, the Court required the federal government to demonstrate both a compelling interest and a "narrowly tailored" legislative remedy.

At issue in *Monterey Mechanical Co. v. Sacramento Regional County Sanitation District*[112] was whether Public Contract Code § 2000 contained the exclusive criteria for assessing good-faith efforts to satisfy affirmative action goals. Section 2000(b) lists 10 criteria for assessing good faith in the contractor's attempt to comply with the local agency's affirmative action goals and requirements. Compliance with those 10 criteria creates a rebuttable presumption affecting the burden of producing evidence that a bidder has made a good-faith effort to comply with the goals and requirements relating to participation by MBEs and WBEs. Monterey contended, on appeal, that Public Contract Code § 2000(b) contains the *exclusive* criteria for assessing good-faith efforts to satisfy affirmative action goals. The appellate court agreed. The court stated that, because the district in this particular case had imposed conditions in addition to those set forth in the 10 criteria listed in section 2000, the local agency had exceeded its authority. The court held that section 2000 permits local governmental agencies to impose affirmative action obligations on prime bidders for public works projects. Where such a provision is invoked, only bidders who either meet the affirmative action goals of a local agency or make a good-faith effort to do so will be eligible for award of the contract. The good-faith effort is to be determined strictly in accordance with the criteria enumerated in Public Contract Code § 2000(b). Although section 2000(b) allows local agencies some discretion in imposing specific obligations, it does not allow for the creation of wholly discrete criteria.

On November 5, 1996, the voters of the state of California adopted the California Civil Rights Initiative, adding section 31 to Article I of the California Constitution. The initiative provides, in relevant part, that

---

[111] 515 U.S. 200 (1995).
[112] 44 Cal. App. 4th 1391 (1996).

[t]he state shall not discriminate against, or grant preferential treatment to, any individual or group on the basis of race, sex, color, ethnicity, or national origin in the operation of public employment, public education, or public contracting.

The Ninth Circuit Court of Appeals held in *Coalition for Economic Equity v. Wilson*[113] that Proposition 209 does not violate the United States Constitution.

The case of *Monterey Mechanical Co. v. Wilson*,[114] although not involving Proposition 209, held that the minority and women business enterprise goals and outreach requirements of Public Contract Code §§ 10115 *et seq.* were unconstitutional because the statute treats contractors differently, according to their ethnicity and sex, with respect to the good-faith requirement because "only those firms not minority- or women-owned must advertise to those respective groups and only minority- and women-owned firms are entitled to receive the bid solicitation." The court further held that the statute could not survive strict scrutiny because there was no evidence that the statute was narrowly tailored to remedy past discrimination by the governmental agency. The court did indicate in dicta that "there might be a non-discriminatory outreach program which did not subject anyone to unequal treatment. But this statute is not of that type." The decision did not address, and left intact, the disabled veteran enterprise component of the statute.

In 1999, Congress reauthorized legislation authorizing the DBE program for the U.S. Department of Transportation. On January 29, 1999, the DOT issued a new final regulation[115] to guide the administration of the DBE program, stating three major goals:

1. to create a level playing field on which DBEs can compete fairly,
2. to mend but not end the DBE program, and
3. to make the DBE program more effective and efficient for all participants.

The new regulations were adopted to conform to the narrow tailoring requirements of the *Adarand* decision. The program will continue until the year 2004, at which time it must be reauthorized by Congress.

At present, to the extent a local public entity receives specified federal funding, current regulations mandate that recipients implement disadvantaged business enterprise programs, establish goals for DBE participation, and require bidders to meet the goals or provide evidence that they made good-faith efforts to meet the goals. Proposition 209, by its terms, does not apply to any action that must be taken to establish or maintain eligibility under any federal program where ineligibility would result in a loss of federal funds to the state.

---

[113] 110 F.3d 1431 (9th Cir.), *cert. denied*, 118 S. Ct. 397.
[114] 125 F.3d 702 (9th Cir. 1997), *reh'g en banc denied*, 138 F.3d 1270 (9th Cir. 1998).
[115] DOT 16-99, 1999 WL 38999 (D.O.T.).

For non–federally funded projects, local agencies may either implement a program similar to that reviewed by the *Domar* courts or seek to establish the documentary evidence necessary to withstand strict judicial scrutiny by performing disparity studies to support the use of race- and gender-based classifications in award of contracts. There is, however, no guarantee that disparity studies will produce statistical data evidencing discrimination or that a court would find the study legally sufficient to support the use of such programs. Even if a disparity study does evidence discrimination, a party challenging the constitutionality of an affirmative action program may present evidence to refute a conclusion that the study establishes discrimination, and the trier of fact may decide that there is not sufficient evidence to establish actual discrimination.[116]

The current trend in affirmative action programs is to place less emphasis on ethnicity and gender and instead utilize economic factors in order to increase the participation of small and disadvantaged businesses on public contracts.[117]

[G] Bid Protests

[1] Procedures

Unlike federal government procurements, which are governed by Federal Acquisition Regulations or General Accounting Office procedures, California has no comprehensive bid protest statute, and the procedures adopted by public agencies vary widely throughout the state. Many local agencies and charter cities use informal procedures that are not reduced to writing. The state, on the other hand, has adopted detailed regulations and statutes requiring multiple notices, written submittals, and protest bonds and imposing severe penalties for filing frivolous protests.[118]

Case law provides that "prior to awarding a public works contract to other than the lowest bidder, a public body must notify the low monetary bidder of any evidence reflecting upon his responsibility received from others or adduced as a result of independent investigation and afford him an opportunity to present evidence that he is qualified to perform the contract, however, due process does not compel a quasi-judicial proceeding prior to rejection of the low monetary bidder as a non-responsible bidder."[119] In other words, pleadings, cross-examination of witnesses and formal findings are not required. Where a bidder is rejected on grounds other than non-responsibility, due process does not require a hearing before a contract is awarded to another bidder;[120] however, affirmative action program requirements may specify other procedures.

---

[116] Engineering Contractors Ass'n v. Metropolitan Dade County, 122 F.3d 895 (11th Cir. 1997).

[117] Cal. Gov't Code §§ 4530 *et seq.*, (Target Area Contract Preference Act); *id.* 7070 (Enterprise Zone Act).

[118] *See, e.g.,* Cal. Pub. Cont. Code §§ 10343, 12125 *et seq.*

[119] City of Inglewood-Los Angeles County Civic Ctr. Auth. v. Superior Court, 7 Cal. 3d 861 (1972).

[120] Educational & Recreational Servs., Inc. v. Pasadena Unified Sch. Dist., 65 Cal. App. 3d 775 (1977).

Many written bid protest procedures require protests regarding defects or improprieties in the bidding documents to be submitted as soon as the basis is known but before the deadline for submittal of bids so that agencies can modify the documents and, if necessary, extend the bidding period, in order to avoid the delay and expense of rejecting all bids and readvertising the work. Protests concerning responsiveness of bids or award of contract are usually subject to strict time limitations to prevent delay in award of the contract.

Typically, standing to protest bidding procedures is limited to participating bidders or as otherwise required by law.[121] In some cases, however, not only the disappointed bidder but also any taxpayer or others designated by statute may attempt to prevent a public entity from awarding a contract to anyone other than the lowest responsible bidder.[122] However, standing to void a contract may be limited. The proper procedure is to seek a writ of mandate based upon an abuse of the public entity's discretion.

Often, a contractor who is not the low bidder on a public works project will want to examine the low bid to see whether it was responsive to the bid requirements. A question arises as to whether the public body is required to disclose to unsuccessful bidders the bid documents submitted by the successful bidder in response to a public works invitation for bids. The California Records Act[123] is based upon a legislative finding that access to information concerning the conduct of the public's business is a fundamental and necessary right of every person in the state of California.[124] California Government Code § 6253(a) provides that public records are open to inspection at all times during office hours of the state or local agency and that every person has the right to inspect any public record, except as stated in the statute. Public bodies are authorized to make reasonable regulations setting forth the procedures to be followed for access to public records. Thus, the unsuccessful bidder should have the right to examine the bidding documents of the successful bidder, except where confidential information is identified. However, the courts have recognized that "[i]t certainly would amount to a disservice to the public if a losing bidder were permitted to comb through the bid proposal or license application of the low bidder after the fact and cancel the low bid on minor technicalities, with the hope of securing acceptance of his, a higher bid. Such construction would be adverse to the best interest of the public and contrary to public policy.[125]

[2] Remedies

After an unsuccessful bidder has exhausted its administrative remedies, and is not satisfied with the agency's decision, it may seek a writ of mandate from

---

[121] *See, e.g.,* Cal. Pub. Cont. Code § 20104.70.
[122] *Id.*
[123] Cal. Gov't Code §§ 6250–6268.
[124] *See id.* § 6250.
[125] *See* Judson Pacific-Murphy v. Durkee, 144 Cal. App. 2d 377 (1956).

the court and may seek injunctive relief.[126] A disappointed bidder has no right to compel the acceptance of its bid or to prevent rejection of all bids. Even if a writ issues, the project may be rebid, and the protester may lose the rebid.

While some courts have refused to hold a public entity liable in a tort cause of action for monetary damages for the mis-award of a public contract to other than the lowest responsible bidder on the grounds of discretionary tort immunity established by the California Government Code,[127] other courts have suggested a low bidder may recover damages based upon a theory of promissory estoppel for its implied promise to award a contract to the lowest responsible bidder.[128]

[3] Standard of Review

The precise standard of review in evaluating the award of a contract or rejection of a bidder by a public agency on a writ of mandate is not entirely clear. Some California authority indicates that a bidder must show fraud or collusion by the public agency.[129] However, the majority of the California bid protest cases have expressed the view that a writ of mandate will be granted if the contract award constitutes an abuse of a public agency's discretion.[130]

An appellate court's review of the award of a public contract is to determine whether its findings and actions are supported by substantial evidence. The court's review is limited to an examination of the proceedings to determine whether the agency's actions were arbitrary, capricious, entirely lacking in evidentiary support, or inconsistent with proper procedure. There is a presumption that the agency's actions were supported by substantial evidence, and the petitioner has the burden of proving otherwise. The court does not reweigh the evidence and must view it in the light most favorable to the agency's actions, indulging all reasonable inferences in support of those actions.[131]

One court has even held that a charter city's award of a contract and its decision to reject protests were legislative actions requiring an exercise of discretion guided by consideration of the public welfare.[132] The court identified a judicial reluctance to inquire into legislative thought processes and concluded that once exercised, legislative discretion is, absent special circumstances, not subject to

---

[126] *See, e.g.,* Rubino v. Lolli, 10 Cal. App. 3d 1059 (1970); Baldwin-Lima-Hamilton Corp. v. Superior Court, 208 Cal. App. 2d 803 (1962); Miller v. McKinnon, 20 Cal. 2d 83 (1942); Cal. Civ. Proc. § 1085; Mike Moore's 24-Hour Towing v. City of San Diego, 145 Cal. App. 4th 1294 (1996); Old Town Dev. Corp. v. Urban Renewal Agency, 249 Cal. App. 2d 313 (1967); Baldwin-Lima-Hamilton Corp. v. Superior Court, 208 Cal. App. 2d 803 (1962).

[127] Rubino v. Lolli, 10 Cal. App. 3d 1059 (1970).

[128] Kajima/Ray Wilson v. Los Angeles County Metro. Transp. Auth., 69 Cal. App. 4th 1458 (1999), *review granted, depublished.*

[129] West v. City of Oakland, 30 Cal. App. 566 (1916).

[130] *See, e.g.,* Old Town Dev. Corp. v. Urban Renewal Agency, 249 Cal. App. 2d 313 (1967); Baldwin-Lima-Hamilton Corp. v. Superior Court, 208 Cal. App. 2d 803 (1962).

[131] *See, e.g.,* Ghilotti Constr. Co. v. City of Richmond, 45 Cal. App. 4th 897 (1996); MCM Constr., Inc. v. City & County of S.F., 66 Cal. App. 4th 359 (1998).

[132] Mike Moore's 24-Hour Towing v. City of San Diego, 45 Cal. App. 4th 1294 (1996).

judicial control and supervision. As a consequence, the steps to be undertaken, the method selected, and the decision reached in the course thereof by the city, in the absence of fraudulent or arbitrary action, would not be interfered with by the courts. The decision of the legislative body is to be evaluated on its face, without an inquiry into the underlying motive or thought processes of the legislature.

Chapter 3
CONSTRUCTION CONTRACTS

§ 3.01   Basic Considerations
§ 3.02   Analysis of Owner-Contractor Relationship
    [A]   Basic Owner-Contractor Agreement
        [1]   Identity of the Parties
        [2]   Description of the Work
        [3]   Time for Performance
        [4]   Contract Price
        [5]   Payment Terms
    [B]   General Conditions
        [1]   Integration
        [2]   Interpretation of Contract Documents
        [3]   Contractor's Review of Documents and Field Conditions
        [4]   Warranties, Guarantees, and Correction of Work
        [5]   Subcontractor's Adoption of Contract Documents
        [6]   Design Delegation
        [7]   Indemnity
        [8]   Claims, Damages, and Dispute Resolution
        [9]   Mediation and Arbitration
        [10]   Delays and Time Extensions
        [11]   No Damages for Delay
        [12]   Concealed or Unknown Conditions
        [13]   Insurance
        [14]   Notice Requirements
        [15]   Termination or Suspension of the Contract
§ 3.03   Analysis of Contractor-Subcontractor Relationship
    [A]   Adoption of Contract Documents and Flow-Down Provisions
    [B]   Payment Terms and Pay-When-Paid Clause
    [C]   Indemnity Obligations
    [D]   Bonding Requirements
    [E]   Obligation to Continue Work
    [F]   Contractor's Right to Terminate Subcontractor
§ 3.04   Material Supplier Contracts

§ 3.05　**Analysis of Owner-Architect Relationship**
　　　　[A]　**Limitation of Liability**
§ 3.06　**Analysis of Design-Build Relationship**

§ 3.01 BASIC CONSIDERATIONS

The written contract forms the basis of the various relationships on a construction project. All parties to the construction process—owners, architects, general contractors, subcontractors, laborers, and material suppliers—have much at stake in the process of writing and executing their respective contracts.

It is standard practice in the construction industry for the "superior" party in a given relationship (the owner vis-à-vis the general contractor or architect, the general contractor vis-à-vis the subcontractor, and so on) to present the cosignatory with a form of agreement for the latter's signature. Perhaps the most widely used standard forms are those published by the American Institute of Architects (AIA) and the Associated General Contractors of America (AGC). Whether the parties use a standard form agreement or one especially drafted for a particular project, a careful review of contract provisions is essential for all parties. No form, however well crafted, may be applied to all situations. Standard forms typically require modification to identify or acknowledge certain facts (e.g., time constraints), to address specific circumstances (e.g., a developer who commences construction but wants the right to assign to a new owner), and to address legal issues that are of particular concern to a party (e.g., limitation or waiver of damages). Reviewing the contract's terms to ensure they clearly define the parties' risks, rights, and duties need not take a great deal of time and can save considerable trouble and expense down the road.

§ 3.02 ANALYSIS OF OWNER-CONTRACTOR RELATIONSHIP

On most projects, and particularly with respect to large and/or complex projects, the relationship, rights, duties, obligations, and allocation of risks between the owner and the general contractor (and sometimes the architect and/or construction manager) are defined by a number of documents and drawings collectively referred to as the contract documents. The contract documents may include any or all of the following:

1. Bidding documents (invitation and instructions)

2. Owner-contractor agreement (identifies parties, work, time, price and payment terms)

3. General conditions (provides non-project-specific items in detail)

4. Supplementary and/or special conditions (provide project-specific information)

5. Drawings (graphic and pictorial documents showing the design, location, and dimensions of the work)

6. Specifications (written requirements for materials, equipment, systems, standards, and workmanship for the work)

7. Addenda (reflect changes prior to execution of the agreement)

8. Modifications (reflect substantial changes to the agreement terms or conditions).

Although each is usually physically separate, the contract documents are connected and cross-referenced to one another so that they collectively form a single contract between the parties.

[A] Basic Owner-Contractor Agreement

The basic owner-contractor agreement is usually separate from the other contract documents and is often the only contract document actually signed by the parties. The agreement, which typically cross-references the other contract documents, may be quite brief and usually covers the five most basic elements of the contract for construction: (1) identity of the parties; (2) description of the work; (3) time for performance; (4) contract price; and (5) payment terms. The following is a brief description of the major elements of the basic owner-contractor agreement:

[1] Identity of the Parties

The basic agreement should use the full legal and/or corporate names of both the owner and the contractor. Although they are not parties to this agreement, the names of the architect/engineer and/or the construction manager are frequently stated. The individuals signing the contract should indicate their legal capacities as individuals, partners, or corporate officers.

[2] Description of the Work

The basic agreement should clearly state the scope of the work to be performed under the contract. This is particularly important when the contractor is to construct either one of several projects or only a portion of a particular project. In such cases, it is vital that there be a clear division between the work of the various contractors in order to avoid future jurisdictional disputes. It is important to define the scope of the work clearly so that it may be determined in the future whether change orders are within the scope of the work originally contemplated by the contract.

[3] Time for Performance

The basic agreement usually establishes a time for commencement and completion of the work. This is a particularly important matter inasmuch as delays and failure to complete the work on time are among the most common and difficult owner-contractor disputes to settle.

When the basic owner-contractor agreement is drafted, the time for performance is usually stated as the number of days from commencement in which the work must be substantially completed. This is usually stated in calendar days, in order to eliminate any disagreement over how to count holidays and weekends, or it is stated by reference to a specific date. Failure to grant normal and reasonable time extensions can lead to a claim for constructive acceleration, so both parties should be wary of agreements prohibiting extensions to the contract time.

Many construction contracts contain a provision stating that "time is of the essence." Such a provision has significant legal effect and is frequently enforced through a provision for liquidated damages. Liquidated damages are a fixed amount, set by contract, to quantify compensation due for unexcused contract delays.

[4] Contract Price

When the contract price is stated as a lump sum, the contract is referred to as a fixed-price contract. The fixed price is usually the bid price of the low bidding contractor, although the actual price is frequently adjusted through the owner's decision to incorporate additive or deductive alternates submitted with the base bid.

The basis for determining the fixed-price contract amount should be clearly set forth in the agreement, which is often done by incorporating the contractor's bid into the agreement by reference. Owners should take note of any exceptions, exclusions, or qualifications contained in a contractor's bid before including it as a binding contract document. The agreement also may contain unit prices or specific amounts for certain categories of anticipated extra work. Because the purpose of such provisions is to simplify and, if possible, avoid future disagreements, the agreement may also set fixed rates for overhead and profit on change orders.

Some agreements call for the owner to pay the actual cost of the work plus a fee to the contractor, which fee may be either a set amount or some percentage of the contractor's cost. Such agreements are known as cost-plus contracts and may include a "not to exceed" provision setting a guaranteed maximum price for the work. Cost-plus agreements should clearly spell out how the contractor will be compensated for change order work. It is most important that reimbursable and nonreimbursable costs be specifically set forth and clearly defined.

[5] Payment Terms

The basic owner-contractor agreement will probably refer to some provision in the general or supplementary conditions containing more detail regarding payment, but it is also common practice for the basic agreement to include a statement of the requirements the contractor must meet in applying for progress payments. A reasonable time should be allowed for the owner to review the contractor's progress payment requests and issue checks.

Most construction agreements contain a provision for retention, which is a percentage (usually 10 percent) of the amounts due the contractor but held by the owner as a type of security for the contractor's full and complete performance and to satisfy potential lien claims at the end of the project.

Final payment is made at the end of the project when the work has been fully performed and consists of the entire unpaid balance of the contract amount, including the retention. In fact, the release of the retention usually constitutes most or all of the final payment.

[B] General Conditions

The general conditions spell out, very specifically, the rights, duties, and responsibilities of the owner and the contractor. In addition, when applicable, they advise the contractor as to the duties and responsibilities of the architect and/or the construction manager. The general conditions are usually standard forms that would apply to most projects of the type being constructed. Project-specific information is usually identified in the special or supplementary conditions.

The AIA's General Conditions of the Contract for Construction (Document A201) is one of the oldest and most commonly used forms. The AIA forms have been used to define construction terms and set legal precedent, and are generally accepted as establishing an industry-wide norm for contracting. The AIA periodically updates and revises A201. This section addresses many of the key provisions typically included in the General Conditions and identifies unique and, in some cases, controversial aspects of the 1997 edition of AIA Document A201.

[1] Integration

An integration clause provides that the contract documents constitute the entire agreement between the owner and the contractor. The most important aspect of an integration clause is that it excludes all written or oral negotiations and representations between the parties if they are not specifically made a part of the agreement. A201 ¶ 1.1.2 further provides that the contract documents are not to be interpreted to create a contractual relationship between the architect and the contractor, between the owner and subcontractor or sub-subcontractor, between the owner and architect, or between any persons of entities other than the owner and contractor.

[2] Interpretation of Contract Documents

The duty and responsibility of interpreting the requirements of the contract documents is usually assigned to the owner's representative in charge of administering the project. AIA Document A201 names the architect for this role. When the architect has a vested interest in a particular interpretation—as in deciding whether a particular detail should have been included as an element of the design or left to the contractor's discretion as a construction method—there is possibly a conflict of interest in allowing the architect's interpretation to govern. Clearly, however, some party must be named as the arbiter of interpretational disputes or a project may be in danger of becoming bogged down in debate between the various interested parties.

Another important consideration is the amount of time allowed for interpreting the contract documents. A201 ¶ 4.2.11 requires the architect to render interpretations "with reasonable promptness and in accordance with any time limit agreed upon." A persistent failure to meet either a standard of reasonableness or an agreed-upon period of time can become a significant cause of disruption and delay to the contractor's progress and lead to disputes over damages for delay.

When no agreed-upon period has been set, then delay by the architect is usually not recognized until 15 days after the owner or contractor has made a written request for the architect's interpretations.

[3] Contractor's Review of Documents and Field Conditions

Most contracts require that contractors certify that they have visited the construction site and familiarized themselves with the local conditions under which the work is to be performed. A201 ¶¶ 3.2.1 through 3.2.3 address the competing goals of architects and contractors. Architects continually look for a remedy against contractors who exacerbate design problems by failing to carefully review and compare the contract documents and site conditions before starting work. Contractors seek to avoid assuming any responsibility for detecting errors or omissions in the architect's plans and specifications. While the language of ¶ 3.2.1 limits the purpose of the contractor's review obligations to exclude detection of errors and omissions or inconsistencies in the contract documents, it expands the scope of the review as well as the grounds for and nature of the penalty imposed for failing to detect errors and violation of building codes.

The contractor must compare drawings, take field measurements, and observe conditions at the site before starting each portion of the work. Thus, the contractor's obligation for review and coordination continues throughout the life of the project.

The purpose of the contractor's review is not to discover errors, omissions, or inconsistencies nor to ascertain compliance with applicable laws, statutes, ordinances, building codes, rules, and regulations. However, the contractor must promptly notify the architect of all discovered errors, omissions, or inconsistencies and variances with applicable laws and codes. The contractor, in essence, is now liable to the owner for damages resulting from such errors, inconsistencies, omissions, and variances that it knows or "should have known" exist. If the contractor fails to perform the review and comparison of documents and site conditions, the contractor must pay such costs and damages to the owner as would have been avoided if the contractor had performed such obligations. Given the expanded penalty, contractors should establish procedures, schedules, and checklists for performing their duties under this section. Contractors should document the activities they undertake and transmit the documentation to the owner and architect as evidence that these contractual duties have been performed.

Contractors should be wary of contract provisions that make them, in essence, "guarantors" of the plans. Signing a contract containing such a provision may bar the contractor from making claims for delays or added costs caused by defects in the plans or specifications. Although it is completely reasonable for the owner to expect the contractor's cooperation in detecting and correcting minor errors in the contract documents, it would be very poor policy indeed for the owner to rely on the contractor's diligence as a substitute for a thorough review of the contract documents.

[4] Warranties, Guarantees, and Correction of Work

A warranty is, in essence, a pledge that something is what it is claimed to be. A201 ¶ 3.5 provides three express warranties: (1) that materials and equipment furnished under the contract will be of good quality and new unless otherwise required or permitted by the contract documents; (2) that the work will be free from defects not inherent in the quality required or permitted; and (3) that the work will conform to the requirements of the contract documents. Unless otherwise limited by contract, the foregoing warranties are enforceable, and the contractor remains liable for any breach until expiration of the applicable statutes of limitation.[1]

In addition to the foregoing warranties, A201 ¶ 12.2, "Correction of Work," contains a guarantee, i.e., a pledge to replace work if it does not meet the specifications or is not as represented. The contractor guarantees that if, within one year of the date of substantial completion, any work is found to be not in accordance with the requirements of the contract documents, the contractor shall return to the site and correct it promptly after receipt of written notice by the owner to do so.

Subparagraph 12.2.5 of Document A201 makes clear that the one-year correction period does not establish a period of limitations with respect to the contractor's other obligations under the contract documents, but only with respect to the specific obligations of the contractor to correct the work. Beyond the specific one-year correction period of ¶ 12.2, the contractor, while not necessarily required to return with its own forces, must nonetheless assume financial responsibility for correction of work during the period set by the applicable statute of limitations.

[5] Subcontractor's Adoption of Contract Documents

For the protection of both the owner and the prime contractor, a contract provision almost always requires each of the subcontractors on a project to adopt the basic owner-contractor agreement and to be bound by the terms of the contract documents or to designated flow-down provisions.[2] As discussed in § 3.03 below, this is a critical provision to ensure the prime contractor's obligations to the owner will be performed by its subcontractors.

[6] Design Delegation

A201 ¶ 3.12.10, added in 1997, provides that delegation of design to the contractor is prohibited unless such services are specifically spelled out by the contract documents or unless the contractor needs to provide such services in order to carry out its responsibilities for construction means and methods. While

---

[1] *See, e.g.,* Cal. Civ. Proc. Code §§ 337, 337.1, 337.15, 338; A201 ¶ 13.7 (Commencement of Statutory Limitation Period).

[2] *See* A201 ¶ 5.3.1.

the first clause puts the burden on the owner and its architect to enumerate the areas of design delegation, the second clause is arguably so broad as to encompass design responsibility for the contractor's entire scope of work.

If professional design services or certifications are required by the contract documents, the owner and architect must specify the performance and design criteria that such services must satisfy. The contractor will not be held responsible for the adequacy of the performance or design criteria required by the contract documents. The architect must review, approve, or take other appropriate action on the design submittals made by the contractor. This provision raises a variety of questions. Since California law requires architects and engineers to be licensed, can an unlicensed contractor sign a contract that requires him to provide, directly or indirectly, some kind of design service? Is it good practice to have the design function diffused and fragmented? Will the contractor be able to provide adequate insurance to cover its new design risk? Is design delegation consistent with competitive bidding laws?

While a contractor may eagerly or begrudgingly accept design responsibility, it is incumbent upon the owner to ensure that there are no gaps in responsibility and to understand that the architect is being compensated for providing less than the complete design of the project. During the proposal or contract negotiation stage, the architect should be required to include a designation of all portions of the design that it intends to delegate as well as the qualifications necessary for those services to be performed. The owner can then require the contractor or its specialty subcontractors to provide evidence of appropriate insurance coverage for the design services that have been delegated.

[7] Indemnity

An *express indemnity* is a contractual agreement by which one party (the indemnitor) engages to protect another (the indemnitee) from the legal consequences of the conduct of one of the parties or of some other person.[3] Although an indemnity agreement is not the same as an insurance contract, its effect is to place on the indemnitor the burden of insuring against any loss within its scope.[4] An indemnity contract is in the nature of a promise to stand behind specifically identifiable work; insurance policies, on the other hand, cover generally defined risks in exchange for a premium having some relationship to those risks. Indemnity is usually unlimited in amount but limited to a given project.[5]

California Civil Code § 1668 makes unlawful any contract that exempts a person from responsibility for his fraud, willful injury, or violation of the law, and therefore may serve to prohibit indemnification for intentional torts. With respect to construction contracts, an indemnity agreement that purports to indemnify against liability arising from the *sole negligence or willful misconduct* of the indemnitee or his agent, or for defects in design furnished by such persons, are

---

[3] Cal. Civ. Code § 2772.
[4] *See* Rossmoor Sanitation, Inc. v. Pylon, Inc., 13 Cal. 3d 622 (1975).
[5] *See* Herrick Corp. v. Canadian Ins. Co., 29 Cal. App. 4th 753 (1994).

against public policy, void, and unenforceable.[6] Furthermore, a provision in a construction contract with a public agency, relieving it from liability for its active negligence, is void and unenforceable.[7]

"Sole negligence" means that, as between the parties to the indemnity agreement, the indemnitee was solely negligent. The negligence of the plaintiff or of other third parties is immaterial.[8] The statute does not prohibit indemnification for a loss that is not due to the indemnitee's sole negligence or sole willful misconduct, but the amount to be indemnified may not include damages paid because of willful misconduct.[9] The provisions of Civil Code § 2778 defining the scope of indemnification rights and obligations are implied into each and every indemnity agreement unless the agreement expresses a contrary intent.[10] When the contract provides for "indemnity against liability," the person indemnified is entitled to recover only upon becoming liable and having made payment.

An "indemnity against claims, demands, or liability" embraces the cost of defense incurred in good faith and in the exercise of reasonable discretion. Importantly, the person indemnifying is bound, on request of the person indemnified, to defend actions or proceedings brought against the indemnitee, and the indemnitee has the right to conduct his own defense if he chooses to do so. If the indemnitor refuses or declines to defend, a recovery against the indemnitee suffered "in good faith" is conclusive such that the indemnitor cannot relitigate the indemnitee's liability.

Special rules of construction have been adopted by the courts with regard to indemnity agreements that relieve the indemnitee of responsibility for his own negligence. If the contract contains a "general" indemnity agreement, which does not provide for indemnity against liability or loss caused by the indemnitee's "active" negligence, the indemnitor has no duty to indemnify for the independent negligence of the indemnitee. Language requiring the indemnitee to indemnify "from all claims for damages to persons" "in any suit of law" or "from any cause whatsoever" has been interpreted as a general indemnity agreement.[11] The California Supreme Court has on several occasions reaffirmed the rule that an indemnity clause that does not precisely address itself to the issue of an indemnitee's negligence is a general indemnity clause, and that a general indemnity clause may be construed to provide indemnity for *passive* negligence, but not *active* negligence of the indemnitee.[12] By comparison, contract language that provides for indemnity "except for loss caused solely by the negligence of the indemnitee" is not a general indemnity clause.[13]

---

[6] Cal. Civ. Code § 2782(a).

[7] *Id.* § 2782(b).

[8] Southern Pac. Transp. Co. v. Sandy Land Protective Ass'n, 224 Cal. App. 3d 1494 (1990).

[9] Smoketree-Lake Murray, Ltd. v. Mills Concrete Constr. Co., 234 Cal. App. 3d 1724 (1991).

[10] Gribaldo, Jacobs, Jones & Assocs. v. Agrippina Versicherunges A.G., 3 Cal. 3d 434, 442 (1970).

[11] Rossmoor Sanitation, Inc. v. Pylon, Inc., 13 Cal. 3d 622 (1975).

[12] Gonzalez v. R. J. Novick Constr. Co., 20 Cal. 3d 798 (1978); Rossmoor Sanitation, Inc. v. Pylon, 13 Cal. 3d at 628; Markley v. Beagle, 66 Cal. 2d 951 (1967).

[13] Ralph M. Parsons Co. v. Combustion Equip. Assocs., Inc., 172 Cal. App. 3d 211 (1985).

A person who participates in some manner in the conduct or omission that caused injury, beyond mere failure to perform a duty imposed by law, is actively negligent and therefore is not entitled to indemnity under a general indemnity agreement.[14] Active negligence may consist of a connection with the act or omission of another by knowledge of it or by acquiescence.[15] Passive negligence is the failure to perform a duty of care, but in which there is no participation in any affirmative act of negligence.[16] Whether negligence is active or passive must be determined by the circumstances of each case.[17]

Typically general contractors require "Type I" indemnity agreements from their subcontractors. Such agreements require the subcontractor to indemnify the general contractor, even where the general contractor may be actively, but not solely, at fault.

Although Civil Code § 2778 provides that the indemnitor also has a duty to defend, that duty does not arise until there is a finding of liability against the indemnitor. In *Regan Roofing Co. v. Superior Court*,[18] a general contractor had been sued for construction defects in its development of a condominium, and sought indemnity from multiple subcontractors pursuant to Type I indemnification provisions in the subcontracts. The trial court held that the provision would cover the general contractor's own negligence in the construction of the project, and also ruled that the subcontractor defendants had a current duty to defend the general contractor. The court of appeal reversed with respect to the finding of a "current" duty to defend, stating that such a finding was premature since "no determination has yet been made as to whether the subcontractors were negligent in the performance of their work, giving rise to a duty to indemnify and a related duty to defend."[19]

Disputes that arise regarding (1) whether the indemnity provision is a "general" versus a "Type I" specific indemnity clause, (2) as to which contractor was actively at fault, passively at fault, or solely at fault, and (3) whether the indemnitor may have no duty to defend until liability is established can create construction risk management problems. A contractor cannot assume that indemnification will occur and, under the *Regan Roofing* decision, would have to defend itself until a finding of liability against one indemnitor. For these reasons, it is important that *in addition to* the contractual indemnity provisions, the indemnitor should be required to have contractual liability coverage under a commercial general liability policy and name the indemnitee as an additional insured on the indemnitor's policy.

The contractor's duty to indemnify and hold harmless the owner under A201 ¶ 3.18.1 (1997 edition) is a general indemnity clause and is limited to the extent claims, damages, losses, or expenses are not covered by a new form of insurance

---

[14] Morgan v. Stubblefield, 6 Cal. 3d 606 (1972).
[15] Jimenez v. Pacific W. Constr. Co., 185 Cal. App. 3d 102 (1986).
[16] Rossmoor, 13 Cal. 3d at 629.
[17] *Id.*
[18] 24 Cal. App. 4th 425 (1994).
[19] 24 Cal. App. 4th at 436.

known as "Project Management Protective Liability Insurance"; however, this new insurance product may not be readily available to all contractors. Indemnity for loss of use damages has been deleted to be consistent with the waiver of consequential damages.[20] This clause provides for limited contractor indemnity obligations to the owner and may prompt owners to simply substitute an entirely new and broader indemnity provision in place of ¶ 3.18.1.

[8] Claims, Damages, and Dispute Resolution

Currently consolidated in A201 ¶¶ 4.3 through 4.6, the procedure for presentation of claims is very specific (in fact, some would argue that it is too specific). All claims arising during the construction project are governed by these paragraphs.

The term "claim" is comprehensively defined and includes all disputes between the owner and the contractor arising out of or relating to the contract. All claims must be initiated by written notice, and the burden of proof rests with the claimant.

It is imperative that the contractor be fully aware of all time limitations so as not to waive a claim inadvertently. A201 ¶ 4.3.2 sets a time limit on claims, requiring all claims to be initiated in writing within 21 days of either the event giving rise to the claim or the claimant's initial recognition of the condition giving rise to the claim, whichever is later.

In an apparent attempt to limit the expense and unpredictability of construction contract litigation, A201 ¶ 4.3.10 provides for a mutual waiver of consequential damages by the owner and contractor. In general, the owner waives damages for loss of use, and the contractor waives damages for home office overhead. The waiver is applicable to all consequential damages due to either party's termination of the contract; however, it is not intended to preclude an award of "liquidated direct damages." The effect and enforceability of this new provision is called into question by inherent ambiguities and conflicts with other AIA document provisions.

The fundamental problem with this provision is that there is no generally accepted definition of the term "consequential damages." Similarly, there is no clear definition of the term "liquidated direct damages." In order to be enforceable, liquidated damages must be the result of the parties' good-faith effort to estimate the actual damages that likely will result from breach, where the exact amount of damages is uncertain and not readily capable of calculation. Direct damages, by contrast, are those that flow naturally or ordinarily from a contract breach and may be easily quantified. Thus, the term "liquidated direct damages" seems to merge two concepts that may be mutually exclusive or at least contain contradictory elements.

Additionally, subparagraph 4.3.10 conflicts with a number of other AIA document provisions. For example, AIA Document A111 (Owner-Contractor Agreement) permits the contractor to recover for certain costs of the work,

---

[20] *See, e.g.,* Regional Steel Corp. v. Superior Court, 25 Cal. App. 4th 525 (1994).

including cost of home office personnel and data processing; however, ¶ 4.3.10 of Document A201 appears to prohibit recovery of these costs. The waiver also appears to conflict with California Public Contract Code § 7102, which voids any public contract provision intended to limit a contractor's right to recover damages for delay.

Most owners will simply strike ¶ 4.3.10. Presuming the waiver is enforceable and is not stricken, contractors and owners will undoubtedly use creative accounting techniques to recharacterize indirect costs as direct costs and thus avoid the waiver.

The architect's role in resolving claims and disputes is detailed in A201 ¶ 4.4.1, providing that all claims, including those alleging error or omission by the architect, but excluding those arising from hazardous materials, shall be referred initially to the architect, who shall within 10 days take one or more of the following actions, as stated in A201 ¶ 4.4.2: request additional data from the claimant or a response with supporting data from the other party; reject the claim totally or partially; approve the claim; suggest a compromise; or advise the parties that the architect is unable to resolve the claim or concludes that it would be inappropriate to resolve the claim. Approval or rejection of claims by the architect will be by written decision, which shall state the reasons therefor. This approval or rejection of a claim will be final and binding but subject to mediation and arbitration.[21]

Pursuant to ¶ 4.4.1, a decision by the architect is a condition precedent to mediation, arbitration, or litigation of any claim between the contractor and owner, including those alleging an error or omission by the architect but excluding those dealing with hazardous waste, if the dispute arises before final payment is due. However, an exception is made if the architect has failed to take action within 30 days after the claim was made.

[9] Mediation and Arbitration

As discussed in detail in Chapter 14, nonbinding mediation and binding arbitration are common methods utilized to resolve construction disputes. A201 ¶ 4.5 requires that any claim related to the contract, with the exception of those dealing with aesthetic effect and those waived pursuant to ¶ 4.3.10, 9.10.4, or 9.10.5, must be submitted to mediation following a written decision by the architect or 30 days following submission of the claim to the architect,[22] as a condition precedent to arbitration. This mandatory provision, however, ignores the fact that the success of a mediation is due, in large part, to the voluntary and nonbinding nature of the process, as well as timing and the prior experience of the parties. The parties should consider modifying this provision to provide more flexibility to the process and thus increase the probability of success of the mediation.

A201 ¶ 4.6 outlines the process for initiating arbitration. Subparagraph 4.6.4 specifically limits the consolidation or joinder of parties other than the

---

[21] *See* A201 ¶ 4.4.5.
[22] *See* A201 ¶ 4.5.1.

owner, contractors, or persons "substantially involved in a common question of fact or law whose presence is required if complete relief is to be accorded in arbitration."[23] For example, an architect may not be joined in any arbitration, except by the architect's written consent.

[10] Delays and Time Extensions

The failure of a contractor to complete construction within the allotted time constitutes a breach of contract entitling the owner to damages. Under normal contract law, the measure of damages would be the reasonable rental value or loss of income the owner could have otherwise collected but for the period of delay, or alternatively, legal interest on the owner's invested funds for the period of time during which no income was available from the project due to the delay.[24]

On rare occasions, a contract may state that time extensions will not be available under any circumstances. This is an onerous provision, and an owner should impose such a policy only in extreme circumstances. Most contractors are reluctant simply to add a contingency factor to their bid for fear that such a contingency will render their bid uncompetitive. For this reason, the owner who insists on such a provision is likely to receive inflated bids or no bids at all on its project.

Even on projects not allowing extensions of time, it is likely that delayed completion due to causes within the owner's control would nonetheless excuse the contractor from a breach of contract. California Civil Code § 1511 provides that failure to perform an obligation is excused when such performance "is prevented or delayed by the act of the creditor . . . even though there may have been a stipulation that this shall not be an excuse." That section does go on to state, however, that the parties may agree that the party claiming prevention or delay is required to provide notice of such a claim "within a reasonable time after the occurrence of the event excusing performance."

Most contracts provide that the contractor be given time extensions for delays beyond the contractor's control if the critical path of the work as a whole is delayed.[25]

Conflicts often occur regarding weather delays. Contractors should be aware that the typical construction contract (including the AIA form) provides that extensions of time for bad weather will be given only for "adverse weather not reasonably anticipated." Document A201 goes so far as to require contractors requesting a time extension for bad weather to document the request with data substantiating that weather conditions were abnormal for the period in question and could not have been reasonably anticipated.[26]

---

[23] See id. ¶ 4.6.4.
[24] See Bird v. American Sur. Co., 175 Cal. 625 (1917), and waiver of such damages under A201 ¶ 4.3.10.
[25] See A201 ¶ 8.3.1.
[26] See id. ¶ 4.3.7.2.

[11] No Damages for Delay

In the event the contractor is delayed in completion of the project by fault of the owner or the owner's agents, the contractor may be entitled to an extension of time but not to additional compensation. This is commonly referred to as a no damages for delay provision. Whether such a provision will be enforced by a court depends on the factual circumstances involved in the delay. These clauses have been extensively litigated throughout the country with inconsistent results.[27]

In 1984, the California legislature added California Public Contract Code § 7102. This provision (as amended in 1987) states:

> Contract provisions in construction contracts of public agencies and subcontracts thereunder which limit the contractee's liability to an extension of time for delay for which the contractee is responsible and which delay is unreasonable under the circumstances involved, and not within the contemplation of the parties, shall not be construed to preclude the recovery of damages by the contractor or subcontractor.
>
> No public agency may require the waiver, alteration, or limitation of the applicability of this section. Any such waiver, alteration, or limitation is void. This section shall not be construed to void any provision in a construction contract which requires notice of delays, provides for arbitration or other procedure for settlement, or provides for liquidated damages.

This statute significantly decreases the availability of a no damages for delay clause as a defense for public bodies. However, in the opinion of the authors, section 7102 does not prevent public entities in California from including a no damages for delay clause in their contracts; it merely prevents the enforcement of such a clause under certain circumstances.

It should be noted that Document A201 specifically provides that recovery of damages for delay by either party is not excluded by ¶ 8.3.1;[28] however, such damages would arguably be waived pursuant to ¶ 4.3.10.

[12] Concealed or Unknown Conditions

Even when the contractor has made a prior independent site inspection, the full details of the true jobsite conditions usually become known only during the course of construction. The site inspection clause bars the contractor from receiving extra compensation because of site conditions that could or should have been detected as a result of such prior inspection. It is expected that the contractor will make adequate provision in its bid for any such known conditions. Moreover, the

---

[27] *See, e.g.,* K&F Constr. v. Los Angeles City Unified Sch. Dist., 123 Cal. App. 3d 1063 (1981); Hawley v. Orange County Flood Control Dist., 211 Cal. App. 2d 708 (1963); McGuire & Hester v. City of S.F., 113 Cal. App. 2d 186 (1952); Milovich v. City of L.A., 42 Cal. App. 2d 364 (1941).

[28] *See* A201 ¶ 8.3.3.

contractor may be precluded from recovering extra compensation for site conditions revealed in soil reports or other information provided by the owner.

This still leaves the problem of site conditions that were unknown and could not have been reasonably anticipated. A contractor may be entitled to rescission under the legal doctrine of impossibility of performance if the jobsite conditions actually encountered prevent performance as required by the contract.[29] This doctrine requires an actual objective impossibility rather than a mere subjective inability of the particular contractor to perform.[30] Extreme impracticability, such as performance costing 10 or 12 times the anticipated cost, has also been held to constitute impossibility.[31] However, unforeseen difficulties of lesser magnitude, even though significant, have been held not to excuse performance.[32] A contractor may also be entitled to rescission if the unanticipated jobsite condition amounts to a mutual mistake of such magnitude that the party seeking rescission would not have entered into the contract if it had had full knowledge of the facts.[33]

Assuming that the actual site conditions are not sufficient to give the contractor a right of rescission, recovery of additional compensation will depend upon the allocation of risk set forth in the agreement with the owner. If the agreement does not assign the risk of unforeseen conditions to the contractor, compensation will be allowed for additional costs.[34]

Under A201 ¶ 4.3.4, the contractor must make a claim for concealed or unknown and unusual conditions promptly, before the conditions are disturbed, and in no event later than 21 days after first observation. The architect must promptly investigate and recommend the appropriate equitable adjustments. If the architect concludes that no adjustment is appropriate, then she must notify the owner and contractor in writing, and any claim must be made within 21 days thereafter.

If the express terms of the contract assign the risk of unknown jobsite conditions to the contractor, its claim for extra compensation may be denied.[35] The contractor will also be entitled to additional compensation when the owner or its agent has misled the contractor concerning concealed conditions.[36]

[13] Insurance

Construction contracts often contain an insurance provision requiring that either the contractor or the subcontractor (1) maintain public liability insurance/ workers' compensation insurance with specified limits, and (2) ensure that the owner

---

[29] *See* Cal. Civ. Code § 1511.
[30] *See* Hensler v. City of L.A., 124 Cal. App. 2d 71, 83 (1954).
[31] Mineral Park Land Co. v. Howard, 172 Cal. 289 (1916).
[32] Kennedy v. Reece, 225 Cal. App. 2d 717 (1964).
[33] *See* Cal. Civ. Code §§ 1577, 1689; Hannah v. Steinman, 159 Cal. 142 (1911).
[34] Warner Constr. Corp. v. City of L.A., 2 Cal. 3d 285 (1970).
[35] Wunderlich v. State, 65 Cal. 2d 777 (1967); Thomas Kelly & Sons v. City of L.A., 6 Cal. App. 2d 539 (1935).
[36] *See* E.H. Morrill Co. v. State, 65 Cal. 2d 787 (1967); Welch v. State, 139 Cal. App. 3d 546 (1983).

and general contractor are named as "additional insureds" on their insurance policies. This insurance provision is therefore used in conjunction with and for purposes of enforcing and securing the benefits of the contractual indemnity/hold harmless provisions. The provision is intended to make sure the subcontractor has sufficient liability insurance protection for itself, and through the requirement of "additional insured endorsements" affords the general contractor and owner additional protection by making them insureds to the subcontractor's policies. This is particularly important if the contractor (additional insured) has a significant "self-insured retention" or "deductible" under its own general liability policy, since the contractor can tender its defense to the insurance carrier for the subcontractor when liability arises out of the subcontractor's work. The cost of the additional insured endorsement is generally passed on to the general contractor as part of the subcontractor's bid and ultimately passed on to the owner.

A "certificate of insurance" is generally a one- to two-page document provided by the insurance broker certifying that certain coverages (general liability, workers' compensation, etc.) are in place for the named insured on the certificate. The entity identified as the "CERTIFICATE HOLDER" on the document is *not* necessarily an additional insured. Generally, the certificate must also state under the category "description of the operations/locations/vehicles/special items," or elsewhere, that the certificate holder is in fact an additional insured under the policy. To ensure compliance, a request should be made for an actual copy of the additional insured endorsement as well.

Even if an actual additional insured endorsement is never issued by the company, a certificate identifying the holder as an additional insured is the same as an endorsement and is enforceable against the insurance company.[37] If the broker issuing the certificate has an agency appointment agreement on file with the state Department of Insurance listing him as an agent of the insurance company, the representations set forth on the certificate are binding upon the insurance company.[38]

To avoid problems with the issuance of a certificate of insurance, the following measures are suggested:

1. Insist upon receiving an actual insured endorsement.
2. Obtain a copy of the insurance policy.
3. Make sure the agent issuing the certificate is licensed in California and has an agency appointment with the state's insurance department or a written agency appointment agreement.
4. To avoid cancellation without notice, have the agent providing the certificate modify language on the certificate to require that the insurer notify all insureds of cancellation.
5. Never destroy insurance certificates since they are most valuable many years after they have been issued.

---

[37] Dart Equip. Corp. v. Mack Trucks, Inc., 9 Cal. App. 3d 837 (1970).
[38] Loehr v. Great Republic Ins. Co., 226 Cal. App. 3d 727 (1990).

In the insurance context, the owner often insists that it be listed as an additional insured under the general contractor's policy. The general contractor insists that it be named as an additional insured under the CGL policy for each subcontractor. In most cases, the endorsement circumscribes in some way the scope of the additional insured's coverage, usually limiting the additional insured's status to a specific length of time, or only while the additional insured is operating in a certain capacity. Subject to whatever restrictions appear in the additional insured endorsement, an additional insured has the same coverage under the policy as do the named insureds.

The availability of defense coverage is one of the primary motives in an entity's requirement for an additional insured status. In most cases, there is an existing express indemnity agreement or provision in which the named insured promises to hold the additional insured harmless for certain losses. A standard CGL policy provides "contractual liability coverage" that will respond in many cases to pay whatever amounts the named insured becomes liable to pay under the indemnity agreement,[39] but status as an indemnitee will not guarantee a defense under the indemnitor's insurance policy. As an additional insured, however, the indemnitee can demand a defense directly under the named insured's general liability policy.

This is particularly important since as an additional insured, the indemnitee may not only seek coverage for its vicarious liability for the acts of the indemnitor (i.e., subcontractor) but also obtain coverage for the indemnitee's own independent negligence.[40] Under these decisions, as long as the indemnitee's (general contractor's) liability arises from the work being performed by the indemnitor (subcontractor), coverage under the subcontractor's CGL policy will respond to both the vicarious liability and the independent liability or fault of the indemnitee/general contractor.

The 1997 edition of AIA Document A201 adds a new ¶ 11.3, which permits the owner to require the contractor to purchase and maintain "Project Management Protective Liability Insurance." Since this new hybrid policy is untested and may not be readily available, it is not clear what claims will be covered. The owner, contractor and architect waive all rights against each other for damages covered by the policy, except such rights as they may have to the proceeds of the insurance. This paragraph also provides that the owner shall not require the contractor to include the owner, architect, or other persons as additional insureds on the contractor's liability insurance coverage under ¶ 11.1. However, if the new project management protective liability policy is *not* procured, the additional insured endorsements should be required.

---

[39] The term "Contractual Liability" can be misleading. Contractual liability coverage only applies to an indemnification/hold harmless agreement to indemnify based upon another's tort—it does not provide coverage for breach of contract.

[40] *See* Shell Oil Co. v. National Union Fire Ins. Co., 44 Cal. App. 4th 1633 (1996); Gulf Oil Corp. v. Mobil Drilling Barge, 441 F. Supp. 1 (E.D. La. 1975); Price v. Zim Israel Navigation Co., 616 F.2d 422 (9th Cir. 1980).

Paragraph 11.4, regarding property insurance, has been revised to include coverage for earthquake, flood, and windstorm. Additionally, the owner, rather than the contractor, now is responsible for costs not covered by deductibles.

[14] Notice Requirements

Proper notice is absolutely essential for the preservation of a party's rights. It is therefore extremely important that the parties to the contract be aware of any and all notice requirements in the contract documents. The notice clause itself[41] usually states nothing more than the requisite formalities involved in the giving of notice. The entire contract should be reviewed for the many specific notice requirements likely to be sprinkled throughout the General Conditions.[42]

Perhaps the most important notice requirement in Document A201 is that pertaining to the claimant's giving of notice of claims against any party.[43] Failure to give the proper notice within specified time limits may preclude the contractor from recovery upon an otherwise valid claim.

If claimants cannot adequately document a potential claim within the stated time limit, they should serve notice apprising the owner of the nature of the claim, in as much detail as possible at the time, and reserving their right to amplify and document the claim further at such time as the necessary facts become known.

[15] Termination or Suspension of the Contract

Many owners routinely added a termination for convenience provision to the 1987 edition of Document A201. Although the 1997 version now includes such a provision at ¶ 14.4, owners, as well as contractors, should modify or at least clarify its terms.

A201 ¶ 14.4 provides for termination of the contract but not performance of the contractor. Contractors should add a clause stating that upon termination, all contractual rights and remedies cease, such as warranty, correction of work, and indemnity obligations. Owners should be aware that ¶ 14.4 provides that upon termination, the contractor is entitled to receive payment for work completed, costs incurred by reason of the termination, *and reasonable overhead and profit on work not executed,* although some of these costs were arguably waived under ¶ 4.3.10. Recovery of the costs enumerated in ¶ 14.4 will put the contractor in a better position than if it had completed the work, i.e., it receives profits without having to perform the work. Owners should eliminate the clause permitting compensation for overhead and profit on work not executed or substitute a graduated termination fee.

---

[41] *See* A201 ¶ 13.3.

[42] Each of the following paragraphs in AIA Document A201 contains a notice requirement of some sort: 2.3, 2.4, 3.2.1, 3.2.2, 3.7.3, 3.7.4, 3.9, 3.12.8, 3.12.9, 3.17.1, 4.3, 4.4.4, 4.6, 5.2.1, 5.3, 5.4.1.1, 8.2.2, 9.4.1, 9.5.1, 9.6.1, 9.7, 9.10, 10.2.6, 10.3.1, 10.3.2, 11.1.3, 11.4.1.2, 12.2.2, 13.3, 13.5.1, 13.5.2, and all of Article 14.

[43] *See* A201 ¶ 4.3.1.

§ 3.03 ANALYSIS OF CONTRACTOR-SUBCONTRACTOR RELATIONSHIP

On a typical construction project, most of the work is performed not by the prime contractor but by numerous subcontractors, each of whom enters into a separate written agreement with the prime contractor. These subcontracts are of vital importance to both parties. Because the subcontractor has no direct contractual relationship with the project owner, its rights, duties, and liabilities depend entirely upon the terms and conditions of its subcontract. Because the prime contractor is dependent upon the performance of its subcontractors in order to meet its obligations to the owner, it is essential that the subcontract terms be comprehensive and consistent with the requirements of the prime contract.

As is the case with prime contracts, several subcontract forms are available. Perhaps the most widely used forms are those drafted by the American Institute of Architects (AIA) and the Associated General Contractors of America (AGC). The following discussion of specific contract terms draws extensively on AIA Document A401, the Standard Form Subcontract.

Because many of the terms of the basic agreement between the owner and the general contractor are either incorporated in the subcontract or have substantially analogous clauses in the subcontract, this discussion of subcontract terms focuses on certain key terms that pose special problems in the subcontract context.

[A] Adoption of Contract Documents and Flow-Down Provisions

Care should be taken by the prime contractor to ensure that the appropriate terms adopting the contract documents and flow-down provisions are contained in each of its subcontracts. This is the prime contractor's best method of ensuring that there is no inconsistency between its obligations to the owner and the subcontractor's obligations to the prime contractor. The contractor should review both agreements to make sure they are consistent with one another.

For subcontractors, a subcontract provision adopting the contract documents of the owner and the prime contractor means that subcontractors cannot merely read the subcontract to determine their rights, obligations, and duties. It is imperative for subcontractors to review the general and special conditions between the owner and the prime contractor to determine how matters such as scheduling, time extensions, and change orders will be handled. This is particularly important on large projects.

[B] Payment Terms and Pay-When-Paid Clause

The most critical element of the subcontract is the clause addressing how the subcontractor will be paid. If an owner requires certain information on the prime contractor's applications for payment and requires those applications by a given date, the subcontracts let by that prime contractor must require subcontractors to supply the prime contractor with similar information far enough in advance of the due date to enable the prime contractor to prepare the application.[44] Document A401 also gives the subcontractor leverage to force the prime contractor to discharge progress payments received from the owner by allowing the subcontractor to stop work should the prime contractor refuse to pay (provided, of course, that the failure to pay the subcontractor is not the subcontractor's fault).[45]

Some payment clauses also require subcontractors to submit mechanic's lien releases for themselves and any of their sub-subcontractors or material suppliers as a condition of receiving payment. Such provisions usually parallel similar requirements in the prime contract documents.

The subcontractor's right to be paid and the prime contractor's duty to pay in the event of nonpayment by the owner is a major issue that must be considered in the drafting process. A401 ¶¶ 11.3 and 12.1 are also noteworthy because they require the prime contractor to pay the subcontractor whether or not the prime contractor itself has been paid, unless otherwise provided.

It should be noted, however, that in the absence of a dispute as to the quality of the subcontractor's work, one California case[46] has held that the subcontractor is entitled to be paid within a reasonable time even if the owner has withheld retention from the prime contractor and even if the subcontract provides that the subcontractor is to be paid when the owner pays the prime contractor. The subcontract involved provided that "Contractor agrees to pay to the subcontractor upon receipt of each payment received from the owner the portion of said payment allowed to contractor on account of subcontractor's work to the extent of subcontractor's interest therein less any percentage retained under said general contract."[47]

The subcontractor's work in that case had been approved by all parties. The prime contractor and the owner were in dispute about matters other than the subcontractor's work, and the owner had withheld money from the prime contractor, who, in turn, had withheld payment from the subcontractor. The subcontractor sued and prevailed. The court stated:

> Defendant's interpretation . . . would postpone payments earned by a subcontractor, itself without fault, until a dispute between the contractor and the owner is resolved, perhaps months or even years later. Indeed, it gives no reasonable assurance that such a dispute would ever be resolved. While the question is unsettled, the contractor continues unobligated to the subcontractor. On the other hand, if the dispute be lost because of the contractor's fault, then surely the contractor must pay his subcontractor creditor from other funds; if won, he must apply all or a substantial part of the money he receives toward his subcontractual obligations. His interest would seem more likely to benefit from avoidance of any settlement with the owner. It is unlikely that such a result was intended by the contracting parties. The rules announced in *Hertzka*

---

[44] *See* A401 ¶ 11.3.
[45] *See id.* ¶ 4.7.1.
[46] Yamanishi v. Bleily & Collishaw, Inc., 29 Cal. App. 3d 457 (1972).
[47] 29 Cal. App. 3d at 461.

& *Knowles v. Salter supra* and *Hawley v. Orange County Flood, etc., Dist., supra*, are applicable.[48]

Some subcontract agreements contain stronger language making payment by the owner to the prime contractor a "condition precedent" to the prime contractor's obligation to pay the subcontractor. The prime contractor inserts this language to avoid financing the job out of its own funds. Any doubt about how such provisions would be interpreted was resolved in *Wm. R. Clarke v. Safeco Insurance Co.*[49] when the California Supreme Court, in a four to three decision, ruled that a "pay-if-paid" provision in a prime contractor's agreement with its subcontractor was void because it resulted in an indirect forfeiture of the subcontractor's constitutionally protected mechanic's lien right. As a result of the *Clarke* decision, it is clear that a prime contractor is obliged to pay its subcontractors within a reasonable time whether or not the prime contractor received payment from the owner.[50]

[C] Indemnity Obligations

The following case summaries illustrate the manner in which California courts interpret contractual indemnity provisions contained in subcontracts.

The court of appeal determined that the duty of a subcontractor to defend a developer/general contractor under a subcontract indemnity clause is significantly narrower, and arises later, than the duty of an insurer to defend under an insurance policy in *Regan Roofing Co. v. Superior Court.*[51] The developer/general contractor for a condominium project was sued by the homeowners' association for alleged construction defects and cross-complained against 24 of its subcontractors for several causes of action, including express contractual indemnity. The trial court granted the developer's motion for summary judgment on two specific issues: (1) that each of the subcontracts had a specific, Type I indemnity provision requiring the subcontractor to defend and indemnify the developer, even against claims based upon the developer's own negligence; and (2) that the subcontractors' duty to defend the developer was independent of and preceded their duty to indemnify.

The court of appeal issued a writ of mandate requiring the trial court to deny the motion for summary judgment as premature. Unlike an insurance policy, which is a contract of adhesion and requires the insurer to defend the insured regardless of actual liability, a contractual indemnity clause does not give rise to a duty to defend unless and until a determination is made that the subcontractors

---

[48] *Id.* at 463.

[49] 15 Cal. 4th 882 (1997).

[50] The contractor can limit its liability pursuant to the 10-day stop work order procedures set forth in California Civil Code § 3260.2. Additionally, many general contractors are adding provisions to their contract forms defining "reasonable time" as a period not shorter than that which is necessary to prosecute to conclusion litigation against the owner. Whether such provisions will be upheld by courts remains to be seen.

[51] 24 Cal. App. 4th 425 (1994), *review denied.*

were negligent in the performance of their work and thereby owed a duty to indemnify the developer/general contractor. The court noted that California Civil Code § 2778(4) does not require an indemnitor to defend an indemnitee without a showing that the indemnitor was itself negligent. Furthermore, it was unclear whether the subcontractors could be required to defend the developer against claims for which the developer, but not the subcontractors themselves, might be strictly liable.

A specific indemnity clause in a subcontract is controlling and nullifies a party's common-law right to implied equitable indemnity and contribution.[52] In the *Regional Steel Corp.* case, the plaintiff, a subcontractor's employee who was unloading rebar, was knocked from a truck by the general contractor's crane. The prime contractor filed a cross-complaint against the subcontractor, seeking implied equitable indemnity and contribution. The subcontract, which originally had a standard indemnity clause requiring the subcontractor to indemnify the prime contractor against claims caused in whole or in part by the acts and omissions of the subcontractor, had been amended to provide that the subcontractor's indemnity obligation was limited to damages caused in whole by the subcontractor's negligence.

An indemnitee may recover the attorney's fees incurred in defense of the indemnified claim but, in the absence of an express contract provision, cannot recover attorney's fees expended to enforce the indemnity agreement itself.[53] Because the subcontract in the *Otis* case did not have an attorney's fees provision for enforcement of the indemnity provision, such fees were not recoverable under the basic rule that attorney's fees are not available to a prevailing party unless provided for in the contract or permitted by statute. The fact that the indemnity clause included attorney's fees as an item of loss did not convert that provision into one for attorney's fees in an action to enforce the subcontract. The attorney's fees for defense of the underlying claim would also have been authorized by statute, per Civil Code § 2778(3), but that statute does not extend to actions for prosecution of the indemnity clause.

In *Hernandez v. Badger Construction Equipment Co.*,[54] the court of appeal applied the tort principle of comparative fault in an action between an indemnitor and an indemnitee who were each partly at fault. This case involved a personal injury action by a shipyard worker and his wife, brought after the injured employee had already received workers' compensation benefits under federal law. The lengthy opinion in this case not only dealt with the workers' compensation issues but also unraveled a very complicated network of indemnity and insurance obligations between the employer, the crane manufacturer, and the crane's lessor.

In *United States Elevator Corp. v. Pacific Investment Corp.*,[55] the court of appeal held that the indemnitor under a general indemnity clause had an obligation to defend the indemnitee. The plaintiff, United States Elevator Corp. (Elevator),

---

[52] *See* Regional Steel Corp. v. Superior Court, 25 Cal. App. 4th 525 (1994).
[53] *See* Otis Elevator Co. v. Toda Constr., 27 Cal. App. 4th 559 (1994).
[54] 28 Cal. App. 4th 1791 (1994).
[55] 30 Cal. App. 4th 122 (1994), *review denied.*

was sued by a third-party business invitee who was injured when she tripped and fell into an elevator compartment that was below floor level when the doors opened. The defendant building owner (Pacific) refused the elevator company's tender of defense but later settled with the third party (Hanold), who dismissed her lawsuit with prejudice as to all parties, including Elevator.

There followed a trial between Pacific and Elevator on an indemnity contract, which was part of a services contract between them. The indemnity contract provided that Pacific would indemnify Elevator for all "claims, demands, and liability for damages for death or bodily injury to persons . . . arising out of or connected with the use, installation or maintenance of the elevator."[56] The trial court held that "Pacific had no obligation to indemnify Elevator for these costs [of defense] in the absence of either a judgment against Elevator or payment by Elevator on Hanold's claim."[57]

The court of appeal reversed, holding that "an indemnity provision in a contract is to be construed under the same rules governing contracts in general, with a view to determining the intent of the parties. In addition, Civil Code § 2778 provides rules of interpretation of a contract of indemnity, unless a contrary intention appears in the document."[58]

First, the court of appeal rejected Elevator's attempt to rely upon a line of cases in which broad general indemnity clauses were held to impose a duty to defend and/or to pay attorney's fees incurred in defense. These cases were distinguished on the grounds that the indemnity provisions either involved inclusive general language such as "hold harmless" or specific references to attorney's fees. Then, however, the court of appeal noted that Civil Code § 2778 "imposes rules of interpretation for indemnity contracts, unless a *contrary* intention appears."[59] Because there was no *contrary* clause in the agreement between Elevator and Pacific, the matter was determined by section 2778(4), which provides: "The person indemnifying is bound, on request of the person indemnified, to defend actions or proceedings brought against the latter in respect to the matters embraced by the indemnity, but the person indemnified has the right to conduct such defense, if he chooses to do so."

Next, the court of appeal distinguished the California Supreme Court's holding in *Gribaldo, Jacobs, Jones & Associates v. Agrippina Versicherunges A.G.*[60] on the grounds that (1) the contract in that case did contain a provision that was contrary to section 2778(4); and (2) "even if the contract did not exclude the duty to defend, the indemnitee waived it by failing to request the indemnitor to provide a defense."[61]

Finally, the court of appeal found "no merit to Pacific's argument that by settling with Hanold, without liability or contribution by Elevator, Pacific can

---

[56] 30 Cal. App. 4th at 124.
[57] *Id.* at 125.
[58] *Id.*
[59] *Id.* at 126 (emphasis in original).
[60] 3 Cal. 3d 434 (1970).
[61] *United States Elevator Corp.*, 30 Cal. App. 4th at 128.

avoid any duty" under Civil Code § 2778(4).[62] Such a result, the court reasoned, would amount to requiring Elevator to indirectly contribute to Pacific's settlement, as section 2778 does not "afford that advantage to an indemnitor."[63]

In summary, the foregoing cases establish that (1) attorney's fees are available only with respect to defending the underlying action and, in the absence of an express contractual provision, cannot be recovered by the prevailing party in a suit to enforce the indemnity provision; (2) the courts will apportion fault between an indemnitor and an indemnitee who are both at fault; and (3) the indemnification obligation carries with it a duty to defend the indemnitee.

[D] Bonding Requirements

Material and labor suppliers to subcontractors may file mechanics' liens and stop notices under California law. The disputes giving rise to such liens are often unknown to the owner and the prime contractor. In order to protect the owner and prime contractor from such surprise claims (which most often arise from misapplication of funds by the subcontractor), many subcontracts include a requirement that the subcontractor furnish not only a performance bond (one that guarantees that the work will be completed satisfactorily) but also a labor and materials bond, which provides for payment of all proper labor and materials bills incurred by the subcontractor on the given project should the subcontractor fail to pay those bills himself. The amount of bonding required on a particular project will vary, depending on the size and reputability of the subcontractor, the concerns of the owner, and the requirements of the project. Accordingly, no general rule can be stated, and Article 13 in AIA Document A401 provides no concrete language on this subject. The AGC form requires such bonding whenever requested by the prime contractor, thus using bonding as a means of obtaining further assurances of a subcontractor's performance should any doubts arise.

[E] Obligation to Continue Work

Suppose a subcontractor and prime contractor enter into a subcontract for $100,000 and a dispute ensues over whether a particular item of work valued at $20,000 is or is not part of the subcontract. Is the subcontractor obligated to perform the work, or can the subcontractor refuse? The answer to this question depends upon the provisions of the subcontract.

Subcontracts often contain provisions identical or similar to the following:

> Subcontractor, in the event of any dispute or controversy with contractor or any other Subcontractor over any matter whatsoever, shall not cause any delay or cessation in or of Subcontractor's work or the work of any other Subcontractor or of the contractor but shall proceed under this Subcontract Agreement with the performance of the work required thereby.[64]

---

[62] *Id.*
[63] *Id.* at 129.
[64] B.C. Richter Contracting Co. v. Continental Cas. Co., 230 Cal. App. 2d 491, 500 (1964).

If the subcontract contains such a provision and the subcontractor refuses to perform, she will be in breach of contract under the ruling in *B.C. Richter Contracting Co. v. Continental Casualty Co.*,[65] which enforced the clause just quoted. If the subcontractor agrees to perform, the cost may exceed any amount the subcontractor can afford, and she may go broke. In any event, the subcontractor will effectively be "financing" the job for the prime contractor, owner, and lender.

On the other hand, if the subcontract does not contain such a clause, or contains a clause providing that the subcontractor may stop the work in the event payments are not made for work performed, the subcontractor is in a much stronger legal position. A401 ¶ 4.7 is very favorable to subcontractors in this regard.

Obviously, a provision obligating a subcontractor to continue work in the event of a dispute is a key clause that should be reviewed and negotiated between prime contractors and owners and prime contractors and subcontractors.

[F] Contractor's Right to Terminate Subcontractor

One of the most difficult questions to arise on any construction project is when the general contractor has the right to terminate a subcontractor for nonperformance. Generally speaking, the answer to that question will depend on the facts of the particular case. If, in fact, the subcontractor has "materially" breached the subcontract, then the prime contractor will be justified in terminating the subcontractor. It is a question of fact in each case, however, as to when a subcontractor has materially breached the subcontract. The contractor must look at its rights as set forth in the subcontract, and at the conduct of the subcontractor, and then act in good faith and comply with the termination provisions of the subcontract.

A case that discusses the right of a general contractor to terminate a subcontractor and the type of facts that will justify such a termination is *Call v. Alcan Pacific Co.*[66] In that case, the plaintiff was the drywall and painting subcontractor on a project being performed by the defendant as the prime contractor. On July 25, 1963, the plaintiff withdrew his employees from the job. On the same day (a Thursday), Alcan, the prime contractor, sent Call, the subcontractor, a telegram stating:

> Your repeated failure to prosecute work diligently, including, but not limited to, the failure to maintain full crews and also pulling your crews off the Richland Housing Project, the most recent [of] which occurred today, you are hereby notified that unless you resume work with a full crew by Monday, July 29, 1963, we consider your contract terminated and take such action as authorized by our contract with you.[67]

On Monday, July 29, 1963, Alcan, the prime contractor, terminated the subcontract and took possession of Call's materials, tools, and appliances at the jobsite. The subcontractor filed a suit against the prime contractor, seeking an

---

[65] *Id.*
[66] 251 Cal. App. 2d 442 (1967).
[67] 251 Cal. App. at 445.

accounting and an injunction. The prime contractor filed an answer and a cross-complaint suing the surety on the subcontractor's performance bond. The subcontractor then filed a new action against the prime contractor, seeking damages for breach of the subcontract and conversion of his equipment. The two lawsuits were consolidated.

A jury trial resulted in a judgment in favor of the subcontractor against the prime contractor, awarding the subcontractor $16,785 for breach of contract and $750 for conversion, less $1,657.78 found to be due from the subcontractor to the prime contractor. The prime contractor appealed. The trial court had found that the subcontractor had performed his subcontract until July 25, 1963, when he withdrew his employees from the site, and that only a small amount of the subcontract remained to be performed as of that date. Earlier disputes and minor breaches had been waived or settled by negotiation and change orders and the parties had, by their actions, waived arbitration provisions. The subcontractor's cessation of work on July 25, 1963, and failure to work between then and July 29, 1963, did not constitute a "material breach" of the subcontract. The project owner's inspection procedures were found unreasonable. In choosing the date of July 29, 1963, on which to terminate the job, the prime contractor refused further performance in the face of the subcontractor's minor breach. Finally, the prime contractor's conduct on July 29, 1963, in terminating the job and seizing the subcontractor's equipment and materials was held to be a "material breach" of the contract by the prime contractor.

On appeal, the prime contractor contended that the trial court had erred in ignoring the special requirements of the painting subcontract. The subcontract provided that "time was of the essence" and that the subcontractor would conduct the work diligently and "supply a sufficiency of properly skilled workmen at all times to the satisfaction of the contractor." Article 10 of the subcontract gave the prime contractor an option to terminate the subcontract under specified conditions. Specifically, Article 10 provided that should the subcontractor, at any time during the progress of the work, refuse or neglect to prosecute the work in accordance with the provisions of the subcontract until fully completed and accepted, the contractor would immediately have the power to terminate the employment of the subcontractor and to enter upon the premises and take possession of all of the subcontractor's materials, tools, and appliances and to take charge of the work for the purpose of carrying out the terms of the subcontract, and further that the prime contractor might adopt such other means as it deemed necessary to complete the work and protect the prime contractor's interests. Paragraph 11 of the subcontract provided for arbitration of disputes and an agreement that "a dispute shall not interfere with the progress of construction."

The court of appeal noted that a contract may contain valid provisions giving one or the other party an option to terminate it on specified conditions. In this case, the subcontractor was required to supply a sufficiency of properly skilled workers at all times, to the satisfaction of the contractor. The court noted that a provision requiring performance to the promisee's satisfaction does not make the contract illusory, as the promisee cannot act arbitrarily. Rather, according to the character of the contract, the promisee will be held either to the standard of satisfaction in good faith or to the standard of a reasonable person. The court further noted that the time-of-the-essence clause meant that the subcontractor's timely performance must be observed unless the prime contractor excused it by waiver or other action. The court also noted that when a contract provides that the contractor may cancel it by written notice upon the subcontractor's failure of timely performance, the contractor has an option to terminate by giving the prescribed notice. Even without an express termination option, the contractor or owner would have a common-law right to cancel the contract if the contractor or subcontractor, without the fault of the owner or contractor, failed to comply with the time-of-essence clause. The court warned, however, that the option to terminate must be exercised in good faith. The court also noted that when performance has a continuous character, such as a construction subcontract, waiver of one breach does not preclude the other party from invoking its contract rights upon a later breach. Thus, when time is of the essence, strict compliance may be waived and then reinstated by a definite notice. The court stated the issue in the case as follows:

> In the present case the issue was not whether Call, the subcontractor, had been guilty of a minor or material breach by the standards applicable to contracts ungoverned by express conditions on the subject; nor, whether Alcan's consequential action correlatively became a minor or material breach. Rather, the issues were whether the subcontractor's conduct has been such as to activate the termination clause of the contract; whether Alcan had exercised its option to terminate; if so, whether Alcan had acted in good faith; and whether earlier waivers or other conduct abrogated Alcan's right to insist on continuing timely performance in compliance with the written contract.[68]

The court concluded that the trial court did not declare findings on these essential issues. Thus, because the trial court failed to find on those material issues, it was reversible error, and the judgment of the trial court was reversed. The foregoing case clearly indicates that the question of whether a general contractor can terminate a subcontractor is dependent upon (1) the terms and conditions of the contract; (2) whether the general contractor has complied with the contract with regard to its termination rights; (3) the facts alleged to activate the termination clause; and (4) whether the general contractor acted in good faith.

§ 3.04 MATERIAL SUPPLIER CONTRACTS

If a material supplier and its customer, whether an owner, subcontractor, or general contractor, enter into a written agreement, the terms and conditions of that written agreement will govern their relationship. Sometimes that written agreement merely takes the form of a credit application, or in some cases a purchase order. What often occurs in transactions between material suppliers and their customers is an exchange of forms, with neither party signing the other party's form.

---

[68] *Id.* at 448.

Most often, the material supplier issues its purchase order form to the owner, general contractor, or subcontractor and requests that customer to sign the purchase order and return it. Instead of doing so, the owner, general contractor, or subcontractor sends back a form of "acknowledgment" setting forth different terms and conditions. Those were the precise facts in *Frank M. Booth, Inc. v. Reynolds Metals Co.*[69] In that case, the subcontractor's purchase agreement had numerous provisions, including the following:

1. An agreement by the seller to repair or replace, at the construction site, at the seller's expense, all defects of materials or workmanship in the material.

2. An agreement by the seller to indemnify the subcontractor and prime contractor against and hold them harmless from any and all loss, damage, expenses, or attorney's fees incurred on account of a breach by the seller of any of the seller's obligations under the agreement.

3. An agreement that the material would conform to the requirements of the subcontractor's contract with the prime contractor that were applicable to the material or, if subcontractor's contract with the prime contractor did not have such requirements, that the material would be of the quality specified in the purchase agreement or of the best grades of their respective kinds, if no quality was specified, and would be fit for the purpose intended.

4. An agreement that no variance from the terms of the purchase agreement would be accepted by the subcontractor unless the variance was approved in writing.

The supplier sent an acknowledgment and sales order to the subcontractor but did not sign the subcontractor's purchase agreement. The acknowledgment and sales order issued by the supplier provided that the supplier's acceptance was expressly conditioned on the subcontractor's agreement to the terms and conditions on the front and reverse side of the sales order. It further provided in bold print, "you will be deemed to have agreed to such terms and conditions and in reliance thereon, this sales order will be placed in schedule unless written notice of any errors or objections to this sales order is received by us within ten days from the date shown below." It further warranted that the product would be free from defects in material and workmanship and would be in conformance to the subcontractor's standard specifications and those stated on the order. It further stated, in bold print, "seller makes no other warranties, express or implied, written or oral, including but not limited to warranties of merchantability or fitness for any particular purpose." It also provided that there was a limitation of liability on the supplier's part, namely, that the supplier's exclusive liability for breach of warranty would be to replace nonconforming products that would not exceed the buyer's purchase price. In no event would the supplier be liable to the

---

[69] 754 F. Supp. 1441 (E.D. Cal. 1991).

subcontractor for loss of profits or revenue or any incidental, consequential, special, or punitive damages. Finally, the supplier's acknowledgment and sales order provided for attorney's fees in the event of suit.

Neither party attempted to communicate with the other regarding the terms of the contract as set forth in their respective forms. The material was delivered and paid for. The subcontractor stored the materials outside. After a nine-month project delay, the subcontractor found many items were damaged and would have to be replaced. The material supplier failed to warn the subcontractor that the materials should not be stored outside and had knowledge that the subcontractor was storing the materials outside. Each party made a motion for summary judgment. The first question before the court was the application of California Commercial Code § 2207, which provides:

> (1) A definite and seasonable expression of acceptance or a written confirmation which is sent within a reasonable time operates as an acceptance even though it states terms additional to or different from those offered or agreed upon, unless acceptance is expressly made conditional on assent to the additional or different terms.
>
> (2) The additional terms are to be construed as proposals for addition to the contract. Between merchants such terms become part of the contract unless:
>
> (a) The offer expressly limits acceptance to the terms of the offer;
>
> (b) They materially alter it; or
>
> (c) Notification of objection to them has already been given or is given within a reasonable time after notice of them is received.
>
> (3) Conduct by both parties which recognizes the existence of a contract is sufficient to establish a contract for sale although the writings of the parties do not otherwise establish a contract. In such case the terms of the particular contract consist of those terms on which the writings of the parties agree, together with any supplementary terms incorporated under any other provision of this code.

Both parties agreed that a contract had been formed between them, that the California Commercial Code governed the dispute, and that the rules of contract formation set forth in Uniform Commercial Code § 2207 applied. The dispute was the legal effect of Uniform Commercial Code § 2207 when applied to the facts of the case. Specifically, the parties disagreed as to whether subsection 3 of section 2207 was triggered by the circumstances of the document exchange and subsequent conduct that occurred here. Specifically, the court stated, "[T]his dispute involves a classic instance of the battle of the forms in which contracting parties exchange preprinted forms that attempt to cast liability for certain categories of damage on the other party to the transaction."[70] The court stated that the issue before the court was whether Uniform Commercial Code § 2207(3)

---

[70] *Booth*, 754 F. Supp. at 1445.

applied or whether the supplier's acknowledgment should be viewed, under Uniform Commercial Code § 2207, as a counteroffer that was later assented to by the subcontractor by its performance. The court stated that the application of section 2207 to undisputed facts is a matter of law.

The court relied heavily on the Ninth Circuit's analysis in *Diamond Fruit Growers, Inc. v. Krack Corp.*[71] The parties exchanged inconsistent forms and then proceeded to perform, with neither party signing the other party's form. The seller expressly conditioned its assent on the buyer/offeror's assent to the additional terms in the seller's acknowledgment. The *Diamond Fruit* opinion noted that to treat contract performance as "assent" to the additional terms in the acknowledgment would be to revert to the common-law "last shot rule" that Uniform Commercial Code § 2207 was designed to abolish.

The court noted that one of the principles underlying section 2207 is neutrality. If possible, the section should be interpreted so as to give neither party to a contract an advantage simply because it happened to send the first, or in some cases the last, form. The *Frank M. Booth, Inc.* court specifically held that section 2207(3) applied. As previously noted, subdivision 3 of section 2207 provides that conduct by both parties that recognizes the existence of a contract is sufficient to establish a contract for sale, although the writings do not otherwise establish a contract. In such cases, the terms of the contract consist of those terms on which the writings of the parties agree. The court stated that in this case, the material supplier was most responsible for the ambiguity, because it inserted a term in its form that required the subcontractor to assent to those additional terms and then took no action to enforce that requirement. The court stated that if the material supplier had truly wanted the subcontractor to be bound by the terms and conditions set forth in its acknowledgment and sales order, then it could have protected itself by not shipping the coils until it expressly obtained the subcontractor's assent to those additional terms.

Under the reasoning of the *Diamond Fruit* case, subsequent performance did not constitute assent by the subcontractor to different or additional terms contained in the acknowledgment. Therefore, the supplier's acknowledgment may not be viewed as a counteroffer assented to by the subcontractor's subsequent conduct. Because Uniform Commercial Code § 2207(3) applies to delineate the terms of a contract under that section, the contract terms here included those upon which the parties' two forms agreed—specifically, the express warranties of good title, freedom from defects of materials and workmanship, and conformity with standard specifications. The parties' forms conflicted with regard to further express warranties, implied warranties, or limitation of liability provisions, so the provisions contained in the forms on those topics dropped out of the contract. However, as both parties' forms permitted recovery of attorney's fees, that would be a part of the contract.

The court then turned to a discussion of the subcontractor's claim that the supplier negligently breached its duty by failing to inform or warn the subcontractor that the materials required special storage and handling procedures to

---

[71] 794 F.2d 1440 (9th Cir. 1986).

prevent damage. The court noted that several courts applying California law have denied tort claims solely for economic loss and have limited the remedies through which the aggrieved commercial litigants may recoup their economic losses to those provided for by the contract and the Uniform Commercial Code. The court further noted that California courts define *economic loss* as damages for inadequate value, costs of repair and replacement of defective product, or consequent loss of profits, without any claim of personal injury or damages to other property. The court went on to note, however, that other California courts have permitted a negligence claim between commercial litigants by finding that when the harm to the plaintiff is foreseeable, a manufacturer has a duty not to impair the commercial operations of the purchasers with a faulty product, citing *Ales-Peratis Foods v. American Can Co.*[72] and *J'Aire Corp. v. Gregory.*[73] The court noted that those cases permit a negligence claim exclusively for economic losses only when there is no privity of contract between the parties and therefore no remedy is available under the UCC or by contract. The court went on to say that California law is not entirely clear as to the viability of a claim in negligence for economic damages between commercial entities that are in privity of contract. In light of that uncertainty, the federal judges must use their own best judgment in predicting what the California Supreme Court would decide in such a case. This judge held that the prevailing view of the California courts denies recovery between contracting commercial parties when the claim is based either upon strict products liability or negligence and is asserted solely for economic losses, because in such instances the parties are protected by the California Commercial Code and the contents of the contract between them. Therefore, the court dismissed the subcontractor's negligence claim.

Although this opinion is only by a district court judge, it sets forth very important concepts relating to dealings between commercial entities in a construction setting, holding that the Uniform Commercial Code applies and the "last form" in the "battle of the forms" may not necessarily be the terms of the contract. In this type of situation, the relationship between a material supplier and its customer should be reduced to a writing signed by both parties so that there is no ambiguity. If, however, the parties merely exchange forms, with neither party signing the other's form, then Uniform Commercial Code § 2207, as interpreted by this decision, will apply.

§ 3.05 ANALYSIS OF OWNER-ARCHITECT RELATIONSHIP

On a traditional design-bid project, the architect's role is to provide a design, in the form of plans and specifications, to meet the owner's needs. Additionally, the architect is usually required to perform construction administration to ensure that the contractor constructs the work in conformance with the plans and specifications. Typically, the owner-architect agreement is negotiated and executed well before the owner-contractor agreement. However, the terms and conditions

---

[72] 164 Cal. App. 3d 277 (1985).
[73] 24 Cal. 3d 799 (1979).

of the owner-architect agreement may limit the flexibility the owner has in drafting the agreement with the general contractor. For example, AIA Document B151 stipulates that the architect will provide construction administration in accordance with the A201 General Conditions. If the owner fails to utilize Document A201 as part of the agreement with the contractor, there may be major conflicts or gaps in responsibility for project administration.

[A] Limitation of Liability

Although a detailed analysis of the basic owner-architect agreement is beyond the scope of this chapter, the reader should be aware that one of the most frequently contested points of negotiation and the subject of litigation is the architect's desire to limit its liability. This issue is addressed in *Markborough California, Inc. v. Superior Court*,[74] wherein the court of appeal upheld a summary judgment enforcing a provision in an agreement between a developer and an architect, which limited the architect's liability for negligent design to the amount of compensation for the design services.

In the *Markborough* case, the defendant, Glenn, received $67,640 to design a man-made lake. The design failed and cost the plaintiff developer, Markborough, more than $5 million in repairs. However, the contract had a provision limiting any damages recovery to the amount of Glenn's fee. Markborough contended that the clause, to be enforceable under California Civil Code § 2782.5, had to have been specifically negotiated and expressly agreed to by the parties. Glenn did not dispute that the clause was never discussed and that he never disclosed the risks involved in the artificial lake system, nor did he deny that the failure cost more than $5 million to repair.

According to section 2782, all provisions that purport to indemnify the promisee against liability for damages arising out of the promisee's sole negligence, or for defects in a design furnished by the promisee, are against public policy and are void and unenforceable. At the same time section 2782 was enacted, in 1967, the legislature added section 2782.5. In 1980, the sections were amended "to clarify the ability of the parties in construction contracts to negotiate agreements that apportion liability."[75] In its present form, § 2782.5 states, in relevant part, that "nothing contained in Section 2782 shall prevent a party . . . from *negotiating and expressly agreeing* with respect to the allocation, release, liquidation, exclusion or limitation as between the parties of any liability (a) for design defects, or (b) of the promisee to the promisor arising out of or relating to the construction contract" (emphasis added).

The court in *Markborough*, after determining that the issue turned upon the intended meaning of the term "negotiating," concluded: "[A]ll that reasonably can be required for 'negotiation' is a fair opportunity for both parties to accept, reject or modify the other's offers or demands."[76] The court went on to distinguish

---

[74] 227 Cal. App. 3d 705 (1991), *review denied*.
[75] 227 Cal. App. 3d at 711.
[76] *Id.* at 714.

other cases in which such clauses were invalidated either because they were imposed on one party "on a take-it-or-leave-it basis" by another party with disproportionate bargaining power[77] or because the provision itself was "not conspicuous, not clear and explicit, and not comprehensible."[78]

This case illustrates the dangers inherent in failing to read and thoroughly understand every provision in any design service or construction agreement. Under this decision, section 2782 offers little public policy protection against unambiguous, explicit disclaimers and other exculpation clauses beyond the common-law protection against adhesion contracts.

§ 3.06 ANALYSIS OF DESIGN-BUILD RELATIONSHIP

The use of design-build as a project delivery method has been steadily increasing over the past three decades. Unlike under the traditional design-bid-build method of project delivery, many private owners and public agencies are now retaining one entity for both design and construction. Typically, the design-builder is a licensed general contractor that employs its own architects and engineers or retains them as subconsultants or is a joint venture between a general contractor and a design firm. Although less frequent on some private works, the design-builder may be an architect-engineer that subcontracts the construction phase.

By using this method of project delivery, owners find that they can avoid the conflicts that arise between architects and contractors when there are errors or omissions in the plans or where the contractor fails to adequately inspect the site, take field measurements, or review and coordinate the plans and specifications. Most design-build contracts also seek to expedite the schedule, and thereby save costs, for construction by utilizing fast-track methods, whereby early phases of construction commence before the design is completed on later phases.

The design-build method does not reduce the tasks required to complete a project but rather allocates responsibilities and risks in a different manner. Although design-build contracting can be utilized for almost any project, the risks and benefits must be evaluated before selecting it as a method of project delivery. The primary advantages to the owner in utilizing design-build are single-point contractual liability for designing and construction; reduction of owner's management time; lower overall project costs; and faster completion. The risks to the owner in utilizing design-build include fewer responsible parties with assets and insurance to satisfy owner claims for defective design or construction, less control over design aesthetics and function, the design-builder's motivation to reduce its cost of performing rather than protect the owner's interest, and fewer checks and balances on design and construction. Many owners seek to minimize the risks of design-build by actively monitoring design and construction, approving design submittals and selection of subcontractors, or conducting peer review of the design-builder. Such active involvement by the owner, however, may blur the line of sole-source responsibility and undercut the advantages sought to be

---

[77] Conservatorship of Link, 158 Cal. App. 3d 138, 141 (1984).
[78] Windsor Mills, Inc. v. Collins & Aikrnan Corp., 25 Cal. App. 3d 987, 993–994 (1972).

gained by utilizing design-build. Thus, owners should incorporate exculpatory language or disclaimers in the design information they give to the design-builder.

Other considerations in determining whether design-build is appropriate involve the nature of the project and the degree to which the owner can accurately describe the scope of work and the end product. In general, if a project is repetitive or does not require detailed owner input into the design—e.g., retail chains, convenience stores, and fast-food restaurants—it is well suited to design-build. Additionally, complex projects with performance-based rather than aesthetic design criteria, such as manufacturing plants or utility projects, are well suited to design-build. Moreover, if the owner does not have a detailed program of requirements for the work or does not adequately describe the scope of the work, it may be dissatisfied with the end product but have no recourse because the design was delegated to the builder.

From the design-builder's perspective, while the team approach to design and construction may result in certain efficiencies in cost and time, by assuming responsibility for the design, the builder loses the most common ground for seeking additional compensation from the owner—i.e., the *Spearin* doctrine of implied warranty that the plans and specifications are free of defects.

Due to the unique nature of the design-build process, it is critical for the owner–design-builder agreement to accurately reflect the scope of work and the levels of risk that each party agrees to assume, and to address foreseeable contingencies. There are a variety of standard forms published by the American Institute of Architects, the Associated General Contractors of America, the Engineer's Joint Contract Document Committee, and the Design-Build Institute of America. However, because design-build projects are scope driven and each has unique circumstances, standard forms are difficult to use and require significant modification. Some of the most important project-specific provisions include payment, changes, differing site conditions, scheduling, and limitations of liability. Moreover, risk allocation, through insurance, must be carefully reviewed and structured to ensure coverage is in place and that both design sources and construction are covered through project-specific insurance with extended reporting/discovery periods covering faulty workmanship or design error claims.

CHAPTER 4
BREACH OF CONTRACT BY OWNER

§ 4.01 Types of Owner Breach
§ 4.02 Failure to Make Progress Payment
§ 4.03 Contractor's Right to Stop Work in the Event of Nonpayment: Private Works of Improvement
§ 4.04 Prompt Payment Legislation in California
§ 4.05 Delay
§ 4.06 Defective Plans and Specifications
§ 4.07 How Courts Interpret Plans and Specifications
    [A] Basic Rule: Implied Warranty of Accuracy of Plans and Specifications
    [B] Contractor Not Responsible for Defective Plans and Specifications
    [C] Enforceability of Exculpatory Clauses
    [D] Rule of Contra Proferentem: Ambiguities in Plans and Specifications Will Be Construed Against the Drafter
    [E] Obligation of Contractor to Proceed with the Work
§ 4.08 Delay in Start of Performance
§ 4.09 Material Change of Contract
§ 4.10 Acceleration
§ 4.11 Recovery of Consequential Damages
§ 4.12 Recovery of Damages for Reduction of Bonding Capacity
§ 4.13 Contractor's Claim for Right to Finish Early
§ 4.14 Delay Damages on Multi-Prime Projects
§ 4.15 Recoverability of Interest for Changed Work

§ 4.01 TYPES OF OWNER BREACH

Both the contractor and the owner can breach a construction contract. This chapter discusses only those breaches of contract that may be committed by an owner and the rights of the contractors and subcontractors in relation to the breach. Because not every type of breach can be covered here, only the major categories of breach are set forth.

Some of the typical owner breaches are

1. Failure to make payment
2. Delay
3. Defective plans and specifications
4. Material breach resulting in abandonment, thereby allowing recovery of reasonable value
5. Failure to grant jobsite access
6. Acceleration
7. Interference with the contractor's performance
8. Failure to approve shop drawings
9. Failure to approve and process change orders and change order requests
10. Failure to inspect and approve work on the critical path
11. Failure to deliver owner-furnished equipment
12. Wrongful termination and ejection of the contractor from the job.

Often the conduct of the owner may not rise to the level of a "breach" of contract, and, as a result, the contractor will present its claim in a request for change order for any additional costs or time impact that the contractor has suffered as a result of improper conduct by the owner.

§ 4.02 FAILURE TO MAKE PROGRESS PAYMENT

The law in this area is fairly easy to state but very difficult to apply. In the leading case of *Integrated, Inc. v. Alec Fergusson Electrical Contractor*,[1] the court laid out the rules as to when a contractor can or cannot cease to perform upon failure of the owner to make progress payments. The rules are spelled out in the following portions of the opinion:

> The proposition that in building contracts failure to make progress payments is not such a breach as will authorize a contractor to abandon the work and sue for damages, but that it does constitute such a breach as will justify rescission and recovery of the reasonable value of labor and materials furnished, was

---

[1] 250 Cal. App. 2d 287 (1967).

first announced in this state in *Cox v. McLaughlin,* 52 Cal. 590; 54 Cal. 605; 63 Cal. 196; 76 Cal. 60 [18 P. 100, 9 Am. St. Rep. 164] and has thereafter been repeatedly stated to be the settled law of this state. (*American-Hawaiian Eng, etc. Co. v. Butler,* 165 Cal. 497, 516 [133 P. 280]; *San Francisco Bridge Co. v. Dumbarton Land & Improv. Co.,* 119 Cal. 272 [51 P. 335]; *Fairchild etc. Co. v. Southern etc. Co.,* 158 Cal. 264, 273–274 [110 P. 951]; *Porter v. Arrowhead Reservoir Co.,* 100 Cal. 500, 502 [35 P. 146]; *Golden Gate Lbr. Co. v. Sahrbacher,* 105 Cal. 114 [38 P. 635]; *Barris v. Atlas Rock Co.,* 118 Cal. App. 606 [5 P.2d 670]; *Monson v. Fischer,* 118 Cal. App. 503 [5 P.2d 628]. *See Big Boy D. Corp., Ltd. v. Etheridge,* 44 Cal. App. 2d 114, 119 [111 P.2d 953]; *Smoll v. Webb,* 55 Cal. App. 2d 456, 458 [130 P.2d 773].)

The mechanical application of the rule as phrased in the cases heretofore cited would lead one to the conclusion that any failure to make progress payments would entitle the contractor to rescind. A careful examination of the facts in those cases in which the rule has been applied reveals, however, that the failures to make progress payments which were held to justify rescission involved either an extended and unreasonable delay, imposition of new and onerous conditions to payment, outright refusal, or a total repudiation of the contract. They did not involve a minor deviation from the covenant. Significantly in *Porter v. Arrowhead Reservoir Co., supra,* 100 Cal. 500, the court discussed the holdings in the *Cox* case and concluded at p. 504: "From the foregoing views it follows that, to entitle the plaintiffs [contractors] to recover in the present action, it is only necessary that they prove a substantial failure upon the part of the defendant to comply with its agreement as to the payment of an installment upon the contract price, and in addition thereto that they rescinded the contract by reason of such failure . . ." In stating the rule, subsequent cases have omitted the word substantial and merely state generally that failure to make progress payments entitles a contractor to rescind. In our opinion, *Porter v. Arrowhead Reservoir Co., supra,* expresses a more precise statement of the rule. (*Guerini Stone Co. v. P. J. Carlin Constr. Co.,* 248 U.S. 344, 344–345, [63 L. Ed. 275, 285, 39 S. Ct. 102, 106]; *United States v. Southern Constr. Co.,* 293 F.2d 493, 498, *reversed in part on other grounds,* 371 U.S. 57 [9 L. Ed. 2d 31, 83 S. Ct. 108].) A slight deviation either in time or amount of progress payments should not justify rescission or abandonment. *See United States v. Southern Constr. Co., supra.*

In the instant case, therefore, if the court should find that Integrated failed to make progress payments within the time required by the contract, the court must determine whether such failure constituted a substantial failure to comply. That is a question of fact to be determined by the trial court. (*Associated Lathing etc. Co. v. Louis C. Dunn Inc.,* 135 Cal. App. 2d 40, 49 [286 P.2d 825]; *Smith v. Empire Sanitary Dist.,* 127 Cal. App. 2d 63, 73 [273 P.2d 37].)

If Integrated's breach be found to be substantial, the court must also determine whether Fergusson was itself in default in any material respect. One who is himself in default in the performance of a dependent or concurrent obligation or where his default is so related to the obligation in which the other has failed that it affects the performance thereof or the duty of the other to perform may not unilaterally rescind because of the other's breach. (*Fairchild etc. Co. v. Southern etc. Co., supra,* 158 Cal. 264, 273; *Nelson v. Spence,* 182 Cal. App. 2d 493, 499 [6 Cal. Rptr. 312]; *American-Hawaiian*

Eng, etc. Co. v. Butler, supra, 165 Cal. 497, 516; *Eade v. Reich,* 120 Cal. App. 32, 38 [7 P.2d 1043].) Where, however, the delinquency of the rescinding party has no relation to the other's obligation in respect of which the right of rescission is asserted, it will not preclude rescission. (*American-Hawaiian Eng etc. Co. v. Butler, supra; Eade v. Reich, supra.*)[2]

In summary, the rules are as follows:

1. The owner's failure to make progress payments may give the contractor the right to rescind and sue for reasonable value.

2. The contractor can rescind and sue for reasonable value only when the failure to make progress payments is a substantial or material breach of contract.

3. If the contractor itself is in breach of contract, then the owner may withhold progress payments and is not in breach of contract.

4. A slight deviation either in the time or amount of progress payments will not justify rescission.

Thus, in each case, it must be determined whether the failure to make the progress payment was a material breach. If the court holds that it was, the contractor is entitled to rescission and can recover the reasonable value of the work performed. If the court finds that the failure to make progress payments was not a material breach or that the owner was justified in withholding payment because the contractor was in breach of contract, the owner is not in breach and may sue the contractor for breach. Thus, the contractor's decision to pull off a job because of nonpayment of progress payments is an extremely risky decision.

A corollary to this rule is the fact that even if the owner's breach is material, the contractor cannot delay in making the decision to rescind to see whether the contract turns out profitably. The contractor must promptly rescind upon material breach. This was clearly noted in *B.C. Richter Contracting Co. v. Continental Casualty Co.*:[3]

> [3] The general rule in California is "One who has been injured by a breach of contract has an election to pursue any of three remedies, to wit: 'He may treat the contract as rescinded and may recover upon a quantum meruit so far as he has performed; or he may keep the contract alive, for the benefit of both parties, being at all times ready and able to perform; or, third, he may treat the repudiation as putting an end to the contract for all purposes of performance, and sue for the profits he would have realized if he had not been prevented from performing.'" (*Oliver v. Campbell,* 43 Cal. 2d 298, 302 [273 P.2d 15]; *Alder v. Drudis,* 30 Cal. 2d 372, 281 [182 P.2d 195]; 12 Cal. Jur. 2d, Contracts, Section 253; Rest., Contracts, Section 347.)
>
> [4] When, after partial performance, the innocent party elects to disaffirm or rescind, there is no longer any contract which conclusively fixes a limit upon

---

[2] 250 Cal. App. 2d at 296–298.
[3] 230 Cal. App. 2d 491 (1964).

his recovery; hence, it is said, he may sue upon a quantum meruit as if the special contract had never been made and may recover the reasonable value of the services performed, even though recovery exceeds the contract price. (*Oliver v. Campbell, supra,* 43 Cal. 2d at p. 304; *Lessing v. Gibbons,* 6 Cal. App. 2d 598, 607 [45 P.2d 258]; 5 Williston on Contracts (rev. ed.) Section 1485, p. 4146; Rest., Contracts, Section 347; *see* Comments, 23 Cal. L. Rev. 313, 7 So. Cal. L. Rev. 338.)

[5] If the innocent party chooses to rescind, he must do so promptly upon discovery of the breach. (Civ. Code, Section 1691; *Beckett v. Kaynar Mfg. Co., Inc.,* 49 Cal. 2d 695, 700 [321 P.2d 749]; 12 Cal. Jur. 2d, Contracts, Section 194.) He may not wait to see whether the contract turns out to be profitable or unprofitable, good or bad. (*Sogg v. Harvey,* 134 Cal. App. 2d 116, 121 [285 P.2d 104].)

[6] Where his performance is not prevented, the injured party may elect instead to affirm the contract and complete performance. If such is his election, his exclusive remedy is an action for damages. (*House v. Piercy,* 181 Cal. 247, 251 [183 P. 807]; *Ely v. Bottini,* 179 Cal. App. 2d 287, 296 [3 Cal. Rptr. 756].)

[7] His measure of damage is the unpaid contract price plus any additional expense attributable to the breach. (Civ. Code, Section 3300; *Buxbom v. Smith,* 23 Cal. 2d 535, 541 [145 P.2d 305]; *Higgins v. Desert Braemer, Inc.,* 219 Cal. App. 2d 744, 751 [33 Cal. Rptr. 527]; *Standard Iron Works v. Globe Jewelry & Loan, Inc.,* 164 Cal. App. 2d 108, 116 [330 P.2d 271]; Rest., Contracts, Sections 329, 346(2)(a); 5 Williston on Contracts (rev. ed.) Section 1338, pp. 3762–3764.)

[8] Affirmation of the contract, on the one hand, and rescission and restitution on the other, are alternative remedies. Election to pursue one is a bar to invoking the other. (*Alder v. Drudis, supra,* 30 Cal. 2d at p. 383; *Lenard v. Edmonds,* 151 Cal. App. 2d 764, 768 [312 P.2d 308]; Rest., Contracts, Section 384; 5 Williston on Contracts (rev. ed.) Section 1464, pp. 4096–4097.)[4]

A clause that contractors should attempt to negotiate and include in their contracts to address this issue would read as follows:

Contractor shall have the right to stop work and keep the job idle if any payment is not made to the contractor under the agreement. Contractor shall not be obligated to perform further until all payments due are received by the contractor.

One clause that contractors should be aware of is the clause that requires the contractor or subcontractor to continue to perform even where there is a controversy between the parties. There was such a clause in the *B.C. Richter* case, which reads as follows:

Subcontractor, in the event of any dispute or controversy with contractor or any other subcontractor over any matter whatsoever, shall not cause any delay

---

[4] 230 Cal. App. 2d at 499–500.

or cessation in or of subcontractor's work or the work of any other subcontractor or of the contractor but shall proceed under this Subcontract Agreement with the performance of the work required thereby.[5]

The court found that this clause obligated the contractor to continue working notwithstanding the dispute over payment. Specifically, the court stated:

> The quoted clause bound Richter to finish its work regardless of any dispute with Hayes-Cal Builders. In effect, the clause was an advance waiver of any right to rescind after partial performance. The net result of the clause was to make a breach of contract action the subcontractor's exclusive remedy. (*Nelson v. Spence,* 182 Cal. App. 2d 493, 497 [6 Cal. Rptr. 312]; 5A Corbin on Contracts, Section 1227; 17A C.J.S., Contracts, Section 422 (1), p. 521, fn. 62.) Having committed itself to complete performance, Richter Contracting was confined to the remedy and to the scale of damages available to one who has completed his contract notwithstanding a breach by the other party—suit on the contract and recovery by the scale of damages which the law applies in such suits. (*Canepa v. Sun Pacific, Inc.,* 126 Cal. App. 2d 706, 712–713 [272 P.2d 860].)[6]

In *Slavin v. Borenstein,*[7] Slavin, a contractor, agreed to build a custom home in Bel Air for his costs plus a fee of 10 percent of the costs of construction. Borenstein, the owner, asked for the cost-plus fee because, at the time of contracting, the project plans were still incomplete. Slavin rejected the use of a standard AIA form contract, so the parties proceeded on the basis of a memorandum from Slavin to Fink, the owner's agent.

From April through November 1983, Slavin worked on the construction and submitted periodic payment requests to Borenstein for both construction costs and the agreed 10 percent fee. The first eight requests were paid in full, request number 9 was paid in two installments, and request number 10 was paid in three installments. Beginning with request number 11, no construction fees were paid, and Borenstein fell behind on payment of construction costs. At the end of July, Slavin was owed $369,983 for construction costs and fees.

When Slavin approached Fink about payment, he was told that Borenstein had cash flow problems and that he would not pay more than $300,000 in fees regardless of the total construction costs. Fink further stated, "You can go take it or leave it. If you don't like it, then you can sue me." Being hard pressed for money, Slavin said he would accept the $300,000 maximum, but only on the condition that Borenstein bring the unpaid balance current. Slavin confirmed this in a letter dated August 10, 1982, stating the amount owed and agreeing to the $300,000 fee limit "payable in 18 equai payments." Slavin's letter concluded: "Please advise me if the above meets with your approval."

---

[5] *Id.* at 500.
[6] *Id.* at 501.
[7] 25 Cal. App. 4th 713, 30 Cal. Rptr. 2d 745 (4th Dist. 1994).

At trial, Slavin testified that he received no response to his letter; Fink stated that he had contacted Slavin and told him Borenstein would not pay in 18 equal installments, but would only pay on a "pro rata percentage as to the completion of the house." Slavin testified that he never agreed to accept the pro rata fees. Most importantly, Slavin never received any further payments on his contractor's fees, though he did receive payments for construction costs "from time to time."

At the conclusion of a trial in November 1991, the trial court awarded Slavin judgment of $875,644 for unpaid construction costs, contractor's fees, and lost profits, plus interest, less an offset of $36,430. In a statement of decision, the trial judge concluded that "Fink did not comply with the conditions of the modification agreement, and therefore, the original agreement was never effectively modified."

Borenstein appealed, contending that the August 10 memorandum was an effective modification of the parties' agreement that limited Slavin's fee to $300,000. Placing its reliance on California Civil Code § 1697 (modification of an oral contract), § 1580 (consent), and § 1584 (performance of the conditions of a proposal), the court of appeal upheld the judgment. Specifically, the court held that the condition stated in the August 10 memorandum regarding payment in 18 equal installments was a counteroffer that Fink testified was never accepted by Borenstein. Moreover, "the evidence reflects that no further payments were made on the contractor's fees, therefore, there was no acceptance by performance."

§ 4.03 CONTRACTOR'S RIGHT TO STOP WORK IN THE EVENT OF NONPAYMENT: PRIVATE WORKS OF IMPROVEMENT

The California Legislature has enacted a new statute giving contractors the right to stop work on a construction project in the event of nonpayment. The legislation adds section 3260.2 to the California Civil Code. This statute is a direct response to the case of *William R. Clarke Corp. v. Safeco Insurance Co.*,[8] which held that "pay-if-paid" clauses are unenforceable. The California Supreme Court, in that case, in a four to three decision, held that contractors could not condition their obligation to pay their subcontractors on receipt of payment from the owner. The effect of the decision was to obligate contractors to pay their subcontractors within a "reasonable" time even though the owner had not paid the contractor. The result of this decision was to require contractors to finance the project when the owner defaulted in payment to the contractor. In an attempt to limit prime contractors' responsibility for guaranteeing payments to their subcontractors even if the owner delays, refuses, or is unable to make payment, this legislation was enacted.

The 10-day stop work order remedy of Civil Code § 3260.2 is available to original contractors who are not paid all monies owed, pursuant to a written contract for a private work of improvement, within 35 days from the date payment is due under the contract, provided there is no dispute as to the satisfactory

---

[8] 15 Cal. 4th 882 (1997).

performance by the contractor. Unless all amounts then due are paid within 10 days of the date of the notice, the original contractor has the right to stop work on the project. In California, an "original contractor" is one with a direct contractual relationship with the owner.[9] The 10-day stop work order statute applies to all contracts entered into on or after January 1, 1999. Retentions withheld by a lender in accordance with the construction loan agreement are not subject to Civil Code § 3260.2.

Before an original contractor can actually stop work under Civil Code § 3260.2, there are certain conditions and specific procedures that must be satisfied. First, the original contractor must be owed monies pursuant to a written contract and 35 days must have elapsed since payment was due. Second, there must be no dispute as to satisfactory performance of the work or materials furnished by the contractor. Third, at least 5 days prior to service of the 10-day stop work order, the prime contractor must post, in a conspicuous location at the jobsite and at the jobsite's main office, if one exists, a notice that the prime contractor intends to file a 10-day stop work order. Fourth, a copy of the written notice must be served upon all subcontractors with whom the original contractor has a direct contractual relationship at the same time the notice is served upon the owner. Specific requirements to effectuate service are set forth in the statute.

Within 5 days of receipt of a 10-day stop work order, the owner must forward a copy of the notice to the construction lender by first-class mail. If the original contractor is not paid within 10 days of the date of the notice, then the original contractor has the right to stop work on the project. Once the dispute is resolved or the 10-day stop work order is canceled, the original contractor must post a notice in a conspicuous location at the job and the main office and serve a copy of the notice on all subcontractors with whom the original contractor has a direct contractual relationship informing them that the dispute has been resolved or the notice canceled.

The original contractor's right to stop work under this statute is in addition to any other rights that the original contractor may have under the law. If all of the requisite notice requirements are met, the original contractor, its surety, subcontractors, and their sureties are not liable for any delays or damages that the owner may suffer as a result of the notice and subsequent work stoppage. A prime contractor's liability to a subcontractor or material supplier resulting from the cessation of work is limited to the amount of monetary damages the subcontractor or material supplier could recover under the mechanic's lien law for goods and services provided up to the date the subcontractor stops furnishing work and/or materials, including the 10-day notice period; however, this provision does not apply to limit monetary damages for custom work, including materials that have been fabricated, manufactured, or ordered to specifications that are unique to the job. If payment is not made within 10 days from the date the notice was served, the original contractor or its surety may seek a judicial determination of liability for the amount not paid in an expedited proceeding in the superior court in the county in which the project is located.

---

[9] Cal. Civ. Code § 3095.

One of the key provisions of the foregoing statute is that which states that this right to stop work order can be given with respect to amounts due over which there is no dispute as to the satisfactory performance of the original contractor. Coupled with this is the right, set out above, of the original contractor, or his or her surety, to seek a judicial determination of liability for the amount not paid in an expedited proceeding in the superior court in the county in which the work of improvement is located. It is anticipated that there will be many occasions where the owner refuses to make a progress payment or retention payment to the original contractor on the grounds that the contractor is in breach of the contract, and therefore, the money is not due. Examples would be claims by the owner that the contractor is behind schedule, has performed defective workmanship, has failed to repair defective workmanship, has failed to perform the work in accordance with the plans and specifications, has wrongfully substituted inferior or defective materials for those called for in the plans and specifications, and other such disputes. It is anticipated that substantial litigation will arise pursuant to this new code section. The statute provides that the superior court shall have an "expedited proceeding," but it sets forth no guidelines whatsoever as to how the courts are supposed to "expedite" the proceedings. The superior courts in most counties in the state of California have very crowded court calendars, and it remains to be seen as to whether the contractors and owners can really "expedite" the resolution of a dispute regarding payment. One could foresee a circumstance where every month the owner and the contractor would be back in court fighting over whether a payment was due. One can also imagine the reluctance of a superior court judge to get involved in an expeditious fashion in these types of disputes, particularly in light of their crowded calendars and particularly in those circumstances where the amount in dispute would be rather small. For example, one can hardly imagine a superior court giving a great deal of time or thought to a dispute over a $1,000 progress payment on a small project. It is anticipated that this new legislation will catch many owners and developers by surprise.

It is also important to note that only the contractor has the right to seek resolution of the dispute in superior court. The statute does not confer that right on the owner as well. If the contractor does not instigate proceedings in superior court to get a resolution of the dispute, the project will come to a standstill. The owner would then be forced to terminate the contractor and employ a new contractor. It is hoped that the service of the notice will force the parties to meet and resolve the matter.

As noted above, the bill was an attempt to overcome the *Clarke v. Safeco* decision so that contractors would not be compelled to continue working and financing the job without payment from the owner. But if substantial disputes arise between the contractor and the owner as to whether payments are due, construction projects could be severely hampered, delayed, and disrupted by having to go to court on a periodic basis to try to determine whether an owner had properly or improperly withheld a payment. It remains to be seen, in actual practice, whether this legislative "solution" to the *Clarke v. Safeco* decision will prove to be effective as a practical matter. Only time will tell whether cash flow on private construction projects will improve as a result. The legislature has seen fit to write into every contract on private works of improvement a right to stop work for nonpayment.

§ 4.04 PROMPT PAYMENT LEGISLATION IN CALIFORNIA

For many years in the construction industry in California, the time for payment under either a prime contract or a subcontract depended upon the terms and provisions of the contract documents themselves. Furthermore, if an owner, contractor, or subcontractor failed to pay its bills within the time set forth in the contract documents, the only penalty was interest from the date the obligation became due; if the contract called for attorney's fees in the event of litigation, the party not making payment would also be obligated to pay attorney's fees. California has now enacted a comprehensive scheme governing payments from owners to prime contractors, from prime contractors to subcontractors, and from subcontractors to sub-subcontractors on construction projects.

Under California Civil Code § 3260, an owner of a private work of improvement must release retention withheld from the contractor within 45 days after the "date of completion." *Date of completion* means any of the following:

1. Date of issuance of any certificate of occupancy covering the work by the public agency issuing the building permit

2. Date of completion indicated on a valid notice of completion recorded pursuant to Civil Code § 3093

3. Date of completion as defined in Civil Code § 3086.

If there is a good-faith dispute between the owner and the contractor, the owner may withhold from the retention 150 percent of the disputed amount. The contractor must, in turn, pay its subcontractors their proportional share of the retention within 10 days after the contractor receives payment from the owner. This statute allows the general contractor to withhold from the subcontractor up to 150 percent of any disputed amount if there is a "bona fide" dispute. If either the owner or the general contractor wrongfully fails to make payment in accordance with these provisions, there is a penalty of 2 percent per month on the amount due "in lieu of any interest otherwise due." In an action to collect the amount due, the prevailing party is entitled to attorney's fees and costs. The statute declares that it is against public policy for any party to require another party to waive the provisions of the statute. The statute does not apply to retention withheld by a construction lender in accordance with a construction loan agreement. The statute also provides that within 10 days after receipt of a written notice by the owner from the original contractor, or by the original contractor from the subcontractor, stating that any work in dispute has been completed in accordance with the terms of the contract, the owner or original contractor shall advise the notifying party of the acceptance or rejection of the disputed work. Within 10 days of acceptance of the disputed work, the owner or original contractor shall release the retained portion of the retention proceeds.

For public works of improvement, California Public Contract Code § 7107 requires public agencies to release retention within 60 days after completion. This period may be increased to 90 days in the case of state agencies that choose to withhold "an amount equal to or less than 125 percent of the estimated value of the work yet to be completed." Section 7107(c) also defines *completion* to include (1) occupation and beneficial use beyond that needed for testing and start-up; (2) acceptance of the completed work by or on behalf of the public agency; (3) a cessation of labor for more than 100 continuous days due to factors beyond the contractor's control; or (4) a cessation of labor for 30 continuous days if the public agency files a notice of cessation or a notice of completion. Section 7107(d) requires the original contractor to pay its subcontractors within seven days of receipt of funds from the owner. The original contractor may withhold 150 percent of amounts in dispute. The wrongful withholding of retention subjects the original contractor or the public entity to a charge of 2 percent per month and reasonable attorney's fees to the prevailing party in any action to recover the retention.

Several statutes address prompt release of progress payments. California Business and Professions Code § 7108.5 applies to progress payments by contractors and subcontractors on both private and public works of improvement. Contractors or subcontractors must pay subcontractors within 10 days after receipt of payment from the owner or contractor; unless otherwise agreed in writing, however, California Public Contract Code §§ 10262 and 10262.5 prohibit a written waiver of the prompt payment requirements on state contracts. The statute provides for a "penalty"' of 2 percent per month on the amount not paid. The 2 percent per month penalty for failure to make prompt payment is paid to the subcontractor rather than to the Contractors State License Board. If, however, there is a bona fide dispute between the contractor and subcontractor, or a subcontractor and a sub-subcontractor, then the contractor or subcontractor may withhold 150 percent of the disputed amount. Violation of the statute is a ground for a disciplinary action under the contractor's license law.[10] If progress payments are wrongfully withheld, the prevailing party is entitled to attorney's fees and costs in an action to collect the progress payments.

California Public Contract Code § 10261.5 provides that state agencies that fail to make progress payments within 30 days after receipt of an undisputed progress payment request must pay interest to the contractor at 10 percent per annum, as provided in section 685.010 of the California Civil Procedure Code.

The Public Contract Code extends the same prompt progress payment requirements to the California State University (section 10853) and to local government agencies (section 20104.50), subject to certain conditions. Section 10853 gives the state university 39 days within which to make a progress payment after receipt of "an undisputed and properly submitted payment request," and requires that the trustees must return any disputed requests to the contractor within seven days of receipt, along with a written statement of "the reasons why the payment request is not proper." This section gives the comptroller 14 days,

---

[10] *See* discussion in § 1.14.

or the trustees 25 days, to process the request following receipt. Each day in excess of the seven-day turnaround for improper requests is deducted from the university's 39 days.

Section 20104.50 expressly preempts the field of public policy with respect to prompt payments by local governments. It allows only 30 days within which local governments must make payments of undisputed and properly submitted requests, while imposing the same requirement of a seven-day turnaround with a written explanation of improprieties in the request, again with the 30 days being reduced day-for-day by any delay in meeting the seven-day return requirement.

From the foregoing, it appears that there is now a comprehensive statutory scheme regulating progress payments and retention on both public and private works of improvement. The purpose of these statutes is, of course, to require prompt payment on construction projects, and all owners, contractors, and subcontractors must be aware of the fact that these provisions are, in effect, written into any general contract or subcontract on public or private construction projects in California.

California Government Code § 926.19 provides that any state agency that fails to pay any undisputed payment shall be liable for interest on the undisputed amount, beginning the 31st day after the agency receives notice that an undisputed payment is due. Interest accrues at a rate equal to interest accrued in the "Pooled Money Investment Account," minus 1 percent. The court must award costs and reasonable attorney's fees to the plaintiff in an action brought under section 926.19, if the state agency violates the section.

§ 4.05 DELAY

Delay by the owner, the most litigated breach of contract action in the construction industry, is discussed in detail in Chapter 7.

§ 4.06 DEFECTIVE PLANS AND SPECIFICATIONS

The courts have long held that damages for defective plans and specifications are recoverable. Generally, the measure of damages is the increased cost of labor and materials necessary to overcome the defect in the plans and specifications. For many years, attorneys practicing in the construction industry proceeded on the assumption that if you could prove that there were defective plans and specifications, the contractor could recover for the extra work and expenses incurred in overcoming the defect. The theory was that in issuing the plans, the owner impliedly warranted that the plans were free from defect and that the building could be constructed from those plans and specifications. A long line of cases, beginning with the early case of *United States v. Spearin*,[11] has stated the foregoing proposition of law.[12]

---

[11] 248 U.S. 132 (1918).
[12] *See also* Gogo v. Los Angeles Flood Control Dist., 45 Cal. App. 2d 334 (1941); Souza & McCue Constr. Co. v. Superior Court, 57 Cal. 2d 508 (1962); Wunderlich v. State, 65 Cal. 2d 777 (1967).

One California case, *Jasper Construction, Inc. v. Foothill Junior College District*,[13] held that in order to recover damages for incorrect plans and specifications, the contractor must prove an affirmative misrepresentation or concealment of material facts in the plans and specifications in order to recover. Specifically, in *Jasper*, the jury was instructed on the theory of implied warranty as follows:

> When a public entity, such as defendant Foothill District in this case, issues plans and specifications to a contractor, such as plaintiff Jasper in this case, the District impliedly warrants that the plans and specifications are free from defect, are complete and will, if followed by the contractor, result in construction of the project intended.[14]

On appeal, the defendant Junior College District contended that the jury instruction constituted prejudicial error because it permitted the jury to find Foothill Junior College District liable in the absence of any evidence of misrepresentation or intentional concealment which was contrary to the law. The appellate court agreed. The court analyzed several cases[15] and held that there could be no liability for a public entity because of extra work caused by plans and specifications that are merely "incomplete" and that recovery on that theory cannot be maintained upon a showing of a "defect" unless that defect consists of an intentional concealment or positive assertions of material facts that prove to be false. It is the opinion of the writers of this book that *Jasper* was not correctly decided and is contrary to the weight of authority in California and elsewhere. Therefore, *Jasper* should be viewed in light of its particular facts.

As set forth above, the general rule of construction law is that when the owner issues a set of plans and specifications, the owner impliedly warrants that they are free from defect and that the building can be constructed from those plans and specifications. Although the opinion of the authors that *Jasper* was not correctly decided and is contrary to the weight of authority was not directly confirmed, the later case of *Tonkin Construction Co. v. County of Humbolt*[16] seems to confirm the general rule as set forth in the *Spearin* case and the cases cited in footnote 12, and at least impliedly overrules the *Jasper* case. Tonkin was seeking money damages against Humbolt County for failure to perform certain representations and agreements implied in the parties' written contract. Tonkin was awarded damages in the trial court, and the county appealed. The facts were that in September 1983, the county and the United States Army Corps of Engineers (Corps) undertook to restore the shoreline in a portion of Humbolt Bay. The county and the Corps agreed that the county would be responsible for having a

---

[13] 91 Cal. App. 3d 1 (1979), *reh'g denied*.
[14] 91 Cal. App. 3d at 7.
[15] United States v. Spearin, 248 U.S. 132 (1918); Robert E. McKee, Inc. v. City of Atlanta, 414 F. Supp. 957 (N.D. Ga. 1976); Wunderlich v. State, 65 Cal. 2d 777 (1967); Souza & McCue Constr. Co. v. Superior Court, 57 Cal. 2d 508 (1962); Gogo v. Los Angeles Flood Control Dist., 45 Cal. App. 2d 334 (1941).
[16] 188 Cal. App. 3d 828 (1987).

seawall built and that the Corps would be responsible for having backfill work performed behind the seawall. That backfill work was necessary for the support and protection of the seawall. The county submitted the contract for construction of the seawall to various bidders, and Tonkin was the low bidder. The Corps contracted with a dredging company, Osburg, for performance of the backfill work. The seawall was to be approximately 1,200 feet in length. The county and the Corps determined that construction of the first 500 feet of the seawall would be completed by October 15, 1983, and Osburg would then perform the necessary backfill work on that portion. Thereafter, Tonkin would complete construction of the remainder of the seawall.

The contract with Tonkin provided that Tonkin was to begin work within 10 days of its receipt of the county's notice to proceed and complete the seawall within 40 days. The contract also provided that Tonkin was obligated to pay the county $500 per day in liquidated damages for each day of delay beyond the 40th day. The contract's progress schedule specified that at least 500 feet of the wall were to be constructed by October 15, 1983, and that Tonkin was to pay the county $2,000 in liquidated damages beginning October 16, 1983, until the first 500 feet of seawall were completed. The contract also provided that "[i]t will be necessary to coordinate scheduling of the various phases of work on this project with the United States Army Corps of Engineers since dredge material necessary for backfill of the wall and support fill for the anchor footing will be provided by their contractor." The contractor received the county's notice to proceed and commenced work. By October 16, 1983 Tonkin had completed the first 500 feet of seawall, as required by the contract. Despite constant communication with the Corps, Osburg did not arrive with its dredge to perform the necessary backfill work. As a result, Tonkin had to haul rock and gravel out to the seawall to protect it from damage by tidal action and storms, had to have its personnel regularly check the wall for damage, and had to keep equipment stationed at the construction site so that it could make any necessary repairs. The dredge finally arrived in December 1983 but was not in full operation until April 1984. The seawall was completed after April 1984. The county took the position that liability for the extra work done by Tonkin could not be established, because the contract for construction of the seawall contained no actual misrepresentation with respect to the availability of the dredge. The county contended that implied representations cannot be the basis of liability and that the contractual provision referring to the necessity of coordinating scheduling constituted a disclaimer as to the exact time of the dredge's arrival. In response to these contentions, the court stated, "These contentions are not borne out by law." The court, after summarizing the law in this area, stated that a contract includes not only the express terms but also any implied provisions that are indispensable to effectuate the intention of the parties. Without commenting upon or expressly overruling the *Jasper* case, the court stated the rule for establishing a public entity's liability as follows:

> The rule establishing a public entity's liability for misrepresentation was set forth by our Supreme Court in *Souza & McCue Constr. Co. v. Superior Court,* supra 57 Cal.2d at pages 510–511: (3) "A contractor of public works who,

acting reasonably, is misled by incorrect plans and specifications issued by the public authorities as the basis for bids and who, as a result, submits a bid which is lower than he would have otherwise made may recover in a contract action for extra work or expenses necessitated by the conditions being other than as represented. [Citations omitted.] This rule is mainly based on the theory that the furnishing of misleading plans and specifications by the public body constitutes a breach of an implied warranty of their correctness." (See also *Warner Constr. Corp. v. City of Los Angeles* (1970) 2 Cal.3d 285, 293–294 [84 Cal.Rptr. 444, 466 P.2d 996]; *Welch v. State of California* (1983) 139 Cal.App.3d 546, 550 [188 Cal.Rptr. 726].)[17]

The court stated that an implied term of the contract between Tonkin and the county was that, once the notice to proceed was issued, the dredge would be available for work on the project. The court stated that the intention of the parties was that completion of the seawall would occur within 40 days of the issuance of the notice to proceed. That interpretation was supported by the fact that the contract provided that the work would start within 10 days and that Tonkin would be liable to the county for liquidated damages. The court found that the intention of prompt completion of the seawall could not be effectuated absent an implied term that the county would ensure the dredge's availability for work on the project. The contract itself provided that the backfill work to be performed by the dredge was necessary to the completion of the seawall. Tonkin, acting as a reasonable public works contractor, was misled by that incorrect implied representation in submitting its bid. Tonkin justifiably relied upon that implied representation in determining the cost of constructing the seawall. As a result, Tonkin did not include in its bid the cost of maintaining the seawall for an indefinite period of time while awaiting the arrival of the dredge. The county impliedly warranted the correctness of that implied representation and was therefore liable for the cost of extra work necessitated by the dredge's failure to arrive. Thus, the appellate court affirmed the judgment of the trial court.

The authors believe that the *Tonkin* case correctly states the law in this area. Even though it did not specifically comment upon or discuss the *Jasper* case, it impliedly overruled that case.

The United States Supreme Court, in *Hercules, Inc. v. United States*,[18] confirmed three lower federal court decisions (two by the Court of Claims, which were consolidated on appeal to the Federal Circuit Court of Appeals) and sustained summary judgments dismissing lawsuits brought against the United States by the manufacturers of the chemical defoliant known as "Agent Orange." The manufacturers had sought to recover their litigation expenses and settlement costs incurred in Agent Orange litigation brought against them by Vietnam War veterans and their families, based on two legal theories: contractual indemnification and warranty of specifications provided by the government. The Supreme Court held that neither the government's implied contractual warranty of specifications

---

[17] 188 Cal. App. 3d at 832.
[18] 516 U.S. 417, 116 S. Ct. 981 (1996).

nor *United States v. Spearin*,[19] the seminal case recognizing a cause of action for breach of such a warranty, extends so far as to render the United States responsible for costs incurred in defending and settling the veterans' tort claims. Under the *Spearin* doctrine, when the government provides specifications directing how a contract is to be performed, as it did in this case, the government only warrants that the contractor will be able to perform the contract satisfactorily if it follows the specifications. Such a warranty does not extend beyond the contractor's performance to third-party claims against the contractor. Thus, the *Spearin* doctrine does not extend to a contractor's post-performance third-party costs, as a matter of law.

In the case of *Howard Contracting, Inc. v. G.A. MacDonald Construction Co.*,[20] the contractor claimed that the owner had failed to inform the contractor of restrictions on access to the project and restrictions on the time for performance of the work, both of which caused delay and increased costs to the contractor. The court reconfirmed the doctrine of "breach of the implied warranty of correctness of plans and specifications," stating the following:

> Under general principles of contract and tort law, a party who conceals or fails to disclose material information to another is liable for fraud. In the public construction contract context, however, the conduct of a public agency which would otherwise amount to a tortious misrepresentation is treated as a breach of contract. The underlying theory is that providing misleading plans and specifications constitutes a breach of the implied warranty of correctness. (*Souza & McCue Construction Co. v. Superior Court* (1962) 57 Cal. 2d 508).

In *City of Salinas v. Souza & McCue*,[21] the Supreme Court stated the applicable law when a public agency fails to disclose information relevant to the performance of a contract:

> It is the general rule that by failing to impart its knowledge of difficulties to be encountered on a project, the owner will be liable for misrepresentation if the contractor is unable to perform according to the contract provisions.

§ 4.07   HOW COURTS INTERPRET PLANS AND SPECIFICATIONS

[A]   Basic Rule: Implied Warranty of Accuracy of Plans and Specifications

As can be expected, many times the fundamental issue before the trier of fact (judge, jury, or arbitrator) is whether the plans and specifications do, in fact, contain errors and omissions. Typically, the issue arises when the contractor encounters

---

[19] 248 U.S. 132, 39 S. Ct. 59 (1918).
[20] 71 Cal. App. 4th 38 (1999).
[21] 66 Cal. 2d 217 (1967).

what it perceives to be an error or omission in the plans and specifications and submits a request for information (RFI) to the architect to resolve the problem. The architect typically responds either by denying that there is any problem or by issuing a "clarification" for the contractor's use. The contractor then contends that there has been a change in scope, for which it is entitled to additional costs and/or time; the owner and architect contend that there has been no change in scope, but instead merely a clarification of the design intent. It then becomes the function of the court to determine whether there is in fact an error or omission in the plans and specifications. As stated above, the courts have held that when the evidence establishes that there is an error or omission that impacts the contractor's cost or time of performance, the contractor is entitled to additional time, compensation, or both.

The courts thus must address the issue of interpretation of plans and specifications. First of all, it must be understood that the plans and specifications on a construction project serve as a "road map"[22] pursuant to which the contractor builds the project. The plans and specifications advise the contractor about the conditions that will be encountered in performing the work and further prescribe the design and construction methods or techniques that the contractor should follow to build the building. Thus, if the plans and specifications do not perform those functions, the implied warranty of accuracy and suitability has been breached, so the contractor may seek extra compensation. This is the basic rule established by the cases cited in footnote 15 of § 4.06. Specifically, the California Supreme Court, in *Souza & McCue Construction,* stated the rule as follows:

> A contractor of public works who, acting reasonably, is misled by incorrect plans and specifications issued by the public authorities as the basis for bids and who, as a result, submits a bid which is lower than he would have otherwise made may recover in a contract action for extra work or expenses necessitated by the conditions being other than as represented. [Citations omitted.] This rule is mainly based on the theory that the furnishing of misleading plans and specifications by the public body constitutes a breach of an implied warranty of their correctness. [Citations omitted].[23]

In numerous cases, contractors have recovered additional compensation as a result of a breach of the implied warranty. In *Warner Construction Corp. v. City of Los Angeles,*[24] the owner (City of Los Angeles) issued a set of plans and specifications that included boring logs of two test holes that had been performed by the city on the project. Those boring logs showed the soil to be "coarse sand with clay binder" when, in fact, the soil was "coarse sand with minute binder," which increased the contractor's cost of performance. The court confirmed the contractor's right to recover additional expense as a result of that error in the plans

---

[22] J.L. Simmons Co. v. United States, 412 F.2d 1360, 1362 (Ct. Cl. 1969).
[23] Souza & McCue Constr. Co. v. Superior Court, 57 Cal. 2d 508, 510–511 (1962).
[24] 2 Cal. 2d 285 (1970).

and specifications. Similarly, in *Welch v. State*,[25] the state of California had certain information in its files concerning tidal conditions in the area of a bridge that was to be repaired by the contractor, which the state failed to disclose in its bid documents. That made the information regarding tides that was disclosed in the plans and specifications misleading. Because the tides were different from what the contractor expected, its cost of performance increased. The court held that the contractor was entitled to recover the additional costs of construction.[26]

Subcontractors have likewise been allowed to recover extra costs incurred because of defective plans and specifications. In *Keller Construction Corp. v. George W. McCoy & Co.*,[27] a subcontractor installed a sewer line in accordance with the plans and specifications. The sewer line broke, and the subcontractor was directed to repair the breaks. The subcontractor sued the general contractor and was allowed recovery. The Louisiana court also held that the owner was obligated to indemnify the general contractor for the amounts the general contractor had to pay the subcontractor. In *Macomber v. State*,[28] a subcontractor built a staircase in accordance with the plans and specifications, but when it went to install the staircase, the unit did not fit. The contractor was allowed to recover from the owner the additional costs incurred by the subcontractor in repairing the staircase to make it fit. Recovery was permitted in this case despite the fact that the architect had denied the claim and no written change order had been issued for the repairs.

[B] Contractor Not Responsible for Defective Plans and Specifications

A corollary to the rule that the contractor is entitled to recover additional compensation for errors and omissions in the plans and specifications is the rule that if the contractor constructs a project in accordance with the plans and specifications and the finished product is defective, the contractor has no liability.[29] A specific example of this concept is *Sunbeam Construction Co. v. Fisci*,[30] a case in which the framing subcontractor and the roofing contractor installed a roof according to the owner's plans and specifications, which called for a flat roof. After the roof had been framed and installed, puddles formed, and the roof leaked. The owner sued the roofing contractor and the framer for its damages. The roofing contractor and the framer prevailed in that action.

It should be noted, however, that even though the contractor installs work in accordance with the plans and specifications, the contractor may still be held liable for any damages caused if it violates industry standards in installing the work. This point is illustrated in *C.C. California Plaza Associates v. Paller &

---

[25] 139 Cal. App. 3d 546 (1983).

[26] *See* the discussion of Howard Contracting, Inc. v. G.A. MacDonald Construction Co. in § 4.06 above.

[27] 119 So. 2d 450 (La. 1960).

[28] 250 Cal. App. 2d 391 (1967).

[29] *See* Kurland v. United Pacific Ins. Co., 251 Cal. App. 2d 112 (1967); Patrick J. Ruane, Inc. v. Parker, 185 Cal. App. 2d 488 (1960); and the *Spearin* case.

[30] 2 Cal. App. 3d 181 (1969).

Goldstein,[31] in which Cal Plaza employed a prime contractor to construct a 10-story office building in 1983. When water intrusion occurred, the alleged source was a parapet cap assembly that had been installed by Paller & Goldstein (P&G). Cal Plaza sued the prime contractor, which cross-complained against P&G for contractual and equitable indemnity.

On the eve of trial, all parties except P&G settled; as part of the settlement, the prime contractor assigned its indemnity rights against P&G to Cal Plaza. The prime contractor had admitted, in requests for admissions, that the installation of the parapet cap flashing at the project was done in accordance with the architect's specifications; however, an expert witness testified that P&G had violated industry standards by neglecting to seal the bolt hole penetrations left by the handrail subcontractor. As a result, there was conflicting evidence as to the liability of P&G. After two days of testimony, the trial court granted a nonsuit and dismissed the action against P&G.

The court of appeal reversed, noting that even though P&G had installed the parapet cap per the plans and specifications, P&G could still be held liable if it failed to follow industry standards by not sealing the bolt hole penetrations. Thus, the granting of the nonsuit was improper and the case was remanded for trial.

In *City Street Improvement Co. v. City of Marysville*,[32] the contractor installed a sewer that was not in accordance with the plans and specifications. However, when the sewer was installed, the city's representatives inspected and approved the contractor's installation. Later, when the sewer failed, the court held that the contractor had no liability. In *Marin Municipal Water District v. Peninsula Paving Co.*,[33] a contractor constructed a highway for the state of California following the plans and specifications and performing the work under the direction and supervision of the State Engineer. The construction work caused damage to a water pipeline on an adjacent property. The court held that the contractor had no liability to the adjacent property owner.

[C] Enforceability of Exculpatory Clauses

Many times the owner will attempt to shift the risk of loss to the contractor by including disclaimers or exculpatory clauses in the contract documents. General disclaimers (for example, clauses requiring the contractor to "examine carefully the site of the work" or stating that "it is mutually agreed that the submission of a proposal shall be considered prima facie evidence that the bidder has made such an examination") will usually not be enforced by the court.[34] On the other hand, if the owner inserts a very specific exculpatory clause that clearly and definitely shifts the risk to the contractor, it will be upheld. In *Wunderlich v.*

---

[31] 51 Cal. App. 4th 1042 (1996).
[32] 155 Cal. 419 (1909).
[33] 34 Cal. App. 2d 647 (1939).
[34] *See, e.g.,* City of Salinas v. Souza & McCue Constr. Co., 66 Cal. 2d 217 (1967), and the cases cited therein. *See also* Robert E. McKee, Inc. v. City of Atlanta, 414 F. Supp. 957 (N.D. Ga. 1979).

State,³⁵ the California Supreme Court refused to allow a contractor to recover for breach of warranty when there was no positive misrepresentation or concealment of facts in the plans and specifications and the state had specifically disclaimed any guarantee of the accuracy of the test results that had been submitted to the bidders. For a contrary case, see *E.H. Morrill Co. v. State*.³⁶

In the case of *Howard Contracting, Inc. v. G.A. MacDonald Construction Co.*,³⁷ the owner contended that the contractor was not entitled to recover for changed conditions by reason of certain exculpatory language in the contract documents. The owner contended that by virtue of the contract documents, the contractor was required to comply with any permits issued on the project. The court, citing the case of *City of Salinas v. Souza & McCue*,³⁸ stated that when a public agency fails to disclose information relevant to the performance of the contract and fails to impart its knowledge of difficulties that the contractor may be expected to encounter on the project, the owner will be liable for misrepresentation if the contractor is unable to perform according to the contract provisions. Following the *City of Salinas* case, the court noted that a contract provision specifically requiring bidders to carefully examine the construction site and jobsite conditions and providing that submission of bid constituted prima facie evidence of the bidders' site inspection cannot excuse a governmental agency from its active concealment of conditions. The court held that the public owner of the project was not absolved from liability by reason of general clauses requiring the contractor to examine building permits and the site of the work.

[D] Rule of Contra Proferentem: Ambiguities in Plans and Specifications Will Be Construed Against the Drafter

One of the rules that comes into play in this area is the rule of *contra proferentem*. The owner, through its architect, is the drafter of the plans and specifications. If there is an ambiguity in the plans and specifications, it will be construed against the owner. This rule is based on the familiar principle of contract interpretation that a court will generally construe any ambiguity against an owner as the drafter of the documents. Even if both the owner's and the contractor's interpretations are reasonable, the contractor's interpretation will prevail. This was clearly stated in *Bennett v. United States*,³⁹ wherein the court said:

> To prevail . . . it is not essential that [the contractor] demonstrate his position to be the only justifiable or reasonable one. A specification susceptible to more than one interpretation, each interpretation found to be consistent with the contract's language and the parties' objectively ascertainable intentions, becomes convincing proof of an ambiguity; the burden of that ambiguity falls solely upon the party who drew the specification.⁴⁰

---

³⁵ 65 Cal. 2d 777 (1967).
³⁶ 65 Cal. 2d 787 (1967).
³⁷ 71 Cal. App. 4th 38 (1999).
³⁸ 66 Cal. 2d 217 (1967).
³⁹ 371 F.2d 859 (Ct. Cl. 1967).
⁴⁰ 371 F.2d at 861.

On the contrary, if the ambiguity in the plans and specifications is patent, then the rule of *contra proferentem* does not apply, and the contractor has a duty to inquire of the owner and architect as to the true meaning of the contract before submitting a bid. This prevents the contractor from taking advantage of the owner, protects other bidders by ensuring that all bidders bid on the same specifications, and helps avoid litigation over ambiguities.[41] Whether a set of plans and specifications is patently ambiguous is a question of contract interpretation, to be decided by the court on a case-by-case basis. First, the court asks whether the ambiguity was patent. This is not a simple yes-or-no proposition, but rather involves placing the contractual language along a spectrum to determine whether the ambiguity is so glaring as to create a duty of inquiry on the contractor's part. If the court decides that the ambiguity was not so patent, then the court reaches the second issue, which is whether the contractor's interpretation was reasonable. A case that examines this set of rules in detail is *Salem Engineering & Construction Corp. v. United States*,[42] in which a contractor built an addition to a federal government office building. The contract provided that the contractor was to remove the existing built-up roof and that the existing structures were to remain in place. The drawings provided "shading" to indicate that roof work was required only on the new addition. The government's clerk of the works demanded that the contractor replace the roof on the existing building and tie it into the roof on the new addition. The contractor did so under protest and filed a claim for the additional work. The court stated that it first had to determine whether the contract specifications and drawings were ambiguous—that is, whether they sustained the interpretations advanced by both parties (citing *Max Drill, Inc. v. United States*[43]). The court explained that, when construing contract documents, the court must assume the position of a reasonable and prudent construction contractor (citing *Norcoast Constructors, Inc. v. United States*[44] and *Firestone Tire & Rubber Co. v. United States*[45]). The court must then interpret the contract language as would a reasonably intelligent contractor who is acquainted with the circumstances surrounding the contract (under *Hol-Gar Manufacturing Corp. v. United States*[46]).

The contractor contended that, because the primary purpose of the contract was to construct the new addition, it was reasonable to interpret the contract to require removal and replacement of the existing roof only to the extent necessary to join the existing roof with the new roof. The government contended that the contract clearly required the contractor to remove and replace the entire roof on the existing building. The government stated that to adopt the contractor's interpretation would give no meaning to a provision in the contract that stated that the contractor was to "remove and replace the existing built-up roof." The

---

[41] *See* Newsom v. United States, 676 F.2d 647 (Ct. Cl. 1982).
[42] 2 Cl. Ct. 808 (1983).
[43] 427 F.2d 1233 (Ct. Cl. 1970).
[44] 488 F.2d 1400 (Ct. Cl. 1971).
[45] 444 F.2d 547 (Ct. Cl. 1971).
[46] 351 F.2d 972 (Ct. Cl. 1965).

court ultimately concluded that the contract could reasonably be interpreted either way and therefore was ambiguous, so it ruled in favor of the contractor. The court stated that although the contract somewhat shifted the burden of resolving ambiguities to the contractor, by requiring that the contractor seek clarifications of ambiguities (citing *W.P.C. Enterprises, Inc. v. United States*)[47] the government could not expect the contractor to notify it of every possible ambiguity or potential interpretation of the contract, as to do so would make the contractor an insurer against all government mistakes (citing *Mountain Home Contractors v. United States*[48]). The court held that the contractor assumes only the burden of interpreting the contract reasonably (citing *Corbetta Construction Co. v. United States*[49]) and need notify the government only of major discrepancies in the plans and specifications (citing *Blount Bros. Construction Co. v. United States*[50]). The court stated that, as the drafter of the contract documents, the government was responsible for ensuring that the words used in the contract reasonably conveyed the intended meaning (citing *John M. McShain, Inc. v. United States*[51]). If the government had wanted the contractor to remove and replace the entire roof, it could have done so by an explicit provision to that effect (as per *L. Rosenman Corp. v. United States*[52] and *Schweigert Tire, Inc. v. United States*[53]). As a result of the foregoing findings, the court concluded that the contractor's interpretation was not based on any major discrepancy or patent ambiguity, and therefore the contractor had no duty to seek clarification. Thus, it was entitled to extra compensation for removing and replacing the entire existing roof.

[E] Obligation of Contractor to Proceed with the Work

A question that often arises is whether the contractor is obligated to proceed if the owner will not grant a change order for additional time and money. There is no easy answer to this question. The contractor has two options: it may continue performance under protest, while reserving the right to claim additional compensation, delay damages, and extensions of time or acceleration damages; or it may cease performance, contending that the refusal to grant a change order is a breach of contract excusing the contractor's further performance. Each case must be analyzed on its own facts. In some cases, if the problem created by the errors and omissions in the plans and specifications is a "material" change, then the contractor can cease performance. On the other hand, if the problem is "minor" or "trivial," the contractor must continue to perform. If the contract contains a clause requiring the contractor to continue to perform disputed work, the contractor must do so. *See* the *B.C. Richter* case in § 4.02.

---

[47] 323 F.2d 874 (Ct. Cl. 1963).
[48] 425 F.2d 1260 (Ct. Cl. 1970).
[49] 461 F.2d 1330 (Ct. Cl. 1972).
[50] 346 F.2d 962 (Ct. Cl. 1965).
[51] 462 F.2d 489 (Ct. Cl. 1972).
[52] 390 F.2d 711 (Ct. Cl. 1968).
[53] 388 F.2d 697 (Ct. Cl. 1967).

A case that sheds some light and guidance in this area is *Coleman Engineering Co. v. North American Aviation.*[54] In *Coleman,* the contractor submitted a bid to engineer and manufacture certain missile trailers for the government. With its bid, the contractor submitted its equipment specifications and its interpretation of the owner's specifications. The owner gave the contractor "go ahead" telegrams to start the design and engineering and delivered a series of purchase orders totaling $527,632. The purchase orders provided that if the owner made changes, the contract price would be equitably adjusted in a manner satisfactory to both parties. After a dispute arose concerning the interpretation of the specifications, the contractor requested additional time and money. The owner issued a specification that the contractor contended would cause a minimum increase of $257,000 in engineering, design, and construction costs. The contractor requested an increase in the contract price. The owner rejected the contractor's request, contending that the contractor had underbid the job and was looking to the owner to bail it out. The owner then offered a mere 10 percent increase in the contract price. The parties negotiated, but the negotiations broke down, and the owner claimed that the contractor was in breach of contract. The owner terminated the contractor and hired another contractor to complete the job, at an increase in price. The contractor sought the reasonable value of what it had performed to the date of termination of the contract, and the owner sought recovery of the excess costs that it had paid to the other contractor to complete the job. The court held that the change order clause left to future determination any adjustment in the contract price due to changes, and that when the parties found themselves unable to agree on a price, the contractor was justified in refusing to perform. Specifically, the court stated:

> Undoubtedly, if the subsequent changes are minor or of not great magnitude the contractor must perform and obtain a subsequent judicial determination as to the price of the changes. However, where the changes are of great magnitude in relation to the entire contract, the contractor must negotiate in good faith to settle the price [citation omitted], and where he had done so, he is not required to continue performance in the absence of an agreement as to the price.[55]

Bear in mind that the foregoing is subject to the caveat mentioned in § 4.02 relating to *B.C. Richter Contracting,* in which the court held that the contractor had an obligation to continue to perform by virtue of a clause in the contract documents that required the contractor to perform notwithstanding disputes.

§ 4.08 DELAY IN START OF PERFORMANCE

The courts have held that a contractor can recover for damages when the construction project is unreasonably delayed in the start of performance. In *Stark*

---

[54] 65 Cal. 2d 396 (1966).
[55] 65 Cal. 2d at 406.

v. Shaw,[56] the roofing subcontractor submitted a bid to roof some houses on March 12, 1954. The bid stated on its face that it was based upon present cost of labor and material and, in the event that there were increases, the prices were subject to revision. The bid was accepted on the same day it was tendered, and a contract was signed. The contract had no date of commencement of work. The prime contractor orally advised the subcontractor that the job would start on April 1, 1954. The starting date was orally changed first to April 15, 1954, then to May 17, 1954. On May 17, 1954, the job had not started. The roofing subcontractor sued for breach of contract, seeking his lost profits. The prime contractor cross-complained for breach of contract. The trial court held that May 17, 1954, was a reasonable time for starting construction and held in favor of the roofing subcontractor. The prime contractor appealed. On appeal, the judgment was affirmed, and the court stated that if no time for performance is stated, then a reasonable time will be inferred.

§ 4.09 MATERIAL CHANGE OF CONTRACT

In the leading case *Daugherty Co. v. Kimberly Clark Corp.*,[57] in which the project in question was a paper mill, the plaintiff was a subcontractor who entered into an original subcontract for $2.9 million and claimed extras of $1.6 million for a total job claim of $4.5 million. The subcontractor was in fact paid $775,000 over the original contract price. The authorized change orders amounted to $875,000. The subcontractor contended that because the original contract had been so vastly changed, there had been a material change in the scope, resulting in an abandonment, thereby allowing the subcontractor to seek recovery on a reasonable value theory.

Summary judgment was granted in favor of the owner. On appeal, that judgment was reversed. The court held that there was a triable issue of fact as to whether the vast number of change orders resulted in an abandonment of the original contract. The appellate court recognized the principle of law that when there is a material change in scope, it may result in an abandonment of the contract, thereby allowing the contractor to recover on a reasonable value theory.

In the case of C. *Norman Peterson Co. v. Container Corp. of America*,[58] the plaintiff/contractor contracted with the defendant/owner for the performance of a mill modernization project. The contractor sued the owner, claiming that it was entitled to recover its total costs expended in completing the project because of numerous errors in the plans and specifications and changes in the construction plans, which resulted in extra work for the contractor. The owner contended that both the contractor and the owner were aware at the outset of the project that revisions would have to be made to the plans and specifications, that there was no

---

[56] 155 Cal. App. 2d 171 (1957).
[57] 14 Cal. App. 3d 151 (1971).
[58] 172 Cal. App. 3d 628 (1985).

material change in the scope of the work to be performed, and that the contract set forth detailed procedures for compensation for performing extra work.

The trial court found that the owner had breached and abandoned the contract when it spent an unreasonable amount of time redesigning the facility after the contractor had started the project, that the project was poorly engineered, that the owner had made an excessive number of changes beyond the contemplation of the parties at the time the contract was executed, and that, therefore, the contractor was entitled to recover on an abandonment theory and could use total cost of the work as the measure of damages. Specifically, the contractor was allowed to recover its total costs expended plus a markup for overhead and profit, less the payments actually received from the owner.

On appeal, the court held that the change orders and extra work imposed by the owner on the contractor were of such a magnitude as to materially change the scope of the work originally contemplated under the contract and that the facts supported the trial court's holding that the parties had abandoned the original contract and that the plaintiff was entitled to the reasonable value of the work it performed on a quantum meruit basis, without being limited by the original contract amount. The court held that, in the context of a construction contract, the imposition of an excessive number of changes may support a finding of abandonment even if the contract is, in fact, completed.

The court further held that the trial court had properly made a finding that the owner had both breached and abandoned the contract because there was no double recovery by the contractor. The appellate court affirmed that the trial court had properly based its judgment on the contractor's total costs because the contractor was unable to keep accurate cost records regarding the numerous, complex changes it was asked to perform by the owner, many of which were based upon oral instructions from the owner. Some pertinent quotations from the opinion are appropriate.

> In the specific context of construction contracts, however, it has been held that when an owner imposes upon the contractor an excessive number of changes such that it can fairly be said that the scope of the work under the original contract has been altered, an abandonment of contract properly may be found. (See *Daugherty Co. v. Kimberly-Clark Corp.*, 14 Cal. App. 3d (1971), at p. 156; *Opdyke & Butler v. Silver*, 111 Cal. App. 2d 912 (1952), at 916–919.) In these cases, the contractor, with the full approval and expectation of the owner, may complete the project. (See *Daugherty, supra*, at pp. 156–157.) Although the contract may be abandoned, the work is not. Under this line of reasoning, the trial court was well justified in determining that, by their course of conduct, the parties had abandoned the terms of the written contract while proceeding to complete the mill restoration project. Abandonment of a contract may be implied from the acts of the parties. . . . Abandonment of the contract can occur in instances where the scope of the work when undertaken greatly exceeds that called for under the contract. In *Opdyke* the written contract provided that all change orders should be approved by the architect, should be in writing and the contract should be adjusted accordingly before the work was done. Yet, because the parties consistently

ignored this requirement in material part, the court allowed a recovery on quantum meruit.[59]

§ 4.10 ACCELERATION

The elements of an acceleration claim are

1. A completion date set forth in the original contract
2. The occurrence of an excusable delay (strikes, inclement weather, and so on)
3. A timely request for an extension by the contractor
4. An improper denial of that request; and
5. Damages to the contractor resulting from the denial.

The damages recoverable in an acceleration claim are basically the increased costs of supervision and the lack of efficiency in having to do a job in a shorter period of time than is reasonably justified. The theory of acceleration was acknowledged in *Huber, Hunt & Nichols, Inc. v. Moore*.[60] The concept has been clearly recognized in many construction cases outside California.[61] The concept of acceleration is discussed in more detail in Chapter 7.

§ 4.11 RECOVERY OF CONSEQUENTIAL DAMAGES

It is not unusual for contractors and subcontractors to seek recovery of what are commonly referred to as "consequential damages," over and above the direct costs incurred because of a breach of a construction contract by the owner. The types of consequential damages that contractors sometimes seek are

1. Loss of profits
2. Loss of business
3. Loss of an advantageous competitive position in the industry
4. Impairment of bonding capacity.

The California Supreme Court has held that these types of damages are indeed recoverable. In *Warner Construction Corp. v. City of Los Angeles*,[62] the plaintiff, Warner Construction, had performed a project for the city of Los Angeles. The contractor brought an action against the city for the balance due on the contract and also damages for delay and breach of contract. The jury returned a verdict for the contractor in the sum of $150,000. The supreme court reversed

---

[59] 172 Cal. App. 3d at 640.
[60] 67 Cal. App. 3d 278 (1977).
[61] *See* Contracting & Materials Co. v. City of Chicago, 314 N.E.2d 598 (Ill. App. Ct. 1974).
[62] 2 Cal. 3d 285 (1970).

the judgment of the trial court insofar as it awarded damages in excess of the amount due on the contract, the amount claimed for delay, and the amount of additional costs of construction attributable to the city's misrepresentation of conditions at the jobsite. The supreme court took the view that the plaintiff had presented no evidence to support the balance of the award (the contractor had claimed compensation for further damages in the nature of lost profits due to impairment of its capital by the increased costs of construction), and the cause was remanded for a new trial on the issue of damages. The supreme court noted that the contractor undoubtedly should recover the additional costs of construction attributable to the city's misrepresentation as to site conditions, plus a reasonable profit on that additional work. At the trial, the contractor testified that it was compelled to bear the additional costs of construction out of its own resources and that loss of capital forced a curtailment in the contractor's construction operations and research, and also led to a reduction of bonding capacity, as well as destruction of the plaintiff's former advantageous competitive position in the industry. In regard to that claim, the court stated the following:

> For an established firm such as Warner Construction Corporation, an award for lost profits could not be criticized as speculative. (See *Lucky Auto Supply v. Turner* (1966) 244 Cal.App.2d 872, 882 [53 Cal.Rptr. 628].) Plaintiff, however, presented no evidence to show the loss of profits resulting from this alleged impairment of capital. Loss of business, restriction of research, reduction of bonding capacity, and destruction of a former advantageous competitive position comprise imponderable factors which may affect different companies to differing extents and amounts. Measurement of the damages required proof of the effect of these factors upon the profits of plaintiff. (*Dallman Co. v. Southern Heater Co.* (1968) 262 Cal. App. 2d 582, 591–592 [68 Cal.Rptr. 873].)[63]

Thus, the supreme court expressly acknowledged that for an established construction company, an award of lost profits could not be characterized as speculative. It reversed the verdict of the jury and the judgment of the trial court because it felt that this plaintiff had not submitted competent evidence to support those types of damages. However, the supreme court did acknowledge that loss of business, restriction of research, reduction of bonding capacity, and destruction of a former advantageous competitive position are damages that could be recovered if properly proved.

The Wisconsin Court of Appeals held that a subcontractor whose performance schedule was accelerated could recover lost profit on other projects from which its resources were diverted in order to meet the accelerated schedule. In *Downey, Inc. v. Bradley Center Corp.*,[64] an owner awarded a contract to Huber, Hunt & Nichols, Inc. (HH&N) to construct a sports arena. HH&N subcontracted the mechanical work to Downey, Inc. During the course of the project, design changes necessitated numerous stop work orders issued by HH&N to Downey.

---

[63] 2 Cal. 3d at 301.
[64] 524 N.W.2d 915 (Wis. Ct. App. 1994).

Because certain athletic events were scheduled for the facility, HH&N was under pressure from the owner to meet the original completion date. HH&N therefore refused to extend Downey's performance period and insisted that Downey do whatever was necessary to stay on schedule. Downey pulled workers and additional resources off its other current projects in order to meet the accelerated schedule. Downey also worked its crews overtime and performed work out of sequence to meet the schedule. HH&N compensated Downey for the overtime labor and other direct costs of the acceleration but refused to allow a claim by Downey for the profit that was lost on other contracts as a result of reallocating its resources. HH&N contended that lost profit on separate contracts was too speculative and therefore not a foreseeable consequence of accelerating the work on the project in question. HH&N also defended the claim on the basis that Downey had failed to comply with notice provisions contained in the subcontract.

The Wisconsin Court of Appeals held that Downey's lost profits on separate contracts was a foreseeable result of the acceleration, pointing to evidence that HH&N had told Downey to do whatever was necessary to stay on schedule despite the stop work orders. The court further noted that HH&N was aware that Downey was pulling crews from its other projects and made no objection. The resulting lost profit was therefore a recoverable cost of the acceleration. Furthermore, the court noted that Downey had informed HH&N that it would send an additional bill for the "impact" that completing the project had caused on Downey's other projects and that HH&N made no objection to that notice at the time. With regard to the notice requirement, the court noted that on 21 separate occasions, Downey had made oral or written complaints regarding suspensions, delays, and out-of-sequence work. Therefore, this was sufficient to comply with the notice provision of the subcontract.

§ 4.12 RECOVERY OF DAMAGES FOR REDUCTION OF BONDING CAPACITY

Although the contractor lost the case, a 1994 appellate decision clearly acknowledged that a contractor can recover consequential damages consisting of lost profits based upon impairment of its bonding capacity. The court, in *S.C. Anderson, Inc. v. Bank of America National Trust & Savings Ass'n*,[65] carefully analyzed the evidence presented by the contractor and concluded that it was insufficient; however, the court did clearly recognize that such damages were recoverable.

First of all, the court acknowledged the fact that a contractor can recover lost profits based upon impairment of its bonding capacity: "In addition to general and punitive damages, Anderson claimed consequential damages, consisting of lost profits, based upon an alleged impairment of its bonding capacity. The California Supreme Court has recognized this ground of recovery in construction cases [*Warner Construction Corp. v. City of Los Angeles*, 2 Cal. 3d 285, 301 (1970)]."[66]

---

[65] 24 Cal. App. 4th 529 (1994).
[66] 24 Cal. App. 4th at 534.

The contractor in this case, Anderson, contended that, because it was not timely paid for its work on two major construction projects it was performing, it was deprived of the opportunity to bid on—and therefore be awarded—contracts for several other construction projects. On two contracts being performed for the same owner, the amount unpaid totaled $248,004.

At trial, the defendant construction lender (the bank) filed a motion in limine to limit the evidence that the contractor could introduce relating to the type of damages being claimed. The trial court ruled that the contractor could present evidence regarding lost profits as to individual projects, provided it established (1) that it had actually prepared a bid for a specific project; and (2) that it was precluded from submitting the bid because its bonding company refused to issue a bid bond.

Anderson had worked on the jobs for which it was not paid in 1985 and early 1986. In May 1986, its bonding capacity was $5 million for a single project, with a $10 million aggregate exposure. The bonding company (Travelers Insurance Company) calculated a contractor's bonding capacity by assessing, among other things, the contractor's working capital. By mid-1986, financial statements disclosed that the lack of payments on the two projects financed by the bank had affected Anderson's working capital such that the bonding company reduced Anderson's aggregate bonding capacity from $10 million to $5 million.

In February 1987, Travelers approved a bid bond to allow Anderson to bid on an elementary school project. Anderson was the low bidder, but all bids were rejected, and the project was scheduled for rebid in March 1987. By the time of the rebid, Travelers refused to provide another bid bond because the project would take Anderson over its newly lowered $5 million bonding capacity. Without the bid bond, Anderson was unable to submit a rebid.

In 1986 and 1987, Anderson had used a 5 to 8 percent profit and overhead figure in its bids. Anderson's rebid on the school project was $2.98 million, which included a 5 percent profit margin of $140,588. The low bid on the job was $3,027,036. Thus, Anderson contended that, but for its inability to obtain the bid bond, it would have been the low bidder for the project and would have realized a profit of $140,588.

After the foregoing evidence had been presented, the bank moved for nonsuit on the ground that the contractor had failed to present facts sufficient to support an award of lost profits. Because the contractor did not present evidence of its past profitability, it failed to establish that it would have earned the 5 percent profit margin projected in its rebid. In addition, the bank argued that because the 5 percent figure represented gross profits, the jury had no basis upon which to calculate lost net profits, which is the proper measure of damages.

Anderson responded to the motion by offering to produce additional evidence to show the company's profitability from 1983 through 1987. Anderson's attorney proposed to recall the bookkeeper to testify as to the "gross margins" during those years, defined as the difference between gross revenue on a job and the job costs. Anderson argued that a contractor incurs certain fixed costs every year, and that once those costs are covered by earlier projects, the profits earned on any later projects are net profits. The bookkeeper had prepared a schedule showing Anderson's gross revenues, gross costs, and gross margins for the five years at issue. Arguing that he had concluded that, based upon the judge's earlier ruling on the bank's in limine motion, he was precluded from introducing the gross margin evidence, Anderson's attorney moved to reopen the case to present the proffered evidence.

The bank opposed the motion to reopen, arguing that there was nothing in the court's in limine ruling to prevent the contractor from presenting the gross margin evidence, and that the bank would be prejudiced because its counsel had revealed a lot of his strategy in his arguments, and a potential witness had remained in the courtroom during the argument on the nonsuit motion. The trial court granted the nonsuit and denied the contractor's request to reopen its case.

The court of appeal, first of all, noted that lost anticipated profits may be recovered only if the evidence shows, with reasonable certainty, both their occurrence and their extent. It is enough to demonstrate a reasonable probability that the profits would have been earned but for the defendant's conduct. The contractor has the burden to produce the best evidence available, under the circumstances, to attempt to establish a claim for loss of profits. The court noted that, in this particular case, the contractor had offered no evidence that would have enabled the jury to conclude that it was reasonably probable that the company would in fact have earned a profit of $140,588 had it been awarded the project. Specifically, the court noted that Anderson did not offer any evidence that its bid was accurate or reasonable, or that it could have done the work required under the plans and specifications for the cost projected in its rebid.

The court noted that many factors might have operated to reduce or eliminate Anderson's anticipated 5 percent profit had it been awarded the job. As an example, the court noted that it is entirely possible that the bid contained errors that resulted in an inaccurate statement on the low side of the cost of the work. It is also possible that the bid was based on overly optimistic takeoffs or was for some other reason an unreasonable assessment of the costs likely to be incurred during actual construction of the project. The court noted that Anderson's proffered evidence addressed none of those subjects. In addition, the court stated that performance inefficiencies during the course of a job can inflate a contractor's projected cost. Such inefficiencies may arise from a variety of causes, including inexperience of the contractor in performing work of the type required by a particular job. The court noted that, in this case, the evidence disclosed during the course of the trial provided little information on that subject.

The court acknowledged that the contractor was only obligated to demonstrate its loss with reasonable certainty and not with absolute precision; but the law also compels a person seeking damages to present the best evidence of its damages which the nature of the case is capable of producing. In this case, Anderson made no showing that it was impossible or impractical to produce evidence relating to the accuracy of its bid, its ability to competently and efficiently perform the project, or its likely net profit.

The court then turned to Anderson's attempt to reopen its case to present additional evidence from its accountant. The court acknowledged that it is error for a court not to permit a plaintiff to reopen its case and introduce further evidence if it is possible to rectify the deficiencies in its original proof. In contrast, denial of a motion to reopen, following the granting of a nonsuit, will not be overturned when the additional evidence offered to be produced is either irrelevant to the issues involved in the case or insufficient to overcome the deficiencies in the party's proof.

Again, the court turned to the offered testimony of the bookkeeper about Anderson's "gross margins" for 1983 through 1987, in order to support the reliability of the 5 percent profit projection in its rebid. The court noted that it was unclear whether the information and testimony to be supplied by the bookkeeper would have cured the deficiencies in Anderson's proof, by establishing, directly or inferentially, that the reasonably anticipated costs of performing the job would have been no greater than 95 percent of the amount of Anderson's rebid. The offer of proof did not explain whether the revenues, costs, and margins to be presented were separately calculated and listed for each job performed during each of the years specified, or instead were nothing more than the cumulative totals segregated only by year.

The court could not say, as a matter of law, whether the evidence about Anderson's business history and its revenues, costs, and margins, for some reasonable time frame, would have been sufficient to support favorable inferences concerning the accuracy and reasonableness of its bid, its ability to efficiently perform the job in accordance therewith, or the reasonableness of its expected net profits. The court did state, however, that it could say, with considerable confidence, that a collection of year-end summaries of one or more categories of general corporate financial data would likely have been inadequate to establish the specific lost profits claimed by Anderson on the particular job in question.

The court concluded that, in the absence of a more precise offer of proof, the appellate court was in no position to arrive at conclusions about the subjects favorable to the contractor, who must bear the consequences of all defects and ambiguities in its offer of proof. Thus, concluded the court, Anderson failed to demonstrate that the evidence it intended to introduce, if its motion to reopen had been granted, would have been sufficient to allow the jury to find that Anderson had lost profits by reason of its inability to rebid the project due to lack of a bid bond.

In dicta, the court of appeal did recognize various methods that are available for a contractor to prove lost profit damages. In a footnote, the court stated:

> 11. Traditionally, three approaches have been utilized to attempt to prove damages resulting from extra or additional work performed by a contractor: (1) the "actual cost" method, where the actual costs of the work are proved by reference to separate records kept for the specific purpose of tracking such damages (see *New Pueblo Constructors, Inc. v. State* (Ariz. 1985), 697 P.2d 185, 194); (2) the "total cost" method, where the contractor subtracts the estimated cost or bid of the entire project from the final actual cost of the entire project in order to arrive at a damage figure. (See *New Pueblo Constructors, Inc. v. State, supra,* 696 P.2d at p.194; *Servidone Const. Corp. v. U.S., supra,* 931 F.2d at p.861–862; and *Huber, Hunt & Nichols, Inc. v. Moore* (1977), 67 Cal. App. 3d 278, 307–309); and (3) the "jury verdict" method, where the jury is presented with whatever actual data and records exist, along with relevant estimates and opinions by experts and comparisons to similar projects,

and invited to "guesstimate" (*Dawco Const., Inc. v. U.S.* (Fed. Cir. 1991), 930 F.2d 872, 881) the amount of the contractor's damages (see *New Pueblo Constructors, Inc. v. State, supra* 696 P.2d at p. 194; *Dawco Const., Inc. v. U.S. supra*, 930 F.2d at p. 881; and *State of California ex rel. Department of Transportation v. Guy F. Atkinson Co.* (1986), 187 Cal. App. 3d 25, 32–33). This means of proving damages may be resorted to only if the plaintiff "can demonstrate a justifiable inability to substantiate the amount of [its] resultant injury by direct and specific proof." (*Joseph Pickard's Sons Co. v. United States* (Fed. Cir. 1976), 532 F.2d 739, 742; see also *Dawco Const., Inc. v. U.S.*, supra, 930 F.2d at p. 881; and State of California ex rel. *Dept. of Transportation v. Guy F. Atkinson Co., supra,* 187 Cal.App.3d at p. 33.)[67]

In conclusion, although there are advantages and disadvantages to proving damages by any method other than the actual cost method, it is clear that California courts will allow contractors to recover lost profits due to impairment of bonding capacity. The contractor will have to produce adequate evidence to support the claim, including that (1) its bonding capacity was impaired as a result of its losses on the job in question; (2) it was precluded from bidding on a specific project as a result; (3) had it bid that project, it would have been the low bidder; (4) it had the knowledge, experience, and ability to perform the project it was precluded from bidding on; (5) its bid was accurate; (6) it could have performed the project for the cost anticipated in its bid; and (7) what its likely net profit would have been.

§ 4.13 CONTRACTOR'S CLAIM FOR RIGHT TO FINISH EARLY

A claim is often made by a prime contractor that it has the right to finish the job early. If the owner wrongfully interferes with that right, the contractor is entitled to additional compensation. Construction contracts traditionally stipulate a completion date. If the contractor goes beyond that date without justifiable excuse, the contractor incurs liability to the owner. Most owners believe, however, that the contractor is not entitled to any additional compensation if the contractor completes ahead of the contract completion date. It is not uncommon, however, for a contractor to plan and bid a contract on the assumption that it can complete the project in less time than allowed by the contract documents. If the contractor can, in fact, complete the project earlier, the contractor then stands a chance of making additional profit by not having to sustain its general conditions over the full length of the contract time for performance. If the project owner delays or otherwise interferes with the contractor's performance, preventing the contractor from meeting a reasonable scheduled completion date, which is ahead of the contract completion date, the question is whether the contractor can recover delay damages and, if so, from what date. Owners, of course, argue that if the contractor was able to complete the work within the contractually authorized period of time, then the contractor suffered no damages. Another subsidiary

---

[67] *Id.* at 538 n.11.

issue in this scenario is whether the contractor is obligated to notify the owner if it does intend to complete the job early, as a condition precedent to the contractor's right to recover damages for interference with its scheduled early completion date.

In the case of *Maryland Department of General Services v. Cherry Hill Sand & Gravel Co.*,[68] a Maryland court ultimately found that the contractor had failed to achieve substantial completion within the contractual performance period and that therefore, the owner had properly refused to accept the project. In arriving at that conclusion, though, the court stated:

> Even where a contractor has completed his project within the time allotted under the contract, the other contracting party may be liable in damages for additional expenses realized where it causes delay in completion such as where the owner breaches its implied warranty to provide accurate and reliable plans and specifications.[69]

The court cited, as support for the foregoing proposition, *Housing Authority v. E.W. Johnson Construction Co.*,[70] and *District Concrete Co. v. Bernstein Concrete Corp.*[71] In the *Housing Authority* case, the contractor completed its work within the contractual performance period and was still awarded damages as a result of delay caused by the owner's faulty specifications. The Arkansas Supreme Court stated:

> The appellant owners' basic and fundamental defense to the contractor's action for damages is essentially that the contractor completed the project within the allotted time under the contract, namely 350 days after receipt of notice to proceed with construction. However, we are not impressed with appellant's argument. Indeed, an owner may not prevent a contractor's early completion of his assignment with impunity.[72]

In the *District Concrete Co.* case, a subcontractor was seeking recovery of delay damages from a supplier, notwithstanding the fact that the subcontractor had completed its work early. The supplier argued that the sub's allegations of a scheduled early completion prior to the onset of winter was "self-serving and speculative." The court allowed the subcontractor to recover from the material supplier. Specifically, the court stated:

> Had the defective concrete not been poured [the subcontractor] estimated completion of the major labor intensive portion of the project by December. As it was, this status was not achieved until the following June. There is no question that the defect caused a delay and that these expenses were incurred

---

[68] 443 A.2d 628 (Md. 1982).
[69] 443 A.2d at 634.
[70] 573 S.W.2d 316 (Ark. 1978).
[71] 418 A.2d 1030 (D.C. 1980).
[72] 573 S.W.2d at 323.

as a result. The damage award need not be absolutely exact; a reasonable estimate based on relevant data is sufficient to support an award.[73]

In *Shore Bridge Corp. v. State*,[74] the contractor completed its work five days before the contractual completion date but was allowed to recover for a one-month delay because the state had forced the contractor's equipment to remain idle. The court rejected the state's argument that if the contractor completed its work within the contractual time period, the contractor could not possibly have incurred any delay damages. Specifically, the court stated:

> Claimant herein was entitled to employ all of the agencies and forces which in its judgment could wisely and advantageously be used in the performance of its work under the contract. A construction which would enable the defendant to retard the prosecution of the work by the claimant or disable it from employing all of the agencies or forces which in its judgment could wisely and advantageously be used in the performance of the contract herein, would operate as a fraud upon the claimant. . . . The claimant was to be enabled to prosecute its work to the utmost advantage and economy.[75]

In *E.J. Albrecht Co. v. New Amsterdam Casualty Co.*,[76] the Court of Appeals for the Seventh Circuit, applying Illinois law, held that the contractor could not be allowed recovery on the right-to-finish-early theory in the absence of an express contractual authorization for compensation for "idle time."

The concept of a contractor's receiving damages by reason of the owner's wrongful interference with the contractor's scheduled early completion date has been recognized for many years in federal government contracting cases. In the case of *Weaver-Bailey Contractors, Inc. v. United States*,[77] the Corps of Engineers awarded a contract to Weaver to build breakwaters, parking lots, and other improvements in a recreation area. The original contract completion date was February 13, 1985. Weaver planned and scheduled the work to be completed on December 3, 1984, prior to the onset of winter weather. Erroneous government design documents prevented Weaver from meeting its planned schedule completion date, as a result of which Weaver was required to maintain a crew at the site throughout the unproductive winter season. Weaver, in fact, completed the work prior to the contract completion date, which the government had extended by 68 days. Weaver filed a claim for delay damages, but the government argued that it could not be held liable because the contractor had never informed it that the contractor intended to complete the work ahead of the contractual completion date. The court rejected that argument and awarded Weaver $469,041 in delay damages. Specifically, the court stated:

> The proper focus is not whether Weaver-Bailey informed the government at the inception of the contract that it intended to complete the project before

---

[73] 418 A.2d at 1038.
[74] 61 N.Y.S.2d 32 (Ct. Cl. 1946).
[75] 61 N.Y.S.2d at 38–39.
[76] 163 F.2d 16 (7th Cir. 1947).
[77] 19 Cl. Ct. 474 (1990).

winter; rather, the focus should be on whether Weaver-Bailey would have completed the project early but for the government-caused delay.[78]

Weaver-Bailey is consistent with prior federal cases on this subject: *Metropolitan Paving Co. v. United States,*[79] *Morrison-Knudsen Co. v. United States,*[80] and *Henry Ericsson Co. v. United States.*[81] The federal Boards of Contract Appeals have likewise allowed these types of claims. The federal Board of Contract Appeals courts have held that a contractor may recover if the contractor can prove

1. The contractor intended to complete the work prior to the contract completion date.
2. The contractor's intended schedule was reasonable.
3. The contractor would have met its intended completion date but for the delay or interference of the owner.[82]

Based upon the foregoing, it appears that if a contractor intends to complete the work prior to the contract completion date, and if the contractor issues a reasonable schedule that would allow it to complete within that time, and if it can prove that it would have met that completion date but for the owner's delay or interference, then the contractor will probably have a claim for breach of its right to finish early.

Several cases in California have addressed the concept of damages being awarded to a contractor for not being able to finish earlier than the contract completion date. In *Maurice L. Bein, Inc. v. Housing Authority,*[83] the court of appeal did not expressly discuss the concept of a contractor's right to finish early; however, the court did confirm an award of damages to a contractor based on a date earlier than the completion date set forth in the contract. The plaintiff (Bein) entered into a contract with the Housing Authority of the City of Los Angeles to construct 62 apartment buildings containing a total of 601 dwelling units. In order for the work to proceed, it was necessary for the Housing Authority to complete the annexation of certain land that made up a portion of the project site. This was referred to as the "County Strip" on which 11 of the 62 apartment buildings (containing 104 units) were to be built.

Although the contract provided 390 days to complete the work, Bein submitted a schedule, which the Housing Authority approved, showing that the work would be completed in 276 days. In fact, it took 654 days to complete the project.

---

[78] 19 Cl. Ct. at 479.
[79] 325 F.2d 241 (Ct. Cl. 1963).
[80] 84 F. Supp. 282 (Ct. Cl. 1949).
[81] 62 F. Supp. 312 (Ct. Cl. 1945).
[82] In that connection, see *Appeal of CWC, Inc.*, ASBCA No. 26432, 82-2 B.C.A. (CCH) ¶ 15,907 (June 19, 1982); and *Appeal of John F. Burke Engineering & Construction*, ASBCA No. 8182, 1963 B.C.A. (CCH) ¶ 3,713 (Mar. 29, 1963).
[83] 157 Cal. App. 2d 670 (1958).

There were two reasons for the delay. One was the failure of the Housing Authority to obtain the "County Strip," and the other was delay in obtaining approval of the project plans and specifications from the city of Los Angeles.

Bein submitted a claim for damages for 378 days—to wit, the difference between the 276 days provided in Bein's approved schedule and the 654 days that it took to complete the project. The trial court found that the schedule submitted by the contractor was reasonable and that the contractor could have completed the work within the 276 days but for the delays caused by the Housing Authority. The trial court also found that the Housing Authority was responsible for the entire 378 days of delay. As a result, the court awarded the contractor damages in the sum of $209,510.77 ($115,413.86 sustained by Bein and $94,096.91 sustained by its subcontractors). The trial court thus allowed the contractor to recover damages sustained by its subcontractors, without even addressing the issue as to whether that was appropriate.

The trial court also found the Housing Authority guilty of misrepresentation because, in order to induce Bein to sign the contract, it had represented that the "County Strip" could be obtained within a few days in spite of the fact that the Housing Authority knew that it had not completed annexation of the County Strip at the time it made that representation.

Although the court of appeal did not expressly discuss the doctrine of the "contractor's right to finish early," it effectively confirmed the concept by awarding damages from the date set forth in Bein's approved schedule to the actual completion date. Another interesting fact was that the contract contained the following provision:

> No payment or compensation of any kind shall be made to the contractor for damages because of hindrance or delay from any cause in the progress of the work whether such hindrances or delays be avoidable or unavoidable.

This was, in effect, a "no damages for delay" clause. However, the court refused to enforce the clause because of the Housing Authority's misrepresentation.

In the case of *Howard Contracting, Inc. v. G.A. MacDonald Construction Co.*,[84] the court impliedly recognized the doctrine of the contractor's right to finish early. In that particular case, the contractor had used bar chart schedules with critical path information to prove delay. The owner had contended that the contractor had not properly proved its case by reason of the fact that it used bar charts rather than critical path method (CPM) schedules to prove delay. In discussing the evidence, the appellate court stated that the unrefuted evidence revealed that the contractor could have completed the contract within its initial schedule, which showed completion earlier than the 330 days allowed under the contract. The court further noted that, but for the delays caused by the owner, the contractor could have completed early. The court acknowledged therefore that the contractor's "right to finish early" claim would be a valid claim under the particular facts of the case.

---

[84] 71 Cal. App. 4th 38 (1999).

As a result of these decisions, it is at least arguable that California courts will accept the concept of the contractor's right to finish early.[85]

§ 4.14 DELAY DAMAGES ON MULTI-PRIME PROJECTS

One project delivery system in common use is the award of separate prime contracts by the owner for various aspects of the work. Owners award multi-prime contracts to be able to "fast-track" the construction process by awarding contracts for the initial phases of the work before the design has been completed for subsequent phases. If a fast-track project is successful, it can reduce the total time for design and construction and thereby reduce the costs of the project. One of the primary risks for contractors in participating in multi-prime projects is delay, as there is no "general contractor," who typically coordinates the various trades on the project. Some owners are not properly prepared to provide the necessary coordination; as a result, the contractors on multi-prime projects often will be delayed by other contractors on the job. The courts have, in some instances, allowed contractors to recover delay damages from the owner as a result of the owner's failure to properly coordinate separate prime contractors.

In *United States Steel Corp. v. Missouri Pacific Railroad*,[86] the owner awarded separate contracts for construction of a bridge substructure and superstructure. The owner issued a notice to proceed to the superstructure contractor before the substructure work had been sufficiently completed. The superstructure contractor was compelled to immediately respond to the notice to proceed because of compulsory contract language and began its preconstruction activities. The superstructure contractor was awarded $364,232 for delay. On appeal, the owner argued that a "no damages for delay" clause precluded recovery. The court of appeal, however, upheld the trial court's application of the principle of active interference, a judicially recognized exception to no damages for delay provisions.

In *Phoenix Contractors, Inc. v. General Motors Corp.*,[87] the owner gave an equipment contractor priority access to the jobsite as a result of which the work of the mechanical contractor was impacted. The mechanical contractor was

---

[85] For additional cases that have discussed this concept of the contractor's right to finish early, see *Barton & Sons Co.*, ASBCA Nos. 9477, 9764, 65-2 B.C.A. (CCH) ¶ 4874 (1965); *Vec, Inc.*, ASBCA No. 35988, 90-3 B.C.A. (CCH) ¶ 23,204 (1990); *Sun Shipbuilding & Dry Dock Co. v. U.S. Lines, Inc.*, 439 F. Supp. 671 (E.D. Pa. 1977); *Owen L. Schwam Construction Co.*, ASBCA No. 22407, 79-2 B.C.A. (CCH) ¶ 13,919 (1979); *Beco Corp.*, ASBCA No. 27090, 82-2 B.C.A. (CCH) ¶ 16,124 (1982); *Lloyd H. Kessler, Inc.*, AGBCA No. 88-170-3, 91-2 B.C.A. (CCH) ¶ 23,802 (1991); *Frazier-Fleming Co.*, ASBCA No. 34537, 91-1 B.C.A. (CCH) ¶ 23,378 (1990); *Gardner Displays Co. v. United States*, 346 F.2d 585, 171 Ct. Cl. 497 (1965); *Montgomery-Ross-Fisher*, PSBCA No. 1033, 84-2 B.C.A. (CCH) ¶ 17,492 (1984); *Schmid v. United States*, 173 Ct. Cl. 302 (1965); *G.M. Shupe, Inc. v. United States*, 5 Ct. Cl. 662, 3 F.P.D. ¶ 11 (1984); *Green Builders, Inc.*, ASBCA No. 35518, 88-2 B.C.A. (CCH) ¶ 20,734 (1988); *Sydney Construction Co.*, ASBCA No. 21377, 77-2 B.C.A. (CCH) ¶ 12,719 (1977); *Rob Glo, Inc.*, VABCA Nos. 2879, 2884, 91-1 B.C.A. (CCH) ¶ 23,357 (1990).
[86] 668 F.2d 435 (8th Cir. 1982).
[87] 355 N.W.2d 673 (Mich. Ct. App. 1984).

allowed to recover delay damages from the owner, again despite a no damages for delay clause in the contract. In *Commonwealth of Pennsylvania State Public School Building Authority v. Noble C. Quadel Co.*,[88] the owner was held liable to the general building contractor for its costs in continuing to provide temporary heat because the owner failed to require the prime HVAC contractor to provide temporary heat upon "enclosure" of the building in accordance with the terms of the prime contract.

Recognizing this potential for liability on multi-prime contracts, some owners have utilized contract language disclaiming liability for delay damages and disclaiming coordination responsibilities. In some cases, these provisions have been enforced. For example, in *M.A. Lombard & Son Co. v. Public Building Commission of Chicago*,[89] the general building contractor was ordered to proceed before the excavation and site preparation work had been completed. The general building contract included a clause that provided that the contractor's only remedy, in the event of delay, regardless of the cause of delay, was an extension of time. The contractor filed a claim for $1.3 million in delay damages, which was denied on the basis of the no damages for delay clause. The court concluded that the owner had bargained for the right to delay its multiple prime contractors.

Owners have also been successful in disclaiming the responsibility for coordinating multiple prime contracts. The state of Mississippi uses a contract clause that makes each contractor responsible for coordinating its work with the work of all the other prime contractors. This clause has been enforced against contractors, resulting in denial of delay claims against the state of Mississippi.[90]

Owners have also been successful in disclaiming coordination responsibilities by delegating the coordination duties to one of the multiple prime contractors, usually the general building contractor. This kind of contract provision has been successfully used to absolve the owner for liability for delay or disruption.[91]

In light of the fact that owners, by contract, can shift the risk of responsibility for delay and coordination, contractors will often file delay claims against the other prime contractors on the project. AIA Document 201, General Conditions of the Contract for Construction, ¶ 6.2 provides that the owner can award separate prime contracts and requires the contractor to coordinate its work with any work being performed by separate prime contractors. These provisions have been interpreted to make each contractor a third-party beneficiary of the other prime contracts. In *Moore Construction Co. v. Clarksville, Department of Electricity*,[92] a site contractor was allowed to sue the general building contractor. In *Barth Electric Co. v. Taylor Bros., Inc.*,[93] an electrical contractor was allowed to

---

[88] 585 A.2d 1136 (Pa. Commw. Ct. 1991).

[89] 428 N.E.2d 889 (Ill. App. Ct. 1981).

[90] *See* Hanberry Corp. v. Mississippi State Bldg. Comm'n, 390 So. 2d 277 (Miss. 1980); CIG Contractors, Inc. v. Mississippi State Bldg. Comm'n, 510 So. 2d 510 (Miss. 1987).

[91] *See* Broadway Maintenance Corp. v. Rutgers, 434 A.2d 1125 (N.J. Super. Ct. App. Div. 1981).

[92] 707 S.W.2d 1 (Tenn. Ct. App. 1986).

[93] 553 N.E.2d 504 (Ind. Ct. App. 1990).

sue the mechanical contractor and the general building contractor for delay. In *Tonn & Blank, Inc. v. Board of Commissioners of LaPorte County*,[94] the general building contractor was allowed to sue an equipment contractor. Each of those cases was based upon an interpretation and application of the AIA Document 201 General Conditions.

A different contract disclaimer clause was used in *Bolton Corp. v. T.A. Loving Co.*,[95] wherein the general building contractor was designated as the "project expeditor" in the contract documents, and the public owner disclaimed coordination responsibility and delegated those duties to the "project expeditor." Under those circumstances, the general building contractor could be sued for delay by the HVAC contractor.

If, however, the contract documents do not establish an affirmative duty to coordinate the work with the other contractors, a delay damages claim against a co-prime may not be maintained.[96] Furthermore, if one of the prime contractors defaults and delays a fellow prime contractor, the delayed contractor will not be able to recover against the performance bond if only the owner is named as an obligee on that bond.[97]

Often, an owner will employ a construction manager if it is awarding multiple prime contracts. The construction manager, as one of its duties, coordinates the work of the multiple prime contractors. In this situation, it logically would appear that the construction manager should be responsible for delay; however, there are few cases in which delay claims against construction managers have been successfully maintained. In *Bagwel Coatings, Inc. v. Middle South Energy, Inc.*[98] the owner, but not its construction manager, was held liable to a contractor when the construction manager ordered the contractor to perform work out of sequence. Typically, no damages for delay clauses are used to protect the owner's agents, including the construction manager, under these circumstances.

However, in the absence of a no damages for delay clause or some other type of disclaimer, multiple prime contractors may be allowed to sue and recover against construction managers for their negligent direction of the work on multiple prime projects.[99]

In *Dynamic Construction Co. v. Barton Malow Co.*,[100] a Michigan court ruled that a construction manager could not be held liable to a general building contractor for delays and field changes. The University of Michigan retained Barton Malow Company as construction manager for a hospital expansion project. The construction management agreement stated that it would not be superseded by

---

[94] 554 N.E.2d 827 (Ind. Ct. App. 1990).
[95] 380 S.E.2d 796 (N.C. Ct. App. 1989).
[96] *See* J.F., Inc. v. S.M. Wilson & Co., 504 N.E.2d 1266 (Ill. App. Ct. 1987).
[97] *See* M.G.M. Constr. Corp. v. New Jersey Educ. Facilities Auth., 532 A.2d 764 (N.J. Super. Ct. Law Div. 1987).
[98] 797 F.2d 1298 (5th Cir. 1986).
[99] *See* Gateway Erectors v. Lutheran Gen. Hosp., 430 N.E.2d 20 (Ill. App. Ct. 1981).
[100] 543 N.W.2d 31 (Mich. Ct. App. 1995).

provisions of contracts for construction and that nothing contained in the construction management agreement should be deemed to give any third party any claim or right of action against the owner or the construction manager that does not otherwise exist without regard to the construction management agreement.

The University awarded the general building construction contract to Dynamic Construction Company, which filed claims against the University alleging differing site conditions, defective plans and specifications, field changes, and delay and disruption. Thereafter, Dynamic filed suit against the University and the construction manager. Dynamic argued that the supervisory and coordination responsibilities of Barton Malow, which were assumed under the terms of the construction management agreement, created a duty owed to the construction contractors, and therefore, to the extent that the construction manager's faulty performance caused increased construction costs, the construction manager should be held liable to the contractor. The court of appeal of Michigan said there was no Michigan precedent regarding a construction manager's liability to construction contractors with whom the construction manager had no contractual relationship. The court reviewed cases from Minnesota and Florida and concluded that in the absence of an express contractual provision to the contrary, a construction manager assumes its supervisory and inspection duties for the sole benefit and protection of the owner and therefore has no duty to protect the interests of the contractors. In this case, the construction management agreement expressly disclaimed any liability to the construction contractors. As a result, the court held that Dynamic could not hold Barton Malow responsible for the increased costs.

In *Amp-Rite Electric Co. v. Wheaton Sanitary District*,[101] a sanitation district hired three prime contractors for a wastewater treatment plant. The plaintiff was an electrical contractor. Another prime contractor was responsible for plumbing, heating, and ventilation, and a third "general" contractor was responsible for the remainder of the work. The prime contract addressed the issue of coordination. The supplemental conditions required that each contractor and its subcontractors coordinate their work with adjacent work, and after approval of the project's construction by the engineer, the general contractor was required to coordinate all work by all contractors on the project. In the section dealing with delays, the contract stated that delays that would allow recovery of damages were "delays beyond the control of contractor . . . including acts of neglect by any separate contractor employed by the owner." The plaintiff electrical contractor was delayed and incurred labor costs three times higher than anticipated. There was loss of productivity caused by having to work simultaneously throughout the construction site, resulting in additional travel time; work was required during the winter months, which was not anticipated; and the electrical contractor was required to work in tandem with the general contractor and the HVAC contractor, both of whom failed to cooperate with the electrical contractor. The owner attempted to expedite performance of the work by attending progress meetings

---

[101] 580 N.E.2d 622 (Ill. App. Ct. 1991).

and by meeting with the contractors to help them develop schedules. On one occasion, the owner attempted to help by converting the HVAC contractor's schedule into a bar chart and forwarding it to the general contractor.

A jury awarded the electrical contractor damages against the owner, which the appellate court affirmed. The court held that the owner was liable for breach of its implied duty to coordinate if it either actively created or passively permitted the continuance of a situation over which it had control that made performance by the electrical contractor more difficult. The court noted that while the contract required the contractors to coordinate among themselves and the general contractor to coordinate the work, the general contractor had no power to enforce coordination. The owner alone could enforce coordination by withholding payments from an offending contractor, by taking over the work of an offending contractor, or by terminating that contractor and replacing it. Thus, the owner had an implied obligation to coordinate, including utilizing its powers to enforce coordination, and whether or not it did so was a question for the jury.

According to these non-California cases, whether a contractor can recover delay damage in a multi-prime project from either the owner or the other contractors will depend upon the language of the contract documents and the facts relating to who caused the delay.

§ 4.15 RECOVERABILITY OF INTEREST FOR CHANGED WORK

Changes in scope are a common occurrence on construction projects. When a change in scope occurs, the contractor typically submits a request for extra compensation, but it usually takes a substantial period of time for the owner to authorize payment; or the owner may not authorize payment at all. The contractor may ultimately recover on its claim, but in the interim the contractor must "finance" the change order work. In other words, the contractor must advance the necessary capital to pay for the labor, services, equipment, or material attributable to the changed work. The contractor finances the extra work either out of its own funds (equity capital) or out of borrowed funds (borrowed capital). The question arises as to whether the costs associated with either of those two types of economic impacts to the contractor are recoverable. Although there are no California cases on the subject, the following decisions provide some guidance on the issue.

In *Appeal of Dawson Construction Co.*,[102] the Veterans Administration issued a change order completely redesigning the radiology rooms in a hospital construction project. The change required substantial payment out of the contractor's working capital. The contractor submitted a price proposal, but the VA did not respond for several months. The change order price was not established until more than a year after the directive had been issued. The contractor was denied recovery of any imputed interest for the use of its equity capital to finance the extra work performed for the owner. Likewise, in *District of Columbia v. C.J. Langenfelder & Son, Inc.*,[103] the government withdrew the availability of a

---

[102] VABCA Nos. 2000, 2016, 86-3 B.C.A. (CCH) ¶ 19,322 (1986).
[103] 558 A.2d 1155 (D.C. Ct. App. 1989).

designated government-owned waste disposal site on a highway construction project. The dispute was not resolved for several years, and the cost of the change was eventually determined to be $320,877. A dissenting judge argued that the "cost" (that is, lost availability of a contractor's equity capital) should be a legitimate item in a change order, but the majority of the District of Columbia Court of Appeals denied the contractor's recovery of imputed interest. The rationale for this rule was stated by the U.S. Court of Appeals for the Federal Circuit in *Gevyn Construction Corp. v. United States*,[104] wherein the court stated that the distinction for recovery of interest on the use of debt capital versus the recovery on the use of equity capital is supportable on the grounds that the costs to the contractor of borrowing capital are clearly determinable, while the value to the contractor of the use of equity capital is not so readily ascertainable. This rationale for the holding appears to be questionable. It is true that interest on borrowed capital is much easier to measure than the value of the loss of use of equity capital. It is not difficult, however, to measure the "imputed interest" on the principal amount of the contractor's equity that is used to finance the changed work.

The courts have held that, when a contractor borrows money to perform changed or extra work, the contractor may recover interest paid on the loan as an element of the change order price adjustment.[105] There are some requirements that must be met in order to recover interest on money borrowed to finance change order work. First, the change order work must result from an owner's directive. If the extra work is caused by other factors, working capital is the sole responsibility of the contractor. As an example, in *Servidone Construction Corp. v. United States*,[106] unanticipated site conditions caused a $23,000,000 overrun on a dam construction project. The contractor was not entitled to recover interest on funds borrowed to finance the additional work.

Even where there has been an owner's directive calling for the extra work, the contractor must be able to prove that the loan was necessitated by that directive. In *Appeal of Gevyn Construction Co.*,[107] the prime contractor was sponsoring a claim on behalf of a subcontractor for interest on money borrowed to perform change order work. The claim was denied because the subcontractor's loans had been necessitated not by change orders but by the improper withholding of funds by the prime contractor. The court stated that, by analyzing the cash flow borrowings and cost of changes, it was clear that the contractor severely restricted the flow of contract earnings to the subcontractor and therefore, the need for borrowing by the subcontractor was traceable to excessive withholding of contract earnings.

In addition, if a contractor borrows money to perform change order work, the contractor should segregate the loan proceeds into a separate account. If the loan proceeds are deposited into a general account, or otherwise commingled

---

[104] 827 F.2d 752 (Fed. Cir. 1987).
[105] *See* Joseph Bell v. United States, 404 F.2d 975 (Ct. Cl. 1968).
[106] 931 F.2d 860 (Fed. Cir. 1991).
[107] ENGBCA No. 3031, 83-1 B.C.A. (CCH) ¶ 16,428 (1983).

with other funds, the contractor will be denied recovery of interest.[108] The Veterans Administration Contract Appeals Board, in the *Bateson* case, stated that the problem was that by commingling the borrowed money with receipts from other jobs, in a general fund from which all jobs were financed, the contractor effectively precluded any possibility of tracing the borrowed money to the financing of the extra work. Even if the loan proceeds are segregated into a separate account, the contractor must also be able to prove that the funds were used exclusively for the performance of the change order work. If any of the funds are used for a different purpose, the contractor will be denied recovery of interest.[109]

As shown by the above cases, it is very difficult for a contractor to recover imputed interest on account of equity capital used to finance extra work. A contractor may recover interest paid on funds borrowed to perform extra work if (1) the loan was necessitated by a directive from the project owner; (2) the loan proceeds were segregated at all times from other contractor funds; and (3) the loan proceeds were used exclusively for the performance of the change order work.

---

[108] *See* Appeal of J.W. Bateson Co., VACAB No. 1148, 86-1 B.C.A. (CCH) ¶ 18,585 (1985).
[109] *See* Appeal of Ed Goetz Painting Co., DOT BCA No. 1168, 83-1 B.C.A. (CCH) ¶ 16,134 (1982).

Chapter 5
BREACH OF CONTRACT BY CONTRACTOR

§ 5.01 Contractor's Performance
    [A] Failure to Perform Work per Plans and Specifications
    [B] Failure to Complete the Work (Abandonment)

§ 5.02 Liquidated Damages for Delay

§ 5.03 Loss of Rents for Delay

§ 5.04 Loss of Profits for Delay

§ 5.05 Fraud

§ 5.06 Offsetting Backcharges

§ 5.07 Latent and Patent Construction Defects
    [A] In General
    [B] Latent Defects
    [C] Patent Defects
    [D] Indemnity Actions Involving Latent Defects

§ 5.08 Recovery of Damages for Violation of the Building Code

§ 5.09 Termination of Contractor

§ 5.10 Impact of Construction Contract Terms

§ 5.11 Conclusion

§ 5.01 CONTRACTOR'S PERFORMANCE

The performance of the contractor is measured by two criteria: (1) whether the contractor performed the work according to the plans and specifications; and (2) whether the contractor completed the work within the time specified in the contract plus any extensions of time authorized by change order.

[A] Failure to Perform Work per Plans and Specifications

The contract agreement typically requires the contractor to perform the work in accordance with the plans and specifications and in a good and workmanlike manner. Failure to do so makes the contractor liable to the owner for the damages the owner sustains. Generally, the measure of damages in this area is the cost of remedying the defect.[1] If, however, the cost of remedying the defect is excessive or would require rebuilding the entire structure, the alternative measure of damages is *diminution in value* (the difference between the value of the building as improperly constructed and what its value would be if properly constructed).[2] In *Bayuk,* the architect's defective plans and specifications resulted in a single-family residence that was poorly constructed. The owner's expert testified that the cost of repair would be economically unfeasible, and the court applied the diminution-in-market-value measure of damages, following *Shell v. Schmidt.*[3]

Often there is a major dispute between the contractor and the owner on this issue, when the costs of remedying the defect are excessive and the diminution in value is less. The contractor will argue that the diminution-in-value rule should apply, and the owner will argue that the costs of remedying the defect rule should apply. A 1990 California decision carefully weighed those two contentions. In *Orndorff v. Christiana Community Builders,*[4] the owners had lived in their home since 1977 and testified that they had no plans to leave it. The defendant developer stipulated that the home was built on defectively compacted soil. The owners presented evidence that it would cost $243,539.95 to repair the defects in the home and to relocate while the necessary repairs were being completed. The owner's appraiser testified that after the home was repaired, it would be worth $238,500. The trial court awarded the homeowners $243,539.95. The developer argued at the trial level that the diminution-in-value rule should apply to this case, because the diminution in value was less than the cost to repair. The developer then appealed to the district court of appeal. Further evidence before the trial court was presented by the developer that the costs of repair should be $118,355, rather than the $221,792.68 testified to by the owners' expert. The owners' expert appraiser also testified that the home without repairs would be worth $67,500,

---

[1] Jones v. Kvistad, 19 Cal. App. 3d 836, 843 (1971).
[2] Bayuk v. Edson, 236 Cal. App. 2d 309, 316 (1965).
[3] 164 Cal. App. 2d 350, 359, 360 (1958). *See also* Amerson v. Christman, 61 Cal. App. 2d 811 (1968).
[4] 217 Cal. App. 3d 683, 266 Cal. Rptr. 193 (1990).

but with repairs would be worth $238,500. The developer's appraiser testified that without the repairs, the house would be worth $160,500, and that with the repairs, it would be worth only $225,500. Thus, there was a sharp conflict in testimony between the owner's experts and the developer's experts.

On appeal, the district court of appeal affirmed the trial court's judgment. The appellate court stated that the law was not as rigid as the developer suggested, and that when the plaintiffs have personal reasons to repair and the costs of repair are not unreasonable in light of the damage to the property and the value after repair, costs of repair that exceed diminution in value may be awarded (citing *Heninger v. Dunn*[5]).

The court in the *Heninger* case had stated that the rule precluding recovery of restoration costs in excess of diminution in value is not of invariable application, and that restoration costs may be awarded even though they exceed the decrease in market value if there is a reason personal to the owner for restoring the property to its original condition, or when there is reason to believe that the owner will in fact make the repairs. The *Orndorff* court stated that in this case, there was such a personal reason. The owners had testified that they had paid a premium for this house because it was located immediately adjacent to an open space easement and that they would stay in the house and repair it if they were awarded the costs of repair. The court also noted that, by virtue of the testimony of the owners' appraiser, the repaired home would be worth $238,500. Because the cost of repair in the sum of $243,539.95 (admittedly more than the value of the home as repaired) was only 2.5 percent greater than the undamaged value of the property, the court held that that was well within reason.

In *Schaffer v. Debbas*,[6] the court of appeal applied the traditional lesser measure rule in choosing whether to award the owner of a defectively constructed home the cost of repair or the diminution in value of the property. Alleging defects in the heating and air-conditioning and drainage systems, the plaintiff homeowners sued a joint venture with which they had contracted to construct a custom home for $1.3 million. After four years of unfulfilled promises to repair the defects, the owners moved and later sold the house for $950,000.

By special verdict, the jury found the cost of repairs to be $130,000 and the diminution in value of the property to be $585,000. The trial court entered judgment on the cost-of-repair amount, and the court of appeal confirmed that decision. Although the general rule is to award the lesser amount, this rule will not always be applied, as noted in *Orndorff v. Christiana Community Builders,* cited *supra,* wherein it was reasoned that a rigid application of the lesser measure rule could force many plaintiffs to abandon their personal residences rather than repair them.

---

[5] 101 Cal. App. 3d 858 (1990).
[6] 17 Cal. App. 4th 33, 21 Cal. Rptr. 2d 110 (1993).

[B] Failure to Complete the Work (Abandonment)

If the contractor fails to "substantially complete" the work, then the contractor has abandoned the work. "Substantial completion" has a definitive meaning in the construction industry. An owner cannot sue a contractor for abandoning the work if the job has been substantially completed. In *Lowy v. United Pacific Insurance Co.*,[7] the contractor stopped work when the job was 98 percent complete. The court held that the job was substantially complete, and therefore the contractor was entitled to the balance of the contract price less the cost of completion. In other words, when the contractor has completed the work and there are only minor or trivial imperfections or defects in the work, the job is substantially complete, and therefore the contractor is entitled to the balance of the contract price less the cost of completing or correcting the minor items that have not been corrected or completed. The doctrine of substantial performance has been well defined in United *States Industries, Inc. v. Vadnais*,[8] as follows:

> Recently the doctrine of substantial performance was once again set forth by our Supreme Court: "Where a person agrees to do a thing for another for a specified sum of money to be paid on full performance, he is not entitled to any part of the sum until he has himself done the thing he agreed to do, unless full performance has been excused, prevented or delayed by the act of the other party." (*Lowy v. United Pac. Ins. Co.*, 67 Cal. 2d 87, 92 [60 Cal. Rptr. 225, 429 P.2d 577].) Applied to building contracts, "where the owner has taken possession of the building and is enjoying the fruits of the contractor's work in the performance of the contract, if there has been a substantial performance thereof by the contractor in good faith, if the failure to make full performance can be compensated in damages to be deducted from the price or allowed as a counterclaim, and if the omissions and deviations were not willful or fraudulent and do not substantially affect the usefulness of the building or the purpose for which it was intended, the contractor may, in an action upon the contract, recover the amount of the contract price remaining unpaid, less the amount allowed as damages for the failure of strict performance." (*Lowy v. United Pac. Ins. Co., supra,* pp. 92–93.)

The key part of the foregoing quotation is that the owner has taken possession of the building and is enjoying the fruits of the contractor's work. This is sometimes referred to as *beneficial use and occupancy.* If the owner can use and occupy the building for its intended use, even though it may not be absolutely complete, it can be said that there has been substantial completion and the contractor is entitled to the contract price less the cost of remedying any defects and the costs to complete.

It should be noted, however, that if the contractor's omissions and deviations are willful or fraudulent and substantially affect the usefulness of the building or

---

[7] 67 Cal. 2d 87 (1967).
[8] 270 Cal. App. 2d 520, 525, 526 (1969).

the purpose for which it was intended, then performance cannot be said to be substantial. *United States Industries,* following *Lowy,* put it this way:

> In *Lowy, supra,* 67 Cal. 2d 86, 92–93, the court pointed out (after reference to other elements) that substantial performance has been achieved "if the omissions and deviations were not wilful or fraudulent and do not substantially affect the usefulness of the building or the purpose for which it was intended. . . ." There must thus be a combination of these two factors: (1) substantial omissions or deviations resulting from (2) wilful or intentional action (or inaction) on the part of the party claiming that he substantially performed.[9]

When the court finds that the contractor has breached the contract by willfully and intentionally deviating from the plans and specifications and those deviations substantially affect the building's usefulness for the purpose for which it was intended, then there can be no finding of substantial performance, and the measure of damages is the cost of correcting the deviations. When there is a material and intentional major deviation from the plans and specifications, the contractor cannot invoke the difference in value rule.

In *Shell v. Schmidt,*[10] the contractor installed stucco rather than wood sheeting on exterior walls, one 35,000 BTU floor furnace rather than two 30,000 BTU wall furnaces, and drywall rather than lath and plaster on interior walls. The contractor offered evidence that these deviations from the plans and specifications were necessary because what was called for in the plans and specifications was unavailable, and expert testimony showed that what was furnished was equal to or better than what was specified; therefore, the value of the house had not been affected by the deviations. The appellate court ruled that the measure of damages was the cost of making the houses comply with the plans and specifications and the difference-in-value rule would not be invoked when there had been an intentional major deviation from the plans and specifications.

Obviously, substantial performance is a question of fact in each case. In *Tolstoy Construction Co. v. Minter,*[11] the appellate court reversed, as a matter of law, a trial court finding of substantial performance because a building addition and carport were so poorly constructed as to require substantial amounts of money to make the building tenantable; a water heater was installed in violation of the building code; the carport was visibly out of plumb; areas of wood were left unpainted; and doors, windows, and light switches were misaligned, and nailheads were left showing.

If the contractor has not "substantially completed" the work (as determined under the rules referred to above) and has terminated its construction work, then the contractor has abandoned the project.[12] *Abandonment* results when the contractor ceases work on the project before "substantial completion," without factual

---

[9] United States Indus., Inc. v. Vadnais, 270 Cal. App. 2d 520, 531.
[10] 164 Cal. App. 2d 350 (1958).
[11] 78 Cal. App. 3d 665 (1978).
[12] Lowy v. United Pac. Ins. Co., 67 Cal. 2d 87; Tolstoy Constr. Co. v. Minter, 78 Cal. App. 3d 665.

and legal justification. Whether the contractor has abandoned the work is a question of fact in each case.

When the contractor abandons the work, the owner's measure of damages is the difference between the contract price and the price of completing the work the contractor had agreed to perform.[13] The owner may also recover any delay or other consequential damages resulting from the contractor's abandonment.[14] The same rule applies to an action by a prime contractor against its subcontractor.[15]

§ 5.02 LIQUIDATED DAMAGES FOR DELAY

The types of damages recoverable by an owner vary from case to case. If the contract has a liquidated damages clause for delays, that clause will establish the measure of damages for delay. In California, liquidated damages clauses are presumed valid, and the burden of proving that the liquidated damages clause is unenforceable is on the party attacking the clause.[16]

In *Pacific Employers Insurance Co. v. City of Berkeley*,[17] the court ruled that a contractor's abandonment of the work did not render the liquidated damages provision in the prime contract inoperable. In *Pacific*, Emerald Builders contracted to construct a community recreation center for the City of Berkeley. Emerald provided the city with a performance bond issued by Pacific Employers. The contract called for liquidated damages of $250 per day. Emerald failed to complete the project by the contractually established deadline and eventually abandoned the job. In response to the city's demand, Pacific Employers arranged for another contractor to finish the work. Upon completion, Pacific Employers requested payment of the contract balance from the city, and the city claimed an offset of $82,500 in liquidated damages, which had accrued between the stipulated completion date and the substitute contractor's actual completion date.

Pacific Employers argued that once the contractor abandoned the project, the liquidated damages clause become inapplicable and unenforceable. The court acknowledged that several prior California cases appeared to stand for that proposition, which had been repeated as the "California Rule" in at least two well-known legal treatises. The court noted that in those cases, the owner had never even made a claim for liquidated damages, and therefore the cases could not stand for the proposition that liquidated damages are unenforceable after the contractor abandons the work. The court felt that the better-reasoned rule would be to allow recovery of liquidated damages regardless of whether the contractor abandons the work; therefore, the city was entitled to retain liquidated damages. The court stated that it is not for delay in completion of the contract but rather

---

[13] Amerson v. Christman, 261 Cal. App. 2d 811, 823 (1968). *See also* Glendale Fed. Sav. & Loan Ass'n v. Marina View Heights Dev. Co., 66 Cal. App. 3d 101, 123 (1977).

[14] Cal. Civ. Code § 3300; Amerson v. Christman, 261 Cal. App. 2d 811.

[15] *See* University Casework Sys., Inc. v. Superior Court, 41 Cal. App. 3d 263 (1974); Associated Lathing Co. v. Louis C. Dunn, Inc., 135 Cal. App. 2d 40 (1955).

[16] *See* Cal. Civ. Code § 1671.

[17] 158 Cal. App. 3d 145, 204 Cal. Rptr. 387 (1984).

for delay in completion of the work that the damages are provided. The liquidated damages clause did not, in terms or by implication, empower the contractor to put an end to the accrual of liquidated damages by the mere process of refusing to go ahead after the time for completion had passed. The rule stated in this case is the majority rule across the United States.

§ 5.03 LOSS OF RENTS FOR DELAY

The owner may also recover loss of rents. In *Bird v. American Surety Co.,*[18] the owner sued a contractor and its surety when the contractor abandoned the job. The judgment of the trial court included loss of rent. The California Supreme Court stated:

> The rentals lost by delay in completion and the attorneys' fees incurred in defending against suits were proper charges in favor of the owner against the contractor, and against the Surety Company, which had given a bond to indemnify the owner against any loss or damage arising by the failure of the contractor to faithfully perform his contract, and for the payment in full of the claims of all persons performing labor upon, or furnishing materials to be used in, the work. (*Hampton v. Christensen,* 148 Cal. 729 [89 Pac. 200]; *Tally v. Ganahl,* 151 Cal. 418, [90 Pac. 1049].)[19]

In *Camper v. McDermott,*[20] the owner of a 50-unit apartment building recovered a judgment for loss of rent by reason of a delay in the completion of the swimming pool. Although the judgment for the owner was reversed on appeal, it was reversed because the owner had not properly proved its loss of rent. Specifically, the court held that such a loss of rent would be recoverable if properly proved. The court stated:

> However, when we consider the issues as to damages awarded on the cross-complaint a different conclusion is required. [1] We do not take the position that no damages with respect to loss of rental by reason of the failure of the plaintiff and cross-defendant to comply with the requirements of the contract would be proper; our conclusion on that branch of the case is only that the evidence in the record does not justify the award of $6,000 or any other amount of specific substantial damages as lost profits. It cannot be questioned that, particularly in a university neighborhood where young renters predominate, the presence of a swimming pool is an asset to an apartment house. Whether or not evidence may be adduced on a retrial which would justify an award of substantial damages is another question and depends on factors which we are in no position to pass upon in view of the present condition of the record.[21]

---

[18] 175 Cal. 625 (1917).
[19] 175 Cal. at 631.
[20] 266 Cal. App. 2d 41 (1968).
[21] 266 Cal. App. 2d at 44.

In *McDevitt & Street Co. v. Marriott Corp.*,[22] the court found that the contractor breached its contract by completing performance 132 days late. The contractor sued the owner, and the owner filed a counterclaim seeking damages for delay in the opening of the hotel. The owner contended that (1) it was deprived of the use of sales proceeds (the hotel was to be sold immediately upon completion to another facet of Marriott); (2) it lost lease payments; (3) it lost management fees; and (4) it incurred additional administrative expenses. The court noted that in the case of delay by a contractor, the measure of damages to the owner is either the rental value of the completed structure or a reasonable return for the structure treated as an investment. The owner sought recovery under the latter formula. The rate of return could be measured by the lost use of the sale proceeds. The owner, who was deprived of the sale proceeds for 132 days, would have used that money to pay down corporate borrowings and to reduce its interest expense. On a sales price of $6,250,000, Marriott was deprived of $203,000. That sum was calculated by using its average weighted borrowing rate, the standard rate charged by Marriott for carrying cost to capitalize a project. Marriott would also lose interest income. The court found that damages assessment to be reasonable.

In addition, during construction, the owner had entered into a ground lease with a restaurant. The court also awarded the owner the lost rent payments for the 19 weeks that completion was delayed. A subsidiary of Marriott entered into a management agreement to manage and operate the hotel after it was sold to the buyer. The management company's fees were a percentage of gross revenues and profits. Marriott had projected the hotel's anticipated gross revenues using average room and occupancy rates. The court concluded that, although it was a close question, Marriott was not entitled to recover such lost management fees. Calculation of the fees was based on the hotel's revenues and profits, calculations of which, the court held, were too speculative because the hotel, although part of an established chain, was a new business. The court awarded most of the owner's claims based upon administrative expenses incurred during the delay period, consisting of site administration, travel, site security, temporary telephone, field office supplies, and bonds and insurance. The court denied a request for the project's contribution to the regional and national overhead allocation of Marriott.

§ 5.04 LOSS OF PROFITS FOR DELAY

Owners have also been awarded loss of profits. In *Burnett & Doty Development Co. v. Phillips*,[23] the owner hired a contractor to do the site work (clearing and grubbing, rough earthwork, sewers, storm drains, initial base rock, curbs, gutters, sidewalks, base rock on streets, street paving, fire hydrants, street monuments and signs, and the placing of riprap, bank planting, and fencing along a flood control channel). On the date set for completion under the contract, the work was not completed. This delay caused a discontinuance of the owner's business (developing and selling residential properties) for five months.

---

[22] 713 F. Supp. 906 (E.D. Va. 1989).
[23] 84 Cal. App. 3d 384 (1978).

The court found that during the period in question, the owner was able to produce houses at an average rate of 3.3 houses per month at an average profit of $2,156 per house. As a result of the five-month delay, the owner was prevented from producing 16.5 houses, for a loss of profit of $35,574. In addition, the delay caused the construction cost of two houses in the subdivision to increase by $4,974. The court also found that the owner suffered increased interest expense on the construction loan of $1,259. Finding that the owner realized a profit of $8,100 upon the ultimate sale of the 42 homes in the subdivision, the court deducted that from the total damages. The damages of the owner were therefore calculated as follows:

1.	Loss of profit	$35,574
2.	Increased construction costs	4,974
3.	Increased interest expense	1,259
	TOTAL	$41,807
4.	Less profit made on sale of homes	8,100
5.	Net amount of judgment	$33,707

The appellate court affirmed the judgment for the owner. The court stated:

> (3) An injured party may recover "for a breach of contract the amount which will compensate it for all the detriment proximately caused [by the breach], or which, in the ordinary course of things, would be likely to result [from the breach]." (Civ. Code Section 3300.) The damages awarded should, insofar as possible, place the injured party in the same position it would have been had the contract properly been performed, but it may not be awarded more than the benefit which it would have received had the promisor performed. (Civ. Code Section 3358; *Steelduct Co. v. Henger-Seltzer Co.,* 26 Cal. 2d 634, 648–649 [160 P.2d 804] (1945); *Glendale Fed. Sav. & Loan Assn. v. Marina View Heights Dev. Co.,* 66 Cal. App. 3d 101, 123 [135 Cal. Rptr. 802] (1977).)
>
> (4) Damages may be awarded for breach of contract for those losses which naturally arise from the breach, or which might reasonably have been foreseen by the parties at the time they contracted, as the probable result of the breach. (*Glendale Fed. Sav. & Loan Assn. v. Marina View Heights Dev. Co., supra,* 66 Cal. App. 3d at p. 125.) Damages must be reasonable, however, and the promisor is not required to compensate the injured party for injuries that it had no reason to foresee as the probable result of its breach when it made the contract. (*Coughlin v. Blair,* 41 Cal. 2d 587, 603 [262 P.2d 305] (1953); *Ely v. Bottini,* 179 Cal. App. 2d 287, 294 [3 Cal. Rptr. 756 (1960).)
>
> Where the injured party shows that, as a reasonable probability, profits would have been earned on the contract except for its breach, the loss of the anticipated profits is compensable. (*Nelson v. Reisner,* 51 Cal. 2d 161, 171–172 [331 P.2d 17] (1958); *Fisher v. Hampton,* 44 Cal. App. 3d 741, 747 [118 Cal. Rptr. 811] (1975).) Where business activity has been interrupted by a breach of contract, damages for the loss of prospective profits that otherwise might

have been made from its operation are generally recoverable where such damages are shown to have been foreseeable and reasonably certain. (*See Grupe v. Glick,* 26 Cal. 2d 680, 692 [160 P.2d 832] (1945); *Guntert v. City of Stockton,* 55 Cal. App. 3d 131, 143 [126 Cal. Rptr. 690, 127 Cal. Rptr. 602] (1976). See also *S. Jon Kreedman & Co. v. Meyers Bros. Parking-Western Corp.,* 58 Cal. App. 3d 173, 184–185 (130 Cal. Rptr. 41) (1976); *Mahoney v. Founders' Ins. Co.,* 190 Cal. App. 2d 430, 436 [12 Cal. Rptr. 114] (1961); *Jegen v. Berger,* 77 Cal. App. 2d 1, 16 [174 P.2d 489] (1946).) "[W]here the operation of an established business is prevented or interrupted, as by a . . . breach of contract . . . damages for the loss of prospective profits that otherwise might have been made from its operation are generally recoverable for the reason that their occurrence and extent may be ascertained with reasonable certainty from the past volume of business and other provable data relevant to the probable future sales" (Grupe v. Glick, supra).

Burnett & Dotty presented evidence that its sole business from August 1971 through November 1974 was that of the construction of single family residences. During the 38-month period commencing in August of 1971 and ending in November of 1974, Burnett & Dotty operated actively for 33 months. Only during the period of mid-November 1972 to mid-April 1973—the forced shutdown period caused by Phillips' breach of contract—was the company unable to produce homes. During the 33 working months, Burnett & Dotty constructed 109 houses at an average rate of 3.3 houses per month. The working experience of the business thus supports the trial court's finding that during the five-month interruption in its business caused by the breach of the contract Burnett & Dotty was prevented from developing an average of 16.5 homes. There was evidence that during the 38-month period beginning in August of 1971 and ending in November of 1974, the lowest average profit realized on any Burnett & Dotty project was $2,156 per home in the Oak Creek Park Subdivision. Multiplying this lowest average profit figure by the 16.5 homes which Burnett & Dotty would have constructed during the forced 5-month shutdown period yields the $35,375 loss of profits found by the trial court to have occurred during the interruption of Burnett & Dotty's business. (*See Hacker etc. Co. v. Chapman V. Mfg Co.,* 17 Cal. App. 2d 265, 267–268 [61 P.2d 944] (1936).)

The parties planned on the contract being fully performed by October 31, 1972. C. S. Phillips, Sr., personally wrote into the contract at the time of its execution, "work to be completed by October 31, 1972." This evidence supports the determination that the parties contemplated or should have contemplated at the time of contracting that the interruption of Burnett & Dotty's business, and attendant loss of profits, was likely to result from Phillips' failure of timely performance. Such losses were a foreseeable result of any extensive delay in performance of the agreement.[24]

It should be noted that the award of increased construction costs and interest was not even challenged on appeal.

---

[24] 84 Cal. App. 3d at 390.

§ 5.05 FRAUD

In an extreme case, *Walker v. Signal Cos.*,[25] a husband and wife were awarded punitive damages for fraud from the builders and sellers of a single-family residence. The plaintiffs had signed a contract with the defendants whereby the defendants were to construct the house and sell the completed product to the plaintiffs. The house had to be completed by April 1, 1973, in order for the plaintiffs to defer capital gains on the sale of their prior personal residence. This fact was known by all parties to the transaction. The house was not, in fact, completed by April 1, 1973. Notice of Completion was finally signed on September 4, 1973. The court first stated the general rule regarding the measure of damages for breach of a construction contract as follows:

> (4) The measure of damages for breach of contract to construct improvements on real property where the work is to be done on plaintiff's property is the reasonable cost to the plaintiff to finish the work in accordance with the contract. (*See, e.g., Jones v. Kvistad,* 19 Cal. App. 3d 836, 843 [97 Cal. Rptr. 100] (1971); *Kitchel v. Acree,* 216 Cal. App. 2d 119 [30 Cal. Rptr. 714] (1963).) Plaintiffs, however, were not required to terminate their contract with the defendants and complete the project themselves. They had every right to expect the defendants would complete the contract in the manner as required by its terms.[26]

The plaintiffs also recovered punitive damages of $100,000 against one of the defendants and $115,000 against another defendant, for a total of $215,000, by reason of fraud—to wit, entering into a construction contract calling for completion on a date certain with no intent to perform. The appellate court affirmed the award of punitive damages, stating:

> Defendants contend plaintiffs are not entitled to punitive damages because their action is one for breach of contract and not for fraud. Punitive damages are not recoverable in an action for breach of contract no matter how wilful, malicious or fraudulent the breach. (Civ. Code Section 3294; *Crogan v. Metz,* 47 Cal. 2d 398, 405 [303 P.2d 1029] (1956).) They may, however, "[b]e awarded where a defendant fraudulently induces the plaintiff to enter into a contract. [Fn. omitted.] (*Kuchta v. Allied Builders Corp.,* 21 Cal. App. 3d 541, 549 [98 Cal. Rptr. 588]; *Horn v. Guaranty Chevrolet Motors,* 270 Cal. App. 2d 477, 484 [75 Cal. Rptr. 871].) The words 'oppression, fraud, or malice,' in Civil Code section 3294 being in the disjunctive, fraud alone is an adequate basis for awarding punitive damages. (*Miller v. National American Life Ins. Co., supra* 54 Cal. App. 3d 331, 336 [126 Cal. Rptr. 731]; *Horn v. Guaranty Chevrolet Motors, supra,* 270 Cal. App. 2d 477, 484 [75 Cal. Rptr. 871].)" (*Glendale Federal Sav. & Loan Assn. v. Marina View Heights Dev. Co.,* 66 Cal. App. 3d 101, 135 [135 Cal. Rptr. 802] (1977).) Although the contract

---

[25] 84 Cal. App. 3d 982 (1978).
[26] 84 Cal. App. 3d at 993.

may fix the compensation, it does not prevent recovery of exemplary damages. (*See Esparza v. Specht,* 55 Cal. App. 3d 1 [127 Cal. Rptr. 493] (1976); *Fletcher v. Western National Life Ins. Co.,* 10 Cal. App. 3d 376, 400 [89 Cal. Rptr. 78, 47 A.L.R.3d 286] (1970); *Wetherbee v. United Insurance Co. of America,* 265 Cal. App. 2d 921, 928 [71 Cal. Rptr. 764] (1968); *Sharp v. Automobile Club of So. Cal.,* 225 Cal. App. 2d 648 [37 Cal. Rptr. 585] (1964); *Acadia California Ltd. v. Herbert,* 54 Cal. 2d 328 [5 Cal. Rptr. 686, 353 P.2d 294] (1960); *Chelini v. Nieri,* 32 Cal. 2d 480, 487 [196 P.2d 915] (1948).) The application of this rule which permits punitive damages where both fraud and breach of contract exists is only consistent with the underlying policy for the application of punitive damages. (Civ. Code Section 3294). . . .

At the time he denied the motion for new trial, the trial judge said the jurors had heard evidence that the "Walkers were put upon by these corporations . . . they weren't treated fairly . . . ." and "that the collective conscience [of the community] speaking through that jury said that this kind of conduct is going to be punished by an award of punitive damages."

When we consider the reprehensibility of the defendants, we cannot overlook the comments of the trial judge to which reference has previously been made. What may now appear from the record to be a mere delay in a construction contract was perceived by 12 jurors and a trial judge as a planned fraudulent scheme resulting in actual damages to the Walkers. An evaluation of the events must be made in human terms giving proper respect to those who listened to the evidence and observed the witnesses.[27]

§ 5.06 OFFSETTING BACKCHARGES

In the construction industry, a general contractor will often use the same subcontractor on several projects. Many times the general contractor will backcharge against subcontractors on a particular project. The question then arises as to whether the general contractor can offset those backcharges against money it owes to the subcontractor on another project. There is no California authority directly on point on this issue. California Civil Procedure Code § 428.10 provides that a defendant can file a cross-complaint against a plaintiff for any claim that it has against the plaintiff, whether or not it is related to the transaction out of which the plaintiff's complaint arises. Thus, as a practical matter, if the subcontractor were to sue the general contractor for the amount due on Job A, the general contractor could file a cross-complaint for the amount that it claims on Job B.

The Supreme Court of Virginia, in *Piland Corp. v. League Construction Co.,*[28] held that a prime contractor could offset a backcharge on one job against a payment due a subcontractor on another job. In that case, League Construction was the subcontractor who had provided work to Piland on two projects. League Construction filed a suit against Piland for $25,300 due on one of the projects. The general contractor, Piland, admitted that it had withheld retention on that project because it was entitled to an offset for $85,000 worth of backcharges for

---

[27] *Id.* at 996.
[28] 238 Va. 187, 38 S.E.2d 652 (1989).

faulty work performed by the subcontractor on another project. The trial court had awarded a summary judgment to the subcontractor. The Virginia Supreme Court reversed. Referring to court rules, the court held that a defendant could plead as a counterclaim any cause of action that it had against the plaintiff, whether or not it grew out of the same transaction mentioned in the complaint and whether or not the claim was liquidated. It is the opinion of the authors that California courts would reach a similar result.

§ 5.07 LATENT AND PATENT CONSTRUCTION DEFECTS

[A] In General

The time within which a plaintiff may bring legal action against a contractor, developer, architect, or others involved in the design or construction of improvements to real property is established by statute and varies according to the legal theory upon which the action is based. Thus, California Civil Procedure Code § 337 imposes a three-year limitations period for actions to recover for damage to property, whereas section 340(3) establishes a one-year period for actions for personal injury or wrongful death. Section 338 requires that actions for breach of written contract must be brought within four years. Such statutes of limitation begin to run as soon as the plaintiff suffers injury or incurs damage. When injury to persons or damage to property is caused by a latent defect, however, the limitations period does not begin to accrue until the plaintiff discovers both the injury and its negligent cause, or could have discovered them through reasonable diligence.[29] This raises a problem for contractors, developers, architects, and others involved in construction, because injury or damage arising from defects caused by negligent performance of their work may not appear for many years.

The legislature has therefore placed outside limits on the time within which suit may be brought for injury or damage arising from construction defects. Civil Procedure Code § 337.1 requires actions arising from patent defects to be brought within four years; section 337.15 allows 10 years to file suit based upon latent construction or design defects. Most significantly, these time limits begin to run from the time of substantial completion of the construction, rather than from the time the injury was or should have been discovered. For this reason, sections 337.1 and 337.15 are technically not statutes of limitation, but statutes of repose, because they determine the time within which a cause of action can arise at all. Such statutes of repose are intended to protect defendants from the legal and financial uncertainties of liability that would otherwise be potentially unlimited in time.

Section 337.1 is quite specific about the four-year time limitation on actions for property damage, personal injury, or wrongful death based upon a *patent defect,* which is defined in section 337.1(e) as "a deficiency which is apparent by reasonable inspection." Subsection (b) provides that if an injury to person or

---

[29] *See* Leaf v. City of San Mateo, 104 Cal. App. 3d 398 (1980).

property, or injury causing wrongful death, occurs during the fourth year after substantial completion, the action must be brought within one year after the injury occurs, irrespective of the date of death, "but in no event may such an action be brought more than five years after the substantial completion of construction of such improvement."

The 10-year limitations period on actions arising from latent defects (section 337.15) does not apply to personal injury actions. Thus, in *Martinez v. Traubner*,[30] a plaintiff was allowed to sue for personal injuries 20 years after project completion.

[B] Latent Defects

Most limitations disputes involving construction defects concern hidden deficiencies. Suits arising from such latent defects pose more difficult legal problems, most of which involve the issues of whether the defect was truly latent and when the defect became known, or patent. Unlike section 337.1, section 337.15 does not specifically address its interaction with other statutes of limitation.

In *North Coast Business Park v. Nielsen Construction Co.*,[31] Defendant Erreca, Inc. acted as excavation and grading subcontractor on a business park project of North Coast, which was completed in 1982. In 1984, North Coast employed a civil engineer to inspect a crack in a floor slab. The engineer opined that drainage problems between Building 9 and a retaining wall had caused erosion that contributed to damage and deflection of the retaining wall. North Coast filed its action in December 1988 against the prime contractor, Nielsen, who in turn filed a cross-complaint against Erreca. Erreca obtained summary judgment based on evidence that the defect in design and construction of the retaining wall was known to North Coast in early 1984, and therefore the three-year and four-year limitations periods had expired.

The court of appeal affirmed, rejecting the contention of North Coast that a claim based on a latent defect is timely so long as the lawsuit is filed within the 10-year period established by section 337.15. The 10-year period does not apply when suit was filed more than 3 years after the defect became manifest and was discovered. The shorter periods established by sections 337 and 338 begin to run when damage is sufficiently appreciable to give a reasonable person notice of a duty to pursue legal remedies. In addition, section 337.15 imposes an absolute 10-year bar based on the date of substantial completion, regardless of the time of discovery.

The statutes therefore set up a two-step process for determining the timeliness of actions to recover damages arising from latent construction defects. First, actions must be filed within three years (section 337 for damage to property) or four years (section 338 for breach of written contract) of discovery; and, regardless of the date of discovery, such actions must be filed within 10 years (section 337.15) of substantial completion of the project.

---

[30] 32 Cal. 3d 755 (1982).
[31] 17 Cal. App. 4th 22 (4th Dist. 1993).

The question of just what constitutes a latent defect within the meaning of section 337.15 may depend upon the circumstances in which the issue arises or upon the court deciding the case. This point is illustrated by two significant 1993 cases, each of which considered the question in the context of homeowners' insurance policies in which coverage was excluded for damages arising from latent construction defects.

In *Chadwick v. Fire Insurance Exchange*,[32] the court of appeal held that "the considerations that have led to a broader definition of latent defect for statute of limitations purposes are inapplicable to determining the scope of a homeowners insurance exclusion."[33] The opinion calls into question expansive interpretations of the latent defect exclusion in insurance policies, which have held such exclusions to apply whenever defective construction could be discovered only by expert examination.

The expansive definition of the term "latent defect" grew, in part, out of the opinion in *Acme Galvanizing Co. v. Firemans Fund Insurance Co.*,[34] a decision that interpreted the term as used in Civil Procedure Code § 337.15, which sets out the limitations period for actions arising from latent defects in construction. That case held that, for purposes of the statute of limitations, a defect is latent if it is not apparent by reasonable inspection. A similar definition was then picked up by other courts for purposes of defining the scope of the latent defect exclusion in insurance policies. *Chadwick* holds that "a standard that classes as latent or inherent all defects whose discovery requires expert examination or analysis sweeps too broadly."[35]

In *Chadwick*, the plaintiff homeowners noticed cracks in the interior sheetrock of their house and reported the loss to the defendant, Fire Insurance Exchange (FIE). Engineers employed by FIE attributed the cracking to defective design and construction of the framing system, including substitution of gang-nailed floor trusses for 2 × 12 floor joists, which created an unusual connection with no solid framing between the bearing and support beams. Another expert employed by the homeowners found numerous poor framing practices that were plainly visible from the crawlspace and concluded that the loss was primarily due to those poor framing practices. The homeowners alleged that they learned of the framing deficiencies only through the experts' inspection. In depositions, they stated they would not have been able to detect or recognize the framing deficiencies without such expert assistance.

The trial court granted summary judgment for the defendant FIE, which was reversed by the court of appeal. The court first rejected the homeowners' primary contention that the defective framing was the result of a nonexcluded peril—namely, "negligent construction," which was alleged to be the efficient proximate cause of the loss.[36] Such analysis is appropriate only when there are

---

[32] 17 Cal. App. 4th 1112 (1st Dist. 1993).
[33] 17 Cal. App. 4th at 1123.
[34] 221 Cal. App. 3d 170 (1990).
[35] 17 Cal. App. 4th at 1123.
[36] *Chadwick*, 17 Cal. App. 4th at 1116.

simultaneous multiple causes for the resulting damage. The court found, under the facts of this case, that negligent construction was not a peril distinct from creation of the defective framing. Here the loss was in fact occasioned by only a single cause: defective framing. An insured may not avoid an express exclusion merely by affixing an additional label to the act or event causing the loss.

The court then turned to the proper definition of latent defect according to the context in which the issue arises. The court rejected FIE's request to apply the standard used in statute of limitations cases—namely, whether "an average homeowner would not have observed or recognized the structural defects . . . without expert assistance."[37] Such an expansive view would be inconsistent with the reasonable expectations of the insured. Under section 337.15, "even actual observation of a deficiency or its immediate effects will not make it patent if the average consumer would not have fully understood its cause or known it to be a deficiency."[38] The court concluded that "[h]owever sensible such a broad view of latency may be in the statute of limitations context, it is inappropriate for interpreting an exclusion in an all-risk homeowners insurance policy."

Next, the court rejected FIE's contention that a broad interpretation was necessary to avoid making the insurer "the guarantor of the quality of construction of all insured property." The insurer could avoid such a result either by itself conducting an expert investigation or by writing into its policies an appropriate exclusion phrased in clear and unmistakable language, which, the court noted, FIE had already done in later editions of the policy in question.

Finally, to the extent they could be interpreted to define latent defect as including defects discoverable only by expert examination, the court of appeal in *Chadwick* disapproved *Carty v. American States Insurance Co.*,[39] *Winans v. State Farm Fire & Casualty Co.*,[40] and *Tzung v. State Farm Fire & Casualty Co.*[41] Still, the court concluded that a reasonable inspection "for purposes of the latent or inherent defect exclusion may, under some circumstances, include appropriate expert assistance or analysis."[42]

Here, the court noted, both parties' experts agreed that much of the loss was due to the substitution of trusses for joists, and the homeowners' expert noted five additional poor framing practices visible from the crawlspace. Summary judgment for the insurer was therefore improper, inasmuch as "this testimony creates a triable issue of fact as to whether a reasonable expert inspection, undertaken at the time the policy issued, would have revealed the defects causing homeowners' loss."[43]

---

[37] *Id.* at 1122.
[38] *Id.* at 1123.
[39] 7 Cal. App. 4th 399 (1992).
[40] 968 F.2d 884 (9th Cir. 1992).
[41] 873 F.2d 1338 (9th Cir. 1989).
[42] *Chadwick*, 17 Cal. App. 4th at 1126.
[43] *Id.*

Another important appellate court case is *City of San Diego v. United States Gypsum Co.*[44] In this case, the court of appeal upheld a lower court's grant of summary judgment and judgment on the pleadings in favor of multiple defendants, all manufacturers and distributors of asbestos products, on the grounds (1) that the plaintiff city's tort claims were time-barred, (2) that its cause of action for nuisance was really a products-liability action, and (3) that its equitable indemnity action could not be maintained in the absence of the city's having sustained damages to a plaintiff by way of judgment or settlement.

The city of San Diego filed its action on May 25, 1988; however, the evidence showed that the city, more than three years earlier, had undertaken an extensive investigation and evaluation of the asbestos in many city-owned buildings and had determined that degrading and deterioration were taking place. At that time, the city had taken certain remedial actions and had brought a federal suit against another defendant in which it alleged $50 million in damages.

Supported by amicus curiae briefs filed by 66 California cities and counties, the city of San Diego contended, on appeal, that (1) the three-year statute of limitations does not begin to run until a particular building is actually contaminated by release of asbestos fibers or dust; (2) defendants created a nuisance on city property; (3) the city has an equitable indemnity cause of action against the defendants even though the city is not presently liable to any injured third party; (4) the city is not bound by the rule barring recovery of economic losses; and (5) public policy allows public entities to be compensated under such circumstances.[45]

The court of appeals affirmed the dismissal of the city's suit, holding that the three-year tort statute of limitations for "injury to real property"[46] began to run as soon as "appreciable and actual damages had occurred." The court found that the city's own allegations of its damages demonstrated that asbestos had been contaminating the buildings from the time of installation, while its earlier federal suit further demonstrated that the city had suffered more than nominal damages prior to May 25, 1985. In light of the extensive studies it had performed from 1978 through 1985, the city, in its summary judgment opposition, offered to stipulate that "it knew asbestos was a dangerous and defective product . . . [and] it knew asbestos was in many of its buildings well before 1984."[47] Moreover, the fact that the city had sued another defendant in the earlier action could not rescue its present suit, "since the statute of limitations is not delayed because plaintiff does not know the wrongdoer's identity."[48]

The court of appeal also rejected the city's argument that the asbestos created a continuing nuisance, finding the case was really based upon defective products liability. The city argued that the broad definition of "nuisance" in California

---

[44] 30 Cal. App. 4th 575 (1994).
[45] 30 Cal. App. 4th at 581.
[46] Cal. Civ. Proc. Code § 338(b).
[47] 30 Cal. App. 4th at 581.
[48] *Id.* at 584.

Civil Code § 3479 goes beyond the traditional concept of nuisance as "a land-connected tort," while the amicus curiae Association of California Counties urged that the city's damages were continuing and occur anew with each significant release of asbestos fibers or dust. The court of appeal thought these arguments proved too much. "Indeed, under City's theory," the court reasoned, "nuisance 'would become a monster that would devour in one gulp the entire law of tort.'"[49]

On the issue of equitable indemnification, the court of appeal cited two prior California Supreme Court decisions to the effect that "a cause of action for indemnity does not accrue until the indemnitee suffers loss through payment of an adverse judgment or settlement."[50] The court noted (1) that the city had incurred no such liability to any third party, (2) that no statute or regulation made defendants jointly and severally liable with the city to third parties, and (3) that the city's equitable indemnity cause of action was also subject to the three-year statute of limitations in California Civil Procedure Code § 338(b).

This case is significant in that it establishes that the shorter three-year statute of limitations begins to run as soon as a "latent defect" becomes known to the property owner, whose knowledge will be presumed as soon as actual and appreciable damage has occurred. Moreover, the concept of continuing nuisance is not a viable theory for reviving an otherwise time-barred cause of action.

Although the "continuing nuisance" theory holds that injury caused by an abatable nuisance or trespass occurs repeatedly and an action thereon against an owner or trespasser may be brought at any time before the nuisance has been abated or the trespass has been discontinued (subject to the three-year limitations period of Civil Procedure Code § 338(b)), the court of appeal held that the property owner's damages action against the installing contractor was subject to the 10-year latent defect statute,[51] which barred a suit brought 19 years after construction was completed.

The court of appeal, in *Scott v. Continental Insurance Co.*,[52] once again examined the definition of "latent defect" and agreed with the opinion in *Chadwick* that the definition of that term adopted in *Acme Galvanizing*, and later used in *Carty*, is flawed and should not be followed. *Acme* and *Carty* defined a *latent defect* as a defect in construction that is "neither readily observable nor apparent on reasonable inspection." The *Scott* court found that the *Acme* opinion had "plucked" that definition of latent defect from the statute of limitations[53] even though section 337.15 begins with the phrase "as used in this section."

[C] Patent Defects

Another case decided by the court of appeal in 1996, *Tomko Woll Group Architects, Inc. v. Superior Court*,[54] considered the issue of when a construction

---

[49] 30 Cal. App. 4th at 586.
[50] *Id.* at 587.
[51] Cal. Civ. Proc. Code § 337.15.
[52] 44 Cal. App. 4th 24, 51 Cal. Rptr. 2d 566 (1996).
[53] Cal. Civ. Proc. Code § 337.15.
[54] 46 Cal. App. 4th 1326 (1996).

defect is sufficiently patent to trigger the four-year limitations period set forth in California Civil Procedure Code § 337.1. An individual plaintiff brought an action for personal injuries resulting from a slip and fall on a lobby or patio designed and built by the defendants. The trial court denied defendants' motion for summary judgment, finding a triable issue of fact as to when the alleged defect became patent so as to trigger the four-year statute of limitations under section 337.1.

In addition, *Sanchez v. Swinerton & Walberg Co.*,[55] stands for the principle that contractors cannot be held liable for patent defects once a building has been accepted by the owner. In a case in which a plaintiff slipped and fell at a newspaper production facility that had been accepted by the owner, the trial court entered summary judgment in favor of the defendant general contractor and its concrete subcontractor. The court of appeal affirmed, based upon the uncontradicted evidence that the standing water that had caused the plaintiff's fall was obvious to any reasonably observant person; therefore, the defect was patent. In fact, the owner had noticed the condition and failed to call it to the attention of the contractors. The court of appeal noted that the California Supreme Court had stated the rule, in 1857, that a contractor that constructs a defective facility, which causes injury, is not liable if the defect was patent after the project has been completed and accepted by the owner. By accepting and operating the facility, the owner assumes responsibility to the world for its sufficiency, and the contractor is thereafter relieved from any further liability.

[D] Indemnity Actions Involving Latent Defects

Fleck v. Bollinger Home Corp.,[56] demonstrates that indemnity actions present a unique situation in which a party ordinarily protected by the 10-year statute of limitations[57] may nonetheless be held liable. This possibility arises because section 337.15(c) allows an indemnitee who has not yet paid a judgment to file an indemnity cross-complaint, even after the 10 years has run in favor of the contractor (or, in this case, the developer). The court of appeal stated that the policy of § 337.15(c) is to impose liability on the one who actually caused the damages, and California law allows a party to assign "choses in action" for injuries to property.

§ 5.08 RECOVERY OF DAMAGES FOR VIOLATION OF THE BUILDING CODE

Most contracts between the owner and the prime contractor (and the subcontracts issued thereunder) require the contractor to construct the project in accordance with applicable building codes. If work is not constructed in accordance with the building code, the question is whether or not that gives rise to liability.

---

[55] 47 Cal. App. 4th 1461 (1996).
[56] 54 Cal. App. 4th 926 (1997).
[57] Cal. Civ. Proc. Code § 337.15.

In *Huang v. Garner*,[58] Mr. and Mrs. Huang, purchasers of an apartment building, brought an action alleging that the building had been defectively designed and constructed. The defendants included the real estate brokers, sellers of the property, various lenders, intermediate owners of the property, and the original owner/developer of the property. The suit went forward against Garner (the original owner/developer), Encinal Park, Inc. (the construction company), Connolly (the civil engineer), and Matson (the building designer). The jury returned a verdict in favor of the Huangs against Garner and Encinal Park. The trial court had granted a nonsuit on all theories of liability upon which Huang had sued Connolly and Matson. With respect to Garner and Encinal Park, the trial court had granted a nonsuit as to negligence and strict liability claims, refused to allow economic damages to the Huangs, and limited their recovery to physical damages only.

The apartments had been constructed on land owned by Garner, by his wholly owned construction company, Encinal Park. Garner also hired the building designer, Matson, to prepare plans for the project. Matson retained the civil engineer, Connolly, to do the structural engineering. The Huangs presented evidence that the plans and specifications were defective, including insufficient fire retardation walls, insufficient sheer walls, and inadequate structure, many of which were claimed to be in violation of the 1961 Uniform Building Code. There was additional evidence that deviation from the plans during construction also contributed to faulty construction. Garner sold the building to Bartels, who sold the property to Piper Banning Group, which sold it to Mr. and Mrs. Huang.

When the Huangs employed an engineer to determine what would be needed to convert the apartments to condominiums, the engineer discovered extensive structural damage in the garage area and other structural and design defects. The engineer testified as to both the extent of the damages and what it would cost to repair the building to bring it up to code requirements. The first issue for the court to decide was whether the Huangs could sue Matson and Connolly. The court of appeal held that the evidence regarding violations of the Uniform Building Code was sufficient to allow the case to go to the jury upon an instruction as to "negligence per se." The court of appeal also held that it was error for the trial court to deny the Huangs the right to reopen their case, after they had rested, to provide expert testimony on the standard of care for designers. The court of appeal agreed that proof of professional negligence requires testimony of experts as to the standard of care in the relevant community; and, in this case, there was no such evidence.

The court of appeal held that, under California Evidence Code § 669, proof of a statutory violation entitles the plaintiff to an instruction on the presumption of negligence when the plaintiff is within the class of persons for whose protection the statute or ordinance was adopted. The court held that the Huangs were in the class of persons intended to be protected under the Uniform Building Code. Therefore, the court of appeal held that the trial court had erred in granting a nonsuit against the building designer and the civil engineer as to the negligence

---

[58] 157 Cal. App. 3d 404 (1984).

claims of the Huangs and, further, that the trial court erred in granting a nonsuit against the original owner and the developer of the building limiting the plaintiff's recovery against them to physical damages only. Moreover, even though there was no testimony as to the standard of care of professionals such as a building designer or civil engineer in the relevant community, the Huangs had presented evidence that the building design and the structural engineering did not satisfy the requirements of the Uniform Building Code. As a result, under Evidence Code § 669, proof of a statutory violation entitled the Huangs to an instruction on the presumption of negligence since they were within the class of persons for whose protection the statute was enacted. The *Huang* case was not followed in the case of *AAS v. Superior Court*;[59] however, the California Supreme Court has granted a hearing in the *AAS* case.

In *Morris v. Horton*,[60] a contractor brought an action against the owners of property who refused to pay for additional costs incurred in correcting framing deficiencies in a staircase and for other work the contractor had undertaken on a time and materials basis. The owners cross-complained for the cost incurred to repair, replace, complete, or change work the contractor had agreed to perform. The owners also claimed that the contractor and a subcontractor were liable to them for the cost of repairing and replacing staircase work that was negligently performed by the subcontractor. The jury awarded the contractor $72,406 and awarded the owners $8,100. The jury specifically found that the owners were not entitled to recover any sum as a result of any inadequacies in the work done by the subcontractor on the staircase.

The court of appeal modified the judgment to adjust the award of damages to the contractor but affirmed the judgment in all other respects. The court held that the owners failed to establish their entitlement to a judgment on the cross-complaint on a negligence per se theory under Evidence Code § 669. As owners and occupants of the building in which the staircase was installed, they were within the class of persons for whose protection the City Building Code was enacted; however, the Building Code provisions prescribing the height and width of stairs and the height of a handrail were designed to ensure the safety of persons using the staircase and were not designed to prevent owners from incurring costs associated with their remediation of the Building Code violations.

The *Morris* court distinguished *Huang v. Garner, supra,* on the ground that the plaintiffs in *Huang* were subsequent purchasers of an apartment building who brought a negligence action against the original designers after discovering that the building had suffered extensive structural damages due to design and structural defects. The *Huang* court held that the plaintiffs were entitled to go to the jury on a negligence per se theory, stating without citation to any authority that it was reasonable to conclude that the Uniform Building Code was designed to prevent the types of defects and damages to construction that occurred in that case. The *Morris* court stated that the injury in the *Huang* case was of an entirely different order than the injury suffered by the Morrises in this case. The evidence

---

[59] 64 Cal. App. 4th 916 (1998).
[60] 22 Cal. App. 4th 968 (1994).

in *Huang* indicated that the designer's violations of the Uniform Building Code had resulted in severe structural damage that was likely to lead to the building's collapse. The *Morris* court adopted the *Huang* court's conclusion that structural provisions of the Uniform Building Code were designed to prevent the structural deterioration and collapse or near collapse of buildings, including the conclusion that provisions of the Building Code regarding the height and width of stairs were not intended to prevent building owners from incurring costs to remediate violations of those provisions. Thus, the *Morris* court held that any costs incurred by the owners for the repair or replacement of the staircase were not the type of "injury" that the Building Code was designed to prevent, and therefore, the owners were not entitled to prevail on a negligence per se theory.

It appears from these two cases that the deciding factor as to whether or not a violation of the Building Code is negligence per se will be the type of defect that is involved. It is the authors' opinion, however, that if the owner could prove negligence on the part of the designer, the owner should prevail.

§ 5.09 TERMINATION OF CONTRACTOR

For a complete discussion of the right of an owner to terminate a contractor, *see* § 6.04 in Chapter 6.

§ 5.10 IMPACT OF CONSTRUCTION CONTRACT TERMS

The types of damages discussed in §§ 5.03 and 5.04 above may be modified by the terms of the contract itself. In 1997, the American Institute of Architects promulgated its 1997 version of the AIA Document A201 General Conditions. Subparagraph 4.3.10 is a new section that was added in 1997. It provides the following:

> 4.3.10. Claims for Consequential Damages. The Contractor and Owner waive Claims against each other for consequential damages arising out of or relating to this Contract. This mutual waiver includes:
>
> 1. damages incurred by the Owner for rental expenses, for losses of use, income, profit, financing, business and reputation, and for loss of management or employee productivity or of the services of such persons; and
>
> 2. damages incurred by the Contractor for principal office expenses including the compensation of personnel stationed there, for losses of financing, business and reputation, and for loss of profit except anticipated profit arising directly from the Work.
>
> This mutual waiver is applicable, without limitation, to all consequential damages due to either party's termination in accordance with Article 14. Nothing contained in this Subparagraph 4.3.10 shall be deemed to preclude an award of liquidated direct damages, when applicable, in accordance with the requirements of the Contract Documents.

Its purpose was to limit consequential damages claims in disputes between owners and contractors. Under said provision, the owner would be waiving many of the large damages claims that it might have against a contractor, to wit, (1) rental expenses, (2) loss of use, (3) loss of income, (4) loss of profit, (5) financing losses, (6) business and reputation losses, and (7) loss of management or employee productivity or services. The contractor waives (1) principal office expenses, (2) loss of financing, (3) loss of business and reputation, and (4) loss of profit except anticipated profit arising directly from the work. The new section also provides that nothing set forth therein shall be deemed to preclude an award of "liquidated direct damages," which term is not defined. The authors believe that owners could still recover costs of completion in the event of abandonment and costs of correcting defective work. It is unclear whether "liquidated damages" (if provided for in the contract) could be recoverable. It is the opinion of the authors that liquidated damages would still be recoverable if provided for in the contract.

§ 5.11 CONCLUSION

Contractors failing to perform properly can be held liable to an owner for:

1. Costs of correcting defective work
2. Costs of completion
3. Delay damages such as loss of rent, liquidated damages, interest, and loss of profits.

In a particularly grievous case, punitive damages for fraud may apply.

CHAPTER 6
CONSTRUCTION CLAIMS

§ 6.01 Introduction: The Importance of Early Claims Recognition
§ 6.02 Overview of Owner's Claims
    [A] Owner's Delay Claims
    [B] Liquidated Damages Clauses
    [C] Contractor's Failure to Perform per Plans and Specifications
§ 6.03 Public Works Contractor's Liabilities for False Claims
§ 6.04 Terminating the Contractor
    [A] Damages Recoverable from the Terminated Contractor
    [B] Rescinding the Contract
§ 6.05 Overview of Contractor's Claims
§ 6.06 Contractor's Time-Related Claims
§ 6.07 Contractor's Delay Claims
    [A] Delay and Time Extension Claims
    [B] Categorizing Types of Delay
    [C] Excusable versus Inexcusable Delays
    [D] Compensable versus Noncompensable Delays
    [E] Concurrent Delays
    [F] Causes and Effects of Delay
    [G] Extended Home Office Overhead: The *Eichleay* Formula
    [H] No Damages for Delay Clauses
§ 6.08 Contractor's Disruption Claims
    [A] Disruption versus Delay
    [B] Examples of Disruption
    [C] Proving Disruption Claims
§ 6.09 Contractor's Acceleration Claims
    [A] Actual Acceleration
    [B] Constructive Acceleration
§ 6.10 Overview of Leading Causes of Claims
    [A] Changes (Actual, Constructive, and Cardinal)
    [B] Actual versus Constructive Changes
    [C] Cardinal Changes
§ 6.11 Differing Site Conditions

§ 6.12   Suspension of Work
§ 6.13   Claim Investigation
         [A]   Claim Documentation Checklist
         [B]   Claim Outline and Narrative
§ 6.14   Quantifying and Proving Damages
         [A]   Pros and Cons of the "Total Cost" Method
         [B]   Applications of the Jury Verdict Method

§ 6.01 INTRODUCTION: THE IMPORTANCE OF EARLY CLAIMS RECOGNITION

Timely recognition and proper handling of construction claims are often the key to a successful project for both contractors and owners. To the contractor, early claims recognition may make the difference between realizing a profit or suffering a loss. Knowledge of claims is equally important to an owner, who must realistically evaluate and promptly respond to claims made by a contractor. Indeed, in certain instances the owner may also have claims against the contractor. Given the trouble and expense of litigation, it is in the best interests of both contractors and owners to do whatever is possible to recognize and resolve disputes in order to avoid litigation, if possible, and to prevail in court if necessary.

Furthermore, early claim recognition is of paramount importance to both the contractor and the owner because most construction contracts have notice clauses requiring the contractor to give notice of various types of claims within a specified number of days. Typically, a contractor is required to give notice within 20 to 30 days after the events giving rise to a claim. A contractor may be entirely barred from recovery on a claim not made within such contractual requirements. Another important consideration is that the party asserting a claim bears the burden of proving its entitlement and documenting its extra costs. Early claim recognition allows the claimant (owner or contractor), to the extent possible, to isolate the costs associated with a particular claim. Once the project is completed, it is usually difficult and sometimes impossible to document a claim properly and completely. Such lack of documentation minimizes the chances of an agreeable settlement between the contractor and the project owner and makes it extremely difficult to obtain a favorable result through litigation or arbitration. Some of the procedures and techniques for properly documenting claims are considered in more detail in Chapter 4.

§ 6.02 OVERVIEW OF OWNER'S CLAIMS

Most claims by project owners against contractors are based on a breach of contract theory. Whether a contractor may be liable to the owner for breach of contract depends upon the answer to two questions: First, did the contractor complete the project within the time required under the contract (including any extensions of time authorized by owner-approved change orders)? Second, did the contractor's work comply with the requirements of the project plans and specifications?

In the absence of a liquidated damages provision, discussed below, an owner's damages will be based upon the general principle that governs any action for breach of contract. California Civil Code § 3300 provides:

> For the breach of an obligation arising from contract, the measure of damages, except where otherwise expressly provided by this Code, is the amount which will compensate the party aggrieved for all the detriment proximately caused thereby, or which, in the ordinary course of things, would be likely to result therefrom.

The application of this general principle of damages is part of the discussion below of the particular kinds of claims that typically arise from construction projects.

[A] Owner's Delay Claims

Even a partial failure of performance by the contractor can give rise to an owner's claim for damages for delay. If the owner can prove that the delay in completion of the project was the fault of the prime contractor or of one or more of its subcontractors and/or material suppliers, the owner may be able to recover the value of the loss of use of the facility between the time the project should have been completed under the contract until actual completion or beneficial occupancy becomes available to the owner. For an owner intending to occupy the completed project itself, such damages may be quite difficult to quantify and prove. This is particularly true of public owners, whose use of facilities is generally not expected to generate income or profits. The burden of proof is significantly easier for the owner of income property, who may recover loss of rent resulting from the delay in completion.[1] For the owner who has borrowed money to finance construction of the project, additional and/or increased interest payments may be recoverable if the contractor knew or should have known that the project was being financed through loans.[2]

In appropriate cases, an owner may also be awarded loss of profits.[3] In extreme cases, when fraud can be established, an owner may even recover punitive damages.[4]

[B] Liquidated Damages Clauses

A liquidated damages clause is a contract provision that requires payment by the contractor of a set amount for each day the project completion is delayed beyond the date set forth in the contract. In many instances, liquidated damages may also be required with reference to "interim milestones," which are specific dates for completing certain portions of the work. One of the primary purposes of liquidated damages is to serve as an incentive to motivate the contractor to complete the project, or some particular portion thereof, in a timely manner. Section 1671 of the California Civil Code creates a presumption favoring the validity of liquidated damages clauses. Generally, courts will enforce such provisions and award liquidated damages unless the party seeking to invalidate the clause is able to prove "that the provision was unreasonable under the circumstances existing at the time the contract was made." This standard is also written into the provision of the California Public Contract Code that requires liquidated damages provisions to be included in state construction contracts.[5] However, under

---

[1] *See* Bird v. American Sur. Co., 175 Cal. 625 (1917). *See also* Camper v. McDermott, 266 Cal. App. 2d 41 (1968).
[2] *See* Mendoyoma, Inc. v. County of Mendocino, 8 Cal. App. 3d 873 (1970).
[3] *See, e.g.,* Burnett & Doty Dev. Co. v. Phillips, 84 Cal. App. 3d 384 (1978).
[4] Walker v. Signal Cos., 84 Cal. App. 3d 982 (1978).
[5] *See* Cal. Pub. Cont. Code § 10226 (formerly Cal. Gov't Code § 14376).

the general principle that liquidated damages should not be awarded when their enforcement would constitute a penalty rather than an approximation of actual damages that are difficult to determine, it is questionable whether an owner would be allowed to recover liquidated damages for delays to a project that do not cause any actual damages to the owner. To date, no California case has been decided in which an owner has been awarded liquidated damages when, upon completion, it was unable to use a finished project due to, for example, a lack of funds with which to operate the facility.

Owners should view liquidated damages clauses with caution and should carefully consider their effects in light of the circumstances involved in their particular project. For example, an owner with significant financing costs and/or expectations of substantial revenues (such as rents) that can be easily calculated may have actual damages that far exceed the stipulated amounts for liquidated damages. Moreover, the existence of liquidated damages may motivate the contractor to be especially assertive in contending and documenting that the owner is at fault for any delays that may be encountered in constructing a project.

Nationally, courts are divided as to whether liquidated damages clauses are valid when the underlying delay is only partly caused by the contractor and is contributed to by the owner.[6] Although some jurisdictions will apportion damages according to the degree of fault on the part of both the owner and the contractor, California courts will not attempt such an apportionment but will refuse to enforce the liquidated damages provision in such cases.[7] Thus, in the *General Insurance* case, it was held that "[t]he correct rule is that where such delays are occasioned by the mutual fault of the parties the court will not attempt to apportion them but will refuse to enforce the provision for liquidated damages."[8] This, then, is the rule in California, at least in cases involving concurrent project delays.

Under the predecessor to Public Contract Code § 10226, cited above, at least one California court determined that liquidated damages were enforceable in a situation in which the owner and the contractor *each* caused delays, holding that "[d]amages are not being apportioned. Damages are liquidated. Quantum of delay in terms of time is all that is being apportioned. That is an uncomplicated fact finding process. That is what courts are for."[9] It important to emphasize that the *Nomellini* case did not involve concurrent delays; rather, the owner and contractor each caused separate delays, as to which they were each entitled to recovery against one another. Thus, the kind of "apportionment" approved of by the

---

[6] For example, in *San Ore-Gardner v. Missouri Pacific Railroad Co.*, 496 F. Supp. 1337, 1349 (E.D. Ark., 1980), the U.S. District Court refused to apportion fault or enforce liquidated damages; whereas, in *Aetna Casualty & Surety Co. v. Butte-Meade Sanitary Water*, 500 F. Supp. 193, 197 (D.S.D. 1980), another District Court held the owner's partial fault for delays did not preclude an award of liquidated damages. *See also* the discussion of "concurrent delay" in § 6.07[E] below.

[7] General Ins. Co. of Am. v. Commerce Hyatt House, 5 Cal. App. 3d 460 (1970).

[8] 5 Cal. App. 3d at 472 (quoting Gogo v. Los Angeles County Flood Control Dist., 45 Cal. App. 2d 334, 344 (1941)).

[9] Nomellini Constr. Co. v. State *ex rel.* Department of Water Resources, 19 Cal. App. 3d 240, 246 (1971). *See also* Robinson v. United States, 261 U.S. 486, 488 (1923).

courts in both *Nomellini* and *Robinson* consists of subtracting the days of delay for which the owner was responsible from a greater number of days for which the contractor was responsible, and enforcing the liquidated damages provision as to the difference.[10] Presumably, if the owner caused a greater number of delay days, the contractor would likewise be entitled to recover its damages for the net difference in its favor.

Finally, in order to collect liquidated damages, an owner must not itself be in breach of the contract's delay provisions, as "[l]iquidated damages are a penalty not favored in equity and should be enforced only after he who seeks to enforce them has shown that *he has strictly complied with the contractual requisite to such enforcement.*"[11] This rule invalidating liquidated damages apparently applies when an owner's delays prevent the contractor's timely completion: "By its own act defendant [owner] rendered performance *within the time limited by the contract impossible* and has therefore lost its right to claim the liquidated damages provided in the contract."[12]

The question arises then: Should an owner include a liquidated damages clause in its construction contract? From the owner's perspective, factors favoring a liquidated damages clause include (1) the clause may provide an incentive for the contractor to pursue the work diligently; and (2) if enforceable, the liquidated daily amount will overcome the owner's usual difficulty in proving and quantifying its damages. The factors against including a liquidated damages clause are:

1. If the owner is even partially at fault for the delays, the resulting invalidation of the liquidated damages clause may entirely preclude recovery of any delay damages by the owner;

2. The very existence of the provision may cause the contractor to be particularly aggressive in attempting to establish and document the owner's fault in delaying construction; and

3. The owner may not be able to insert a high enough dollar amount for liquidated damages in the contract and the contractor may actually determine that it is in its best interests to "buy" the time at the liquidated damages rate.

[C] Contractor's Failure to Perform per Plans and Specifications

If the contractor fails to meet its contractual obligation to perform the work in a good and workmanlike manner and as required by the project plans and specifications, it will be held liable to the owner for the resulting damages. Under

---

[10] *Nomellini*, 19 Cal. App. 3d at 245 (quoting *Robinson*, at 261 U.S. at 487–488).

[11] General Ins. Co. v. Commerce Hyatt House, 5 Cal. App. 3d 460, 472 (1970) quoting Aetna Cas. & Sur. Co. v. Board of Trustees, 223 Cal. App. 2d 337, 340 (1963)) (emphasis in original).

[12] Aetna Cas. & Sur. Co. v. Board of Trustees, 223 Cal. App. 2d 337, 340 (1963) (quoting Gogo v. Los Angeles County Flood Control Dist., 45 Cal. App. 2d 334, 344 (1941)) (emphasis in original).

what has been called the *cost of correction rule,* the usual measure of damages for failure to construct in conformity with the plans and specifications is the cost that would reasonably be required to remedy the defect and make the product conform to the contract's requirements.[13]

Under other similar circumstances, the alternate measure of damages may be measured by the diminution in market value as a result of the defects. This is determined by the difference between the value of the building as improperly constructed and what its value would have been if it had been properly constructed. This measure of damages is appropriate when the cost of repair would exceed the loss of market value or is otherwise excessive, when there would be economic waste in making repairs, or where remedying the defect might require rebuilding the entire structure. For example, in *Bayuk v. Edson,*[14] the plaintiff's residence was poorly constructed following defective plans and specifications prepared by the project architect. Based upon expert testimony that the cost of repair would be economically unfeasible, the court awarded damages based on the diminution of market value.[15]

A similar circumstance arises when a contractor has performed the work in good faith but has somehow failed to comply with the technical requirements of the contract documents. Where such failure does not have significant impact on the value or usefulness of the finished project, the contractor is said to have accomplished "substantial performance" of the contract. When the owner has taken possession and is making use of a project, the owner is considered to be the beneficiary of the contractor's substantial performance; thus, the owner's actual damages may be deducted from, or offset against, the contract price otherwise due the contractor. In an action to recover under the contract, the contractor who has rendered substantial performance is entitled to recover the unpaid portion of the contract price, less damages for the lack of strict performance.[16]

Thus, in the most common circumstance, where no specific performance criteria are set forth in the contract, a contractor's work will nonetheless be expected to be of reasonable quality. If the work does not meet this implied "reasonableness standard," the contractor may be denied recovery under the contract and/or the owner may be entitled to recover damages. Where, in addition to the usual detailed drawings and specifications, the contract specifications also provide for the finished work to achieve a specific performance standard, a contractor who constructs the work in strict accordance with the detailed plans will usually not be held in breach of the contract if the specific performance criterion is not achieved.[17]

---

[13] Jones v. Kvistad, 19 Cal. App. 3d 836 (1971).

[14] 236 Cal. App. 2d 309 (1965).

[15] *See also* Shell v. Schmidt, 164 Cal. App. 2d 350 (1958); Amerson v. Christman, 261 Cal. App. 2d 811 (1968).

[16] Tolstoy Constr. Co. v. Minter, 78 Cal. App. 3d 665 (1978).

[17] *See* Sunbeam Constr. Co. v. Fisci, 2 Cal. App. 3d 181 (1969); Kurland v. United Pac. Ins. Co., 251 Cal. App. 2d 112 (1967).

When an owner sues a contractor for failure to complete the project (i.e., partial performance), the usual measure of damages is the owner's cost of completion (e.g., payments to another contractor who finishes the job), plus incidental damages, less any expenses that may have been saved by the breach.[18]

§ 6.03 PUBLIC WORKS CONTRACTOR'S LIABILITIES FOR FALSE CLAIMS

California's False Claims Act[19] ("the State Act") was enacted in 1987 and closely parallels the United States False Claims Act[20] ("the Federal Act") as amended in 1986. Also known as *qui tam* statutes, both the state and federal laws prohibiting and punishing false claims by contractors are coming into ever increasing usage by public owners and by private citizens. A civil action on a contractor's allegedly false claim may be brought by the U.S. or California Attorney General, by the prosecuting authority of a political subdivision of the state, or by a *qui tam* plaintiff. *Qui tam* is a shortened form of the Latin phrase "*qui tam pro domino rege quam pro se ipso in hac parte sequitur,*" which translates: "he who sues on behalf of the King as well as for himself." The *qui tam* provisions in the False Claims Acts authorize a private citizen to bring suit on behalf of the government and receive a percentage of the amount recovered. The existence and growing usage of these laws has significantly increased the risks involved in seeking and performing public works contracts and pursuing claims. Moreover, while most California contractors are more likely to be exposed to liability under the State Act, they should also be alert for possible federal liability on projects that involve any "flow-down" federal funding.

For the reasons explained in more detail below, contractors who are contemplating filing claims on public projects should be wary that many public agencies are aggressively pursuing false claims actions and are also seeking criminal indictments in certain cases. General contractors who are contractually required to "certify" claims of their subcontractors must also keep in mind that the act of certifying such claims may leave them open to a false claims action. General contractors in this position may be well advised to require such subcontractors to indemnify them from such a possibility by an appropriate indemnity clause in their subcontracts.

The Federal False Claims Act was amended and revitalized in 1986, based upon what Congress perceived to be a "severe" problem of the "growing pervasiveness of fraud" in federal programs and procurement. As a result of the 1986 amendments, *qui tam* filings under the Federal Act increased from a mere 33 filings in 1987 to 534 filings in 1997. Following suit, California (in 1987) and several other states enacted their own False Claims Acts, directly modeled after the federal statute.

---

[18] Kennedy v. Reece, 225 Cal. App. 2d 717 (1964).
[19] Cal. Gov't Code §§ 12650 *et seq.*
[20] 31 U.S.C. §§ 3729 *et seq.*

Although California's False Claims Act has been in effect for a dozen years, there is a dearth of case law applying and/or interpreting its provisions.

As a result of the legislation passed in the 1980s, both the Federal and State False Claims Acts have taken on a new level of significance to construction contractors. Target defendants of the original federal law, enacted during the 1860s, were mostly Civil War defense contractors. Defense contractors remained the primary targets for false claims actions following the 1986 amendment of the Federal Act. By the early 1990s, healthcare contractors had become the main target defendants, and, by the late 1990s, the main targets have become public works construction contractors.

Both the State and Federal False Claims Acts impose civil liability on any contractor[21] who presents a false claim for money, property, or services to a governmental entity or representative. The State Act defines a *claim* to include any request or demand for money, property, or services made to any employee, officer, or agent of the state or of any political subdivision, or to any contractor, grantee, or other recipient, whether under contract or not, if any of the funding comes from the state or any of its political subdivisions.[22] This is based upon the definition of "claim" in the Federal Act, which applies whenever the U.S. government either provides all or any portion of the money requested or demanded, or will reimburse the payer for any part of the money to be paid to the contractor claimant.[23]

The specific acts prohibited under both the State and Federal False Claims Acts are stated in substantively identical lists,[24] which include the following very broad categories:

1. Knowingly presenting or causing to be presented to an officer or employee of the state, or of any political subdivision thereof, a false claim for payment or approval

2. Knowingly making, using, or causing to be made or used a false record or statement to get a false claim paid or approved by the state or by any political subdivision

3. Conspiring to defraud the state or any political subdivision by getting a false claim allowed or paid by the state or by any political subdivision

4. Having possession, custody, or control of public property or money used or to be used by the state or by any political subdivision and knowingly delivering or causing to be delivered less property than the amount for which a certificate or receipt is given

---

[21] The False Claims Acts are not limited to contractors but apply as well to a wide variety of claims made to public entities.

[22] Cal. Gov't Code § 12650(a). The term "political subdivision" includes any city, city and county, tax or assessment district, or legally authorized local governmental entity with jurisdictional boundaries (*id.* § 12650(c)), as well as the University of California (*id.* § 12652.5).

[23] 31 U.S.C. § 3729(c).

[24] Cal. Gov't Code § 12651(a)(1)–(7); 31 U.S.C. § 3729(a)(1)–(7).

5. Having authorization to make or deliver a document certifying receipt of property used or to be used by the state or by any political subdivision and knowingly making or delivering a receipt that falsely represents the property used or to be used

6. Knowingly buying or receiving, as a pledge of an obligation or debt, public property from any person who lawfully may not sell or pledge the property

7. Knowingly making, using, or causing to be made or used a false record or statement to conceal, avoid, or decrease an obligation to pay or transmit money or property to the state or to any political subdivision.

To this list, California adds an eighth category, which seems to be especially directed at general contractors who certify a subcontractor's claim and add their own markup:[25]

8. Being a beneficiary of an inadvertent submission of a false claim to the state or a political subdivision, subsequently discovering the falsity of the claim, and failing to disclose the false claim to the state or the political subdivision within a reasonable time after discovery.

Any person who commits one of these prohibited acts will be liable to the governmental entity (i.e., United States, California, or political subdivision) for "treble damages," meaning three times the amount of actual damages that are sustained by the governmental entity, plus court costs and a civil penalty of up to $10,000 for *each* false claim.[26] The Federal Act is essentially the same, except the amount of the civil penalty is at least $5,000 and not more than $10,000.[27] Moreover, such civil liability is *in addition* to any criminal penalties that may apply; and any contractor who has been convicted (or who pleads guilty or *nolo contendere*) in a criminal case based upon the same facts will not be allowed, in the civil action, to deny the essential elements of the civil offense.[28] Otherwise, in prosecuting a civil action, the Attorney General, other public prosecutor, or *qui tam* plaintiff must prove all of the elements of the offense by a preponderance of the evidence.[29] Both the Federal and State Acts require proof that the contractor submitted false claim information "knowingly," which is defined to mean that the person had actual knowledge of the information, or acted in deliberate ignorance or in reckless disregard of the truth or falsity of the information. No proof of specific intent to defraud is required.[30]

The $10,000 civil penalty *per false claim* is especially dangerous because if, for example, the basis for a false claim is an overstatement of overhead

---

[25] Cal. Gov't Code § 12651(a)(8).
[26] *Id.* § 12651(a).
[27] 31 U.S.C. § 3729(a).
[28] Cal. Gov't Code § 12654(d); 31 U.S.C. § 3731(d).
[29] Cal. Gov't Code § 12654(c); 31 U.S.C. § 3731(c).
[30] Cal. Gov't Code § 12650(b); 31 U.S.C. § 3729(b).

expenses or labor rates, each time that overstated cost appears in an invoice or payment request, it is, by definition, a separate false claim and will therefore subject the defendant contractor to an additional $10,000 penalty.

Liability under the State Act is joint and several for any act committed by two or more persons.[31] Thus, regardless of whether defendants are sued together in a single action, or individually in separate actions, they are each liable for full satisfaction of all damages and penalties assessed under the State Act. Furthermore, in light of the very broad language used to describe the prohibited acts, liability may attach to attorneys, CPAs, claims consultants, and virtually anyone else who participates in the preparation of a "false" claim.

Not all false claims against government entities fall within the realm of the False Claims Acts. In general, the following are excluded:

1. The State Act does not apply to any controversy involving an amount less than $500 in value.[32] For purposes of the rule, "controversy" means any one or more false claims submitted by the same person.

2. The State Act does not apply to a defendant contractor's fraudulent act involving a claim, record, or statement made pursuant to the California Tort Claims Act.[33]

3. Claims made under the laws of workers' compensation and insurance are specifically excluded by the State Act.[34]

4. Claims pursuant to state and federal tax laws are not subject to either False Claims Act. Claims under California's Revenue and Taxation Code are excluded by the State Act,[35] and claims under the Internal Revenue Code are similarly excluded by the Federal Act.[36]

5. Actions may not be brought by *qui tam* plaintiffs based upon information already known to the government. Under the State Act, a court may not have jurisdiction over an action for fraudulent claims initiated by a *qui tam* plaintiff against a member of the State Senate or Assembly, a member of the State Judiciary, an elected official in the executive branch of the state, or a member of the governing body of any political subdivision, if the action is based on evidence or information known to the state or political subdivision when the action was brought.[37] This provision is modeled after a similar provision of the Federal Act[38] that denies court jurisdiction, where the government already knows the underlying information, in cases that are brought by *qui tam* plaintiffs

---

[31] Cal. Gov't Code § 12651(c). The Federal Act contains no such provision.
[32] Cal. Gov't Code § 12651(d). The Federal Act contains no such provision.
[33] Cal. Gov't Code §§ 810–996.6.
[34] *Id.* § 12651(e).
[35] *Id.* § 12651(f).
[36] 31 U.S.C. § 3729(e).
[37] Cal. Gov't Code § 12652(d)(1).
[38] 31 U.S.C. § 3730(e)(2).

against members of the U.S. government (i.e., members of Congress, members of the judiciary, and "senior executive branch officials").

6. Under both the State and Federal Acts, a *qui tam* plaintiff may not bring an action that is based on allegations or transactions that are the subject of an existing civil suit or an administrative civil money penalty proceeding in which the government is already a party.[39]

7. Both the State and Federal Acts deny jurisdiction for their courts to hear *qui tam* actions based on publicly disclosed matters. The State Act prohibition[40] includes all such actions that are "based upon public disclosure of allegations or transactions in a criminal, civil, or administrative hearing, in an investigation, report, hearing, or audit conducted by or at the request of the Senate, Assembly, auditor, or governing body of a political subdivision, or from the news media, unless the action is brought by the Attorney General or the prosecuting authority of a political subdivision, or if the person bringing the action is an original source of the information." The Federal Act prohibition[41] is substantively similar with reference to U.S. governmental entities. The term "original source" under the State Act "means an individual who has direct and independent knowledge of the information on which the allegations are based, who voluntarily provided the information to the state or political subdivision before filing an action based on that information, and whose information provided the basis or catalyst for the investigation, hearing, audit, or report that led to the public disclosure." Under the Federal Act,[42] "original source" is similarly defined with respect to the U.S. government but does not include the requirement that the *qui tam* plaintiff's information "provided the basis or catalyst" for the government's activities leading to the disclosure.[43]

8. Both the State and Federal Acts significantly restrict *qui tam* actions based on information discovered by governmental employees. Under the State Act,[44] courts are denied jurisdiction to hear *qui tam* actions for fraudulent claims "based on information discovered by a present or former employee of the State or of a political subdivision during the course of his or her employment, unless that employee first in good faith exhausted existing internal procedures for reporting and seeking recovery of such falsely claimed sums through official channels and unless the State or political subdivision failed to act on the information provided within a reasonable period of time." The analogous provision

---

[39] Cal. Gov't Code § 12652(d)(2); 31 U.S.C. § 3730(e)(3).
[40] Cal. Gov't Code § 12652(d)(3)(A), (B).
[41] 31 U.S.C. § 3730(e)(4)(A).
[42] Cal. Gov't Code § 12652(d)(3)(B).
[43] 31 U.S.C. § 3730(e)(4)(A).
[44] Cal. Gov't Code § 12652(d)(4).

of the Federal Act[45] is much more limited and prohibits actions brought by military personnel against other members of the armed services "arising out of such person's service in the armed forces."

Both the State and Federal Acts contain significant "whistle-blower" provisions, intended to prevent employers from retaliating against employees who report false claims and/or cooperate in enforcement suits.[46] Employees who suffer any such discrimination are empowered to bring legal action to recover "all relief necessary to make the employee whole." Such relief specifically includes reinstatement of employment and restoration of seniority status, double back pay plus interest, and "compensation for any special damages sustained as a result of the discrimination, including litigation costs and reasonable attorneys' fees."

The percentage that a *qui tam* plaintiff can recover under the State Act varies from 15 percent to 50 percent of the "proceeds," depending upon the level of participation, if any, of the attorney general and/or other public prosecuting authority.[47] The comparable percentages recoverable by a *qui tam* plaintiff under the Federal Act range from 15 percent to 30 percent.[48] In addition, under both the Federal and State Acts, a prevailing plaintiff may be compensated for reasonable expenses, costs, and attorneys' fees.[49]

The potential for such recovery is, intentionally, a strong incentive for private citizens to bring *qui tam* actions against claimants who may have submitted false claims to public entities (or owners of projects receiving government funding). While this is, no doubt, sound public policy from the point of view of public owners, there is an obvious danger of abuse in any law that enlists the public as, in effect, "bounty hunters" in search of "windfall profits" of their own. Indeed, there is always a potential for abuse where risks are low and possible rewards are high. The authors are likewise mindful of the potential for abuse on the part of public project owners themselves, who may be tempted to use the threat of False Claims Act enforcement actions to deter or minimize a contractor's legitimate claims. While such intimidation tactics are clearly unethical, and arguably illegal, there is little to deter them in either of the False Claims Acts. The potential for such abuse is quite real, given the high stakes involved in many public works projects and the fact that public entities and their employees are protected by sovereign immunity from liability for bad-faith actions that would subject a private citizen to punitive damages.

The fact is, there appears to be almost no downside risk to an overzealous public owner, attorney general, local prosecuting authority, or private citizen in bringing a false claims action against a contractor submitting a claim. The only deterrent, which appears to be quite remote, is the possibility that a court could

---

[45] 31 U.S.C. § 3730(e)(1).

[46] Cal. Gov't Code § 12653; 31 U.S.C. § 3730(h).

[47] *See* Cal. Gov't Code § 12652(g). The State Act defines *proceeds* to "include civil penalties as well as double or treble damages as provided in Section 12651." *Id.* § 12652(g)(7). The Federal Act contains no similar definition but appears to imply the same.

[48] 31 U.S.C. § 3730(d).

[49] Cal. Gov't Code § 12652(g)(8); 31 U.S.C. § 3730(d)(2).

award the prevailing defendant its reasonable attorneys' fees if the court finds that the claim of a *qui tam* plaintiff was "clearly frivolous, clearly vexatious, or brought solely for the purposes of harassment."[50] This is a high standard and bears a difficult burden of proof (beyond a mere preponderance of evidence), and thus does not appear to pose a significant threat to federal, state, political subdivision, or *qui tam* plaintiffs.

In light of the foregoing, the authors suggest that all contractors with claims on public works construction projects should:

1. Carefully review, for accuracy and reasonableness, all invoices, payment requests, requests for equitable adjustment, and claims prior to submitting them to the public owner.

2. Accumulate and maintain backup documents for all costs and estimates that may potentially result in claims.

3. Carefully distinguish between their actual, estimated, and/or contractual costs and the rates specified in all contracts with public entities.

4. Carefully review all subcontractor pass-through claims, and require appropriate backup documentation and information.

Finally, general contractors who are obligated under their prime contracts to certify subcontractors' pass-through claims should be sure to include in all subcontract agreements a well-drafted indemnification provision for any subcontractor false claim violations.[51]

§ 6.04 TERMINATING THE CONTRACTOR

The most extreme remedy available to an owner is terminating the contractor's performance. Some contracts have a provision that allows the owner to terminate the contract for any reason. Such provisions are called "termination for convenience" clauses and generally require the owner to compensate the contractor for the work actually performed prior to such termination. As this kind of termination is usually triggered by factors (e.g., the owner's project financing) that have nothing to do with the contractor's performance, and such provisions usually define the method of calculating the contractor's compensation, they are typically not a source of contested claims and are therefore not considered in this chapter.

Termination "for cause" is a much different matter and almost always results in claims back and forth between the owner and the terminated contractor. Owners

---

[50] This identical language is in California Gov't Code § 12652(g)(9) and 31 U.S.C. § 3730(d)(4).

[51] This may be ultimately a futile effort in cases where it is alleged that the general contractor and subcontractor are conspirators with respect to the false claim, under California Government Code § 12651(a)(3) or 31 U.S.C. § 3729(a)(3). Also, under federal case law, there is no legal right to indemnity or contribution among participants in a scheme to defraud the government in violation of the False Claims Act. *See* Mortgages, Inc. v. U.S. Dist. Court, 934 F.2d 209 (9th Cir. 1991).

must be extremely careful when making a determination to terminate a contractor for cause. Records must be carefully documented to demonstrate (1) that the contractor is in material breach of contract; (2) that the owner has given the contractor an opportunity to cure the breach; (3) that the contractor has failed to do so; and (4) that the contractor's breach of contract is not excused or otherwise justified by some act or omission on the part of the owner. In other words, the owner must demonstrate that it is not itself in breach of the contract.

The most crucial element of a termination for cause is that the contractor's breach of the contract is material.[52] This element requires much more than a minor deviation in the work or a slight departure from the terms of the contract; it requires a failure by the contractor to perform one or more of the essential terms of the contract. To be deemed material, a breach must go to the very purpose of the contract itself. The factors affecting materiality are set forth in the *Restatement (Second) of Contracts* § 266 as:

1. The extent to which the injured party will be deprived of the benefit which he reasonably expects;

2. The extent to which the injured party can be compensated for the part of that benefit of which he will be deprived;

3. The extent to which the party failing to perform (or failing to offer to perform) will suffer forfeiture;

4. The likelihood that the party failing to perform (or failing to offer to perform) will cure his failure, taking into account all the circumstances, including any reasonable assurances; and

5. The extent to which the behavior of the party failing to perform (or failing to offer to perform) comports with standards of good faith and fair dealing.

Some construction contracts contain termination clauses that allow one party to terminate the other upon the occurrence of a specifically stipulated default or failure of performance. Such contract provisions define the grounds for termination thereunder. Thus, if a termination meets those specific contractual criteria, it need not otherwise satisfy the legal criteria for materiality.

Most construction contracts provide a number of specific notice requirements. In the case of a rescission for cause, such notice requirements should be strictly complied with, and if a surety is involved, it is particularly crucial to give the surety appropriate notice of the contractor's default. A failure to properly notify the surety could exonerate the surety's obligations under its performance bond, which is an important source for the owner's recovery. This is especially true on any project in which the termination was based upon the contractor's failure to meet its financial obligations, or a declaration of insolvency or bankruptcy, in which case recovery against the contractor itself is likely to be impossible.

---

[52] *See* Crofoot Lumber, Inc. v. Thompson, 163 Cal. App. 2d 324 (1958).

Before any decision is made to terminate a contractor's performance, it is extremely important that careful thought be given to how the project will be completed. An owner must take into account (1) how the project will be completed; (2) in what time frame this can be accomplished; and (3) what the cost will be. There is usually a significant economic cost incurred in the process of changing contractors in the middle of a project, as any replacement contractor must (1) mobilize and commence performance where the terminated contractor left off; (2) negotiate arrangements with subcontractors and suppliers to complete the project; (3) assume the risk of deficient work having been performed by the terminated contractor; and (4) reestablish proper coordination and scheduling among the trades for completing their work. For these reasons, a replacement contractor is usually in a position to demand a high price from the owner for taking over a project that was in such trouble to have justified termination of the original contractor. In addition, any future claims based upon deficiencies in the work will be complicated by the original contractor's and the replacement contractor's each contending that the other was responsible for the failure of any particular aspect of the work to comply with the plans and specifications and/or to meet specific performance criteria.

Public works contracts are further complicated by the need to comply with applicable competitive bidding requirements or to declare that the project is being completed on an emergency basis so as to avoid competitive bidding for the completion work. In light of these factors, it will be very difficult or extremely costly to hire a completion contractor, and it is usually better (regardless of the pain and agony) to finish the project with the original contractor and bring legal action for damages for breach of contract *after* completion, rather than terminating the contractor and being unable to complete the project.

[A] Damages Recoverable from the Terminated Contractor

If the owner does terminate a contractor for cause and successfully proves a material breach of contract and the other elements discussed above, the owner may recover monetary damages from the terminated contractor. The measure of such damages is usually the amount by which the actual costs of completing the project exceeds the amount that the owner would have paid to complete the project under the original contract.[53] However, if it is determined that the owner's termination of the contractor was not justified, or that the owner did not act properly in effecting the termination (e.g., failed to give proper notice and/or a reasonable opportunity to cure), not only may the owner not recover from the contractor the excess costs of completing the project, but also the contractor may be able to recover from the owner the value of the work in place at the time of the wrongful termination, as well as lost profits it would have made if allowed to complete the project.[54]

---

[53] *See* Fairlane Estates, Inc. v. Carrico Constr. Co., 228 Cal. App. 2d 65 (1964).

[54] *See* Thomas Haverty Co. v. Pacific Indem. Co., 215 Cal. 555 (1932); McConnell v. Corona City Water Co., 149 Cal. 60 (1906).

[B] Rescinding the Contract

While a termination leaves the contract in place and seeks damages based upon the rights and duties of the contracting parties, the effect of a rescission is to cancel, void, and nullify the contract. Rescission is different from termination in that it repudiates the entire contract, and any recovery thereafter between the parties is based upon the legal principles of restitution—a legal concept that is based upon putting the parties in the position they would have been in if there had never been a contract. Therefore, the owner is obligated to pay the contractor the reasonable value of the work performed as of the rescission, without reference to the contract price or any of the other provisions of the contract. It is rarely in the best interests of an owner to rescind a construction contract, as most construction contracts are written by or on behalf of the owner and contain many provisions that benefit the owner's interests.

§ 6.05 OVERVIEW OF CONTRACTOR'S CLAIMS

A construction claim is a contractor's request for additional compensation beyond the contract price and/or for an extension of the time within which it is contractually required to complete a project. All such claims are governed by the specific provisions of the contract; therefore, it is incumbent upon the contractor to be completely familiar not only with all of the terms of the contract itself but also with all of the contract documents that are typically incorporated therein by reference.[55] There are two primary types of contractor claims against project owners: time-related claims and payment-related claims, both of which are governed by the general principle of damages recoverable for a breach of contract.[56] Perhaps the most obvious contractual obligation of a project owner, and one of the most frequent grounds for contractors' claims, is the owner's duty to pay the contractor for work performed. This obligation and most claims arising therefrom are discussed in detail in Chapter 4. The contractor's right to additional compensation when the owner changes the project, adding to the scope of work required under the contract, is discussed in § 6.10[A] below.

In general, if a contractor's failure to complete a project within the time contemplated by the construction contract is caused, in whole or in part, by factors within the control of the project owner, then the contractor may be able to recover damages for additional costs incurred as a result of being on the project longer than anticipated. Certainly the most complex issues are those involved in time-related claims, several different types of which are discussed in detail below.

The contractor claims discussed in this chapter may be pursued by a subcontractor as well as by a general contractor. If the subcontractor's claims are directly against the general contractor, the subcontractor may bring suit in its own name for breach of the subcontract agreement. In such cases, the general

---

[55] Such incorporated contract documents typically include plans and specifications, general conditions and special conditions, and often a variety of other documents and forms that must be followed or used in constructing the project.

[56] Cal. Civ. Code § 3300, quoted in § 6.02 above.

contractor may assert the same kinds of defenses than an owner has against a contractor claim. But what if the subcontractor's damages are caused by the project owner? Because subcontractors are not in privity of contract with the project owner, they have no legal standing to present their claims directly to the owner in their own name. Moreover, in many cases the subcontractor's claims and the general contractor's claims arise from the same owner-caused problems that are discussed below.

One possible approach is for the subcontractor to assign its claims to the general contractor. More typically, the general contractor and its subcontractor agree to cooperate in presenting the subcontractor's claims against the owner as a "pass-through" claim. This is usually done in either of two ways: (1) the subcontractor may sue the general contractor and the general contractor may sue the owner for indemnity against the subcontractor's claims; or (2) the general and its subcontractor may enter into a pre-litigation agreement,[57] whereby the general agrees to assert the subcontractor's claims, along with its own, on a "pass through" basis in exchange for the subcontractor's agreement to accept any damages the general may recover on its behalf as full resolution of its claim. Each of these methods was validated in the recent case of *Howard Contracting, Inc. v. G.A. MacDonald Construction Co.*[58]

A potential problem with such liquidating agreements is that the settlement or waiver of the subcontractor's claims against the general contractor calls into question whether there then remains a liability to the subcontractor against which the general contractor may seek indemnity from the owner. The *Howard Contracting* decision, by implication, refuted that argument in its summary of its holding on pass-through claims:

> The settlement and litigation agreement was predicated on the availability of the pass-through claim process. If the process is not available to the parties, MacDonald Construction has potential exposure to Soil Retention Systems through an action for recission of the agreement based on a failure of consideration. From a public policy standpoint, the initiation of separate litigation against a general contractor should not be compelled merely to enable a subcontractor to obtain standing when the general contractor previously agreed to pursue the subcontractor's damages claim against the public agency on a pass-through basis. To hold otherwise would be to insist on needless additional and duplicative litigation.[59]

Thus, the *Howard Contracting* decision confirms the validity of both methods of handling pass-through claims. In the case of pass-through agreements, it also offers the cautionary note that the agreement should normally be presented as part of the evidence at trial.

---

[57] Such agreements between general contractors and their subcontractors are commonly referred to as "liquidating agreements," though the Court in *Howard Contracting* does not use that specific term.
[58] 71 Cal. App. 4th 38 (1998).
[59] 71 Cal. App. 4th at 60.

§ 6.06 CONTRACTOR'S TIME-RELATED CLAIMS

Everyone is familiar with the truism that "time is money." This is particularly true in construction, because generally all of the contractor's costs—labor, materials, overhead, and supervision costs—increase with the passage of time. Second only to its obligation to construct the project in accordance with the plans and specifications is the contractor's obligation to complete construction within the period of time set forth in the contract.

In construction contracts, the contractor must complete construction by the completion date or be in breach of the contract. The project completion date may be in the form of a date certain or within a set number of calendar days after receiving a notice to proceed from the owner. The contract may also contain interim milestone dates by which the contractor is required to complete designated portions of the project. Moreover, if the contract itself is silent as to time, the courts will read into the contract, by implication, the parties' intent that the project be completed within "a reasonable time." In other words, the court will make a determination, based upon the individual circumstances surrounding the project, that the contract was to be completed within a certain time.[60]

Moreover, in the context of a construction contract, time is a double-edged sword. Typically, the contractor has based its bid on the assumption that the project can and will be built within the time frame provided in the bid documents and the project specifications. If this assumption proves invalid because of the project owner's fault, or the fault of one for whom the owner is responsible (e.g., the project architect or the owner's construction manager), the contractor may be entitled to recover its extra costs as delay damages. To protect against such a possibility, many owners attempt to limit their liability by including no damages for delay clauses, allowing the owner to grant the contractor time extensions for excusable delays but to refuse to provide additional monetary compensation.[61] On the other hand, delays within the control of the contractor or those for whom the contractor is responsible (e.g., its subcontractors and suppliers) may subject the contractor to damages claims of the owner,[62] such as liquidated damages in the form of a daily monetary penalty for each day that completion is delayed beyond the contractually agreed period.[63]

There are generally three types of time-related claims that may be pursued against an owner by a contractor: delay, disruption, and acceleration.

§ 6.07 CONTRACTOR'S DELAY CLAIMS

Contractor delay claims arise whenever the contractor's performance is delayed by causes beyond its control. Depending upon the classification of the delay (excusable or inexcusable, compensable or noncompensable), the typical

---

[60] *See, e.g.,* Brower Co. v. Garrison, 468 P.2d 469 (Wash. Ct. App. 1970).
[61] *See* discussion in § 6.07[H] below.
[62] *See* § 6.02[A] above.
[63] *See* § 6.02[B] above.

construction contract will provide various forms of relief to the contractor. It should be noted that such relief is entirely dependent upon the terms of the contract itself inasmuch as a contractor has no inherent right to receive time extensions. A contract that provides an absolute requirement for completion by a certain date, with no exceptions for any reason, is not necessarily illegal or unenforceable. Thus, it is extremely important for the contractor to be aware of the contract provisions regarding time of performance.

[A] Delay and Time Extension Claims

A contractor's right to various forms of relief because of specific problems causing delay to its performance is entirely dependent upon the express provisions of the contract itself. Such provisions allow the owner and the contractor to attempt to limit their own liability by placing the risk of delay upon the other party. It is therefore extremely important for the contractor to be completely familiar with all contract terms relating to delay in order to properly assess the risk allocation with regard to that specific project.

It is important to note that most construction contracts provide for time extensions only when delays affect the overall completion of the project or a contractually required milestone date.[64] Most contracts also require that time extensions be requested in writing within a certain period of time (usually quite short) immediately following the events causing the delay. It is all too easy for the contractor to ignore or give scant attention to time extension notice requirements until late in the project, when the completion date is fast approaching. If the contractor fails to request time extensions as the delays occur, there is a significant risk of being foreclosed from its own remedies for delay and/or being exposed to the owner's remedies under the contract.

[B] Categorizing Types of Delay

There are basically three important categories of delay on construction projects, each of which has particular significance with respect to the remedies that may be available under the contract; these are:

1. Excusable or inexcusable delays;
2. Compensable or noncompensable delays; and
3. Concurrent delays.

[C] Excusable versus Inexcusable Delays

Any delay that is the fault of the contractor or any of its subcontractors or material suppliers is classified as *inexcusable*. In other words, the contractor will be entitled to no extra compensation or any extensions of time because of such

---

[64] Such delays are said to be on the project's "critical path" as determined by CPM (critical path method) scheduling, which is discussed in detail in Chapter 7 below.

delays. Moreover, these are the kind of delays that typically entitle the owner to recover damages from the contractor. *Excusable delays,* on the other hand, are usually unforeseen and entitle the contractor to an extension of time within which to complete the project, provided that the cause of delay falls within an appropriate contract provision and the contractor complies with the notice requirement for such delays.[65] The mere fact that a delay is excusable does not necessarily entitle the contractor to any additional compensation for increased costs resulting from the delay. This reflects the fact that, for many excusable delays, neither party is at fault. Still, even without the right to receive extra compensation, it is a significant advantage for the contractor to receive a time extension in order not to be subject to liquidated damages or other penalties in favor of the owner.

Examples of the most commonly encountered types of excusable delays include the following:

Unusually severe weather. Delays occasioned by weather that is unusually severe for the place and season are generally considered excusable delays. Unusually severe weather conditions are considered to have been unforeseeable at the time of contracting, whereas normal weather conditions for the area, no matter how severe, are considered to be foreseeable and therefore are not generally excusable.

Acts of God. In addition to weather, other natural occurrences that could not have been reasonably foreseen or prevented by the contractor or any other party to the contract are generally classified as excusable delays. Act of God occurrences, also known as force majeure, must be entirely natural (i.e., not the result of human intervention) and, like weather, must be of unanticipated force and/or severity.

Strikes. Strikes and other forms of labor unrest are among the most frequent causes of delays on construction projects. They are generally considered excusable delays if they are unforeseen and beyond the contractor's control.

Unavailability of materials. Contractors may be entitled to an excusable delay if they are unable to obtain, in a timely manner, materials necessary to the construction of the project. Again, the unavailability of the materials must be unforeseeable and beyond the control (that is, without fault or negligence) of both the contractor and the subcontractor or material supplier in question. The issue of whether unavailability of materials qualifies as an excusable delay is usually judged by the same criteria as commercial impossibility of performance. In other words, mere difficulty, or even an increased price, will not qualify. Only true impossibility or a huge increase in price will constitute an excusable delay.

---

[65] For typical contract provisions defining excusable delay in public works projects, see Standard Specifications for Public Works Construction (the "Greenbook") § 6-6.1; with respect to private works, see American Institute of Architects Document A201, General Conditions of the Contract for Construction ("AIA Document A201"), ¶ 8.3.1.

[D] Compensable versus Noncompensable Delays

If a delay is classified as *compensable,* it simply means that the contractor, in addition to an extension of time, should also be entitled to additional financial compensation for its increased costs. For a delay to be compensable, it must be caused by the project owner, or one acting on the owner's behalf, such as the project architect or engineer or an owner's representative, without any contributing fault of the contractor or its subcontractors.

The courts have consistently held that the owner has a duty to cooperate and to avoid hindering the contractor's performance, and that such duty is implied by the law in every construction contract. Thus, a contractor has a legal right to recover monetary damages resulting from owner-caused delays, even in the absence of any express contractual provision for such compensation. The owner's liability arises from its breach of the implied duty of cooperation. A project owner may seek to avoid such liability by including a no damages for delay clause in the contract.[66]

Examples of the most frequently encountered owner-caused delays include:

1. Delay in obtaining permits and/or making the site available to the contractor;
2. Failure to respond to a contractor's request for information or instruction;
3. Errors and omissions in the plans and specifications;
4. Failure to timely provide owner-furnished materials or equipment;
5. Delay in returning shop drawings;
6. Delay in investigating and responding to a contractor's requests for change orders and/or time extensions;
7. Failure to make proper and timely inspections of the project;
8. Failure to make proper and timely progress payments;
9. Failure to make proper progress payments, coupled with the issuance of multiple changes and/or stop work orders; and
10. Approval of the contractor's project schedule with actual knowledge that the schedule will result in conflicts between the work of that contractor and other contractors on the same or related projects.

The concept of "reasonableness" is a factor in each of the examples listed above.

---

[66] *See* § 6.07[H] below.

[E] Concurrent Delays

In many instances, especially in large, complex construction projects, several delay causes may occur simultaneously and/or overlap one another. It is not always easy, and is sometimes even impossible, to determine which delays are actually affecting the overall completion date. Such delays are said to impact the project's "critical path." Moreover, both parties frequently are at fault for certain delays, and it may be difficult or impossible to apportion the blame attributable to each party. Such delays are referred to as *concurrent* delays. In such situations, courts have sometimes declined to apportion the delay and have ruled that neither party is entitled to recover delay damages. As noted below, however, there is a trend toward apportionment, based upon the concept of relative fault[67] that has developed in the field of tort law.

To complicate matters still further, on a single project there may be multiple occurrences of concurrent or allegedly concurrent delays. For example, a contractor may contend that its progress has been adversely affected by a delay in the issuance of project drawings and extensive modifications thereafter. This may be complicated by unreasonable delays in review and approval of the contractor's shop drawings, failures by the architect to issue needed clarifications, and delays by the owner in processing related change orders. Each of these problems may cause significant delay and disruption to the contractor's work, and it may be impossible to sort out how much delay is caused by any one of them. So long as these are all owner-caused delays, they are not considered concurrent, even though they may significantly overlap one another.

The issue of concurrency arises when the owner, defending against the contractor's claims, contends that an overriding simultaneous cause of delay was the contractor's own mismanagement of the project. For example, the owner may contend that the contractor employed unqualified personnel, did not adequately supervise its own employees, and/or failed to properly schedule the work or coordinate the activities of the various subcontractors. If it can be shown that these delays contributed to or overlapped the owner-caused delays, then they may be considered concurrent and may be treated as noncompensable delays. For delays to be categorized as concurrent and therefore noncompensable, each side must show that the delay attributable to the other affected work on the critical path. In other words, both parties' delays must have affected the overall completion date for the project.[68] Only if this is shown will the delays truly be categorized as concurrent and therefore noncompensable.

It is often impossible to establish that either party was entirely responsible for a particular delay. This may be due to the unavailability of clear and convincing evidence, or it may simply be that both parties were partly at fault. The trend of court decisions is toward an apportionment of fault for the delay among

---

[67] *See, e.g.,* General Ins. Co. of Am. v. Commerce Hyatt House, 5 Cal. App. 3d 460 (1970); J.A. Jones Constr. Co. v. Greenbriar Shopping Ctr., 332 F. Supp. 1336 (N.D. Ga. 1971), *aff'd,* 461 F.2d 1269 (5th Cir. 1972).

[68] *See* the discussion in Chapter 7 below.

the parties, based on the concept of relative fault that was developed in the area of tort law. When it is possible to segregate causation, responsibility for critical path delays will be allocated between the parties.[69] If it is impossible to make such an allocation, an estimate of relative fault may be derived by the jury verdict method.[70]

It should be quite apparent that critical path network scheduling is a very useful and sometimes necessary tool for proving a delay claim. This is true even though the contract may not have required the contractor to use critical path scheduling during construction of the project.

[F] Causes and Effects of Delay

Delay claims usually arise in conjunction with one or more of the other categories of construction claims. Whenever a contractor encounters deficient plans, extensive changes to the plans, differing site conditions, suspensions of work, or any other kind of owner-caused disruptions, a delay claim will almost certainly result. For this reason, disruption and delay claims are the most frequent types of claims, arising whenever a contractor's performance is delayed or its planned work sequence is interfered with.

In all cases, whether other types of claims lead to a delay claim depends upon whether the result is the inability of the contractor, through no fault of his own, to complete the work within the time required by the contract. Even if intermediate deadlines are delayed, a contractor will have a valid delay claim only when overall completion of the project is affected.

A contractor with a valid delay claim may attempt to recover the following types of damages:

1. Continuing direct job expenses, such as salaries of foremen and a superintendent;
2. Extended home office overhead, if attributable to the additional time on the project;
3. Additional local (jobsite) office expenses;
4. Increased material costs and/or other expenses due to escalating prices;
5. Additional and/or increased insurance costs;
6. Extra bond premium expenses;

---

[69] *See, e.g.*, Chaney & James Constr. Co. v. United States, 421 F.2d 728, 190 Ct. Cl. 699 (1970). *See also* Nomellini Constr. Co. v. State *ex rel.* Department of Water Resources, 19 Cal. App. 3d 240, 246 (1971), in which it is stated that "categorical statements that where delays are caused on both sides there is no way to 'apportion damages' are an absurdity."

[70] *See, e.g.*, State v. Guy F. Atkinson Co., 187 Cal. App. 3d 25 (1986); Raymond Constructors, Ltd. v. United States, 411 F.2d 1227, 188 Ct. Cl. 147 (1969). *See also* the detailed discussion in § 6.14[B] below.

7. Equipment standby costs;
8. Loss of profits from impairment of capital;
9. Impairment of bonding capacity; and
10. Impairment of the contractor's former advantageous competitive position.

[G] Extended Home Office Overhead: The *Eichleay* Formula

The computation of extended home office overhead damages is a particularly difficult problem. Although there are several formulas for computing the allocation of home office overhead to the particular project, the best known is the *Eichleay* formula. In *Eichleay Corp.*,[71] the government allowed a contractor to recover for extended home office overhead due to suspension of its work and specifically approved the following formula:

1. Allocation of a portion of the contractor's total overhead incurred during the contract period, based on the ratio of project billings to total company billings during the contract period;
2. Calculation of a daily overhead cost by dividing the total overhead allocable to the contract (item (1)) by the number of construction days provided in the contract; and
3. Multiplication of the daily overhead rate by the number of days of owner-caused delay.

Although it has been widely criticized and attacked in several government contract cases, in three recent cases, the Court of Appeals for the Federal Circuit has reinstated the *Eichleay* formula and established it as the exclusive means for calculating unabsorbed overhead arising from government-caused project delays. It is therefore probably the most widely recognized formula for calculating delay damages for increased home office overhead expenses.

In 1992, the Court of Appeals for the Federal Circuit confirmed that the *Eichleay* formula may be used to compute home office overhead under certain circumstances. In *C.B.C. Enterprises, Inc. v. United States*,[72] the contractor, C.B.C. Enterprises, had a contract with the Navy to construct improvements at a Marine Air Corps Station in Cherry Point, North Carolina. A change order was issued, adding work with direct costs of $10,846 and extending the contract completion date by 24 days. The Navy calculated home office overhead at 13.94 percent of the change order's direct cost, which was the same rate used in the basic contract. The contractor argued that it should have been allowed to calculate home office overhead according to the *Eichleay* formula, which would have given it

---

[71] 60-2 B.C.A. (CCH) ¶ 2,688 (ASBCA 1960). *See also* Construction Litigation: Representing the Contractor (J. Carter, R. Cushman, & W. Palmer eds., Aspen Law & Business, formerly published by John Wiley & Sons 1992).

[72] 978 F.2d 669 (Fed. Cir. 1992).

additional sums for recovery of home office overhead. When the Navy rejected the contractor's claim, the contractor sought recovery and was supported by the Associated General Contractors of America. The Federal Circuit ruled that the *Eichleay* formula may be used only when a government-caused suspension of work of uncertain duration occurs, which the contractor does not contemplate, interrupting the contractor's cash flow, and thereby resulting in the project in question not properly contributing its fair share to home office overhead. When the disruption, suspension, delay, or interference with the work is uncertain, unknown, and unexpected by the contractor, the contractor has no direct costs to bill, and thus has no income coming in to help absorb the home office overhead. Likewise, when the disruption or suspension is of uncertain duration, the contractor is unable to obtain other work to compensate for the delay, disruption, or suspension of the project in question. Specifically, the court stated: "The Eichleay Formula was devised to calculate reimbursable home office overhead costs in the event of suspension of work on a contract when the suspension decreases the stream of direct costs against which to assess a percentage rate for reimbursement."[73]

The court went on to state that, when additional work is added by a change order, there are, in fact, additional direct costs that the contractor can bill pursuant to the change order. Therefore, it is more appropriate to compute the home office overhead on the basis of a fixed percentage of the direct costs. The court stated that, under those circumstances, there is a known change in scope and additional time. Hence, the contractor is not incurring any disruption or reduction in cash flow. Specifically, the court stated:

> C.B.C. experienced no suspension in work, no idle time and no uncertain periods of delay during the agreed upon extended contract performance period. . . . [I]t is inappropriate to use the Eichleay Formula for contract extensions because adequate compensation for overhead expenses may usually be calculated more precisely using a fixed percentage formula.[74]

The *Eichleay* formula continues to be applied to calculate unabsorbed home office overhead.[75] In 1994, when the Court of Appeals for the Federal Circuit again considered the *Eichleay* formula, it held that it is the exclusive formula by which to calculate such overhead. In *Wickham Contracting Co. v. Fischer*,[76] the contractor, Wickham, contracted to renovate a federal post office and courthouse. The contract allowed 365 days for completion; the government caused a delay of 969 days. The contracting officer awarded Wickham many of its direct costs incurred as a result of the delay and an additional $333,084 on its claim for unabsorbed home office overhead. The court described the *Eichleay* formula as follows:

---

[73] 978 F.2d at 671.

[74] *Id.* at 675.

[75] *See, e.g.*, Interstate Gen. Gov't Contractors, Inc. v. United States, 12 F.3d 1053, 1056 (Fed. Cir. 1993).

[76] 12 F.3d 1574 (Fed. Cir. 1994).

The Eichleay formula requires three steps: 1) to find allocable contract overhead, multiply the total overhead cost incurred during the contract period times the ratio of billings from the delayed contract to total billings of the firm during the contract period; 2) to get the daily contract overhead rate, divide allocable contract overhead by days of contract performance; and 3) to get the amount recoverable, multiply the daily contract overhead rate times days of government-caused delay.[77]

The court then defined the "overhead pool" to be used in connection with the *Eichleay* formula as follows: "General and administrative salaries, rent, insurance, depreciation, hospitalization and medical costs, dues and subscriptions, office expenses, auto and truck maintenance, utilities, plans and specifications, cleaning, protection, taxes and licenses, and officer's salaries."[78]

Wickham argued that the contracting officer should have included specific field costs for travel, business meetings, telephones, professional fees, union fringe benefits, payroll taxes, and equipment rental. Inclusion of those items would have increased the *Eichleay* allocation from 34 to 80 percent. The Federal Circuit held that the General Services Administration Board of Contract Appeals had properly considered those to be direct expenses and not overhead costs. The court noted that, to apply the *Eichleay* formula, the contractor would have to show that (1) a compensable delay occurred; (2) the contractor could not have taken on any other jobs during the contract period; and (3) the contractor could not have avoided the overhead expenses by reducing its staff. When the government requires the contractor to stand by and it is uncertain as to when the work may be resumed, this often precludes the contractor from taking on additional jobs or laying off workers.

The court noted that a contractor's *overhead* is properly defined as the costs expended for the benefit of its business as a whole and that accrue over a period of time. If it is possible to trace a particular cost to a particular job, then that cost is a direct cost and not overhead. Allocation of home office overhead, based upon a pro rata share of billings, is not necessary because the overhead costs cannot be traced to a particular job. The court noted that *Wickham* sought to avoid the *Eichleay* formula by including direct costs in the overhead pool. Most significantly, the court concluded that "[w]hen a contractor satisfies the prerequisites for application of the Eichleay formula, that formula is the exclusive means available for calculating unabsorbed overhead to the delayed contract."[79]

In a recent case in which both parties advanced the *Eichleay* formula as the appropriate calculation, the U.S. District Court, in reliance upon the Federal Circuit decisions in the *C.B.C. Enterprises* and *Interstate* cases, set forth a convenient chart for the *Eichleay* formula calculation.[80]

---

[77] 12 F.3d at 1577.
[78] *Id.* at 1576.
[79] *Id.* at 1579.
[80] Aircraft Gear Corp. v. Kaman Aerospace Corp., 875 F. Supp. 485, 496 (N.D. Ill. 1995).

[H] No Damages for Delay Clauses

Many owners attempt to limit their exposure to a contractor's delay claims by adding a *no damages for delay* clause to their construction contracts. Such clauses typically provide that the contractor will be entitled only to an extension of time for any delay caused by the owner but will not be entitled to recover monetary damages. The basic legality of such clauses has been upheld by the California courts.[81]

It should be noted, however, that courts generally construe such clauses strictly against the owner in order to limit their application and effect, especially when enforcement will cause a hardship to the plaintiff contractor.[82] Thus, such clauses will not be given effect if:

1. There is evidence of fraud, bad faith, or malice on the part of the owner;

2. The delay is so long that it justifies the contractor's abandoning the job; or

3. The owner actively interferes with the contractor's normal progress.[83]

The enforceability of no damages for delay clauses on California public projects is strictly limited by California Public Contract Code § 7102, which provides:

> Contract provisions in construction contracts of public agencies and subcontractors thereunder which limit the contractee's liability to an extension of time for delay for which the contractee is responsible and which delay is unreasonable under the circumstances involved, and not within the contemplation of the parties, shall not be construed to preclude the recovery of damages by the contractor or subcontractor. No public agency may require the waiver, alteration, or limitation of applicability of this section. Any such waiver, alteration, or limitation is void. This section shall not be construed to void any provision in a construction contract which requires notice of delays, provides for arbitration or other procedures for settlement or provides for liquidated damages.

This provision was invoked by the Court of Appeal in a recent case to overturn a "no damages for delay" clause in a public works contract issued by the City of Los Angeles.[84] The Court noted that, while Standard Specifications[85] § 6-6.1 allows only a time extension for delays caused by unforeseen events,

---

[81] *See, e.g.*, K&F Constr. v. Los Angeles City Unified Sch. Dist., 123 Cal. App. 3d 1063 (1981).

[82] *See, e.g.*, Yamanishi v. Bleily & Collishaw, Inc., 29 Cal. App. 3d 457, 463 (1972).

[83] *See, e.g.*, Hawley v. Orange County Flood Control Dist., 211 Cal. App. 2d 708 (1963).

[84] Howard Contracting, Inc. v. G.A. MacDonald Construction Co., 71 Cal. App. 4th 38 (1998).

[85] Standard Specifications for Public Works Construction (Building News, Inc.), commonly known as the "Green Book" of specifications.

§ 6-6.3 creates an exception by specifically providing for payment for unreasonable and unanticipated delays caused by the public entity.[86] The Court of Appeal rejected the City's contention that the contractor had waived its right to delay damages by signing the contract, quoting the language in § 7102 that specifically voids any "waiver, alteration, or limitation of the applicability of this section."[87]

§ 6.08 CONTRACTOR'S DISRUPTION CLAIMS

A contractor may have a claim for *disruption* when an owner causes a change in the method and/or sequence of construction upon which the contractor based its bid. In other words, if the owner does anything to prevent the contractor from actually performing in the manner contemplated in its bid, the contractor has a claim for the resulting additional expenses. Because of the relationship between disruption and the contractor's planned method of performance, disruption is often called "interference" or "hindrance."

Disruption claims are based upon the following assumptions about the rights and obligations of the parties to a construction contract:

1. When bidding for a construction contract, the contractor is entitled to schedule its work so that it can be performed in the most efficient and cost-effective way available to the contractor, consistent with the project's plans and specifications.

2. The parties have an "implied duty to cooperate," meaning that neither party will act or fail to act so as to prevent the other party from exercising its rights and/or performing its contractual obligations. This concept is perhaps the most basic implied obligation in contract law.[88]

3. The contractor, in planning its method and sequence of performance, must act reasonably and may not assume that other parties will make extraordinary efforts to allow it to perform as planned.

[A] Disruption versus Delay

Although they often occur together, there is a significant difference between a disruption claim and a claim for "pure delay." Although disruption of the contractor's planned performance usually causes a loss in efficiency, it does not necessarily follow that the overall contract completion, the critical path, will be delayed as a result of such disruption. In many cases, the contractor may still be able to complete its work within the contractually required time, even though the cost of performing may be increased due to the owner's disruption or interference.

---

[86] *Id.*, at 71 Cal. App. 4th 49.
[87] *Id.*, at 71 Cal. App. 4th 49–50.
[88] *See, e.g.,* Milstein v. Security Pac. Nat'l Bank, 27 Cal. App. 3d 482 (1972); Joanaco Projects v. Nixon & Tierney Constr. Co., 248 Cal. App. 2d 821 (1967); McWilliams v. Holton, 248 Cal. App. 2d 447 (1967).

The significance of the distinction between disruption and delay is that they result in different types of damages recoverable by the contractor. Recoverable damages for disruption are likely to include increased labor costs resulting from inefficiency, increased manpower employed on the project to compensate for inefficiencies, increased labor costs resulting from the need to mobilize and demobilize work crews, and so on. Damages arising from pure delay, as discussed §§ 6.07[F] and [G] above, are those cost increases associated with an extended performance period, including increased jobsite and home office overhead, escalation of wages and material costs, equipment standby costs, extended and/or escalating finance costs, and so on.

Another significant consideration in categorizing a contractor's claims as either delay or disruption is whether the contract contains a no damages for delay clause. Such clauses are generally interpreted as not barring the contractor from recovering damages arising from the owner's interference, hindrance, or disruption. In fact, an owner's disruption of the contractor's performance may result in the non-enforceability of a no damages for delay clause.[89] Thus, even though the contract may contain an otherwise enforceable no damages for delay clause, the contractor may still be able to recover on a claim for disruption.

Another important concept in the consideration of delay and disruption claims is what is called the *ripple effect*. Because of the intricate interrelationship among all of the construction activities on a given project, it is readily apparent that disruption of any particular operation is likely to have an effect upon succeeding activities, even though they may not be directly affected by the cause of the original disruption.

For many years, contractors on federal projects were precluded from recovery of delay damages resulting from the ripple effect.[90] However, a revision modified the *Rice* doctrine by allowing a contractor to recover for delays to unchanged work resulting from changes and disruptions of other work. For this reason, a contractor on federal projects should carefully document its indirect costs caused by disruptions. Such costs may be included in the contractor's request for equitable adjustment, or recovered as damages in a breach of contract action, if they necessarily and directly resulted from performance of changed and/or disrupted work.[91]

[B] Examples of Disruption

A contractor's recovery for disruption damages is usually based upon one or more of the provisions of the construction contract, such as the changes clause, the suspension of work clause, or the changed conditions clause.

Although there are many circumstances that can give rise to a disruption claim, there are only a few contract clauses entitling the contractor to recover its

---

[89] *See, e.g.,* Hawley v. Orange County Flood Control Dist., 211 Cal. App. 2d 708 (1963).
[90] United States v. Rice, 317 U.S. 61 (1942) (established the *Rice* doctrine).
[91] *See* Law v. United States, 195 Ct. Cl. 370 (1971).

resulting damages. Nonetheless, regardless of the work involved, a claim for disruption can arise on virtually any construction project and under most construction contracts. Examples of disruption claims on government projects include:

1. Disruption of planned work sequence due to mis-fabrication of government-furnished steel.[92]

2. A change in the contractor's manner of performance resulting from governmental delay in issuing a notice to proceed on a project the contractor planned to perform concurrently with another project to obtain efficiency and cost savings.[93]

3. Breach of duty to avoid unreasonable interference with the contractor's performance, arising from errors in surveying a proposed road and the failure to designate feasible stockpile sites.[94]

4. Reduced efficiency of the contractor's labor because of noise from an adjacent government power check facility, resulting in a change in the manner or method of performance.[95]

5. Failure to provide adequate site access, constituting a change in the manner of the contractor's performance.[96]

6. Changes amounting to 10 percent of the contractor's estimated cost of work on a post office project, requiring inefficient, out-of-sequence work, directly attributable to government changes. The contractor, who worked as expeditiously as possible, was allowed to recover increased costs on unchanged work.[97]

7. Need to install dryer vent holes out of sequence and in a more costly manner, caused by government delay in furnishing information.[98]

8. Change in the sequence of work because of the government's delay in performing inspection.[99]

---

[92] Power City Constr. & Equip., Inc., IBCA No. 490-4-65, 68-2 B.C.A. (CCH) ¶ 7126. *See also, e.g.,* Appeal of Fluor, Utah, Inc., IBCA No. 1068-4-75, 81-1 B.C.A. (CCH) ¶ 14,876; Electronic & Missile Facilities, Inc. v. United States, 416 F.2d 1345, 1361 n.50 (Ct. Cl. 1969).

[93] Pan Arctic Corp., ASBCA No. 20133, 77-1 B.C.A. (CCH) ¶ 12,514.

[94] Lewis-Nicholson, Inc. v. United States, 550 F.2d 26, 213 Ct. Cl. 192 (1977).

[95] Nichols-Dynamics Co., ASBCA No. 17949, 75-2 B.C.A. (CCH) ¶ 11,556.

[96] Reliance Enters., ASBCA No. 20808, 76-1 B.C.A. (CCH) ¶ 11,831. *But see* Walter Dawgie Ski Corp. v. United States, 30 Fed. Cl. 115, 39 Cont. Cas. Fed. (CCH) ¶ 76,583 (citing *Reliance* but holding for defendant on ground that no express or implied contract language related to site access).

[97] Coley Properties Corp., PODBCA No. 291, 76-1 B.C.A. (CCH) ¶ 11,701, and 75-2 B.C.A. (CCH) ¶ 11,514. *But see* McMillan Bros. Constr. Inc., EBCA No. 328-10-84, 91-1 B.C.A. (CCH) ¶ 23,351 (citing *Coley* but holding that compensable impact is not established solely by showing the number of changes or amount of cost increases but also requires supporting evidence and testimony).

[98] Fred A. Arnold, Inc., ASBCA No. 18915, 75-2 B.C.A. (CCH) ¶ 11,496.

[99] Fullerton Constr. Co., ASBCA No. 11500, 67-2 B.C.A. (CCH) ¶ 6394.

9. Loss of efficiency due to the owner's delays in furnishing information needed by the contractor to perform the work as planned.[100]

The common thread in all of these examples is that the contractor's claims arose from an unanticipated change in the contractor's planned method or sequence of construction, due to causes for which the owner was responsible, that resulted in loss of the normal efficiency upon which the contractor had based its bid.

[C] Proving Disruption Claims

There are two basic ways to prove a disruption claim. First, a historical comparison can be made between the project in question and the contractor's previous projects of a similar nature. The contractor will attempt to demonstrate by this comparison that its labor costs for jobs of the type in question were significantly lower than on the current project and that specific owner-caused events caused the increase in the contractor's labor costs.

The second method often used by contractors is often called the "measured mile" approach. The contractor takes an unimpacted segment of the subject project and uses that as an unimpacted base line. This is compared with the impacted portion of the project to demonstrate that the contractor's labor productivity, which was much greater in the unimpacted period, represents a loss of inefficiency resulting from owner-caused events.

Although disruption claims are among the most common encountered on major construction projects, they are often the most difficult to prove. There is no substitute for contemporaneous project records, such as accurate and informative daily or weekly project reports, indicating where work was taking place, the subcontractors and workers on the job, and where work could not occur because of owner-caused problems.

§ 6.09 CONTRACTOR'S ACCELERATION CLAIMS

The term *acceleration* applies to any situation in which the project owner requires the contractor to complete all or a portion of the project either prior to the completion date set forth in the contract or prior to the adjusted completion date that would have resulted if the contractor had been given time extensions to which it was entitled under the contract. Such acceleration may be either *actual* or *constructive,* depending upon whether the contractor has received a specific directive from the owner or has simply been denied proper adjustments to its progress schedule. In either case, the contractor may be entitled to recovery of additional costs resulting from extra manpower, overtime, increased equipment rentals, or inefficiencies arising from changes in sequence and/or the need for greater manpower on the jobsite.

---

[100] Arvol D. Hays Constr. Co., ASBCA No. 25112, 84-3 B.C.A. (CCH) ¶ 17,661.

[A] Actual Acceleration

Actual acceleration occurs when a specific directive from the owner requires the contractor to complete all or some portion of its work prior to the completion date(s) agreed to in the contract. In most cases, an owner who directs a contractor to accelerate its performance will agree, in advance, to compensate the contractor for the additional cost of the acceleration. In any event, the fact that the contractor has been accelerated is apparent from the owner's express directive, and the contractor's right to additional compensation is likewise clear.

[B] Constructive Acceleration

Constructive acceleration presents considerably more difficulty in establishing the contractor's right to recover damages. Constructive acceleration occurs in any situation in which the contractor is forced to complete its work ahead of a properly adjusted progress schedule, usually because the owner has refused to grant a time extension for an excusable delay. When an owner refuses to grant a time extension to which the contractor is entitled under the contract, the contractor remains contractually obligated to complete construction by the originally specified completion date. The effect is to force the contractor to accelerate its work in order to avoid being subject to the owner's damages for delay.

When a contractor believes it is being accelerated in the absence of a specific acceleration order from the owner, it is particularly important to inform the owner promptly (in writing) that the owner is forcing the contractor to accelerate its performance. The important element in such notice is to emphasize that the acceleration is not voluntary on behalf of the contractor and will likely result in increased costs. Prompt, clear, written notice of a constructive acceleration claim is imperative, or the contractor risks being deemed a volunteer, in which case the acceleration will be noncompensable.

A contractor's claim for constructive acceleration includes the following five elements:

1. An excusable delay affecting the overall completion date that therefore entitles the contractor to a time extension;
2. The contractor's timely request for a reasonable time extension;
3. The owner's unreasonable refusal to grant the requested time extension;
4. The owner's demand (expressed or implied) that the contractor must complete the work within the original performance period; and
5. The contractor's increased or extra costs resulting from the acceleration.

Examples of constructive acceleration include owner delays in issuing design drawings; issuing of numerous change orders; and delays in delivering owner-furnished materials or equipment. Such actions must be coupled with the owner's refusal to issue a proper time extension, which therefore requires the contractor to increase its workforces, or add extra shifts, in an attempt to achieve the original (unadjusted) contract completion date. Constructive acceleration also results when the owner refuses to acknowledge the effects of adverse weather not reasonably anticipated (or of any other excusable delay, even though it may be noncompensable) and expressly or impliedly (by denying a valid time extension request) requires the contractor to complete the project in accordance with the original (unadjusted) schedule.

In *Mobile Chemical Co. v. Blount Bros. Corp.*,[101] both the contractor and the owner were held liable to project subcontractors for constructive acceleration. In that case, Mobile awarded a fixed-price contract to Blount to construct a chemical plant. Blount performed no actual construction work but was responsible for planning, coordinating, and scheduling the work of its numerous subcontracts. The contract required Blount to use its "best efforts" to complete the project by January 1983. The project was delayed in the beginning, in part by the owner's late delivery of owner-furnished equipment and in part by Blount's inadequate scheduling and coordination efforts. Mobile told Blount that a later completion date would be acceptable, but Mobile and Blount agreed to maintain a united front in their dealings with the subcontractors. As a result, Mobile and Blount insisted that the January completion date was crucial and pushed the subcontractors to add crews and accelerate their schedules to meet that date. After the project was actually finished in April 1983, the subcontractors sued both Mobile and Blount for acceleration and were successful at the trial court level.

On appeal, Blount argued that it was inappropriate to hold the general contractor liable to its subcontractors for acceleration, when the decision to accelerate had been made by the owner, citing cases holding that the owner alone would be liable for an acceleration of the schedule. The court stated that this project was different in light of the facts that Blount performed no construction work with its own forces and that the acceleration cost Blount nothing. In fact, a shorter construction period reduced Blount's overhead and increased its profit on its fixed-price contract with the owner. As a result, Blount was a beneficiary, not a victim, of the acceleration. In addition, the court said that Blount actively participated in the decision to accelerate the subcontractors and then implemented that decision. The court held that Blount and Mobile must therefore bear equal responsibility for the subcontractors' cost increases resulting from the acceleration.

§ 6.10 OVERVIEW OF LEADING CAUSES OF CLAIMS

Construction claims do not typically arise in a vacuum. The causes of both payment-related and time-related claims are usually rooted in circumstances that are themselves major claim areas meriting separate attention. Perhaps the most common of these involves changes in the work. Such changes may be initiated by a formal change order issued by the owner or by an oral field directive by an owner's representative. Changes also may be required to correct defects in the plans and specifications, or in response to the acts and omissions of third parties for whom the owner is responsible.

---

[101] 809 F.2d 1175 (5th Cir. 1987).

Another category of claims, for increased costs arising from changed conditions or differing site conditions, most commonly occurs in projects involving excavation and/or tunneling work. Such claims result when the contractor encounters a subsurface condition that is materially different from what was anticipated when the contract was entered into, based upon the original specifications, a soils report, or other information provided by the owner.

A third category of claims, for suspension of work, arises when there is an owner-caused work stoppage. An owner's suspension order may be written or oral and may be caused by a third party's acts or omissions that prevent the contractor's work from proceeding as planned.

[A] Changes (Actual, Constructive, and Cardinal)

Most construction contracts contain a changes clause permitting the owner to unilaterally require the contractor to perform changes in the work that increase or decrease the contract price. Such clauses allow the owner to add or delete work from the project without stopping the progress of the work, pending processing of a change order to the contract, or even resolution of a dispute in the courts. In the absence of such a provision, under normal contract law, a contractor could sue for breach of contract if the owner made significant additions to, or deletions from, the work originally contemplated at the time the contract was made.

Obviously, changes in the work are likely to affect the cost and/or the time required to perform the work. Therefore, changes clauses usually provide an administrative procedure to allow the contractor to receive reimbursement for costs of changes in its work by way of a change order to the contract. Thus, the changes clause reserves the owner's right to make changes to the contract and imposes on the contractor an obligation to proceed with the work as directed, pending later resolution of any claims or disputes arising from those changes.

An important limitation on the changes clauses is that the extra work order thereunder must apply to changes within the scope of the original contract.[102] Therefore, the contractor's duty to proceed does not apply to an order to perform work outside the scope of the changes clause. A contractor will also be excused from performing added work when required to wait for specific directions from the owner, or when there is an impossibility of performance. The courts have struggled with an adequate definition of what constitutes a change within the scope of the contract. The United States Supreme Court has defined such work as that which "should be regarded as fairly and reasonably within the contemplation of the parties when the contract was entered into."[103]

Whether a particular change is within the general scope is often a question of degree, as to which courts consider the following basic criteria:

1. The general nature of the original contract compared to the changed work;

---

[102] Valley Constr. Co. v. City of Calistoga, 72 Cal. App. 2d 839 (1946).
[103] Freund v. United States, 260 U.S. 60 (1922).

2. The comparative cost of the original contract compared with that of the changed work;

3. The comparative time of performance;

4. The origin of the change (e.g., a decision to change the work occurring after contract execution, or a change resulting from an error or omission in the project plans discovered after contract execution); and

5. The number of changes required by the owner and/or the quantity of work affected by owner changes.

[B] Actual versus Constructive Changes

The clearest case of an *actual change* occurs when the owner issues a written change order and request that the contractor submit its proposed costs and/or time impacts for performing the changed work. In such cases, there is usually no doubt that the contract amount will be adjusted. When there is a written change order, the only issue is the reasonableness of the contractor's price quotation and whether the change justifies a time extension. It is much more difficult to recognize a *constructive change,* which results when an owner directs the contractor to perform additional work that the owner, but not the contractor, considers to be within the original scope of work required to be performed under their contract.

In *Len Co. & Associates v. United States,*[104] the court of claims described a constructive change as follows:

> Each of the other elements of the standard "changes" or "extras" clause has been present—the contracting officer has the contractual authority unilaterally to alter the contractor's duties under the agreement; the contractor's performance requirements are enlarged; and the additional work is not volunteered but results from a direction of the Government's officers.[105]

Examples of constructive changes include:

1. Defective or ambiguous contract plans or specifications that require performance of extra work by the contractor;[106]

2. Directives issued by the owner to other trades or subcontractors that affect the contractor's work;

3. An order from an inspector or other governmental authority to correct work to comply with applicable codes or regulations, although the defective work was performed in conformity with the contract documents;

4. An owner's directive to accelerate the work;

5. Added or changed work required by an answer to a contractor's request for information, or a revision to a contract drawing or specification;

---

[104] 385 F.2d 438, 181 Ct. Cl. 29 (1967).
[105] 385 F.2d at 443.
[106] *See, e.g.,* MacIsaac & Menke Co. v. Cardon Corp., 193 Cal. App. 2d 661 (1961).

6. A disputed order to perform corrective work; and

7. Work called for by the specifications that is either impossible or impractical to perform.

One of the most common reasons for constructive changes is a difference in opinion between the owner and the contractor as to the proper interpretation of the contract plans or specifications. Typically, an owner will refuse to issue a formal change order because it believes that the work in question is already required by the terms of the construction documents, and a change order is therefore unnecessary.[107] A constructive change occurs in the event that the owner's interpretation is later proved to be incorrect, thus indicating that a change order should have been issued.[108]

Another point worth noting involves the issue of whether work required by the original plans and specifications was either impossible or impractical to perform. In order to shift the risk of performance from the contractor to the owner, the contractor must show that extreme and unreasonable difficulty, expense, injury, or loss would be involved in performing according to the plans and specifications. The contractor bears the burden of proving that it was not possible to comply with the plans and specifications and still meet the contract's schedule requirements. Finally, in instances where the contractor voluntarily accepts related design responsibilities, the contractor may be deemed to have assumed the risk of nonperformance.

Most construction contracts require that the contractor must give the owner notice of a constructive change within a specified time (typically 20 days) of the events giving rise to the claim. In most federal contracts, the changes clause further requires that the contractor submit to the government, within 30 days of receipt of a written change order, a written statement of the general nature and monetary extent of the contractor's claim.

[C] Cardinal Changes

If the owner requires changes that, by virtue of their nature or the total quantity thereof, make the project to be constructed one that is fundamentally different from that upon which the contractor based its bid, the result is a *cardinal change*. The essence of a cardinal change is that it requires the contractor to build a fundamentally different project than originally planned. For this reason, cardinal changes will free the contractor from the terms of its bid and allow it to recover the reasonable value of the labor and materials provided to the project, plus reasonable markups for overhead and profit. In effect, a cardinal change will convert a fixed-price contract into a time and materials contract.[109]

---

[107] Dawson Constr. Co., GSBCA No. 3820, 75-1 B.C.A. (CCH) ¶ 11,339.
[108] Blake Constr. Co., GSBCA No. 2477, 71-1 B.C.A. (CCH) ¶ 8870.
[109] *See, e.g.,* Allied Materials & Equip. Co. v. United States, 569 F.2d 562, 215 Ct. Cl. 406 (1978); Wunderlich Contracting Co. v. United States, 351 F.2d 956, 173 Ct. Cl. 180 (1965); C. Norman Peterson Co. v. Container Corp. of Am., 172 Cal. App. 3d 628 (1985).

§ 6.11 DIFFERING SITE CONDITIONS

Ordinarily, a contractor's contractual duty to investigate the project site prior to submission of its bid does not require the contractor to investigate hidden subterranean conditions to determine the suitability of the site for the intended project. Unless specifically required by a statute or an express contractual provision, the contractor is entitled to assume that the owner has adequately investigated the contract site and appropriately prepared the contract documents.

The typical differing site conditions clause entitles the contractor to an equitable adjustment in the amount of its contract if it encounters either or both of the following: subsurface or related physical conditions at the site that differ materially from those indicated in the contract documents (Type I); or an unknown physical condition of an unusual nature at the site, differing materially from those ordinarily encountered and generally recognized as inherent in work of the character provided for in the contract documents (Type II).

These provisions are so common that the two types of differing site conditions are usually referred to as Type I and Type II conditions. A *Type I* condition, by its very nature, requires that there be an express statement in the contract documents regarding the subsurface conditions that materially differs from what is actually encountered during construction. There can be no Type I condition in the absence of such a contractual provision.

In contrast, the finding of a *Type II* condition does not require any contractual representation as to the nature of the subsurface conditions of the project site. Rather, a Type II condition exists whenever the contractor encounters subsurface conditions not ordinarily encountered and not generally recognized as inherent in the type of work involved in that specific project.[110]

In the absence of a differing site conditions clause, contractors would likely feel compelled to add a contingency amount to their bids to account for unknown subsurface conditions. Rather than pay this extra premium, owners generally afford the contractor the protection of a differing site conditions clause. Such provisions are a standard part of the construction contracts on most public works projects.[111] By taking the subsurface condition gamble out of the bidding process, contractors will not be required to consider such contingencies in their bids.[112]

One very important requirement affecting both Type I and II differing site conditions is that the contractor, upon encountering the concealed condition, must immediately stop its work and notify the owner. This requirement is intended to allow the owner to make its own assessment of the condition prior to its being disturbed by the contractor's activities. If a contractor proceeds without first giving such opportunity to the owner, it will usually be barred from recovering on a differing site conditions claim. Thus, *Greenbook* § 3-4 provides: "The

---

[110] Warner Constr. Corp. v. City of L.A., 2 Cal. 3d 285 (1970); E.H. Morrill Co. v. State, 65 Cal. 2d 787 (1967); Wunderlich v. State, 65 Cal. 2d 777 (1967).

[111] *See* Standard Specifications for Public Works Construction (the "Greenbook") § 3-4.

[112] Foster Constr. C.A. & Williams Bros. Co. v. United States, 435 F.2d 873, 193 Ct. Cl. 587 (1970).

contractor's failure to give notice of changed conditions promptly upon their discovery and before they are disturbed shall constitute a waiver of all claims in connection therewith."

Typical examples of Type I differing site condition include (1) when, during excavation, the contractor encountered boulders significantly larger than those indicated by the soil test borings included in the contract documents;[113] and (2) when the contractor's excavation encountered substantially more rocks than shown by the logs or borings.[114] Such a finding, moreover, may be based upon the contractor's own interpretation, if reasonable, and does not necessarily depend upon a geologist's determination.[115]

Type II differing site conditions are encountered less frequently and are significantly more difficult to prove.[116] In part, this is because reasonable persons may differ in their subjective judgment as to what constitutes an unusual condition. Type II differing site conditions were found to exist in the following cases:

1. Cattle bones encountered beneath an inlet structure were held to be a highly unusual and unanticipated differing site condition that weakened the structure.[117]

2. The contractor's pre-bid site inspection would not have revealed the necessity to remove 600–700 gallons of subsurface oil during excavation of a prison site.[118]

3. The contractor encountered, in an elevator shaft, extra steel framework that had to be removed.[119]

4. Concrete beams, automobile parts, and railroad ties were held not within the ordinary definition of rubble in an excavation contract.[120]

5. Unforeseeable conditions adversely affected the contractor's dredging operations and precluded its timely completion of the project.[121]

6. Buried gas line, construction debris, and footings constituted changed conditions, in spite of the government owner's ignorance of their presence.[122]

The following types of conditions have been encountered by contractors and held not to constitute differing site conditions within the meaning of the contract

---

[113] E. Arthur Higgins, AGBCA No. 76-128, 79-2 B.C.A. (CCH) ¶ 14,050.
[114] Bick-Com Corp., VACAB No. 1320, 80-1 B.C.A. (CCH) ¶ 14,285.
[115] Betancourt & Gonzales, S.E., DOT CAB Nos. 2785, 2789, 2799, 96-1 B.C.A. (CCH) ¶ 28,033.
[116] Charles T. Parker Constr. Co. v. United States, 193 Ct. Cl. 320 (1970).
[117] Panhandle Constr. Co., DOT CAB No. 77-31, 79-1 B.C.A. (CCH) ¶ 13,576.
[118] Mutual Constr., DOT CAB No. 1075, 80-2 B.C.A. (CCH) ¶ 14,630.
[119] A.D. Herman Constr., GSBCA No. 4823, 78-1 B.C.A. (CCH) ¶ 13,187.
[120] Maverick Diversified, Inc., ASBCA No. 19838, 76-2 B.C.A. (CCH) ¶ 12,104.
[121] D. W. Sandau Dredging, ENGBCA No. 5812, 96-1 B.C.A. (CCH) ¶ 28,064.
[122] R. A. Glancy & Sons, Inc., VABCA No. 2327, 87-3 B.C.A. (CCH) ¶ 20,068.

clause: hurricanes;[123] excessive rainfall;[124] high water levels;[125] flooding due to man-made causes;[126] and excessive snowfall.[127]

If the contract documents require the contractor to investigate the site prior to the submission of a bid, the risk of differing site conditions falls upon the contractor only if such conditions are readily apparent or would have been revealed by a reasonable site inspection. Thus, a claim will be denied only when the concealed condition could have been discovered by a "reasonably prudent contractor."[128] The requirement of a pre-bid site inspection affects the foreseeability of the condition, but it does not void the differing site conditions clause of the contract.[129]

§ 6.12 SUSPENSION OF WORK

Most construction contracts allow the owner to order the contractor to cease working for a period of time, either for the convenience of the owner or because of the contractor's failure to perform the work in accordance with the contract documents. Provisions such as those found in AIA Document A201 and in the Standard Form Federal Government Contract allow the owner to issue a written order to the contractor to suspend all or any part of the work for such time as may be appropriate for the convenience of the owner or the government. If the contractor's work is suspended under these contract terms, the contractor may be entitled to an equitable adjustment of the contract price. However, in order to recover damages under a typical suspension of work clause, the contractor must show that its work was suspended, delayed, or interrupted for an *unreasonable* period of time.

Although it is common practice for an owner to issue the contractor a written work suspension order, it is not unusual for an owner's field representative to orally order the contractor to suspend work. There may also be a *constructive suspension of work,* whereby the acts or omissions of the owner or its representative make it impossible or impractical for the contractor to proceed. Examples of such constructive suspensions would include an owner's failure to act upon a contractor's request for assistance, an architect's failure or unreasonable delay in providing critical information or clarification of the contract documents, or an owner's unreasonable failure to respond to the contractor's request for deviation from the contract requirements. In these situations, the owner's acts or omissions are deemed to have caused the contractor's work to be suspended, even in the absence of an express directive from the owner to suspend the work. When there is a constructive suspension of work, the contractor will ordinarily be entitled to an increase in contract price and a time extension.

---

[123] Norfolk Dredging Co. v. United States, 360 F.2d 619, 175 Ct. Cl. 504, *cert. denied,* 385 U.S. 919 (1966).
[124] McCutcheon-Peterson, IBCA No. 1392-9-80, 81-1 B.C.A. (CCH) ¶ 14,997.
[125] Roen Salvage Co., ENGBCA No. 3670, 79-2 B.C.A. (CCH) ¶ 13,882.
[126] Lee Hoffman v. United States, 340 F.2d 645, 166 Ct. Cl. 39 (1964).
[127] Warren Painting Co., ASBCA No. 18456, 74-2 B.C.A. (CCH) ¶ 10,834.
[128] Powell Gen. Contracting Co., DOT CAB No. 1088, 80-2 B.C.A. (CCH) ¶ 14,680.
[129] Farnsworth & Chambers, Inc. v. United States, 346 F.2d 577, 180 Ct. Cl. 992 (1965).

A contractor should always examine its construction contract to see if it has a suspension of work clause and to note carefully any requirements to timely notify the owner when there has been a work suspension. A typical requirement is that, in the absence of a written directive from the owner, the contractor must notify the owner within 20 to 30 days of any event giving rise to a constructive suspension of work. A failure to comply with the notice requirement may bar the contractor's claim. Once the contractor has encountered a suspension of work, it is important to begin monitoring and documenting all additional costs associated with the suspension of work. Such documentation is vital to recovery of the contractor's extra costs.

§ 6.13 CLAIM INVESTIGATION

The first step in the determination of any claim involves an investigation of the problems that were encountered. Once these problems are identified, the contractor must then determine if it is entitled to recover money and/or a time extension for such claims.

There are essentially three approaches that can be taken in investigating claims: an event-oriented investigation, a plan-oriented investigation, or a loss-oriented investigation. Under the *event-oriented* approach, a particular event or series of events is investigated to determine if it establishes the basis for a claim. This method is best utilized when the contractor suspects or knows of a particular problem or change in the work that may give rise to a claim. Once the particular problem or item of added work is identified, the investigation is directed toward a review of the facts and circumstances surrounding those particular events.

In a *plan-oriented* investigation, the contractor compares the as-bid configuration reflected in the original bid documents with the as-built configuration as shown on the final construction documents. Any discrepancies or variances noted are then investigated to determine whether they were the result of the owner's directives, for which the contractor has not received compensation.

Under the *loss-oriented* claim investigation approach, the contractor is alerted to the possibility of a claim as a result of unanticipated expenditures of money. By pinpointing the source of its cost overruns, the contractor can determine why it incurred a particular loss.

In undertaking these various approaches to claim investigation, a number of categories of information should be explored. Both the contractor and the owner should:

- ____ Review and compare the final working plans and specifications with the as-bid plans and specifications.

- ____ Review the contract specifications with a view toward noting discrepancies and ambiguities.

- ____ Compare actual labor cost accounts against the bid labor or budgeted labor cost accounts.

- ____ Review all material purchase orders.

§ 6.13[A]   CALIFORNIA CONSTRUCTION LAW

___ Review contractor-to-owner correspondence.

___ Review owner-to-contractor correspondence.

___ Review internal contractor correspondence and memoranda.

___ Review the contractor's inspection system and documentation.

___ Interview the contractor's superintendent and other knowledgeable project personnel.

___ Review the contractor's scheduling information.

[A]   Claim Documentation Checklist

Regardless of the approach followed in investigating potential claims, painstaking factual research and documentation are essential. The time spent prior to claim preparation will not only result in better claim preparation but also reveal more evidence and material that can be used in the presentation of the claim in court or before an arbitration panel. Furthermore, proper claim investigation, like preparation of the claim itself, involves a team effort. Accordingly, it is essential that the contractor line up representatives from accounting, engineering, legal, purchasing, and construction administration to assist in the claim investigation.

No construction claim can be properly prepared without adequate underlying project documentation. On any major construction project, at least the following records should be maintained:

___ Site Investigation Reports, including the following:

   (a)   Site condition

   (b)   Access

   (c)   Temporary power

___ Daily Logs and Daily Reports, documenting the following:

   (a)   Weather

   (b)   Number of men on job, by trade and by job (foreman, superintendent, laborer)

   (c)   Tasks being performed

   (d)   Conversations

   (e)   Visitors to site

   (f)   Problems encountered

___ Correspondence

___ Minutes of Meetings, including all the following:

   (a)   Meetings with owner, architect, subcontractors

   (b)   Scheduling meetings

   (c)   Pay request meetings

___ (d) Jobsite meetings

___ Change Order Log, documenting the following:

    (a) Date problems arose

    (b) Dates requests for change orders were made

    (c) Dates change orders were issued

    (d) Dates when work was done

    (e) Dates when work was billed

    (f) Dates when change orders were paid

___ Photographs and/or Videotape

___ Scheduling Documents, showing the following:

    (a) Schedules prepared

    (b) Schedules received

    (c) Schedules distributed

___ Shop Drawing Log, showing the following:

    (a) Dates submitted

    (b) Dates returned

    (c) Dates approved

___ Records of Progress Payments

___ Cost Accounting Records

___ Detailed Records of Disputed Items (particularly records regarding cost tracking)

[B] Claim Outline and Narrative

If it has been determined through the claim investigation process that the contractor has performed additional and/or changed work for which it has incurred added cost, it must then be determined why the contractor was injured and who was responsible for the added costs to perform the work. If it is determined that the costs were the result of formal change orders, constructive change orders, or other acts or omissions for which the owner is directly or indirectly responsible, the contractor should prepare and submit a claim package to the owner. The first step in this phase is to prepare a claim outline that tells, in short form, the story of each claim. At a minimum, this outline should contain:

1. The basic facts of what happened, including a brief summary of either what the owner did, or failed to do, or what it required of the contractor;

2. The rationale as to why the contractor should be compensated (a legal theory for recovery); and

3. A summary of the costs associated with the claim (a general statement of what type of costs and/or delays resulted from the owner's action for which the contractor should be compensated).

At this stage in the claim investigation process, the input of a lawyer becomes essential. The lawyer should evaluate the supporting documents and claim outline to determine if there is legal liability and whether a credible claim can be prepared from the claim outline. He or she also must determine whether the legal theory stated is correct and, if so, whether all essential elements exist to support the claim's proposed legal theory. Furthermore, the lawyer must evaluate all contract modifications and release language to determine whether any of the contractor's rights to recover on the claim have been waived.

Once it has been determined that there is a valid factual and legal basis for the claim, a more detailed narrative of the claim must be prepared with complete legal, technical, and financial support. This task usually requires a team effort by contractor representatives from accounting, legal, engineering, and contract administration. The claim draft should state as succinctly as possible what the contract required, what the owner required the contractor to do over and above the contract requirements, what added contractor effort resulted from the owner's directives over and above the contract requirements, and what added costs and/or time was incurred, for which the contractor makes a claim. Finally, all relevant project correspondence, memoranda, contract provisions, and other supporting documents should be attached to the claim as appendices. The claim narrative should include the following five sections:

1. Summary of Claim(s);
2. Relevant Contract Requirements;
3. Description of Changed, Added, and/or Impacted Work;
4. Summary of Damages; and
5. Supporting Legal Analysis.

By following this procedure, a contractor will be able to maximize its recovery, whether by negotiation of a favorable settlement or by pursuing the claims through the judicial system.

§ 6.14 QUANTIFYING AND PROVING DAMAGES

Many arguments occur in court cases regarding what proof is adequate to quantify a delay, disruption, and/or acceleration claim. The most common methods for determining claims are the actual cost, total cost, modified total cost, and jury verdict methods. Although most courts and commentators give lip service to the principle that the preferred basis for determining an equitable adjustment is actual, historical cost information,[130] a review of cases reveals a very large

---

[130] See, e.g., AKCON, Inc., ENGBCA No. 5593, 90-3 B.C.A. (CCH) ¶ 23,250 (1990).

number of claims and awards based upon total cost, modified total cost, and jury verdict methods.

The actual cost method is typically preferred "because it provides the court, or contracting officer, with documented underlying expenses, ensuring that the final amount of the equitable adjustment will be just that—equitable—and not a windfall for either the government or the contractor."[131] This method attempts to segregate costs associated with a change at or before the time the change occurs. However, if the contractor is unable to segregate the costs because of the complexity of the claim, as with delays or acceleration, the total cost method is a better approach. This method "subtracts the contractor's bid estimate from the total of all project costs incurred, producing a total cost attributable to the owner's breach."[132]

As its name implies, the modified total cost method is a variant of the total cost method, in that it recognizes certain costs as being the responsibility of the contractor or concedes that a bid estimate must be adjusted upward to account for foreseeable but unanticipated expenses.

The jury verdict method, by contrast, is frequently resorted to when the total cost or modified total cost method has been rejected, usually because it fails to differentiate among several possible causes of increased expense, or because of a partial failure or lack of available proof. What is typically emphasized in a jury verdict approach is the fact that contractor injury has been clearly established, but the precise extent of that injury can only be ascertained by approximation. What should be noted here is the frequency with which the jury verdict method is invoked *after* a total cost or modified total cost approach has failed or been rejected. The jury verdict method has thus come to occupy a fall-back position, a posture designed to sustain a verdict of entitlement in the absence of precise documentation.

[A] Pros and Cons of the "Total Cost" Method

California appellate cases provide fairly little guidance as to what types of proof trial courts will accept for quantification of damages for delay, disruption, and acceleration claims. As described above, the total cost method subtracts the contractor's estimated costs, as set forth in its bid, from the total project costs actually incurred to determine the overrun due to the additional work. This method is used when specific project records and cost documents are unavailable.[133] The accuracy of this method is dependent upon the accuracy of the bid estimate, and it will weigh heavily against the contractor's claim if it is shown that its bid was significantly underestimated. To ensure some level of credibility when using this method, the contractor must establish four factors:

---

[131] Dawco Constr., Inc. v. United States, 930 F.2d 872, 882 (Fed. Cir. 1991).

[132] B. Bramble & A. Phillips, Construction Litigation: Strategies and Techniques 135–136 (1989).

[133] S.C. Anderson, Inc. v. Bank of Am. Nat'l Trust & Sav. Ass'n, 24 Cal. App. 4th 529, 538 n.11 (1994); Huber, Hunt & Nichols, Inc. v. Moore, 67 Cal. App. 3d 278, 307–309 (1977); Servidone Constr. Corp. v. United States, 931 F.2d 860, 861–862 (Fed. Cir. 1991); State *ex rel.* Department of Transp. v. Guy F. Atkinson Co., 187 Cal. App. 3d 25, 32 (1986).

1. Nature of losses makes it impossible or highly impracticable to determine them with a reasonable degree of accuracy;
2. Plaintiff's bid or estimate was realistic;[134]
3. Its actual costs incurred were reasonable; and
4. It was not responsible for the added expenses claimed.[135]

The reasonableness of efforts made on the part of the contractor—or the lack thereof—do not go unnoticed by the courts. For example, in *Huber, Hunt & Nichols,* the court was particularly unimpressed with the contractor's efforts, citing its failure to "take the time to plan check the plans and specifications prior to bid."[136]

A modified version of the total cost method is justifiably much more popular. This method basically allows the court more discretion in calculating the damages claim. In *Seattle Western Industries, Inc. v. David A. Mowat Co.,* the Washington court allowed a subcontractor to use the modified method to recover damages from an architect.[137] In presenting its claim for delay damages due to design error, the subcontractor used a modified total cost approach by deducting from the difference between actual cost and bid estimate those cost increases it thought attributable to the contractor or its lower-tier subcontractors. It also discounted the profit markup on certain aspects of the work to allow for the learning curves of its crews. This translates into the following formula:

> total claim = (actual cost incurred, less original bid amount) less (costs due to general contractor and its subs) less (some profit markup to allow for learning curve of crew)[138]

Relying on that evidence, the jury awarded the contractor $180,299 in delay damages. The Washington Supreme Court upheld almost the entire award, because the subcontractor plaintiff demonstrated that no more precise method existed for segregating the increased costs caused by the delay, and the contractor proved that its original estimate for the project was reasonable.

With increasing frequency, boards of contract appeals and courts are availing themselves of a constrained jury verdict method very similar to a modified total cost claim. Three cases illustrate this point. In *AKCON, Inc.,*[139] a claimant engaged to rehabilitate a tunnel presented a total cost claim and came close to losing its case-in-chief by an insistence on the total cost method. Only through the intervention of the Board was the period of costs able to be isolated and identified.

---

[134] The court used this factor in *In re Meyertech Corp.,* 83 F.2d 410 (3d Cir. 1987), to prohibit use of the total cost method. *See* Huber, Hunt & Nichols, 67 Cal. App. 3d at 298.

[135] Servidone Constr. Corp. v. United States, 931 F.2d 860, 861 (Fed. Cir. 1991), (citing WRB Corp. v. United States, 183 Ct. Cl. 409, 426 (1968); Boyajian v. United States, 423 F.2d 1231, 1243 (Ct. Cl. 1970)).

[136] 67 Cal. App. 3d at 298.

[137] 110 Wash. 2d 1, 7, 750 P.2d 245 (1988).

[138] *Id.*

[139] ENGBCA No. 5593, 90-3 B.C.A. (CCH) ¶ 23,250 (1990).

A similar salvage operation was performed by the Claims Court in *Servidone Construction Corp. v. United States*,[140] in which a contractor was successful in obtaining an equitable adjustment based upon a Type II differing site condition.[141] To show the amount of its injury, Servidone presented its claim by the total cost method. Servidone's total cost claim essentially passed muster, except that the Claims Court concluded that it was "plain that Servidone underbid."[142] In fact, it underbid the project by at least $9 million. The Claims Court nonetheless proceeded to refashion Servidone's claim for the benefit of the claimant. The court insisted that the recovery was not based upon the jury verdict method but rather upon a modified total cost calculation of the court's own making that substituted another contractor's bid amount for a more realistic assessment of what Servidone should have bid.[143]

Finally, in *E.W. Eldridge, Inc.*, the court applied the jury verdict method where the defending agency offered a considerably broad range of remedies, charitably described by the Board as "schizophrenic."[144] At the hearing, the government proposed not less than four total cost approaches, ranging in dollar amounts from $231,000 to $342,000. In its subsequent brief, however, the government presented the Board with yet another range, this time spanning from $140,000 to $200,000. The Board weighed the "total lack of any substantive justification for the claim" and "[t]he abandonment by both parties of attempts to prove actual costs of the constructive change" and resorted to the jury verdict method to award the claimant the sum of $342,077, or roughly 37 percent of its total claim.[145]

[B] Applications of the Jury Verdict Method

A legitimate defense to a claimant's request to use the actual or total cost method is to show that the plaintiff's claimed costs are unreasonable. In such a case, if the defending party does not show what amount would be reasonable, the court has three options: (1) to dismiss the claim for failure of proof; (2) to allow the claim on the theory that the respondent has failed to show what would be reasonable; or (3) to utilize all materials in the record and reach a jury verdict.[146]

Use of the jury verdict method is considered appropriate only if the court determines:

1. that clear proof of injury exists;
2. that there is no more reliable method for computing damages; and

---

[140] 19 Cl. Ct. 346 (1990), *aff'd*, 931 F.2d 860 (Fed. Cir. 1991).
[141] *See* discussion of differing site conditions in § 6.11 above.
[142] 19 Cl. Ct. at 385.
[143] *Id.*
[144] ENGBCA No. 5269, 90-3 B.C.A. (CCH) ¶ 23,080 (1990).
[145] 90-3 B.C.A. (CCH) 23,080, at 115,900–115,991.
[146] Tom Shaw, Inc., DOTBCA Nos. 2106, 2108, 2109, 2110, 2131, 90-1 B.C.A. (CCH) ¶ 22,580 (1990).

3. that evidence is sufficient for the court to make a fair and reasonable approximation of damages.[147]

In the underlying case, a prime contractor, Dawco, commenced an action in the U.S. Claims Court for an equitable adjustment based upon differing site conditions encountered by its landscape subcontractor.[148] The Claims Court found that Dawco had, in fact, encountered a Type I differing site condition and was accordingly entitled to an equitable adjustment.[149]

When Dawco submitted its own "total cost" claim of $575,000, the government objected on the ground that the total cost theory of recovery is disfavored under the law and regarded as the "least reliable method of determining quantum."[150] Although Dawco insisted that its claim was not a total cost claim because "it was separated into components of labor, material, equipment," the Claims Court agreed with the government:

> It was a total cost claim because it did not provide a way to determine what costs were incurred under the contract, and which were the result of additional work attributable to the differing site condition. For example, plaintiff claimed labor costs, but within the category of labor there was no way to distinguish what labor costs resulted from the original contract and which costs resulted from the differing site condition.[151]

The Claims Court was more sympathetic to the concerns of Dawco's landscape subcontractor, who did not know how to prepare a total cost claim and apparently did not recognize that it had in fact submitted one. The court's sense of fairness soon prevailed, as it struggled to find a principle to support an award of damages to the subcontractor:

> For the court to find entirely or partially in favor of plaintiff and then not determine the amount of money owed would subject the parties to the burden of further attempts to agree upon quantum, which they have heretofore been unable to do. . . . The court felt a duty to conclude the case and not to remand it to a certain dead end only to have it reappear on the court's docket.[152]

Slowly but surely, the Claims Court wound its way toward the high ground of fairness, finally resorting to the jury verdict method,stating that

> [t]he conceptual touchstone for the use of the jury verdict approach is the existence of conflicting competent evidence calling into question the accuracy and

---

[147] Dawco Constr., Inc. v. United States, 930 F.2d 872, 880 (Fed. Cir. 1991), *overruled on other grounds in* Reflectone, Inc. v. Dalton, 60 F.3d 1572, 1578 (Fed. Cir. 1995) (citing WRB Corp. v. United States, 183 Ct. Cl. 409, 425 (1968)).

[148] Dawco Constr., Inc. v. United States, 18 Cl. Ct. 682 (1989), *aff'd in part and rev'd in part*, 930 F.2d 872 (Fed. Cir. 1991).

[149] 18 Cl. Ct. at 687, 696.

[150] *Id.* at 696.

[151] *Id.* at 696 n.10.

[152] *Id.* at 698.

reliability of the plaintiff's computations. As the name implies, the technique by necessity requires application of the fair and informed discretion of the trial judge. The court believes that the use of the approach is especially appropriate here, where the principal source of guidance is the rather nebulous concept of "reasonableness."[153]

On that basis, the Claims Court awarded the plaintiff an equitable adjustment of $529,935.

On appeal, however, the Federal Circuit wasted no time in reversing the Claims Court's application of the jury verdict method.[154] In the view of the Federal Circuit, the Claims Courts protective concern for the subcontractor's failure to segregate its actual costs improperly relieved plaintiff's counsel of the obligation of proving its damages. According to the Federal Circuit Court, the plaintiff must furnish a detailed and documented cost breakdown.[155] Furthermore, if Dawco as claimant could not submit such actual cost data, then it was incumbent upon Dawco, before the jury verdict could be invoked on its behalf, to "'demonstrate a *justifiable inability to substantiate* the amount of his resultant injury by direct and specific proof'."[156] The court stated:

> Clearly, Dawco was in an ideal position to detail all its costs. Or, at least, it could have, and should have, been. The issuance of a change order request should signal to the prudent contractor that it must maintain records detailing any *additional* work, just as should the encountering of differing site conditions.[157]

The Federal Circuit illustrated what it considered to be the particular perils of the Claims Court's fast and loose application of the jury verdict method when it focused on the tendency of the jury verdict method to swell the scope of a purported claim in the absence of substantiating actual cost data:

> If anything, the twisted road the Claims Court traveled from Edmundson's $8,100 documented additional expense to the court's final determination that Dawco was entitled to $529,935 as an equitable adjustment for the different site conditions underscores the dangers of using the "jury verdict method." Its primary peril, as evidenced in this case, is the risk that unrealistic assumptions will be adopted and extrapolated, greatly multiplying an award beyond reason, and rewarding preparers of imprecise claims based on undocumented costs with unjustified windfalls.[158]

Thus, the court concluded that the general contractor did not suffer any injuries beyond its subcontractor's claim, nor was the general contractor justifiably

---

[153] *Id.* at 699.
[154] Dawco Constr., Inc. v. United States, 930 F.2d 872, 881 (Fed. Cir. 1991).
[155] 930 F.2d at 882.
[156] *Id.* at 881 (quoting Joseph Pickard's Sons Co. v. United States, 532 F.2d 739, 742, 209 Ct. Cl. 643 (1976)) (emphasis added).
[157] *Id.*
[158] *Id.* at 882.

unable to document its losses, and, therefore, insufficient evidence existed to reasonably approximate the damages impact.[159]

In *State ex rel. Department of Transportation v. Guy F. Atkinson Co.*,[160] the California Court of Appeal upheld the trial court's confirmation of an arbitrator's award in favor of the plaintiff contractor, thus approving both the total cost and jury verdict approaches to quantification of damages. Brushing aside the disfavored nature of the total cost method, the court held that "this method of proving damages is properly permitted in cases such as this, where accurate assessments of costs 'as planned' are difficult, if not impossible to ascertain."[161]

The court accepted the arbitrator's findings that the state's changes not only adversely affected the amount of labor and equipment but also seriously disrupted the planned sequence and efficiency of the work.[162] It rejected the state's contention that only a small portion of the work was directly affected by changes, saying that failed to take into consideration the resulting "massive ripple effects."[163] The court also emphasized the state's failure to suggest a viable alternative "to measure the actual cost of a project before any change occurs" and that "even a State claims evaluator conceded that such calculations were difficult at best and that using pre-change costs derived from bid estimates represented the only feasible method."[164]

In contrast, the court was impressed that the contractor had made "numerous comprehensive attempts to document the additional costs encountered" and had requested additional time to make even further attempts.[165] The court concluded that "[s]uch an approximation was eminently reasonable given the inherent complexities of the cost accounting problems posed by the indirect, yet widespread, effect of the State's ordered changes."[166]

Yet another court made use of the claimant's total cost presentation to implement the jury verdict method. In *Tutor-Saliba-Perini*,[167] the Board of Contract Appeals noted that "[i]ndeed, Appellant's total cost presentation, while not acceptable at face value, still may be used where warranted as an acceptable starting point for determining damages [citation omitted]."[168]

In *Delco Electronics Corp. v. United States*,[169] a contractor, Delco, sued to recover an equitable adjustment on behalf of itself and three of its subcontractors. Delco prevailed on behalf of its primary subcontractor, under a modified total cost theory of recovery, when the Board found that the four criteria necessary to prevail on a total cost claim had been satisfied:

---

[159] *Id.* at 881.
[160] 187 Cal. App. 3d 25 (1986), and 67 Cal. App. 3d 278 (1977).
[161] 187 Cal. App. 3d at 32.
[162] State *ex rel.* Department of Transp. v. Guy F. Atkinson Co., 187 Cal. App. 3d 25, 33 (1986).
[163] *Id.*
[164] 187 Cal. App. 3d at 32 (citing Boyajian v. United States, 423 F.2d 1231, 1244 (Ct. Cl. 1970)).
[165] *Id.*
[166] *Id.*
[167] PSBCA No. 1201, 87-2 B.C.A. (CCH) ¶ 19,775 (1987).
[168] 87-2 B.C.A. (CCH) ¶ 19,775, at 100,076.
[169] 17 Cl. Ct. 302 (1989), *aff'd*, 909 F.2d 1495 (Fed. Cir. 1990).

As the Court of Claims explained, the total cost approach will be considered reliable, and the cost figures arrived at thereby will be considered reasonable, if certain indications of reliability are present. These indications are where "(1) the nature of the particular losses makes it impossible or highly impractical to determine them with a reasonable degree of accuracy; (2) the plaintiff's bid or estimate was realistic; (3) its actual costs are reasonable; and (4) it was not responsible for the added expenses."[170]

The government had advocated that the Claims Court should avail itself of the jury verdict method to decide the quantum issue. This method, the Claims Court noted, may be invoked:

> Where a court or a board of contract appeals sitting as the finder of fact is confronted with competent but conflicting evidence on what is a reasonable amount for an equitable adjustment, it may employ what is referred to as the "jury verdict approach." The Court may adopt the jury verdict approach where, (1) there is clear proof that the contractor was injured; (2) there is no more reliable method for computing damages; and (3) the evidence adduced is sufficient for the court to make a fair and reasonable approximation of the damages [citations omitted].[171]

The Claims Court went on to observe that "the conceptual touchstone for the use of the jury verdict approach is the existence of conflicting competent evidence calling into question the accuracy and reliability of the plaintiff's computations."[172] Finally, the Claims Court concluded "that under either a modified total cost approach or a jury verdict approach," the record supported a total recovery of $19,318,624.[173] The Claims Court decision seemed well-reasoned and judicious; despite the large monetary amounts at issue, a modified total cost theory of recovery seemed appropriate. Curiously, the Claims Court felt that a jury verdict method would have produced the same result.

What is most unusual about the *Delco* case is the manner in which the Claims Court elected to deal with the separate total cost claims of the *contractor* ("Delco claims"). Delco had insisted that, for itself, it could not segregate its approximately $3.6 million in additional engineering costs because, unlike its subcontractor, whose work had changed *in its entirety* and who therefore stood to recover all of its costs under a modified total cost theory, Delco's own work had "not changed at all, just increased in amount."[174] The Claims Court was decidedly underwhelmed by Delco's tactical inconsistency, and pointedly repeated Delco's argument: "The Plaintiff maintains that because the nature of the work was not changed, it was not practical for Delco to separately identify the additional work required by the change from the work required under the original contract."[175] In considering the plausibility of Delco's rationale for not segregating

---

[170] 17 Cl. Ct. at 327–328 (citing WRB Corp. v. United States, 183 Ct. Cl. 409, 426 (1968)).
[171] *Id.* at 323.
[172] *Id.* at 323–324.
[173] *Id.* at 329.
[174] *Id.* at 330.
[175] *Id.*

its additional costs, the Claims Court was quick to observe that Delco could and should have made better efforts to support its claims. Thus, the Claims Court chastised Delco that

> [a]lthough Delco's employees may have believed that segregation and documentation [of costs] were not practical or convenient, the court concludes that such actions were prudent under the circumstances. Delco was perfectly capable of segregating costs.... Credible evidence was provided that Delco had the capability to segregate costs and simply chose not to do so. The court concludes that segregation of costs was not so impractical as to be unduly burdensome.[176]

The Claims Court noted that Delco's failure to segregate its actual costs and maintain supporting documentation and data constituted a fatal flaw in its claim. "This flaw brings into question the reliability of the amount sought."[177] The government, not unexpectedly, attacked Delco for this gross failure to segregate its costs, which was seemingly willful and flagrant, and argued for no recovery at all, refusing even to submit proposals for alternative amounts. The Claims Court, in a display of protective sympathy for the contractor, rejected, as a "wildly divergent position," the government's argument that "these shortcomings somehow deprive a contractor of any recovery for effort that competent evidence proves was expended."[178]

On the basis of what the Claims Court characterized as credible evidence, the government insisted upon an absolute denial of Delco's claim and thus provided the court with no "alternative guidance for the quantum of recovery." In response, the Claims Court recited its conclusion: "Faced with conflicting competent evidence as to the appropriate amount of the recovery, the court must resort, once again, to the jury verdict approach."[179]

What is also unusual about the Claims Court's ruling is that the government elected not to furnish the court with any "reasonable alternative guidance for the quantum of recovery."[180] Rather, the government insisted that Delco was not entitled to anything, because of its conscious failure to segregate actual costs. This strategy backfired when, because of the lack of reasonable alternatives, the Claims Court resorted to the jury verdict approach.[181] It is arguable, however, that its own reasoning demonstrates that the Claims Court had no basis to award damages based on the jury verdict method, because there existed no "competent but conflicting evidence." After all, the court had earlier stated that the "conceptual touchstone for the use of the jury verdict approach is the existence of conflicting competent evidence calling into question the accuracy and reliability of the plaintiff's computations."[182]

---

[176] 17 Cl. Ct. at 330–331.
[177] *Id.* at 331.
[178] *Id.*
[179] *Id.*
[180] *Id.*
[181] *Id.*
[182] *Id.* at 323–324.

Chapter 7
SCHEDULING AND PROOF OF DELAY CLAIMS

§ 7.01   Introduction: Who Has the Burden of Proof, and How Is It Met?
§ 7.02   Use of CPM Scheduling Techniques Preferred
§ 7.03   What Is CPM Scheduling?
§ 7.04   How CPM Scheduling Techniques Are Used to Prove Delay
§ 7.05   Type of Delay Must Be Determined
§ 7.06   Conclusion

§ 7.01 INTRODUCTION: WHO HAS THE BURDEN OF PROOF, AND HOW IS IT MET?

One of the major issues tried over and over in construction cases is delay. Attorneys and claims consultants deal with this issue constantly. Typically, each side produces an expert at trial to present its view of who delayed the project and what the impacts of that delay were. This chapter looks at the case law regarding who has the burden of proof on the issue of delay and how that burden of proof can be met.

§ 7.02 USE OF CPM SCHEDULING TECHNIQUES PREFERRED

Courts and boards of contract appeals have criticized the use of bar chart schedules and have shown a strong preference for the use of critical path method (CPM) scheduling techniques in the proof and defense of delay claims. For example, in *Minmar Builders, Inc.*,[1] the General Services Board of Contract Appeals (GSBCA) stated:

> [T]wo of the contractor's construction schedules were . . . nothing . . . more than a bar chart showing the duration and projected calendar dates for the performance of the contractual tasks. Since no interrelationship was shown as between the tasks, the chart cannot show that project activities were dependent on prior performance . . . much less whether overall project completion was thereby affected.[2]

In *Wilner v. United States*,[3] the Court of Claims stated:

> Plaintiff's position . . . is that the critical path was always in the building. Plaintiff . . . failed to diagram these delay episodes or otherwise to depict the precise route of the path. Plaintiff thoroughly described each item of delay, but did not present evidence concerning an overall review of the critical path he sponsored. Undoubtedly, plaintiff does have a position regarding the proper course of the critical path. Unfortunately, he failed to supply a critical path analysis, and the Court is not obligated to attempt to construct one for him. Due to the absence of plaintiff's view of the critical path, the Court cannot assign weight to any concept of the critical path as propounded by the plaintiff.[4]

However, in the recent case of *Howard Contracting, Inc. v. G.A. MacDonald Construction Co.*,[5] the California Court of Appeal upheld an award of damages

---

[1] GSBCA No. 3430, 72-2 B.C.A. (CCH) ¶ 9599. *See also* R.W. Contracting, Inc., ASBCA No. 24627, 84-2 B.C.A. (CCH) ¶ 17,302; Haas & Haynie Corp., GSBCA Nos. 5530, 6224, 6638, 6919, 6920, 84-2 B.C.A. (CCH) ¶ 17,446, at 86,800, wherein the Board stated: "We simply do not understand the contractor's reason for abandoning its CPM. A CPM schedule is never rigid. It has built-in flexibility to permit graphic recognition of changes so they can be managed."
[2] 72-2 B.C.A. (CCH) ¶ 9599, at 44,857.
[3] 23 Cl. Ct. 241 (1991).
[4] 23 Cl. Ct. at 255–256.
[5] 71 Cal. App. 4th 38 (1998).

where a contractor's "bar chart schedule was based on a critical path method analysis." The Court specifically held that: "The bar chart schedule used here identified the project's critical path and demonstrated that the delays constituted critical path delays.[6]

In *Al Johnson Construction Co. v. United States*,[7] the United States Court of Appeal described CPM scheduling as "a favorite device with present day fact finders in contract disputes."

CPM scheduling is just as important in job management and administration as it is in claims presentation and defense. This was recognized in *Continental Consolidated Corp.*[8] as follows:

> The CPM scheduling technique is one which requires a breakdown of the entire project into individual tasks and an analysis of the number of days required to perform each task. The analysis is then programmed into a computer which produces a chart showing the tasks and a line which controls the completion of the overall work. The line through the modes, the function points for completion of essential tasks, is known as the critical path. In addition, there are numerous side paths for subordinate tasks which normally can be performed without affecting the critical path. However, these subordinate tasks, if improperly scheduled or unduly delayed in performance, can on occasions become critical and thus change the critical path for the entire project.
>
> The critical path method of scheduling requires the logical analysis of all the individual tasks entering into the complete job and the periodic review and re-analysis of progress during the performance period. It is essential that any changes in the work and the time extensions due to the contractor be incorporated into the progress analysis concurrently with the performance of the changes or immediately after the delay and thus integrated into the periodic computer runs to reflect the effect on the critical path. Otherwise the critical path chart produced by the computer will not reflect the current status of work performed or the actual progress being attained.[9]

The CPM schedule must be properly prepared if it is to be accepted by the courts. It must also be used in the field; a completely theoretical schedule developed strictly for the purpose of supporting a claim is much less likely to be accepted. In *Chaney & James Construction*,[10] the Board completely rejected the contractor's CPM schedule, stating:

> [T]hese charts cannot be considered evidence of the facts they portray. While we accept the expert's testimony that the charts appear to be technically correct and logical, the work sequence shown was not demonstrated to be the

---

[6] 71 Cal. App. 4th at 52.
[7] 854 F.2d 467 (Fed. Cir. 1988).
[8] ENGBCA No. 2743, 67-2 B.C.A. (CCH) ¶ 6624 (1967). *See also* Fortec Constructors v. United States, 8 Cl. Ct. 490 (1985); Ballenger Corp., DOTBCA Nos. 74-32, 74-32A, 74-32H, 84-1 B.C.A. (CCH) ¶ 16,973, at 84,524 (1983).
[9] 67-2 B.C.A. (CCH) ¶ 6624, at 30,715.
[10] FAACAP No. 67-18, 66-2 B.C.A. (CCH) ¶ 6066, at 28,076, 28,077 (1967).

only possible sequence in which the work could have been accomplished; nor was it demonstrated that the sequence presented in the charts was necessarily the best one. The work sequence shown was not used in estimating and bidding the job since the original chart from which the two exhibits were derived was not in existence until late 1962, near the end of the project. Also, as the contractor's project manager admits, the sequence shown on the critical path charts was not followed in performing the contract work.

Under the circumstances the critical path charts cannot be accepted as establishing either the facts they portray or the reasonableness of the contract's assertions as to the influence of specific incidents on work progress."

The CPM schedule must be kept up to date and must incorporate delays as they occur. If delays are not concurrently inputted into the computer, the effect upon the critical path and job planning cannot be properly maintained. In *Blinderman Construction Co. v. United States*,[11] the Court of Federal Claims stated:

Said contractual provisions acknowledge the principle that accurate, informed assessments of the effect of delays upon critical path activities are possible only if up-to-date CPM schedules are faithfully maintained throughout the course of construction. Otherwise, the critical path produced by the computer will not reflect the current status of the work performed or the actual progress being obtained. *Continental Consol. Corp.*, ENGBCA Nos. 2743, 2766, 67-2 (CCH) No. 6624, at 30, 715, 1967 WL 320 (1967) quoted with approval in *Fortec Construction,* 8 Cl.Ct. at 506.

Obviously, then, if the CPM is to be used to evaluate delay on the project, it must be kept current and must reflect delays as they occur. *Fortec Construction,* 8 Cl.Ct. at 505.

Thus, plaintiff has failed to demonstrate how the data used to prepare its CPM scheduling relates, if at all, to actual conditions prevailing on the construction site at the times of the alleged delays.

Like any other computerized aid to comprehension, CPM analysis is only as good as the underlying information upon which it is based. *Bednar, et al. Construction Contracting* 664 (citing *Lane Verdugo* ASBCA No. 16327, 73-2 (CCH) BCA No. 10, 171, 1973 WL 1896 (1973).[12]

The Court of Federal Claims denied a delay claim because the contractor's bar charts failed to establish the interrelationship between the disrupted tasks and other activities on the schedule's critical path. In *Mega Construction Co. v. United States,*[13] the U.S. Postal Service awarded a contract to Mega Construction to construct a post office in Canoga Park, California. The contract did not require Mega to submit a CPM analysis of its proposed progress schedule; it only required Mega to maintain a "practicable progress chart."

---

[11] 39 Fed. Cl. 529 (1997).
[12] 39 Fed. Cl. at 585.
[13] 29 Fed. Cl. 396 (1993).

Mega claimed that a variety of government acts and omissions delayed its completion by 272 days. In support of its claim, Mega offered a version of its original as-planned schedule, as well as an as-built schedule, both in bar chart form. Each delay event reflected in the as-built schedule was cross-referenced to documents or statements attributing the delaying event to the government. The Federal Court of Claims emphasized that, in order to recover for delay, a contractor must prove that governmental shortcomings interrupted work on the critical path. The court defined the critical path as that sequence of construction tasks for which there is no timing leeway. The tasks must be performed on schedule, or completion of the entire project will be delayed; the separate tasks are inherently interdependent, as one task cannot commence before the prior task is completed. The court, in commenting upon the fact that it considered Mega's bar chart inadequate, stated the following:

> Plaintiff's bar chart depicted its version of numerous work items. However, it failed to prove that the claimed delays occurred along the critical path, because it does not indicate the interdependence of any one or more of the work items. Plaintiff proffered documents prepared solely for use at trial as its estimate of work items that were on the critical path while the project was ongoing, but offered no credible evidence of the interdependence of the project's activities.[14]

The court also noted that Mega's bar charts were based in part on documents that had never been introduced into evidence. Furthermore, Mega's scheduling consultant had relied upon self-serving statements by Mega's officers without attempting to independently verify those representations. For those reasons, the contractor's delay claim was denied.

The *Mega Construction Co.* case was distinguished in the case of *Howard Contracting, Inc. v. G.A. MacDonald Construction Co.*[15] In the *Howard Construction* case, the owner contended that *Mega Construction Co.* stood for the proposition that a contractor was required to use a computer-generated network diagram schedule known as a critical path method schedule to establish a claim for construction delay damages, and therefore, because the contractor in that particular case did not utilize a CPM schedule, they had not proved their delay claim. The court, first of all, noted that in a federal government construction project, delays that prolong activities of a project falling on the critical path may be compensable. The court noted that the *Mega Construction* case held that a contractor's entitlement to delay damages required the presentation of evidence that establishes the critical path of the project and the occurrence of delays along that critical path. The court noted that in the *Mega Construction* case, the contractor's ineffective use of a bar chart at trial failed to identify the critical path or to demonstrate that compensable project delays occurred on that path. The court held that *Mega Construction* does not stand for the proposition that use of a critical path method schedule is required to establish the occurrence of compensable

---

[14] 29 Fed. Cl. at 428.
[15] 71 Cal. App. 4th 38 (1999).

project delays. The court held that the contractor was not required to use a critical path method schedule to establish critical path delays. The court further noted that the contractor in the *Howard Contracting* case, unlike the contractor in the *Mega Construction* case, used a bar chart schedule that was based on a critical path method analysis and identified the project's critical path, and further demonstrated that delays constituted critical path delays; therefore, the contractor had properly proved delay.

Another case illustrating how a contractor's claim can be lost due to inadequate scheduling (even when a CPM is used) is *Appeal of J.W. Bateson Co.*,[16] in which the contractor (Bateson) was employed to construct a physical education center at the United States Naval Academy in Annapolis, Maryland. The project was completed late, and liquidated damages were assessed by the Navy. Bateson filed a claim alleging that defective building design caused a seven-and-one-half month delay and that, but for the design deficiencies, the project would have been finished ahead of the contract completion date. Bateson also alleged that, as a direct result of defective specifications, its costs increased due to labor inefficiency in connection with the concrete work.

The specifications required a CPM schedule to ensure adequate planning and execution of the work and to help the contracting officer appraise the reasonableness of the proposed schedule and to evaluate the work. Although the specifications required that the CPM network was due 45 days after the Notice of Award, Bateson submitted its CPM schedule six months late. By the time the CPM schedule was submitted, the contractor was already alleging project delays, particularly in connection with shop drawing approval. The CPM schedule also omitted shop drawing submittals. Bateson's project manager testified that he was so busy expediting the project that he put the CPM schedule "on the back burner." Although the specifications called for a minimum of 500 and a maximum of 800 activities in the CPM schedule, Bateson's CPM schedule included 2,000 activities. This original CPM schedule also showed the project being completed one month late. When the Navy rejected that, Bateson revised its CPM schedule to show completion within the contract time.

The Board of Contract Appeals found that Bateson's CPM schedule was "long delayed" and was "inexcusably delayed." The Board also found that, after the CPM schedule was submitted and adjusted at the Navy's request, Bateson found that the CPM schedule had lost its significance as a scheduling instrument. As a result, the Board found that the contractor had not shown that its CPM schedule ever had, or deserved, status as a credible schedule for use in fulfilling the purposes of the specifications; rather, it had been used primarily to further the contractor's claims, which began to emerge at some point during the creation of the CPM documents. The Board stated it therefore had no confidence in, and would not rely upon, the as-planned portions of the CPM schedule and considered only the as-built portions as evidence. In effect, the Board held that, where a CPM schedule is used primarily as a claims device rather than to manage the project, it will be entitled to no weight.

---

[16] ASBCA No. 27491, 84-3 B.C.A. (CCH) ¶ 17,566 (1984).

Thus, many courts and boards of contract appeals have shown a preference for the use of CPM scheduling techniques in determining disputes regarding delay, but only when CPM scheduling is also used in managing the project.

§ 7.03 WHAT IS CPM SCHEDULING?

Many books and articles have been written describing how CPM scheduling is performed; the following analysis is just a basic overview of those techniques. If one thinks about building a house, one can ascertain how a CPM schedule is put together. The first thing one has to do is prepare the site (grading and excavation). Trenches are then dug for foundations and any underground utilities (plumbing, sewer, electrical, etc.). One then pours the foundation, frames the house, puts on the roof, installs rough electric and plumbing, wraps the exterior, drywalls or plasters the interior, and installs finishes (finish electrical and plumbing, cabinets, flooring, carpeting, painting, etc.).

As can be logically seen from the foregoing, a CPM schedule must show (1) each of the activities the contractor must perform to construct the project; (2) the durations of those activities, based upon resources (labor and equipment), productivity of the resources, and the estimated quantity of the work; (3) sequencing of those activities in a logical fashion (that is, one can't frame the house before the foundation is poured); and, finally, (4) the interrelationships among all the activities.

The person preparing the foregoing information feeds it into the computer, and specialized CPM software calculates and computes the "critical path," depicting the information in the form of a "network diagram" (see *Continental Consolidated Corp.*[17]). This critical path is what distinguishes a CPM schedule from a bar chart, which cannot show the interrelationships among the construction activities.

The foregoing basic description of how CPM schedules are prepared is not intended to be all-inclusive. Many other technical elements are involved in CPM scheduling, but the foregoing should be sufficient to give the reader a general understanding of the basic elements.

It should be noted here that, as a practical matter, the contractor generally has the obligation to plan, schedule, and coordinate the work.[18] Although some contractors do not involve their subcontractors in preparing the schedules for a project, most subcontracts do require the subcontractor to adhere to the contractor's schedule. The courts have likewise held that there is an implied obligation on the part of subcontractors to follow the instructions of the contractor, so long as the contractor acts in a reasonable fashion.[19] It is to the advantage of the contractor to involve its subcontractors so that they cannot complain about the

---

[17] ENGBCA No. 2743, 67-2 B.C.A. (CCH) ¶ 6624 (1967).
[18] Able Elec. Co. v. Vacanti & Randozzo Constr. Co., 212 Neb. 619, 324 N.W.2d 667 (1982); S. Leo Harmony, Inc. v. Binker Mfg. Co., 597 F. Supp. 1014 (S.D.N.Y. 1984).
[19] Peter Kiewit & Sons Co. v. Iowa S. Util. Co., 355 F. Supp. 376 (S.D. Iowa 1973); McCarty Corp. v. Pullman-Kellogg, 571 F. Supp. 1341 (M.D. La. 1983).

SCHEDULING AND PROOF OF DELAY CLAIMS § 7.03

schedule.[20] If the subcontractors are involved in creation of the schedule, or are at least put on notice so that they can participate if they elect to do so, then it will be more difficult for them to later complain that the schedule was unreasonable.

The contractor's project schedule must be reasonable to the subcontractors, as it relates to duration and sequencing.[21] During performance of the work, the contractor may allow its subcontractors to perform according to the established schedule, subject to reasonable revisions.[22] The contractor may not unreasonably alter the schedule to the detriment of the subcontractors, even if the owner consents to the change.[23] Subcontractors cannot, however, expect exact compliance with the schedule; there will always be some variations on most construction projects. The contractor's performance must be "within the bounds of reason" and in general conformance with the project schedule.[24]

Courts have likewise described CPM scheduling. For example, in *Santa Fe, Inc.*,[25] the VABCA described CPM scheduling as:

> a management technique by which a project can be broken down into a number of identifiable tasks (activities) and assigned various resources (e.g. time/duration, management cost). These tasks are then sequentially interconnected, reflecting the various interdependence of the activities to provide an overall schedule to complete the project. The result of this scheduling process is a path(s) through this schedule, which if postponed, will delay the project completion. All other paths through the project schedule can experience some postponement . . . without delaying the overall project completion. The amount of postponement which a path of activities can experience without delaying the completion of the project is called "total float." . . . The more total float a path of activities has, the longer it can be postponed without delaying project completion.[26]

---

[20] United States v. F.D. Rich Co., 439 F.2d 895 (8th Cir. 1971); Kroeger v. Franchise Equities, Inc., 490 Neb. 731, 212 N.W.2d 348 (1973).

[21] United States *ex rel.* Heller Elec. Co. v. William F. Klingensmith, Inc., 670 F.2d 1227 (D.C. Cir. 1982); Able Elec. Co. v. Vacanti & Randozzo Constr., 212 Neb. 619, 324 N.W.2d 667 (1982).

[22] Illinois Structural Steel Corp. v. Pathman Constr. Co., 23 Ill. App. 3d 1, 318 N.E.2d 232 (1974).

[23] Nathin & Co. v. George A. Fuller Co., 347 F. Supp. 17 (W.D. Mo. 1972) (both owner and contractor were held liable to subcontractor when work had been unreasonably scheduled to benefit owner and contractor, but to detriment of subcontractors).

[24] Southern Fireproofing v. R.F. Ball Constr. Co., 334 F.2d 122 (8th Cir. 1964).

[25] VABCA Nos. 1943 et al., 84-2 B.C.A. (CCH) ¶ 17,341. The VABCA also stated (concerning the use of CPM schedules):

> [W]hen properly utilized, CPM allows the owner and subsequent review bodies to determine with greater exactitude whether, and to what extent, a particular change order affects the critical path and hence delays ultimate performance.

84 2 B.C.A. (CCH) ¶ 17,341, at 84,411.

[26] *Id.* at 86,405.

§ 7.03

In *Fortec Constructors v. United States*,[27] the Court of Claims stated:

> The reason that the determination of the critical path is crucial to the calculation of delay damages is that only construction work on the critical path had an impact upon the time in which the project was completed. If work on the critical path was delayed, then the eventual completion date of the project was delayed. Delay involving work not on the critical path generally had no impact on the eventual completion date of the project.[28]

As can be seen from the foregoing, it is almost always necessary to use CPM scheduling techniques to successfully prove delay. Only items on the critical path that are delayed will be considered to have delayed completion of the project.[29]

In *Appeal of Montgomery-Ross-Fisher, Inc.*,[30] the Board, based upon the testimony of expert witnesses for both the government and the contractor, defined *critical path* as follows:

> The critical path on a CPM diagram is a line depicting the critical activities necessary to complete a project in the shortest time from the beginning activity to completion of the project. It is the longest continuous path in the network from start to finish of the project. Activities on the critical path do not have float or slack time and, theoretically, a delay of a critical path activity should cause a resulting delay in the subsequent critical path activities unless the work is accelerated, assuming the time durations on the schedule are valid.[31]

Citing *United States Fidelity & Guaranty Co. v. Orlando Utilities Commission*,[32] the Board in *Montgomery-Ross-Fisher* further stated:

> The critical path is the longest series of the work activities through the performance of a whole project. If an activity on the critical path exceeds its

---

[27] 8 Cl. Ct. 490 (1985). In this case, the court rejected the government's CPM analysis because it was not properly updated. The court stated:

> The critical path changed from that depicted in the CPM diagram introduced into evidence. The Corps, however, refused to grant timely and adequate time extensions and to authorize revisions to the CPM to reflect the changed performance critical path. As a result, it is impossible to determine from the CPM diagram whether a particular activity was critical or non-critical, on schedule or behind schedule.

8 Cl. Ct. at 505. *See also* Weaver-Baily Contractors, Inc. v. United States, 19 Cl. Ct. 474, 481–482 (1990).

[28] 8 Cl. Ct. at 505.

[29] *See* Blackhawk Heating & Pumping Co., GSBCA No. 2432, 76-1 B.C.A. (CCH) ¶ 11,649, wherein the GSBCA stated:

> Since liquidated (i.e. delay) damages are only imposed for delays in project completion, it is manifest that only those delays should be considered which actually offset project completion. By their nature the delayed activities involved must necessarily lie on the critical path as it was actually completed.

76-1 B.C.A. (CCH) ¶ 11,649 at 55,579.

[30] PSBCA Nos. 1033, 1096, 84-2 B.C.A. (CCH) ¶ 17,492 (1984).

[31] 84-2 B.C.A. (CCH) ¶ 7,492, at 87,120.

[32] 564 F. Supp. 962, 968 (M.D. Fla. 1983).

SCHEDULING AND PROOF OF DELAY CLAIMS § 7.03

scheduled duration, the termination of the project will be delayed unless some other activity on the critical path is performed in less than its scheduled time. A work activity not on the critical path may be completed later than its scheduled time without affecting the termination of the project unless the noncritical activity exceeds its "float" and thereby becomes an activity on the critical path.[33]

In *Blinderman Construction Company v. United States*,[34] the Court of Federal Claims stated the following:

> CPM is a system of project planning, scheduling and control which combines all relevant information into a single master plan, permitting the establishment of the optimum sequence and duration of operation; the interrelation of all efforts required to complete a construction project are shown; an indication is given of the efforts which are critical to timely completion on the project. *Dictionary of Architecture and Construction* at 228. See also *Wilner v. United States*, 23 Ct.Cl. at 244–245 (quoting *Haney v. United States*, 230 Ct. Cl. 148, 167–169, 676 F.2d 584, 595 (1982)).[35]

The *Montgomery-Ross-Fisher* case, addressed above, also discussed the "contractor's right to finish early" concept (*see* § 4.13). Specifically, the contractor contended that the government had ignored its right to complete its contract performance early, and since it was prevented from doing so because of extra work, it was entitled to additional compensation from the government. In response to this argument, the Board stated:

> Appellant argues that respondent ignores its right to complete its contract performance early and that it was prevented from doing so because of the extra work. It is true, as Appellant contends, that where a contractor establishes it would have completed the contract earlier, impact damages may be recovered for changes causing delays preventing early completion although a contract is completed on time. See, e.g., *Schmid v. United States*, 173 Ct. Cl. 302, 351 F.2d 651 (1965); *Gardner Displays Co. v. United States*, 171 Ct. Cl. [*27] 497, 346 F.2d 585 (1965); *Metropolitan Paving Co. v. United States*, 163 Ct. Cl. 420, 325 F.2d 241 (1963); Owen L. Schwam Constr. Co., ASBCA No. 22407, 79-2 BCA ¶ 13,919 . . . . Absent a contract prohibition, a contractor has a right to better his progress and the Government has an implied obligation to cooperate and not to impede or delay the contractor's performance. *Eickhoff Constr. Co.*, ASBCA No. 20049, 77-1 BCA ¶ 12,398. We are persuaded Appellant would have finished the contract earlier than the completion date but for Modification 1 extra work.[36]

The *Montgomery-Ross-Fisher* case also discussed the question of whether or not the number of changes and the percentage of the price of the changes to

---

[33] 84-2 B.C.A. (CCH) ¶ 7,492, at 87,123.
[34] 39 Fed. Cl. 529 (1997).
[35] 39 Fed. Cl. at 579.
[36] 84-2 B.C.A. (CCH) ¶ 7,492, at 87,124.

the original contract price would, in effect, constitute a "cardinal change." (*See* § 6.10[C].) In commenting upon this issue, the Board stated:

> The parties' chief witnesses disagreed [*29] on whether the number of changes (forty-one with numerous separate items) and the percentage of the price of the changes, excluding Modification 1, to the original contract price (5%), were inordinately high (F.O.F. 20). However, impact is not demonstrated solely by showing the number of changes or clarifications to the contract, nor by comparing the cost of the changes to the original contract price. *Coley Properties Corp.*, PSBCA No. 291, 75-2 BCA ¶ 11,514, aff'd in part and rev'd on another claim, *Coley Properties v. United States,* 219 Ct. Cl. 227, 593 F.2d 380 (1979). See also, *Pathman Constr. Co.,* PSBCA No. 444, 79-2 BCA ¶ 14,027, recon. denied (Jan. 17, 1980), aff'd, *Pathman Constr. Co. v. United States,* 227 Ct. Cl. 670 (1981). There is no fixed rule setting forth the number of changes or percentage of price of changes to establish a cumulative impact due to many changes. A 10% increase was not adequate to establish a cardinal change or breach in Coley Properties, supra, at p. 54,940. Moreover, although the Board in Coley found impact liability it did so based upon a showing that disruption and inefficiency resulted from [*30] the changes and not upon their cost or numbers. Appellant has failed to produce persuasive evidence of disruption and inefficiency here.[37]

§ 7.04 HOW CPM SCHEDULING TECHNIQUES ARE USED TO PROVE DELAY

The first step in the usual process of proving a delay claim is to determine the as-planned schedule, which depicts how the contractor planned to perform the project. This usually consists of the original approved schedule for the project. The second step is to prepare the as-built schedule, which depicts how the project was actually built. That schedule is prepared from contemporaneous project records (i.e., schedules, daily reports, minutes of meetings, correspondence, transmittals, photographs, videotapes, payment applications, testing lab reports, etc.). The third step is obvious: one compares the differences between the as-planned schedule and the as-built schedule to illustrate which critical activities were, in fact, delayed.

The final step is to prepare the entitlement schedule, which shows the delays for which the contractor is entitled to recover damages. This is a crucial step, as it requires an analysis of why the activities on the critical path were delayed and who was responsible for those delays. This, likewise, is determined by a detailed review of the project documents, coupled with interviews of the percipient witnesses. Entitlement schedules are usually one of two types: either an as-planned schedule extended by owner-caused delays, or an as-built schedule collapsed by removing owner-caused delays, thereby depicting the date by which the contractor could have completed the work but for the owner's delay.[38]

---

[37] *Id.*

[38] The entitlement schedules can be used to calculate delay damages (Cannon Constr. Co., ASBCA No. 16142, 72-1 B.C.A. (CCH) ¶ 9404), evaluate acceleration claims (Kenneth Reed

In doing a critical path delay analysis, one must also consider concurrent delays; that is, delays by the contractor that are concurrent with delays caused by the owner. If the owner causes a delay to an activity on the critical path, but the contractor is simultaneously delaying the same work, then the contractor will not be entitled to any compensation. The reasoning for this result is that those critical activities would have been delayed anyway by the contractor; therefore, the owner's concurrent delay is immaterial. Several cases can be cited to support this principle.

In *John Murphy Construction Co.*,[39] the Agricultural Board of Contract Appeals (AGBCA) denied the contractor's delay damages claim, stating that "it does not appear from the record that but for the government caused delays the contractor could have completed the work on time . . . . The contractor was at least concurrently responsible for the delay."[40] In *Fischbach & Moore International Corp.*,[41] the ASBCA stated: "[I]t is axiomatic that a contractor asserting a claim against the government must prove not only that it incurred the additional costs making up its claim but also that such costs would not have been incurred but for government action."[42]

The effect of concurrent delay was also explained in *Clive Construction Co.*[43] as follows:

> Concurrent delay does not bar extensions of time, but it does bar monetary compensation for daily fixed overhead costs . . . because such costs would be incurred on account of the concurrent delay even if the government-responsible delay had not occurred.[44]

Of course, there will be occasions when there are concurrent delays, but the delays of the owner or the delays of the contractor are of different, though perhaps overlapping, durations. With modern CPM scheduling techniques, these events can be graphically depicted, so that the party responsible for the longer delay can be made to bear the expense of the extended delay beyond the other party's shorter, concurrent delay. The burden of proof for both the owner and the contractor in such circumstances was explained in *Bigelow, Inc.*[45] as follows:

> The government has the burden of proof with respect to the contractor's failure to perform pursuant to the terms and conditions of the contract. If this proof burden is satisfied and the contractor believes that its failure is excusable, then the contractor has the burden of coming forward with proof that the performance failure was due to causes beyond its control and without its

---

Constr. Corp., ENGBCA Nos. 2748 et al., 72-1 B.C.A. (CCH) ¶ 9407), and assess liquidated damages (Hardeman-Monier-Hutcheson, ASBCA No. 104444, 67-1 B.C.A. (CCH) ¶ 6158).

[39] AGBCA No. 418, 79-1 B.C.A. (CCH) ¶ 73,836.
[40] 79-1 B.C.A. (CCH) ¶ 73,836, at 13,836.
[41] ASBCA No. 18146, 77-1 B.C.A. (CCH) ¶ 12,300.
[42] 77-1 B.C.A. (CCH) ¶ 12,300, at 12,300.
[43] ASBCA No. 28600, 84-3 B.C.A. (CCH) ¶ 17,594.
[44] 84-3 B.C.A. (CCH) ¶ 17,594 at 87,661.
[45] ASBCA No. 24376, 81-2 B.C.A. (CCH) ¶ 15,300.

fault or negligence. When a contractor is seeking extensions of contract time, for changes and excusable delay, which will relieve it from the consequences of having failed to complete the work within the time allowed for performance, it has the burden of establishing by a preponderance of the evidence not only the existence of an excusable cause of delay but also the extent to which completion of the contract work as a whole was delayed thereby.[46]

The court in *Bigelow* went on to state:

> "[T]he government, by establishing that the contract was incomplete on its due date, . . . has satisfied its burden of proof with respect to the contractor's failure of performance. Thus we are left with a legal evaluation of the excuses proffered by the contractor for its untimely performance."[47]

As a result, if a contractor completes a project beyond the scheduled contract completion date, the burden of proof shifts to the contractor to explain why. This was described in *Bell Construction Co.*[48] as follows:

> In order to prevail on a claim of excusable delay . . . a contractor . . . must go beyond mere allegations and demonstrate with credible proof the existence of certain facts and the satisfaction of certain criteria. First, there must be an identification of the work as controlling the overall completion of the contract. Second, it must be established that this controlling work was delayed by the weather. Thirdly, it must be established that the weather experienced was unforeseeable, i.e., unusually severe.

The Department of Transportation Board of Contract Appeals (DOTBCA) described the contractor's burden of proof as follows:

> A contractor must demonstrate that its performance was actually delayed by the government's action before it is entitled to a time extension and commensurate costs. A contractor carries this burden if it demonstrates that work on an item on the critical path to job completion was delayed by an act of the government.[49]

A delay to an activity that is not on the critical path will not delay the project as a whole. Only delays to critical path activities delay the project as a whole. See *Blinderman Co. Company v. United States*,[50] wherein the Court of Federal Claims stated the following:

> Where, as in the case at bar, the parties use CPM to evaluate contract performance, Courts consistently hold that no proven injury results from construction delays unless it is shown that the activities delayed are on the project's

---

[46] 81-2 B.C.A. (CCH) ¶ 15,300 at 75,737.
[47] *Id.*
[48] ASBCA No. 23376, 79-2 B.C.A. (CCH) ¶ 13,908, at 68,272.
[49] Appeal of Ealahan Elec. Co., DOTBCA No. 1959, 90-3 B.C.A. (CCH) ¶ 23,177, at 116,325.
[50] 39 Fed. Cl. 529 (1997).

critical path. *Fortec Constructors v. United States,* 8 Cl.Ct. 490, 595 (1985), aff'd, 804 F.2d 191 (Fed.Cir. 1986); *Wilner v. United States,* 23 Cl.Ct. at 244 (citing *Broome Constr. v. United States,* 203 Ct.Cl. 521, 528, 482 F.2d 829, 833 (1924)); *Youngdale,* 27 Fed.Cl. at 550 (citing *Haney v. United States,* 230 Ct. Cl. at 167–168, 676 F.2d 584; *G.M. Shupe, Inc. v. United States,* 5 Cl.Ct. 662, 728 (1984)); *Mega Constr.,* 29 Fed.Cl. at 424.[51]

§ 7.05 TYPE OF DELAY MUST BE DETERMINED

Once the actual delays to the job have been determined by using the CPM techniques and analyses described in this chapter, the type of delay must be determined to ascertain which party, if any, is entitled to compensation or damages for the delay. As discussed in Chapter 7, all delays fall into one of three categories: (1) excusable; (2) inexcusable; or (3) compensable. Excusable delays are those caused by factors beyond the control of either the owner or the contractor; examples include weather, strikes, and acts of God. Because neither the owner nor the contractor can control this type of delay, the contractor is entitled to a time extension but no additional compensation. This is sometimes referred to as excusable/noncompensable delay.

The second type of delay, inexcusable, includes any delay that is within the control of the contractor, such as failure to properly man the job, properly schedule the work, timely provide equipment or material to the project, and other types of delay for which the contractor is responsible. Inexcusable delays do not entitle the contractor to a time extension or compensation; in fact, they make the contractor liable for damages the owner may sustain, such as lost rents, loss of use, extended finance costs, and the like.

The third type of delay is compensable delay, which includes any delay to the job caused by the owner and/or by agents for whose acts and omissions the owner is responsible, such as the owner's architect or construction manager. Sometimes called owner-caused delay, examples include failure to grant site access; provision of defective plans and specifications; delays in responding to RFIs (requests for information), shop drawings, or submittals; or other acts or omissions of the owner that delay work on the critical path. For compensable delay (i.e., owner-caused delay), the contractor is entitled to both a time extension and additional compensation.

Sometimes the various types of delay will overlap, which is also referred to as concurrent delay. If an excusable or compensable delay occurs concurrently with an inexcusable delay, the overlapping portion of the delays is inexcusable because the contractor would have been unable to work due to its concurrent, inexcusable delay. If an excusable delay occurs concurrently with a compensable delay, the same logic dictates that the delay will be considered excusable, because the contractor would have been unable to work notwithstanding the owner's (compensable) delay, and therefore the contractor is entitled to a time extension.

---

[51] 39 Fed. Cl. at 584.

§ 7.05

Generally, it is owners who attempt to use concurrent delays as a defense. When the contractor submits a claim for additional compensation due to alleged owner-caused (compensable) delay, the owner will attempt to find evidence that there was a concurrent, inexcusable (contractor-caused) delay to avoid liability for delay damages. An example of this is found in *Beckman Construction Co.*,[52] wherein the contractor contended that the owner (government) caused a constructive change in the requirements for a material-handling system, which, in turn, caused a 72-day delay in obtaining that equipment. The owner (government) was able to present evidence to show that the contractor was behind schedule in constructing the space that housed the equipment and, therefore, could not have installed the equipment even if it had been delivered timely.

In *Clive Construction Co.*,[53] inaccurate contract drawings delayed the work by six weeks. The owner (government) was able to show that, concurrent with this delay, the contractor was prevented from working in any event by the failure of one of its suppliers to deliver equipment.[54]

The modern trend is to apportion the resulting damages when one delay overlaps another. For example, in *Toombs & Co. v. United States*,[55] design errors required the owner (government) to stop work on one part of the project. Problems then developed with the performance of the contractor's masonry subcontractor. The initial delay caused by the government was treated as compensable, but when the problems with the masonry subcontractor became concurrent, the government was held to have no further liability for the delay.

When there is no claim by the contractor for delay damages, as when there are concurrent delays, the question arises as to whether the contractor is nevertheless entitled to a time extension so as to avoid the owner's damages claims (e.g., liquidated damages or actual damages). In *Hood Plumbing*,[56] the owner (government) delayed in finalizing the terms of a contract modification. Concurrent with that delay, the contractor was having problems with a supplier; therefore, the AGBCA held that delay to be inexcusable (that is, a inexcusable delay concurrent with an excusable/compensable delay). As a result, the AGBCA held that the contractor was not entitled to a time extension and therefore remained liable to the owner for liquidated damages.

For a contrary case, see *C.G. Norton Co.*,[57] in which an owner's design error was concurrent with a delay caused by the contractor's fabricator. The ENGBCA did not assess liquidated damages against the contractor; rather, it refused to attempt to allocate blame for the delay or to hold either the government or the contractor liable. It was therefore a wash. The contractor could not recover delay damages, and the government could not recover liquidated damages.

---

[52] ASBCA No. 24725 (1983).
[53] ASBCA No. 28600, 84-3 B.C.A. (CCH) ¶ 17,594 (1984).
[54] *See also* Appeal of Rivera Contracting, ASBCA No. 28888 (1985); Appeal of Volpe-Head, Joint Venture, ENGBCA No. 4722, 89-3 B.C.A. (CCH) ¶ 22,105 (1989).
[55] 4 Cl. Ct. 535 (1984).
[56] AGBCA No. 84-181-1 (1987).
[57] ENGBCA No. 5182 (1988).

A case in which the ASBCA did apportion and allocate fault for delay is *B.D. Collins Construction Co.*,[58] in which the government issued 10 change orders that extended project completion by 153 days. During performance of the work, the contractor suffered two delays to the critical path: a 37-day labor strike (excusable/noncompensable) and a 70-day delay caused by the contractor (inexcusable). The ASBCA subtracted those two delays (37 + 70 = 107) and allowed the contractor 46 days of extended field overhead (153 − 107 = 46).

Suppose that both the owner and the contractor concurrently contribute to the *same* delay. Generally, when both parties contribute to the delay, it is impossible to apportion fault and the result will be a wash. Neither the contractor nor the owner will be able to recover delay damages from the other.[59]

Critical path scheduling may come into play in this area as well, as illustrated by *Wilner v. United States*.[60] In that case, there was an owner-caused (compensable) delay concurrent with a contractor-caused (inexcusable) delay. The contractor was able to show that the government's delay was to the critical path, whereas the contractor's delay was to a construction activity off the critical path. As a result, the contractor was allowed to recover delay damages, because the critical path of the project was, in fact, delayed by the owner but not by the contractor.

When delays are sequential and not concurrent and can be apportioned, each party may recover damages against the other in proportion to the net amount of critical delay actually caused by each party. This is illustrated by *Williams Enterprises v. Strait Manufacturing & Welding, Inc.*,[61] a case in which a steel tower collapsed, stopping steel erection for three months. The steel erection sub-subcontractor sued the steel fabrication subcontractor; the steel fabrication subcontractor sought indemnity from the prime contractor; and the prime contractor countersued the steel fabrication subcontractor for the delay. The steel fabrication subcontractor and steel erection sub-subcontractor both contended that they were not liable for delay damages, because the contractor had concurrently delayed the project by late approvals of shop drawings for the exterior precast panels. Their contention was that the project would have been delayed by the precast panels, even if the tower collapse had never occurred. The court found that the delays by the prime contractor in approving the shop drawings did, in fact, delay the job. The prime contractor, in its CPM analysis, showed that the tower collapse caused a 106-day delay. It also showed that only 23 days of that delay were concurrent with the prime contractor's delay in approval of the shop drawings. As a

---

[58] ASBCA No. 42662.

[59] *See* C.G. Norton Co., ENGBCA No. 5182 (1988); Appeal of J.B.L. Constr. Co., VABCA No. 1799, 86-1 B.C.A. ¶ 18,529 (1985); Appeal of Coffey Constr. Co., VABCA No. 3361 (1993); J.A. Jones Constr. Co. v. Greenbriar Shopping Ctr., 332 F. Supp. 1336 (N.D. Ga. 1971). *See also* Beckman Constr. Co., ASBCA No. 24725, 83-1 B.C.A. (CCH) ¶ 16,326 (1983); S.O.G. San Ore Gardner v. Missouri Pac. R.R., 658 F.2d 562 (8th Cir. 1981); Blinderman Constr. Co. v. United States, 695 F.2d 552 (Fed. Cir. 1982); Midstate Constructors, Inc., PSBCA No. 913, 81-1 B.C.A. (CCH) ¶ 14,898 (1991).

[60] 23 Cl. Ct. 241 (1991).

[61] 728 F. Supp. 12 (D.D.C. 1990).

result, the contractor was entitled to recover delay damages for a net critical path delay of 83 days (106 − 23 = 83).

As shown above, one of the key elements in any delay analysis is the classification of the type of delay. This, in turn, determines which way the damages will flow. If it is an excusable delay (force majeure, strike, weather, act of God, etc.), the contractor will be entitled to a time extension but no additional compensation. If it is a compensable (owner-caused) delay, the contractor will be entitled to a time extension *and* compensation. If it is a inexcusable (contractor-caused) delay, the contractor will not be entitled to either a time extension or compensation and may be liable to the owner for liquidated or actual damages.

§ 7.06 CONCLUSION

When a project is completed beyond the contract completion date, both the owner and the contractor will likely suffer damages. For the contractor, such damages will consist of extended jobsite expenses and extended home office overhead, and perhaps other consequential damages, such as impairment of bonding capacity, loss of profit on missed opportunities, increased interest expenses, and other such damages. The owner may suffer loss of rent, loss of profits, increased interest expenses, and other damages resulting from its inability to use the project on the date contemplated.

The owner can establish its *prima facie* case merely by establishing the contract completion date and the actual completion date. The burden of proof then shifts to the contractor to prove that the delay was caused by other factors, such as owner-caused delay (excusable and compensable) or excusable delay (non-compensable). As a practical matter, both the owner and the contractor would be well advised to be able to present competent evidence showing which activities on the critical path were delayed and who was responsible for each delay. To be successful in court, this usually has to be done by employing the CPM scheduling techniques described in this chapter.

Chapter 8
EXPANDING LIABILITY IN THE CONSTRUCTION INDUSTRY

§ 8.01   Introduction
§ 8.02   Strict Liability
§ 8.03   Fraud
§ 8.04   Breach of Implied Warranty
§ 8.05   Third-Party Beneficiary
§ 8.06   Interference with Prospective Economic Advantage
§ 8.07   Owner's Liability to Subcontractor for Defective Plans and Specifications
§ 8.08   Failure to Exercise Reasonable Care
§ 8.09   Current Trends in California
§ 8.10   Certificate of Merit Required by CCP § 411.35
§ 8.11   Liability of Consultants
§ 8.12   Liability for Defective Product That Has Not Yet Caused Damage: The Economic Loss Rule
§ 8.13   "Stigma Damages" Are Not Recoverable

§ 8.01 INTRODUCTION

For many years, it was established law across the United States that in order to sue someone, privity of contract with that person was necessary. That "privity doctrine" has now been substantially eroded in many areas of the law, including construction law. Courts in California and across the country have modified the privity concept and are now more willing to look at the facts of a given case to determine who caused the loss and attempt to place liability on that person, even if there is no privity of contract between the injured party and the party causing the loss.

§ 8.02 STRICT LIABILITY

The first case in California to adopt the concept of "strict liability" was *Kriegler v. Eichler Homes, Inc.*[1] Eichler Homes was a mass producer of single-family residences in which a heating system that was embedded in the concrete turned out to be defective and damaged the home. The court held that Eichler, as the mass producer of the home, could be held "strictly liable" to the ultimate purchaser of the home for the damages.

The reasoning in *Eichler* was extended to a mass producer of residential lots in *Avner v. Longridge Estates.*[2] Then, in *Stewart v. Cressview Mutual Water Co.,*[3] a developer was held strictly liable when the plaintiff's home was destroyed by fire, due to a defective water distribution system that had failed to supply sufficient water. The latest case to hold a developer strictly liable for damages as a result of a defective structure was *Del Mar Beach Club Owner's Ass'n v. Imperial Contracting Co.*[4] Thus, to date, California courts have limited strict liability to mass producers of homes and residential lots.

However, California courts have held that owners cannot sue subcontractors with whom they have no contractual relationship on theories of strict liability and/or failure to disclose defects in the product. In *La Jolla Village Homeowners Ass'n v. Superior Court,*[5] the owners of a condominium project sued subcontractors with whom they had no direct contractual relationship, on theories of strict liability and failure to disclose defects in the product. The court of appeal held that subcontractors could not be held strictly liable for defects in construction and, further, that the homeowners had no cause of action for nondisclosure against the subcontractors. This trend was continued in *Monte Vista Development Co. v. Superior Court (Wiley Tile Co.).*[6]

In *Monte Vista,* the court of appeal held that a tile subcontractor (Wiley), who purchased and installed a soap dish under a subcontract with the subdivider,

---

[1] 269 Cal. App. 2d 244 (1969).
[2] 272 Cal. App. 2d 607 (1969).
[3] 34 Cal. App. 3d 808 (1973).
[4] 123 Cal. App. 3d 898 (1981).
[5] 212 Cal. App. 3d 1131 (1989).
[6] 226 Cal. App. 3d 1681 (1991). *But see* § 8.05, wherein an owner was allowed to sue a subcontractor on a third-party beneficiary theory.

was not strictly liable to a home buyer who was severely lacerated when she leaned on the soap dish. The holding was based upon the fact that a tile subcontractor is not engaged in the business of selling soap dishes. *Restatement (Second) of Torts* § 402A(1)(a) provides that one who sells a defective product is liable to the consumer for physical harm if the seller is engaged in the business of selling such a product. The *Monte Vista* court held that the tile subcontractor was not a "seller" of the soap dish within the meaning of the *Restatement.* The court stated that it was of no matter to Wiley whether the subdivider or someone else supplied the tile fixtures. Wiley's job was only to do the work, and therefore it did not come into the chain of commerce as a supplier of a soap dish in such a way as to be strictly liable for the injury.

Lack of privity of contract likewise defeated the claim of a concrete contractor against the manufacturer of a "slipform" paving machine in *All West Electronics, Inc. v. M-B-W, Inc.*[7] When a concrete contractor (All West) contacted a retailer (C&R) to discuss purchase of a slipform paver, C&R contacted the manufacturer (M-B-W), who conducted a demonstration in which it was stated that the machine would perform as stated in the manufacturer's brochure. The written brochure stated: "STATE OF THE ART SLIPFORM PAVING . . . OUT PERFORMS big machines on small jobs . . . Any jobs with tight radius work." The retailer then leased the machine to the contractor. The retailer was paid by the finance company to which it assigned the lease. The machine totally failed to perform as advertised. The retailer had filed for bankruptcy protection, so the concrete contractor sued the manufacturer (M-B-W). The court of appeal held that the concrete contractor could not recover from the manufacturer since there was no privity of contract between them. In California, in order to sue for breach of implied warranties of fitness for use and merchantability, there must be privity of contract.[8]

§ 8.03 FRAUD

Other cases in the construction industry have continued to impose liability on other theories. For example, in *Walker v. Signal Cos.,*[9] a developer, whose "inexplicable delays" in completing a house for the plaintiffs resulted in their losing deferral of capital gain on the sale of their former residence, was held liable for fraudulently inducing a contract whose terms the developer did not intend to keep.

---

[7] 64 Cal. App. 4th 717 (1998).
[8] The *All West* court followed the cases of *Burr v. Sherwin Williams Co.,* 42 Cal. 2d 682 (1954); *U.S. Roofing, Inc. v. Credit Alliance Corp.,* 228 Cal. App. 3d 1431 (1991), and *Osborne v. Subaru of America, Inc.,* 198 Cal. App. 3d 646 (1988).
[9] 84 Cal. App. 3d 982 (1978).

§ 8.04 BREACH OF IMPLIED WARRANTY

In *Pollard v. Saxe & Yolles Development Co.*,[10] the builder and developer of a new condominium was held liable to the purchasers of the condominiums on a theory of breach of an implied warranty that the completed structure was designed and constructed in a good and workmanlike manner. In *Aced v. Hobbs-Sesack Plumbing Co.*,[11] a subcontractor was held liable to a prime contractor for breach of the warranty of merchantability in the installation of tubing in a radiant heating system.

§ 8.05 THIRD-PARTY BENEFICIARY

In *Gilbert Financial Corp. v. Steel Form Contracting Co.*,[12] an owner was allowed to sue a roofing subcontractor for breach of the implied warranty of fitness, on a third-party beneficiary theory, when the roof proved to be defective. The owner had no direct contractual relationship with the subcontractor.

In *COAC, Inc. v. Kennedy Engineers*,[13] an engineering firm with a direct contractual relationship with an owner was held liable, on a third-party beneficiary theory, to a contractor with whom it had no direct contractual relationship, for delay in issuing an environmental impact report.

In *Del E. Webb Corp. v. Structural Materials Co.*,[14] a material supplier was held liable to a prime contractor with whom he had no direct contractual relationship. The contractor's numerous legal theories, including third-party beneficiary theory, survived on appeal.

§ 8.06 INTERFERENCE WITH PROSPECTIVE ECONOMIC ADVANTAGE

In *J'Aire Corp. v. Gregory*,[15] a tenant who was not in privity of contract with the general contractor was held to have a cause of action against the general contractor on a theory of negligent interference with prospective economic advantage. The general contractor had delayed in completing a restaurant the plaintiff was leasing from the County of Contra Costa.

In *Chameleon Engineering Corp. v. Air Dynamics, Inc.*,[16] a supplier of air-conditioning units to a subcontractor was held liable to the general contractor for delay, on a theory of negligent interference with prospective economic advantage. In that case, the supplier was to provide air-conditioning units to the heating, ventilating, and air-conditioning subcontractor.

---

[10] 12 Cal. 3d 374 (1974).
[11] 55 Cal. 2d 573 (1961).
[12] 82 Cal. App. 3d 65 (1978).
[13] 67 Cal. App. 3d 916 (1977).
[14] 123 Cal. App. 3d 593 (1981).
[15] 24 Cal. 3d 799 (1979).
[16] 101 Cal. App. 3d 418 (1980).

§ 8.07 OWNER'S LIABILITY TO SUBCONTRACTOR FOR DEFECTIVE PLANS AND SPECIFICATIONS

In an unpublished opinion, *Vanlar Construction, Inc. v. County of Los Angeles*,[17] one of the claims made by the mechanical subcontractor, who had sued the prime contractor and the owner (the county of Los Angeles) was that the owner was liable to the subcontractor for errors and omissions in the plans and specifications, which errors and omissions caused damage to the subcontractor. At the conclusion of the evidence, the trial court granted a motion for judgment on the pleadings as to the subcontractor's cause of action against the county for declaratory relief. On appeal, the court stated that the issue, as to the subcontractor, was whether judgment on the pleadings should have been granted on the subcontractor's action for declaratory relief against the county relating to damages for negligent interference with prospective economic advantage.

The court of appeal held that a judgment on the pleadings should not have been granted. There was no reason why the subcontractor could not state a cause of action, under a negligence theory of action, for declaratory relief against the county and that lack of privity of contract between the subcontractor and the county would not foreclose such an action. The court relied upon the *J'Aire* and the *Chameleon Engineering* cases discussed above, and commented upon the *Gilbert Financial* case, cited in § 8.05 above. The court noted that, in the *Gilbert* case, the appellate court held that an owner could state a good cause of action for negligence against a subcontractor for the installation of a defective roof. Commenting upon the *J'Aire, Chameleon,* and *Gilbert* cases, the *Vanlar* court stated the following:

> The foregoing cases do not involve the precise situation of a subcontractor suing the owner on a negligence theory. *J'Aire* is a suit by a lessee against the owner's general contractor. *Chameleon* is an action by a supplier of a subcontractor against the general contractor. *Gilbert* involves an action by an owner against a subcontractor. Despite the difference in relationships, as the case at bar involves a suit for negligence by a subcontractor against an owner, this should make no difference as a matter of logic or law. Surely if an owner can sue a subcontractor as in *Gilbert,* logically a subcontractor should be able to sue an owner under a proper tort cause of action even in the absence of privity of contract.

> Hence the judgment on the pleadings against Hickman is reversed to the extent that its second cause of action for declaratory relief includes a claim for negligent interference with economic advantage. Hickman may be compelled to amend its pleadings in the court below, if it can do so, to include the pleading requirements of *J'Aire, Biakanja* and *Chameleon, supra.*

Because *Vanlar* is an unpublished opinion, it cannot be cited as authority in other cases. It does, however, recognize the point that a subcontractor can sue

---

[17] Case No. B005504 (Ct. App., 2d Dist., Div. 5 May 30, 1985).

an owner with whom it has no contractual relationship for negligent interference with economic advantage.

This concept has been recognized in court opinions outside California. For example, in *Guardian Construction Co. v. Tetra Tech Richardson, Inc.*,[18] the state of Delaware engaged an engineer to prepare the contract documents and specifications for a beach breakwater project. The plans and specifications were part of the bid package issued to contractors bidding on the job. The low bidder was awarded the contract and engaged a subcontractor to provide labor and materials in accordance with the plans and specifications. After work began, the contractor and subcontractor discovered that the tidal heights and projected benchmark had been miscalculated by the engineer. As a result, the contractor and subcontractor were unable to perform the work as planned and, in fact, incurred extra costs. When the contractor and subcontractor brought suit against the architect, alleging negligent misrepresentation, negligence, and breach of contract, the engineer moved for summary judgment. The Delaware Superior Court noted that the issue of an engineer's liability for economic losses to a contractor or a subcontractor, with whom it has no privity of contract, was a question of first impression in Delaware. Accordingly, the court looked to other jurisdictions for guidance. The court noted that, traditionally, attempts by an injured third party to recover for damages arising out of a contractual duty failed for lack of privity, citing *Ultramares Corp. v. Touche Ross & Co.*[19] The court also noted that the traditional rule in those cases was that a contractual relationship was necessary in order to recover for purely economic losses. As a result, architects and engineers were shielded from direct liability to construction contractors for cost increases arising from defective plans and specifications. However, the court noted that, with the passage of time, the privity of contract rule had fallen into disfavor, and courts had begun to permit recovery of purely economic losses when the information was prepared with the intent that it be relied upon by a specific group and a member of that group, in fact, relied upon the information to its detriment (citing *E.C. Goldman, Inc. v. A/R/C Associates*[20]).

The court noted further that, recently, a third and more expansive rule had come into being—to wit, permitting recovery whenever it was foreseeable that the damaged party might have relied upon the information (citing *East River Steamship Corp. v. Transamerica Delaval, Inc.*[21]). After considering those rules, the court adopted the "specific intent" rule and held that the engineer prepared the plans and specifications while knowing and intending that contractors preparing to bid on the project would rely upon those documents in preparing their estimates. Should it be established that the documents were in error, and that the contractor and subcontractor relied on those documents, recovery might be allowed. Therefore, the summary judgment motion was denied. It is clear from

---

[18] 583 A.2d 1378 (Del. Super. Ct. 1990).
[19] 174 N.E. 441 (N.Y. 1931).
[20] 543 So. 2d 1268 (Fla. Dist. Ct. App. 1989).
[21] 476 U.S. 858 (1986).

the foregoing that the trend in the courts is to place liability on the person causing the loss, despite any lack of privity of contract.

In preparing plans and specifications, a design professional acts as an independent contractor. Thus, the design professional may be held liable to a prime contractor and/or other members of the contractor's group for negligence in the preparation of such documents.[22] In addition, the courts have held that a design professional may be held liable for negligent approval of materials that later proved to be defective.[23] In light of the foregoing, contractors and subcontractors should not overlook this area of potential liability when they seek recovery for losses incurred as a result of defective plans and specifications.

§ 8.08 FAILURE TO EXERCISE REASONABLE CARE

In *Shurpin v. Elmhirst*,[24] a downslope property owner sued a soils engineering firm employed by the adjacent upslope property owner to stabilize and rebuild a slope. The engineering firm performed no work on the plaintiff's property and provided only a recommendation for reconstruction of the upslope owner's property. The court of appeal held that the engineering firm had acted in a professional capacity in designing the corrective measures and owed a duty to foreseeable plaintiffs who might be injured if the engineering firm failed to exercise reasonable care. Consequently, the adjacent downslope landowner was a party who might foreseeably be injured if the engineer failed to exercise reasonable care.

§ 8.09 CURRENT TRENDS IN CALIFORNIA

Thus, it can be seen that the trend in California is to place liability upon the party who has caused injury to a third party, even though there may be no contractual relationship between the complaining party and the party causing the injury.

While the cases remain intact that permit plaintiffs to hold third parties liable for damages, regardless of the absence of privity of contract, at least one aspect of expanding liability in California has been retracted. In *Seaman's Direct Buying Service, Inc. v. Standard Oil Co.*,[25] the California Supreme Court held that a tort cause of action might lie "when, in addition to breaching the contract, defendant seeks to shield itself from liability by denying, in bad faith and without probable cause, that the contract exists." Thus, a new tort cause of action was established in California.

The *Seaman's* decision, however, presented the appellate courts with a number of unanswered questions, and those courts reached varying and often inconsistent conclusions in response. Furthermore, subsequent opinions of the California Supreme Court have indicated a reluctance to authorize tort recovery

---

[22] *See* A.R. Moyer, Inc. v. Graham, 285 So. 2d 397 (Fla. 1973).
[23] *See* Shoffner Indus., Inc. v. W.B. Lloyd Constr. Co., 257 S.E.2d 50 (N.C. 1979).
[24] 148 Cal. App. 3d 193 (1983).
[25] 36 Cal. 3d 752 (1984).

for contract breaches except in situations involving insurance. In the face of these conflicting rulings, the California Supreme Court finally held, in *Freeman & Mills, Inc. v. Belcher Oil Co.*,[26] that the *Seaman's* tort of bad-faith denial of a contract is no longer viable. Although the *Freeman* ruling may signal the beginning of the end to at least one aspect of expanding liability in California, the Supreme Court did not slam the door completely shut. The court emphasized, "[N]othing we say here would prevent the Legislature from creating additional civil remedies for noninsurance contract breach, including such measures as providing litigation costs and attorney's fees in certain aggravated cases, or assessing increased compensatory damages covering lost profits and other losses attributable to the breach, as well as restoration of the *Seaman's* holding if the Legislature should deem that course appropriate."[27]

§ 8.10 CERTIFICATE OF MERIT REQUIRED BY CCP § 411.35

In California, filing of a "certificate of merit" in accordance with California Civil Procedure Code § 411.35 is a necessary prerequisite to maintaining an action for damages against an architect, engineer, or land surveyor, arising out of their alleged professional negligence. Section 411.35 provides that, on or before the date of service of the complaint on any defendant, plaintiff's attorney shall execute and file a certificate of merit declaring one of the following:

1. That the attorney has reviewed the facts of the case, consulted with at least one design professional who practices or teaches in an accredited college or university and is licensed to practice in the same discipline as the defendant, and that the attorney has concluded on the basis of such review and consultation that there is a reasonable and meritorious cause for the filing of such action; or

2. That the attorney was unable to obtain the consultation required because a statute of limitations would impair the action (resulting in a 60-day extension for filing the certificate required pursuant to paragraph (1)); or

3. That the attorney has made three separate good-faith attempts with three separate design professionals and none of those contacted would agree to such a consultation.

The statute further provides that a violation of the statute may constitute unprofessional conduct and be grounds for discipline against the attorney, as well as providing grounds for a demurrer or motion to strike made by the defendant design professional.

Although the statute provides that an attorney who submits a certificate has a privilege to refuse to disclose the identity of the design professional consulted and the contents of the consultation, the court may require the attorney to divulge

---

[26] 11 Cal. 4th 85 (1995).
[27] 11 Cal. 4th at 103.

the names of design professionals refusing consultation. The statute also provides that, upon the "favorable conclusion" of the litigation with respect to any party for whom a certificate of merit was filed, the court may require the attorney for the plaintiff who executed the certificate to reveal the name, address, and telephone number of the person consulted at an *in camera* proceeding (i.e., outside the presence of counsel and parties). If the trial judge finds there has been a failure to comply with this section, the court may order a party, a party's attorney, or both to pay any reasonable expenses, including attorney's fees, incurred by another party as a result of the failure to comply with section 411.35.

Section 411.35 was implemented to discourage the filing of frivolous malpractice actions against design professionals. The Legislature intended to strengthen the certificate of merit requirement by allowing the assessment of reasonable expenses, including attorney's fees as sanctions against a noncomplying party pursuant to subdivision (h).[28] However, a settlement in which the defendant design professional waives fees and costs is not considered a "favorable conclusion" even if the design professional makes no monetary contribution toward the settlement. Therefore, in *Korbel v. Sheung-Chi Chou*,[29] the court of appeal held that, under such circumstances, the settling engineer was precluded from invoking subdivision (h) of section 411.35 to obtain disclosure of the name of plaintiff's attorney's consulting engineer as a means of recovering fees and costs.

§ 8.11 LIABILITY OF CONSULTANTS

An entire new business has grown up in the construction industry: that of expert consultants who advise the parties to construction projects as to their rights and liabilities with regard to claims on troubled construction projects. It is very common for owners, contractors, subcontractors, and others, and/or their attorneys, to hire claims consultants to aid in the presentation or defense of claims. A question that often arises is, What is the liability of these claims consultants should they fail to perform their services in a good and workmanlike manner?

In *Murphy v. A.A. Matthews*,[30] an engineering firm provided claims preparation services to a subcontractor that had performed foundation work for nuclear cooling towers and incurred unforeseen expenses as a result of problems on the job. The subcontractor retained the engineering firm to prepare claims for additional compensation and to present those claims at an arbitration hearing. After an unsuccessful arbitration hearing, the subcontractor sued the engineer for negligence in preparing and documenting the subcontractor's claims. The subcontractor alleged that the engineer failed to discover discrepancies among various documents, did not document the claims clearly and cogently so that they could be understood, provided insufficient supporting data on some of the claims, and prepared the claims in such a manner that the data appeared to have been manipulated.

---

[28] Guinn v. Dotson, 23 Cal. App. 4th 262 (1994).
[29] 26 Cal. App. 4th 1427 (1994).
[30] 841 S.W.2d 671 (Mo. 1992).

A trial court granted the engineer's motion to dismiss, based upon witness immunity; however, the Supreme Court of Missouri reversed that decision and remanded the case. The supreme court held that professionals who provide litigation-related services for compensation may be held liable to their clients and that the doctrine of witness immunity does not apply to expert witnesses who are sued by their clients. The court noted that the witness immunity doctrine was based upon the rationale that such immunity is required so that witnesses will not be reluctant to testify; therefore, the doctrine is limited to actions against a witness for defamation or other such causes of action based upon the witness's testimony.

Absolute immunity is applied only when the witness's statements were relevant to judicial proceedings or when they were made directly in the judicial proceeding or in an affidavit or pleading. The policy considerations supporting the witness immunity doctrine do not extend to negligence suits by clients against the experts they retained. Because such experts are paid professionals, there is no necessity to protect them with immunity in order to ensure frank and objective testimony.

Finally, the court in *Murphy* recognized that allowing suits against experts is no different than allowing malpractice suits against trial attorneys. Like trial attorneys, expert witnesses enter into a commercial relationship with their clients and function as advocates. The court noted that its decision was limited to pre-trial litigation support services, and did not reach the issue of whether the witness immunity doctrine would apply to actual trial testimony.

§ 8.12 LIABILITY FOR DEFECTIVE PRODUCT THAT HAS NOT YET CAUSED DAMAGE: THE ECONOMIC LOSS RULE

In a very significant case, *Zamora v. Shell Oil Co.*,[31] the court of appeal held that a product that has "degraded" but has not yet failed cannot form the basis of liability under either a negligence or strict liability theory. Shell, which had no contractual relationship with the plaintiff homeowners, supplied material used in the manufacture of plumbing pipes. The pipes had "micro-cracking," and there was evidence that they would fail in the future; however, the evidence showed that the pipes did not leak and had not yet failed. The court held that those homeowners who had sustained no leakage from their pipes could not recover from Shell because, even though there was degradation to the homeowners' plumbing systems and failure of the pipes in the future was reasonably certain, the homeowners had suffered no cognizable present damage that would support a negligence or strict liability cause of action. This case could have far-reaching implications in construction defect cases where, quite often, there is a claim that some component part of the structure will fail in the future. The *Zamora* case stands for the proposition that, until there is an actual failure, plaintiffs suffer no recoverable damages from defective work and/or materials.

---

[31] 55 Cal. App. 4th 204 (1997).

The *Zamora* case was followed in *AAS v. Superior Court of San Diego*,[32] in which homeowners and their association sued the developer, the prime contractor, and a subcontractor for negligence, alleging that their homes had serious construction defects, including violation of the Building Code. Those defects, however, had not yet damaged property or injured any person. The court held that, under the "economic loss" rule, the homeowners and homeowner association could not sue for negligence unless the defective construction had resulted in actual physical injury to persons or property. The California Supreme Court has granted a hearing in this case; therefore, this decision is not final.

§ 8.13 "STIGMA DAMAGES" ARE NOT RECOVERABLE

Homeowners who have construction defects in their homes may bring an action to recover for damages to persons or property arising from those construction defects. If they recover for the damages and then repair the homes, when attempting to sell those homes, they are obligated to disclose to potential purchasers that there were defects in the home that have been repaired. Such homes may thus have a "stigma" attached to them that makes them less marketable and/or diminishes their value. In *AAS v. Superior Court of San Diego*,[33] the court of appeal held that such stigma damages are not recoverable in California. The court affirmed a ruling of the trial court that denied an offer of proof that homes with construction defects were worth 2.8 percent less than those without defects, even after full repairs. The court held that such evidence was generic and speculative. The California Supreme Court has granted a hearing in this case; therefore, this decision is not final.

---

[32] 64 Cal. App. 4th 916 (1998).
[33] 64 Cal. App. 4th 916 (1998).

Chapter 9

MECHANICS' LIENS, STOP NOTICES, AND BONDS ON PRIVATE WORKS OF IMPROVEMENT

§ 9.01 Remedies Available
    [A] Mechanic's Lien
        [1] Statutory Implementation
        [2] Other Persons Entitled to File Liens
        [3] Design Professionals Lien
        [4] Persons Not Entitled to File Liens
        [5] Property to Which Mechanics' Liens Attach
        [6] Notice of Nonresponsibility
    [B] Stop Notices
        [1] Contents of Stop Notices
        [2] Case Law Interpretation
            [a] *Familian Corp. v. Imperial Bank*
            [b] Negligence in Disbursing Construction Loan Funds
            [c] Equitable Liens
            [d] General Contractor's Stop Notice Rights
            [e] Defending Stop Notice Claims
            [f] Effect of Payment Bond on Stop Notices
        [3] Attorney's Fees on Bonded Stop Notice
    [C] Bonds
        [1] Perfecting Payment Bond Claims
        [2] Off-Site Bonds

§ 9.02 Procedure for Filing Mechanics' Liens and Stop Notices
    [A] Twenty-Day Notice
        [1] In General
            [a] When Given
            [b] By Whom Given
            [c] To Whom Given
            [d] Where Given
            [e] How Given
            [f] Other Requirements
        [2] Contents of the Notice
        [3] Disciplinary Action

[4] Filing Preliminary Notices with County Recorder
[B] Cases Interpreting Preliminary Notice
[1] *Romak v. Prudential*
[2] *Brown Co. v. Superior Court*
[3] *IGA v. Manufacturers Bank*
[4] *James v. Five Points Ranch*
[5] *Industrial Asphalt v. Garrett Corp.*
[6] *Kodiak Industries v. Ellis*
[7] Additional Case Examples

§ 9.03 Recording the Lien
[A] Recording Requirements
[1] Notice of Completion
[B] Multiple Contracts
[C] Contents of Mechanics' Liens
[1] Description of Site Sufficient for Identification
[2] Name of Owner or Reputed Owner
[3] Name of Person by Whom Claimant Was Employed or to Whom Claimant Furnished Its Materials
[4] Signature of Claimant
[D] Time for Filing Suit to Foreclose
[1] Petition to Remove Lien Not Foreclosed
[2] Lien Foreclosure Actions Stayed Pending Arbitration
[E] What Constitutes Completion
[F] Other Mechanic's Lien Issues
[G] Stop Notice
[1] Construction Funds Are Trust Funds
[2] Service of Stop Notice
[3] Contents of Stop Notice
[4] Time for Filing Suit
[5] Statute of Limitations on Filing Suit to Enforce Stop Notice Is Subject to Doctrine of Equitable Tolling
[H] Payment (Labor and Materials) Bonds
[1] Preliminary Notice
[2] Time for Filing Suit
[3] Cases Interpreting Bond Claims

§ 9.04 Special Problems with Mechanics' Liens
[A] Priority
[B] Separate Works of Improvement
[C] Separate Residential Units
[D] Segregation of Lien
[E] Condominiums
[F] Core Work versus Tenant Work

MECHANICS' LIENS, STOP NOTICES, AND BONDS

      [G]   Proof

§ 9.05   Defenses to Mechanics' Liens and Stop Notices
      [A]   Application of Payments
      [B]   Releases
            [1]   The Historical Context
            [2]   The *Halbert's Lumber* Decision
            [3]   *Halbert's Lumber* Developments
            [4]   Analysis of the New Release Forms
                  [a]   Definitions
                  [b]   Exceptions to the Release
                  [c]   Inherent Ambiguity of the Unconditional Release Form
            [5]   Recommendations and Conclusions
      [C]   Effect of Assignment for Benefit of Creditors
      [D]   Restrictive Endorsements
      [E]   Economic Duress
      [F]   Joint Checks
      [G]   Unclean Hands: Fraud by the Lien Claimant

§ 9.06   Miscellaneous Points
      [A]   Removal of Materials: Right to Lien
      [B]   Voidable Preferences
      [C]   Are Wall-to-Wall Carpets Lienable?
            [1]   In General
            [2]   Conclusion
      [D]   Unreasonable Delay in Payment/Pay-If-Paid Clauses Unenforceable
      [E]   Diverting Funds
      [F]   Lien Law Liberally Construed
      [G]   Oral Agreements for Extras
      [H]   Amount of Lien
      [I]   Ninety-Day Foreclosure Limit
      [J]   Interest
      [K]   Jurisdiction
      [L]   Forged Endorsement of Joint Check
      [M]   Lien Forfeiture: Willful Overstatement
      [N]   Slander of Title
      [O]   Lien Contents
            [1]   Description of Property
            [2]   Amount of Liens
      [P]   Implied Warranty of Design
      [Q]   Personal Liability of Owner
      [R]   Parties
      [S]   Release Bond
      [T]   Contractual Waivers of Mechanic's Lien Remedies

§ 9.01 REMEDIES AVAILABLE

[A] Mechanic's Lien

The mechanic's lien in the state of California is constitutional in origin. Article XIV, § 3, of the California Constitution provides as follows:

> Mechanics, materialmen, artisans, and laborers of every class shall have a lien upon the property upon which they have bestowed labor or furnished materials, for the value of such materials, for the value of such labor done and materials furnished; and the legislature shall provide, by law, for the speedy and efficient enforcement of such liens.

The constitution is not self-executing, and therefore the legislature must enact the procedure for enforcing mechanic's lien rights.[1]

The California Supreme Court has held the mechanic's lien remedy constitutional.[2] It has recognized the importance of protecting the constitutionally guaranteed right of lien, stating:

> As to the interest of the owner whose property is subject to a mechanic's lien, we shall explain that he suffers only a minor deprivation by reason of the lien since he retains possession and use of the land; the owner whose account is subject to a stop notice suffers only the encumbrance of the very funds he has previously allocated for the exclusive purpose of paying construction costs. Moreover, the owner enjoys a variety of measures by which he can protect himself against the impact of such a lien, most notably the requirement that the mechanic must file a preliminary notice before filing or recording his lien, thus affording the owner opportunity to take legal steps against any imposition of an improper lien.
>
> As to the worker whose labor has gone into the property, we point out that he would suffer a major deprivation by the abolition of the lien. Without recourse to prevent the owner from the disposition of the property, or to bar the dissipation of loan funds allocated to the payment of construction costs, the worker would be left with only an unsecured and potentially uncollectible claim for compensation for labor that has enhanced the value of the property itself.
>
> The remedy of stop notice is also limited to persons who supply labor or materials (Sections 3158, 3159), and, when served upon the construction lender, attaches only to funds previously committed to finance construction of the improvement (Section 3162). Although the work of the laborer and goods of the materialman do not directly enhance the value of the loan funds, that fund is commonly secured by a trust deed upon the real property where the improvement is constructed. By enhancing the value of that realty, the laborer and materialman also increase the security of the fund. (*H. O. Bragg Roofing, Inc. v. First Federal Sav. & Loan Assn., supra* 226 Cal. App. 2d 24,

---

[1] *See, e.g.,* Frank Curran Lumber Co. v. Eleven Co., 271 Cal. App. 2d 175 (1969).
[2] Connolly Dev. Inc. v. Superior Court, 17 Cal. 3d 803 (1976).

28; *Rossman Mill & Lbr. Co. v. Fullerton S. & L. Assn., supra,* 221 Cal. App. 2d 705, 709.)

The mechanic's lien derives from the California Constitution itself; the Constitution of 1879 mandated the Legislature to grant laborers and materialmen a lien upon the property which they have improved; no other creditors' remedy stems from constitutional command. (*See Martin v. Becker,* 169 Cal. 301, 316 [146 P. 665] (1915).) Indeed this state, from the earliest days, and consistently thereafter has asserted its interest in protecting the claims of laborers and materialmen. In 1850 the first session of the California Legislature enacted a mechanic's lien law. (Stats. 1850, ch. 87, Sections 1–14, at pp. 211–213.) Moreover, the courts have uniformly classified the mechanic's lien laws as remedial legislation, to be liberally construed for the protection of laborers and materialmen. When the practice of recording a construction loan trust deed before the commencement of construction reduced the effectiveness of the mechanic's lien, the courts and the Legislature evolved alternative remedies—the equitable lien and the stop notice—which attached directly to the loan fund.

This protective policy continues to serve the needs of the construction industry. As was pointed out in *Cook v. Carlson, supra* 364 F. Supp. 24, 29: "Labor and material contractors [in the construction industry] are in a particularly vulnerable position. Their credit risks are not as diffused as those of other creditors. They extend a bigger block of credit, they have more riding on one transaction, and they have more people vitally dependent upon eventual payment. They have much more to lose in the event of default. There must be some procedure for the interim protection of contractors in this situation." Without such interim protection, the improvement may be completed, the loan funds disbursed, and the land sold before the claimant can obtain an adjudication on the merits of his claim.

In summary, we conclude that the recordation of a mechanic's lien, or filing of a stop notice, inflicts upon the owner only a minimal deprivation of property; that the laborer and materialman have an interest in the specific property subject to the lien since their work and materials have enhanced the value of that property; and that state policy strongly supports the preservation of laws which give the laborer and materialman security for their claims. In measuring these values, we do not deal in cold abstractions: we take into account the social effect of the liens and interests of the workers and materialmen that the liens are designed to protect. We measure these valued interests against the loss, if any, caused to the owner. The balance tips in favor of the worker and the materialman; we conclude that the safeguards provided by California law to protect property owners against unjustified liens are sufficient to comply with due process requirements. We therefore uphold the constitutionality of the mechanic's lien and stop notice laws.[3]

---

[3] 17 Cal. 3d at 807, 826–828.

[1] Statutory Implementation

The legislature, following the constitutional mandate, has expanded the class of persons entitled to liens. California Civil Code § 3110 provides as follows:

> *Enumeration of persons entitled.* Mechanics, materialmen, contractors, subcontractors, lessors of equipment, artisans, architects, registered engineers, licensed land surveyors, machinists, builders, teamsters, and draymen, and all persons and laborers of every class performing labor upon or bestowing skill or other necessary services on, or furnishing materials or leasing equipment to be used or consumed in or furnishing appliances, teams, or power contributing to a work of improvement shall have a lien upon the property upon which they have bestowed labor or furnished materials or appliances or leased equipment for the value of such labor done or materials furnished and for the value of the use of such appliances, equipment, teams, or power whether done or furnished at the instance of the owner or of any person acting by his authority or under him as contractor or otherwise. For the purposes of this chapter, every contractor, subcontractor, sub-subcontractor, architect, builder, or other person having charge of a work of improvement or portion thereof shall be held to be the agent of the owner.

Civil Code § 3111 provides:

> *Trust fund for payment of fringe benefits supplemental to wage agreement.* For the purposes of this chapter, an express trust fund established pursuant to a collective bargaining agreement to which payments are required to be made on account of fringe benefits supplemental to a wage agreement for the benefit of a claimant on particular real property shall have a lien on such property in the amount of the supplemental fringe benefit payments owing to it pursuant to the collective bargaining agreement.

Civil Code § 3112 provides:

> *Claimant making site improvement at instance or request of owner.* Any claimant who, at the instance or request of the owner (or any other person acting by his authority or under him, as contractor or otherwise) of any lot or tract of land, has made any site improvement has a lien upon such lot or tract of land for work done or materials furnished.

Although Civil Code § 3111 grants to trust funds mechanic's lien rights for fringe benefits, the California Supreme Court has ruled that such liens are invalid. In *Carpenters Southern California Administrative Corp. v. El Capitan Development Co.,*[4] the California Supreme Court held that Civil Code § 3111 is preempted by the Employee Retirement Income Security Act (ERISA) and is

---

[4] 53 Cal. 3d 1041, 282 Cal. Rptr. 277 (1991).

therefore invalid. A subsequent opinion followed the *Carpenters* case and held that trust funds may not invoke the stop notice remedy either.[5]

The courts have continued to prevent trust funds from seeking recovery under state law as a result of their claims being preempted by ERISA. California Business and Professions Code § 7071.5 of the Contractors License Law provides that the license bond that is required to be maintained by a licensed contractor shall be for the benefit of any express trust fund damaged as a result of the licensee's failure to pay fringe benefits for eligible employees represented by a union that is signatory to a collective bargaining agreement.[6] This provision for enforcement of fringe benefit contributions against a contractor's license bond was held to be preempted by reason of the fact that ERISA provides its own collection mechanisms.[7]

In *Operating Engineers Pension Trust v. Insurance Co. of the West*,[8] the Operating Engineers Pension Trust and other fringe-benefit trusts brought an action to collect certain fringe benefits from payment and performance bond sureties, on the theory that the trust was a third-party beneficiary of the bonds. The court of appeal, as it has in the other cases referenced *supra,* held that the trust funds could not recover on the bonds by reason of the fact that theirs was a state law claim barred by ERISA.

[2] Other Persons Entitled to File Liens

The courts, interpreting the California constitution and its statutes, have held the following to be entitled to mechanics' liens:

- Motor transportation brokers[9]
- House movers[10]
- Carpenters[11]
- Licensed land surveyors and civil engineers[12]
- Architects.[13]

In *Golden Eagle Insurance Co. v. First Nationwide Financial Corp.*,[14] Golden Eagle was the surety on a payment bond on a work of improvement. The developer had not paid a subcontractor, who filed a mechanic's lien and brought an

---

[5] Carpenters Health & Welfare Trust Fund v. Developers Ins. Co., 11 Cal. App. 4th 1539 (1992). *See also* Carpenters Health & Welfare Trust Fund v. Tri Corp., 23 F.3d 849 (9th Cir. 1994).
[6] Cal. Bus. & Prof. Code § 7071.5(d).
[7] *See* Carpenters Health & Welfare Trust Fund v. Surety Co., 13 Cal. App. 4th 1406 (1993).
[8] 35 Cal. App. 4th 59 (1995).
[9] Contractors Dump Truck Serv., Inc. v. Gregg Constr. Co., 237 Cal. App. 2d 1 (1965).
[10] Palmer v. Lavigne, 104 Cal. 30 (1894).
[11] Ogram v. Welchoff, 40 Cal. App. 298 (1919).
[12] Myers v. Alta Constr. Co., 37 Cal. 2d 739 (1951).
[13] Design Assocs., Inc. v. Welch, 224 Cal. App. 2d 165 (1964).
[14] 26 Cal. App. 4th 160, 31 Cal. Rptr. 2d 815 (1994).

action to foreclose that lien. Golden Eagle had an obligation on its payment bond to pay that subcontractor; it did so, took an assignment of the subcontractor's mechanic's lien right, and substituted itself as plaintiff in the mechanic's lien foreclosure action.

The court held that Golden Eagle, having paid the subcontractor pursuant to its payment bond obligation, was subrogated to the rights of the subcontractor, including the subcontractor's right to enforce its mechanic's lien. The court noted that, in this particular case, the grading work that the subcontractor had performed had never been paid for by the owner, and thus it was equitable to allow the surety company to be subrogated to the subcontractor's mechanic's lien rights. The court noted, however, that had the owner already paid for the grading work, the result would be different, as it would be inequitable to allow the surety company to enforce the lien right, thereby resulting in a double payment by the owner.

[3] Design Professionals Lien

The "Design Professionals Lien" gives design professionals (architects, engineers, or licensed land surveyors) the right to file a mechanic's lien for unpaid fees even though no construction is commenced on the project. This law is set forth in sections 3081.1 through 3081.10 of the California Civil Code. It applies to claims for services pursuant to a written contract with a landowner for the design, engineering, or planning of a work of improvement.[15]

To record such a lien, the design professional must have a contract with the owner of the property for design services and that owner must also be the owner of the property at the time of recordation of the lien. Furthermore, a building permit or other governmental approval in furtherance of the work of improvement must have been obtained in connection with or utilizing the services rendered by the design professional.[16]

Procedurally, the conditions precedent to the recording of a lien include (1) the owner must have defaulted on its payments to the design professional; (2) at least 10 days prior to recordation of the lien, the design professional must have made written demand for payment on the owner by first-class registered or certified mail; and (3) the lien must be recorded in the county recorder's office and must name the claimant, specify the amount due, identify the current record owner of the property (which must be legally described), and specify the building permit or other governmental approval for the work that was obtained in connection with or utilizing the services of the design professional.[17] The design professional must bring an action to enforce the lien within 90 days of its recordation. If full or partial payment is made to the design professional, the design professional must execute a full or partial release. The lien automatically expires upon commencement of construction under Civil Code § 3081.4, after which the

---

[15] Cal. Civ. Code § 3081.1.
[16] *Id.* § 3081.2.
[17] *Id.* § 3081.3.

design professional's remedy shifts to the normal mechanics' lien remedies.[18] The design professional must record the lien within 90 days of the design professional's actual or constructive knowledge that the owner does not intend to commence construction of the project.[19]

The lien does not take priority over the interests of a purchaser, lessee, or lender who has duly recorded its interest before the recording of the design professional's lien.[20] The design professional's lien does not take priority over the encumbrance of a construction lender.[21] Furthermore, the lien does not apply to the construction of a single-family, owner-occupied residence for which the construction costs are less than $100,000.[22]

[4] Persons Not Entitled to File Liens

The courts have held that certain persons are not entitled to liens:

- *Materialman's materialmen.*[23] The question to be decided in each case is whether the person to whom the material is furnished is another materialman or a subcontractor. *Theisen* addressed the definition of a *subcontractor:*

  > In our opinion the essential feature which constitutes one a subcontractor rather than a materialman is that in the course of performance of the prime contract he constructs a definite, substantial part of the work of improvement in accord with the plans and specifications of such contract, not that he enters upon the job site and does the construction there.[24]

- *Unlicensed contractors.* A person who acts in the capacity of a contractor and is unlicensed may not succeed. California Business and Professions Code § 7031 provides:

  > *Allegation and proof of license in action on contract:* Exception as to licensed contractor acting in joint venture, etc., without additional license therefor. No person engaged in the business or acting in the capacity of a contractor, may bring or maintain any action in any court of this state for the collection of a compensation for the performance of any act or contract for which a license is required by this chapter without alleging and proving that he was a duly licensed contractor at all times during the performance of such act or contract, except that such prohibition shall not apply to contractors who are each individually licensed under this chapter but who fail to comply with Section 7029.

---

[18] *See id.* § 3081.6.
[19] *Id.* § 3081.7.
[20] *Id.* § 3081.9(a).
[21] Cal. Civ. Code § 3081.9(b).
[22] *Id.* § 3081.10.
[23] Theisen v. County of L.A., 54 Cal. 2d 170 (1960); Piping Specialties Co. v. Kentile, Inc., 229 Cal. App. 2d 586 (1964).
[24] Theisen v. County of L.A., 54 Cal. 2d at 183.

There have been many cases interpreting the license law:

1. Contractor could not "lend his license" to another to meet the licensing requirement.[25]

2. Licensed contractor could not recover when he had a heart attack and his bonding company (unlicensed as an electrical contractor) finished the job for him.[26]

3. Unlicensed partnership was not entitled to recover.[27]

4. General contractor was allowed to recover on a contract calling for masonry work only.[28]

5. When the claimant furnished labor on an hourly basis and material at agreed prices and the plaintiff was supervised by others and directed what to do, the claimant was an employee and did not need to be licensed.[29]

6. Contractor was entitled to recover on a contract and for foreclosure of a mechanic's lien even though the contractor was not licensed when the contract was signed on a Friday but was licensed the following Monday before work commenced.[30]

[5] Property to Which Mechanics' Liens Attach

The mechanic's lien attaches primarily to the structure and secondarily to the ground. California Civil Code § 3128 provides:

> *Work and land to which lien attaches: Interest less than fee simple estate.* The liens provided for in this chapter shall attach to the work of improvement and the land on which it is situated together with a convenient space about the same or so much as may be required for the convenient use and occupation thereof, if at the commencement of the work or of the furnishing of the materials for the same, the land belonged to the person who caused such work of improvement to be constructed, but if such person owned less than a fee simple estate in such land then only his interest therein is subject to such lien, except as provided in Section 3129.

In the leading case of *English v. Olympic Auditorium, Inc.*,[31] the California Supreme Court held that the lien attaches primarily to the structure and secondarily

---

[25] Rushing v. Powell, 61 Cal. App. 3d 597 (1976).
[26] Weeks v. Merritt Bldg. & Constr. Co., 39 Cal. App. 3d 520 (1974).
[27] Frank v. Kozlovsky, 13 Cal. App. 3d 120 (1970).
[28] Martin v. Mitchell Cement Contracting Co., 74 Cal. App. 3d 15 (1977).
[29] Jackson v. Pancake, 266 Cal. App. 2d 307 (1968). *See also* Rodoni v. Harbor Eng'rs, 191 Cal. App. 2d 560 (1961); Contractors Dump Truck Serv. v. Gregg Constr. Co., 237 Cal. App. 2d 1 (1965); Dahl-Beck Elec. Co. v. Rogge, 275 Cal. App. 2d 893 (1969).
[30] Vitek, Inc. v. Alvarado Ice Palace, Inc., 34 Cal. App. 3d 586 (1973).
[31] 217 Cal. 631 (1933).

to the land. Therefore, if the work of improvement is being done by a tenant and the landlord posts and records a valid notice of nonresponsibility, then the lien attaches to the leasehold interest of the tenant and upon the structure down to the surface of the ground.[32] This lien on the structure has been enforced when one court ordered that the lessee has a certain time period to pay the lien. If not paid, the building must be sold and the buyers must remove it within a further time period. If not removed, the lien rights cease.[33]

The owner of the fee, however, cannot have his cake and eat it too. The owner cannot obligate the tenant to make the improvements and then turn around and record a notice of nonresponsibility and exempt its fee from the lien. If the improvements are mandatory under the lease, then the notice of nonresponsibility will be held invalid.[34]

In one case, a retirement association tried to defeat a mechanic's lien on its property pursuant to California Government Code § 31452, which provides that any property purchased for investment by a county employees retirement association is not subject to execution or any other process of court. In *Parsons Brinckerhoff Quade & Douglas, Inc. v. Kern County Employees Retirement Ass'n*,[35] Parsons was an architect who designed the Beale Memorial Library and administered the construction. On October 15, 1987, the owners recorded a notice of completion and on the same day sold the library to the Kern County Employees Retirement Association. On November 16, 1987, the architect recorded its claim of lien and got a judgment foreclosing the same. The Kern County Employees Retirement Association appealed the judgment and contended that Government Code § 31452 defeated the mechanic's lien. The court of appeal held that the statutory exemption relied upon by the retirement association must yield to the California Constitution, which clearly guarantees mechanic's lien rights.

In another case, the California court held that a party may not have a mechanic's lien on Native American–owned property. In *Inland Casino Corp. v. Superior Court*,[36] a contractor fabricated and installed kitchen equipment in a bingo hall operated by a corporation on tribal land. The contractor filed a mechanic's lien when it was not paid and brought an action to foreclose the lien. The court held that, under 28 U.S.C. § 1360(b), state courts lack jurisdiction over actions concerning rights in tribal trust property. The court stated that the trial court lacked jurisdiction over the foreclosure cause of action, because the mechanic's lien attached to the structure upon which the labor and material had been bestowed, and the bingo room in which the equipment had been installed was Indian realty. Thus, the claimant had no lien right.

---

[32] *Id. See* Ott Hardware Co. v. Yost, 69 Cal. App. 2d 593 (1945); Jay Bailey Constr. Co. v. Berry Hotel Corp., 221 Cal. App. 2d 135 (1963).

[33] English v. Olympic Auditorium, Inc., 217 Cal. 631 (1933); Ecker Bros. v. Jones, 186 Cal. App. 2d 775 (1960).

[34] Ott Hardware Co. v. Yost, 69 Cal. App. 2d 593; Los Banos Gravel Co. v. Freeman, 58 Cal. App. 3d 785 (1976).

[35] 5 Cal. App. 4th 1264 (1992).

[36] 8 Cal. App. 4th 770 (1992).

In *Red Mountain Machinery Co. v. Grace Investment Co.*,[37] the Ninth Circuit held that a mechanic's lien claim could be asserted against the leasehold interest of a party who had leased tribal lands from the Yavapai-Prescott Indian Tribe. The lease had a term of 65 years and contained an option to renew for 25 years. The purpose of the lease was to develop a multiple-use facility on the tribal land, including retail, office, lodging, medical, residential, and recreational buildings. The lessee was responsible for paying the costs of all construction on the project, which was called "Frontier Village." The lessee obtained a construction loan from a bank to finance the improvements. The bank's security was on the leasehold interest.

When Red Mountain was not paid for equipment it rented for the project, it recorded a mechanic's lien, naming the tribe as the owner of the property and the lessee as the lessee. Red Mountain filed an action to enforce its lien against the leasehold interest of the lessee only; it did not seek to foreclose its lien upon the underlying real estate owned by the tribe. The lease had been approved by the Secretary of the Interior, who has responsibility for approving such leases. The Ninth Circuit held that federal law did not preempt application of the Arizona mechanics' lien law to the leasehold interest held by the lessee.

[6] Notice of Nonresponsibility

The *notice of nonresponsibility* is defined in California Civil Code § 3094:

Notice of nonresponsibility: Posting. Notice of nonresponsibility means a written notice, signed and verified by a person owning or claiming an interest in the site who has not caused the work of improvement to be performed, or his agent, containing all of the following:

a. A description of the site sufficient for identification.

b. The name and nature of the title or interest of the person giving the notice.

c. The name of the purchaser under contract, if any, or lessee, if known.

d. A statement that the person giving the notice will not be responsible for any claims arising from the work of improvement.

Within 10 days after the person claiming the benefits of nonresponsibility has obtained knowledge of the work of improvement, the notice provided for in this section shall be posted in some conspicuous place on the site. Within the same 10-day period provided for the posting of the notice, the notice shall be recorded in the office of the county recorder of the county in which the site or some part thereof is located.

There has been much litigation over the validity of notices of nonresponsibility.[38]

---

[37] 29 F.3d 1408 (9th Cir. 1994).
[38] *See* Jay Bailey Constr. Co. v. Berry Hotel Corp., 221 Cal. App. 2d 135 (1963) (summary of cases addressing notices of nonresponsibility).

When the person causing the work to be done is a vendee who does not acquire title, then the lien is likewise against the building, and the owner may protect its fee interest by posting and recording a notice of nonresponsibility.[39]

A line of cases has developed holding that an owner who is a "participating owner" in the construction project may not protect its interest in the property by posting and recording a notice of nonresponsibility. In the early case of *Ott Hardware Co. v. Yost*,[40] the owner (Yost) leased the property, which had an obsolete and antiquated theater building on it, to Kaplan under a lease that contained the following provisions:

1. That Kaplan would remodel the building to be used as a motion picture show;
2. That the plans and specifications for the remodeling would be presented to and approved by Yost;
3. That Kaplan would furnish a faithful performance and a labor and material bond;
4. That Yost would pay Kaplan $100 per month from rentals for the actual costs of alterations, up to $10,000;
5. That Kaplan would furnish Yost a statement of the labor and materials supplied as the work progressed;
6. That Kaplan would pay the cost of all labor and materials at least five days prior to the time for filing any lien;
7. That Kaplan would not have the right to remove any improvements made by him when the lease terminated;
8. That the lease would remain in escrow until all the improvements were completed and paid for; and
9. That an addenda to the lease provided that Kaplan would pay to Yost $1,000 (to be returned to Kaplan upon the expiration of the time for filing liens) in consideration for waiving the requirement for a labor and material and faithful performance bond.

The lease and the addenda were deposited into escrow. The escrow instructions provided that the escrow agent was to hold the documents until Yost notified the escrow holder that the lease had been complied with. Yost executed a notice of nonresponsibility, posted it on the property, and recorded it on November 2, 1940. More than $12,000 was spent for labor and material in rehabilitating the building. Kaplan failed to pay for the labor and material, and the claimants recorded mechanics' liens against the property. Yost demanded that the escrow agent return the lease and the addenda for Kaplan's failure to comply with its terms and conditions. The escrow agent complied, and Yost served Kaplan with

---

[39] *See* American Transit Mix Co. v. Weber, 106 Cal. App. 2d 74 (1951).
[40] 69 Cal. App. 2d 593 (1945).

a notice of default under the lease. Yost took possession of the property and sold it to a third party.

The trial court rendered a judgment for the lien claimants and provided that if the liens were not paid within 90 days after the judgment became final, the building (down to the surface of the ground) would be sold according to law, with the lien claimants being paid out of the proceeds of the sale. The trial court further held that the lien claimants were not entitled to a lien upon the land or any portion thereof.

That judgment was reversed on appeal. The court of appeal noted that a lessee is not considered the agent of the lessor within the contemplation of the mechanic's lien statutes merely because of the relationship of landlord and tenant; and that an agency relationship is not created merely by the fact that the lessor consents to the making of improvements by the lessee on the property; however, an agency relationship will be created under the mechanic's lien statutes where there is a contractual provision in the lease that requires the lessee to make the improvements.

The court of appeal then examined the California case of *English v. Olympic Auditorium, Inc.*,[41] which held that when the building of the improvements is optional by the lessee, then the posting and filing of the notice of nonresponsibility by the lessor relieves the land from the effect of a mechanic's lien. However, the court went on to state that, if the instrument creating the relationship (the lease) is such that the transaction establishes, in effect, that the lessee is but an agent of the owner in causing the improvements to be made to the owner's property, and if the making of the improvements is not optional with the lessee and the lessee is obligated as a condition or covenant of the lease to make the improvements, and if a breach of that covenant returns the property to the owner greatly enhanced in value, and if the owner promises to repay the lessee the estimated cost of a major portion of the improvements out of future rents, then an agency relationship is created. In view of that statement, the court held that, under the facts of this case, the lease made the improvements mandatory upon the lessee. The court distinguished *English* by stating that in *English* the lessee was merely authorized to construct the auditorium building and was not obligated to construct it. The court held that, when a lease makes it mandatory upon the lessee to make the improvements as a condition of the lease, then a notice of nonresponsibility filed by the owner is ineffectual and will not relieve the owner of liens filed against the owner's property, and said liens will attach not only to the leasehold interest of the lessee but also to the fee interest of the owner. Therefore, the notice of nonresponsibility filed in this case was invalid.

The *Ott Hardware* case was followed in *Los Banos Gravel Co. v. Freeman.*[42] In *Los Banos*, the owner leased the property (undeveloped land) to a tenant, under a lease that provided that the lessee was to start construction of a gas station and restaurant within 120 days or the lease would be rendered invalid. When the

---

[41] 217 Cal. 631 (1933).
[42] 58 Cal. App. 3d 785 (1976).

lessee did start the work, the owner posted and recorded a notice of nonresponsibility. The lien claimants, who had furnished labor and material to the lessee and had served no preliminary notice, sought to foreclose their liens; but the trial court denied the liens by reason of the owner's notice of nonresponsibility. The court of appeal reversed and held that, under *Ott Hardware,* because the lease made the improvements mandatory, the owner was a "participating owner," and therefore the notice of nonresponsibility was invalid. The court further held that a preliminary notice was unnecessary because the claimants had a direct contract with a "participating owner."

In *Ecker Bros. v. Jones,*[43] the owner had posted and recorded a notice of nonresponsibility. The lien claimant obtained a judgment foreclosing its lien on the building down to the surface of the ground. The lessee was given 90 days to pay the lien claim. The trial court held that, if the lien claim was not paid within the 90-day period, the building would be sold and the buyers would be obligated to remove the building within 20 days; and if the building was not removed, the lien rights would terminate. The court of appeal upheld the judgment, holding that the lien claimant has a lien on the building down to the surface of the ground when the lien claimant deals with a lessee and the owner has posted and recorded a valid notice of nonresponsibility. The court stated that this was the equitable result for the noncontracting owners inasmuch as the building would not remain on the land, making it unusable.

In *Baker v. Hubbard,*[44] the owner posted and recorded a notice of nonresponsibility for a project wherein work was being done by its tenant. Because neither the notice of nonresponsibility that was posted on the property nor the notice of nonresponsibility that was recorded was verified, the court of appeal held the notice of nonresponsibility was invalid.

In light of the foregoing cases, it is clear that, if an owner posts and records a valid notice of nonresponsibility within 10 days after obtaining knowledge that work is being done by its tenant, then the lien claimants will have a lien on the leasehold interest of the tenant and on the building down to the surface of the ground. If the building is then sold and removed, the claimant's lien continues in full force and effect. If the building is not removed, then the lien claimant's lien rights cease. If the lease between the landlord and the tenant, under the facts of a particular case, make the improvements by the tenant mandatory, then the notice of nonresponsibility may be held invalid, and the lien claimants will be entitled to claim a lien on the owner's fee interest in the property. Thus, the terms and conditions of the lease must be examined carefully in each case, to determine the extent of the landlord's participation in the construction project.

[B] Stop Notices

Although the constitutional provision in § 9.01 refers only to a mechanic's lien right, the legislature has provided for a stop notice right on private works of

---

[43] 186 Cal. App. 2d 775 (1960).
[44] 101 Cal. App. 3d 226 (1980).

improvement. Mechanics' liens often are wiped out by foreclosure of the construction loan. The supreme court, in *Connolly Development, Inc. v. Superior Court*,[45] recognized this problem when upholding the constitutionality of the mechanic's lien and stop notice and stated:

> Although the mechanic's lien may provide adequate protection for materialmen when the owner finances the improvement from his own funds, such liens can be wiped out by the foreclosure of a lender's trust deed. The value of the stop notice lies in the fact that its lien attaches to the unexpended balance of the loan, not to the land, and thus survives foreclosure of the trust deed. The stop notice claimant also acquires a right to the fund superior to that of any assignee from the owner or contractor (§ 3166), and superior to the lender's contractual right to employ unexpended funds to complete the work of improvement. (*A-1 Door & Materials Co. v. Fresno Guar. Sav. & Loan Assn., supra* 61 Cal. 2d 728, 733–735; *Lumber Co. v. Northwestern S. & L. Assn.*, 255 Cal. App. 2d 490, 496–497 [71 Cal. Rptr. 422] (1968); *Rossman Mill & Lbr. Co. v. Fullerton S. & L. Assn.*, 221 Cal. App. 2d 705, 710 [34 Cal. Rptr. 644] (1963).)

In light of this, the legislature enacted the stop notice law, a very effective procedure whereby funds in the hands of the construction lender may be held up by a stop notice.

[1] Contents of Stop Notices

California Civil Code § 3103 provides as follows:

> *Stop Notice: Service.* Stop notice means a written notice, signed and verified by the claimant or his agent, stating in general terms all of the following:
>
> - The kind of labor, services, equipment, or materials furnished or agreed to be furnished by such claimant.
> - The name of the person to or for whom the same was done or furnished.
> - The amount in value as near as may be, of that already done, or furnished and of the whole agreed to be done or furnished.
>
> Such notice, in the case of any work of improvement other than a public work, shall be delivered to the owner personally or left at his residence or place of business with some person in charge, or delivered to his architect, if any, and, if the notice is served upon a construction lender, holding construction funds and maintaining branch offices, it shall not be effective as against such construction lender unless given to or served upon the manager or other responsible officer or person at the office or branch thereof administering or holding the construction funds. Such notice, in the case of any public work for the state, shall be filed with the director of the department which let the contract and, in the case of any other public work, shall be filed in the office of the controller, auditor, or other public disbursing officer whose duty it is to make payments under the provisions of the contract, or

---

[45] 17 Cal. 3d 803, 809 (1976).

§ 9.01[B]   CALIFORNIA CONSTRUCTION LAW

with the commissioners, managers, trustees, officers, board of supervisors, board of trustees, common council, or other body by whom the contract was awarded. No such notice shall be invalid by reason of any defect in form; it is sufficient to substantially inform the owner of the information required.

Any stop notice may be served by registered or certified mail with the same effect as by personal service.

[2] Case Law Interpretation

The courts in California have interpreted the stop notice law favorably to the claimant.

Basically, the courts have held that when the construction lender receives a properly bonded stop notice, it must set aside funds for that stop notice and cannot use the funds for any other purpose:

- Lender cannot apply the loan funds to the note and trust deed when the owner defaults in the face of a stop notice.[46]

- Assignment of the construction funds from the borrower back to the lender is invalid as against a stop notice.[47]

- Lender cannot use the funds to complete the construction in the face of a stop notice.[48]

- The leading case on the effect of a bonded stop notice stated:

    The fundholder must therefore withhold from funds furnished to pay construction costs or arising out of a construction loan sufficient money to answer bonded stop notice claims regardless of the terms of its contract with the owner. If the terms of that contract determined the rights of the claimants under subsection (h), the parties to the contract could effectively eliminate those rights. They might, for example, condition the lender's obligation to pay on there being no stop notices filed. Subsection (h) requires that funds earmarked for construction purposes be used to pay suppliers of labor and materials who file claims under the subsection and therefore supersedes the private arrangements of borrower and lender.[49]

- Lender could not pay some stop notice claimants without giving the other stop notice claimants a right to adjudicate. Lender should interplead.[50]

- Subcontractor dealing with an owner/builder may file a stop notice.[51]

---

[46] Calhoun v. Huntington Park First Sav. & Loan Ass'n, 186 Cal. App. 2d 451 (1960).

[47] Rossman Mill & Lumber Co. v. Fullerton Sav. & Loan Ass'n, 221 Cal. App. 2d 705 (1963).

[48] H.O. Bragg Roofing, Inc. v. First Fed. Sav. & Loan Ass'n, 226 Cal. App. 2d 24 (1964).

[49] A-1 Door & Materials Co. v. Fresno Guar. Sav. & Loan Ass'n, 61 Cal. 2d 728, 734 (1964).

[50] Idaco Lumber Co. v. Northwestern Sav. & Loan, 265 Cal. App. 2d 490 (1968). *See also* Miller v. Mountain View Sav. & Loan Ass'n, 238 Cal. App. 2d 644 (1965).

[51] Korherr v. Bumb, 262 F.2d 157 (9th Cir. 1958); Miller v. Mountain View Sav. & Loan Ass'n, 238 Cal. App. 2d 644.

[a] *Familian Corp. v. Imperial Bank*

On virtually all construction projects, the construction loan funds two types of costs. The loan pays for *hard costs,* the costs of design and construction of the project; it also funds *soft costs,* including points, interest, and administrative fees to be paid to the lender by the borrower directly from the construction loan. Although it had been previously conjectured and theorized based on prior cases that a construction lender might be held personally liable to stop notice claimants if there were insufficient funds in the construction loan to meet the stop notice because the lender had previously disbursed construction loan funds to itself for soft costs, *Familian Corp. v. Imperial Bank*[52] is the first case in California to expressly so rule.

In *Familian,* the project failed and Imperial Bank (the construction lender) foreclosed on the project. Because Imperial Bank's loan was recorded prior to the commencement of construction, the loan had first priority, and its foreclosure wiped out all mechanic's lien claims. However, more than 20 material suppliers and subcontractors had served stop notice claims on Imperial Bank. As is most often the case on failed projects, the stop notice claims were considerably in excess of the remaining loan funds. Imperial Bank then filed what is known as an interpleader action and deposited in court the unexpended construction loan proceeds, which amounted to approximately $105,000. Imperial informed all the stop notice claimants that as far as it was concerned, the stop notice claimants could claim a pro rata share of the $105,000 it had deposited in court.

Most of the stop notice claimants were willing to accept this proposal, but Familian argued that the unexpended loan funds included not only the $105,000 remaining in the loan proceeds but an additional $528,000 in points, interest, and administrative fees included in the construction loan but deducted by the bank at the very outset of the loan. Familian argued that under California law and specifically pursuant to California Civil Code § 3166, it was improper for the construction lender to deprive claimants of the construction loan proceeds by paying itself or by using the funds to finish the construction project either before or after the receipt of a bonded stop notice.

Both the trial court and the appellate court agreed with Familian and ruled that the stop notices attached not only to the unexpended construction loan proceeds but also to the soft costs the bank had previously deducted from the loan.[53]

In the case of *Steiny & Co. v. City Corp. Real Estate, Inc.,*[54] the court of appeal distinguished and partly overruled the *Familian Corp.* case. The court in

---

[52] 213 Cal. App. 3d 681 (1989).

[53] The *Familian* court based its ruling both on California Civil Code § 3166 and on *Miller v. Mountain View Savings & Loan Ass'n,* 238 Cal. App. 2d 644 (1965), and *Rossman Mill & Lumber Co. v. Fullerton Savings & Loan Ass'n,* 221 Cal. App. 2d 705 (1963). *See also* Commercial Standard Ins. Co. v. Bank of Am., 57 Cal. App. 3d 241 (1976) (construction lender held to owe duty to surety company not to disburse construction loan funds before actual progress of completion of project).

[54] 84 Cal. Rptr. 2d 38 (Ct. App. 1999).

the *Steiny & Co.* case held that payments for interest and loan fees legitimately incurred and disbursed by the lender to itself, prior to the service of a stop notice, did not constitute an assignment within the meaning of Civil Code § 3166 and therefore were not "loan funds" from which a stop notice claimant may attempt to seek recovery. The *Familian Corp.* case held that the stop notice applied to monies that the lender disbursed to itself for pre-allocated loan expenses, including interest, loan fees, document preparation fees, and general administrative expenses. The *Steiny & Co.* case held that monies that the loan company paid to itself for interest and loan fees were not subject to stop notices. Obviously, there is now a conflict at the appellate court level as to what types of soft costs are subject to a stop notice and what types of soft costs are not subject to a stop notice. Reading the two cases together, it would appear that document preparation fees and general and administrative expenses are subject to stop notices. The conflict arises only with regard to interest and loan fees, on which there are conflicting opinions.

[b] Negligence in Disbursing Construction Loan Funds

The courts have protected the rights of stop notice claimants from improper disbursement of construction loan funds. In a case not involving a bonded stop notice, the California court held that a construction lender owes a duty to exercise reasonable care in the disbursement of construction loan funds to a surety.

In *Commercial Standard Insurance Co. v. Bank of America*,[55] an owner brought an action against the building contractor and the surety on the performance and payment bonds as a result of the contractor's default, failure to pay subcontractors and suppliers, and failure to complete the building for the owner. The bonding company on the performance and payment bond cross-complained against the construction lender. In the cross-complaint, the surety alleged that the construction lender had disbursed an excessive amount of loan proceeds to the contractor ahead of the progress payment schedule. The surety sought to recover from the lender any costs that it would incur in completing the work and paying the bills on the project. The surety sued the lender for implied indemnity, equitable subrogation, and negligence. The trial court sustained demurrers to the surety's cross-complaint, and the surety appealed. The appellate court reversed the decision of the trial court and held that the surety company could state a cause of action against the construction lender on the theory of equitable subrogation and negligence.

With regard to the theory of subrogation, the cross-complaint alleged that the bank's negligence caused the loss. Accordingly, the surety would be subrogated to the rights of the owner against the bank. With regard to the negligence

---

[55] 57 Cal. App. 3d 241 (1976).

theory, the court stated the following policy considerations as to whether to impose a duty of care:

1. Foreseeability of harm to the plaintiff
2. Degree of certainty that the plaintiff suffered injury
3. Closeness of the connection between the defendant's conduct and the injury suffered
4. Moral blame attached to the defendant's conduct
5. Policy of preventing future harm
6. Extent of the burden to the defendant and the consequences to the community of imposing a duty to exercise care, with resulting liability for breach
7. Availability, costs, and prevalence of insurance for the risk involved.

In analyzing those elements, the court noted that the bank was aware of and in fact had requested the performance and payment bonds. Therefore, it was foreseeable to a reasonable person in the bank's position that the bank's negligence in disbursing the loan proceeds could cause the surety losses under those bonds. The court stated that it was certain that the surety had suffered injury, because the surety was liable to the owner for completion costs and the unpaid claims of subcontractors and suppliers. The court further stated that it could see no particular moral blame attached to the bank's conduct but that the imposition of a duty upon the bank to exercise reasonable care in the disbursement of loan proceeds, in cases in which it has undertaken to disburse loan proceeds in accordance with the value of the construction work performed, would probably cause the bank to exercise due care in future cases. The court also said that imposing a duty upon the bank to exercise reasonable care under these circumstances would be imposing upon the bank the burden of employing persons of sufficient competency to evaluate the progress of construction; there was no untoward consequence to the community of imposing such a duty. The court admitted that the only policy consideration with which it had any difficulty was the availability, cost, and prevalence of insurance. The court stated that although the bank probably could insure against negligent conduct in disbursing loan proceeds, the spreading of the risk of such failures is the very business of the surety. The surety company was paid a premium for the performance and payment bonds and was in at least as good a position as the bank to spread the risk of improper disbursement equitably throughout the construction industry. However, the court held that, on the whole, the balancing of those policy considerations favored imposing a duty on the bank. Citing Civil Code § 1714, the court held that every person is responsible not only for its willful acts but also for any injury occasioned by its want of ordinary care that results in damage to another.

Looking at this case and the seven elements of duty as set forth therein, it is apparent that a court may impose a duty on a construction lender to exercise reasonable care in the disbursement of construction loan funds running in favor of bonded stop notice claimants. It is foreseeable to a construction lender that if it fails to properly disburse the construction loan funds in accordance with the progress of the work, there may be insufficient funds to complete the project and, therefore, insufficient funds to pay the suppliers of labor and material—and thus result in the filing of bonded stop notices. If the lender can answer, "We ran out of money and therefore we have no liability on your bonded stop notice," then the spirit, purpose, and intent of the stop notice law will be defeated. It is certain that when the lender runs out of money, unpaid subcontractors and suppliers are in fact injured, and that they will resort to their statutory remedy of a bonded stop notice. It is also clear that by disbursing the money ahead of the progress of the job, and thereby depleting the loan funds, the lender causes direct injury to the suppliers of labor and material who do not get paid and whose labor and material have created the structure that stands as the collateral for the construction loan. It cannot be said that any moral blame attaches to the lender's conduct, but imposition of a duty upon banks to exercise reasonable care in the disbursement of loan proceeds would tend to cause banks to exercise due care in future cases. This promotes the policy of preventing future harm to stop notice claimants. The burden on the bank is not substantial. Most good construction lenders do, in fact, employ inspectors to inspect the progress of the work to ensure that the construction loan remains "in balance," that is, that there are adequate funds to ensure completion of the work. If the funds are inadequate, the lender can require the borrower to deposit more funds. Imposing a duty on lenders in the face of potential stop notices would only require them to do what they normally do to protect their own interests. Insurance may or may not be available, but, as indicated in the *Commercial Standard Insurance* case, on balance a duty should be imposed upon the construction lender.

It is the opinion of the authors that if a construction lender disburses construction loan funds ahead of the progress of the work and thereby depletes the construction loan funds, the construction lender could be held liable to bonded stop notice claimants for those excessive disbursements. This could be easily documented by subpoenaing the records of the lender and comparing the percentage of disbursement of the loan funds with the percentage of completion of the project at the time of disbursement. The only way that a construction lender will run out of money before completion of the job is when it disburses the money ahead of the progress of the job.

[c] Equitable Liens

The equitable lien is no longer enforceable in California.[56]

In *Nibbi Bros. v. Brannon Street Investors*,[57] the general contractor, Nibbi, was working toward the completion of the project when payments from the owner of the project became irregular. Nibbi approached the lender, who assured Nibbi

---

[56] Cal. Civ. Code § 3264. *See* Boyd & Lovesee Lumber Co. v. Western Pac. Fin. Corp., 44 Cal. App. 3d 460 (1975); Nibbi Bros., Inc. v. Brannon St. Investors, 205 Cal. App. 3d 1415 (1988).

[57] 205 Cal. App. 3d 1415 (1988).

that "you will be paid for work performed." Based in large part on this representation, Nibbi proceeded to complete the project and spent $66,000 for additional improvements.

In actuality, at the time the lender informed Nibbi, "You will be paid," the lender had already recorded a notice of default on the deed of trust securing the loan. The lender did not inform Nibbi that it had recorded the notice of default. The lender then proceeded to foreclose on its trust deed, which had priority over any mechanics' liens because it had been recorded before construction commenced.

Nibbi could not recover against the owner, who was bankrupt, and sued the lender on an equitable lien theory, as well as on the tort theory that the lender had negligently omitted to inform Nibbi that it had recorded a notice of default on the deed of trust. Nibbi argued that this was a material omission and that Nibbi had relied on the lender's representation "You will get paid."

In the court case that followed, the court ruled that the lender did not have a duty to inform the general contractor that a notice of default had been recorded and further ruled that the representation "You will be paid" was not a promise by the lender to pay the general contractor for work performed. The court also held that based on California Civil Code § 3264, the general contractor could not use an equitable lien theory. The court ruled that the only way that a contractor could get to the lender and the construction loan proceeds was through a stop notice. But the catch-22 was that the court reaffirmed that at the time of the *Nibbi Bros.* case, general contractors did not have stop notice rights; therefore, the general contractor could not recover against the lender. As noted below, general contractors now have stop notice rights.

[d] General Contractor's Stop Notice Rights

Although *Nibbi Bros.* reaffirms that pursuant to section 3264 the equitable lien is "dead," the California Civil Code was amended, effective January 1, 1989, to provide general contractors with stop notice rights. As such, a general contractor who learns that an owner is having financial difficulties or is unable to make payments required under the construction contract should serve a bonded stop notice directly on the lender on behalf of itself, its subcontractors, and material suppliers.

[e] Defending Stop Notice Claims

A plaintiff who lost his suit on a stop notice and the bonding company on the stop notice bond were held liable for the owner's attorney's fees in defending the stop notice suit.[58]

---

[58] Flintkote Co. v. Presley of N. Cal., 154 Cal. App. 3d 458 (1984); Systems Inv. Corp. v. National Auto. & Casualty Ins. Co., 25 Cal. App. 3d 1057 (1972).

[f] Effect of Payment Bond on Stop Notices

If the owner secures and records a payment bond pursuant to California Civil Code § 3235, then the lender may ignore bonded stop notices, except that of the general contractor. The California Supreme Court, in *Connolly Development, Inc. v. Superior Court*,[59] in upholding the constitutionality of the mechanic's lien and stop notice, specifically recognized this, stating, "[T]he owner and lender can protect themselves against stop notices by securing and recording a payment bond from the general contractor."[60] The construction lender must withhold funds pursuant to the bonded stop notice of the general contractor.

Civil Code §§ 3159 and 3162 were amended, effective January 1, 1995, to state in unequivocal terms that a general contractor's stop notice rights are not conditioned on the prior recording of a payment bond. Subsection (a)(1) of both Civil Code § 3159 and § 3162 provide, in relevant part:

> (a)(1) The construction lender shall withhold funds pursuant to a bonded stop notice filed by an original contractor, regardless of whether a payment bond has previously been recorded in the office of the county recorder where the site is located in accordance with Section 3235. See also Building Profit Corp. v. Mortgage Realty Trust, 36 Cal. App. 4th 683 (1995).

[3] Attorney's Fees on Bonded Stop Notice

Civil Code § 3176, relating to stop notices for private works, provides as follows:

> In any action against an owner or construction lender to enforce payment of a claim stated in a bonded stop notice, the prevailing party shall be entitled to collect from the party held liable by the court for payment of the claim, reasonable attorney's fees in addition to other costs and in addition to any liability for damages.

> The court, upon notice and motion by a party, shall determine who is the prevailing party for purposes of this section, whether or not the suit proceeds to final judgment. Except as otherwise provided by this section, the prevailing party shall be the party who recovered a greater relief in this action. The court may also determine that there is no prevailing party. Where an action has been voluntarily dismissed or dismissed pursuant to a settlement of the case, there shall be no prevailing party for purposes of this section.

> Where the defendant alleges in his or her answer that he or she tendered to plaintiff the full amount to which he or she was entitled, and thereupon deposits in court for the plaintiff, the amount so tendered, and the allegation is found to be true, then the defendant is deemed to be a prevailing party.

---

[59] 17 Cal. 3d 803 (1976).
[60] 17 Cal. 3d at 809. *See* Cal. Civ. Code §§ 3161, 3162, 3235.

Civil Code § 3176.5, relating to stop notices for private works of improvement, provides as follows:

> If the plaintiff is the prevailing party in any action against an owner or construction lender to enforce payment of a claim stated in a bonded stop notice, any amount awarded on the claim shall include interest at the legal rate calculated from the date the bonded stop notice is served upon the owner or construction lender pursuant to Section 3172.

These code sections received their initial interpretation by the appellate courts in a very unusual case. In *Villinger/Nicholls Development Co. v. Meleyco*,[61] the contractor entered into a contract with a homeowner to remodel the homeowner's house. The contract had an arbitration clause in it. A dispute arose, and the contractor filed a bonded stop notice with the construction lender. The owners filed a demand for arbitration, and the dispute was submitted to arbitration. In the arbitration proceedings, the contractor was the prevailing party. The contractor then petitioned the superior court to confirm the award in the form of a judgment and also sought prejudgment interest and attorney's fees under the above-referenced code sections. The court held that a petition to confirm an arbitration award was not an "action" as that term is used in Civil Code § 3176 and, therefore, the contractor was not entitled to attorney's fees on its stop notice claim. Had an arbitration not been involved in this case and the matter been decided pursuant to a court trial, it is the opinion of the authors that the contractor would have been entitled to attorney's fees under Civil Code §§ 3176 and 3176.5.

[C] Bonds

The legislature has enacted legislation to allow an owner to bond its job and thereby limit liens to the owner's balance due the prime contractor. California Civil Code §§ 3235 and 3236 provide:

> *Restrictions on recovery where original contract and payment bond are recorded: Place for recording: Deficiency judgment.* In case the original contract for a private work of improvement is filed in the office of the county recorder of the county where the property is situated before the work is commenced, and the payment bond of the original contractor in an amount not less than 50 percent of the contract price named in such contract is recorded in such office, then the court must, where it would be equitable so to do, restrict the recovery under lien claims to an aggregate amount equal to the amount found to be due from the owner to the original contractor and render judgment against the original contractor and his sureties on such bond for any deficiency or difference there may remain between such amount so found to be due to the original contractor and the whole amount found to be due to claimants.

---

[61] 31 Cal. App. 4th 321 (1995).

Limitation on owner's liability: Right of owner to exact necessary bond or security. It is the intent and purpose of Section 3235 to limit the owner's liability, in all cases, to the measure of the contract price where he shall have filed or caused to be filed in good faith his original contract and recorded a payment bond as therein provided. It shall be lawful for the owner to protect himself against any failure of the original contractor to perform his contract and make full payment for all work done and materials furnished thereunder by exacting such bond or other security as he may deem necessary.

[1] Perfecting Payment Bond Claims

Statutory labor and material bonds are seldom found on private works of improvement, but when they are it is the primary source of recovery because the liens are then limited to the balance due from the owner to the contractor and stop notices (except that of the general contractor) can be ignored.[62] If a labor and material bond has been recorded, then in order to recover on it the claimant must do two things: (1) serve a preliminary notice under Civil Code § 3097 (*see* §§ 9.02[A] through 9.02[B][7]) or serve the principal and surety on the bond within 15 days of the notice of completion or 75 days of completion if no notice of completion is recorded;[63] and (2) bring suit on the bond within six months after completion.[64]

It is quite common in the construction industry for general contractors to place in their subcontract agreements provisions stating that the subcontractor shall not be entitled to recover from the general contractor unless the general contractor, in turn, is entitled to recover from the owner. In *Wm. R. Clarke Corp. v. Safeco Insurance Co.*,[65] the subcontract had an extensive clause precluding recovery by the subcontractor from the general contractor unless and until the general contractor, in turn, recovered from the owner. Specifically, the clause provided that receipt of funds by the contractor from the owner was a condition precedent to the contractor's obligation to pay the subcontractor. The clause also contained a provision stating that the subcontractor did not waive its right to enforce its statutory mechanic's lien rights or remedies, and that such mechanic's lien rights would be the sole remedy and means for payment on account of work performed by the subcontractor.

The general contractor obtained a payment bond pursuant to Civil Code § 3235. The construction work began, and the owner stopped making payments to the general contractor; the general contractor, in turn, stopped making payments to its subcontractors. The subcontractors brought an action on the payment bond, and the surety company defended on the basis of the contractual provision precluding recovery by the subcontractors from the contractor unless the contractor had been paid by the owner. Specifically, the surety company relied upon Civil Code § 2809, which provides that the obligation of a surety cannot be larger

---

[62] *See* Cal. Civ. Code §§ 3161, 3162, 3236.
[63] *Id.* §§ 3242(a), (b), 3240, 3227.
[64] *Id.* § 3240.
[65] 38 Cal. App. 4th 1655 (1995).

in amount or in any other respect more burdensome than that of its principal. Essentially, the surety argued that, because its principal, the general contractor, was not contractually obligated to the subcontractors, the subcontractors could not recover on the payment bond.

The court of appeal first noted that the California Supreme Court held in *Bloom v. Bender*[66] that the provisions of Civil Code § 2809 could be waived. The court then concluded that, by reason of the reservation of the right of the subcontractors to pursue their mechanic's lien rights, the subcontractors also had the right to recover on the payment bond. Specifically, the court found that the payment bond in question was issued for the purpose of complying with the mechanic's lien law; therefore, an action to recover payment on the payment bond is an action to enforce the claimant's mechanic's lien rights and thus a remedy expressly reserved by the subcontractors in their subcontracts. The subcontractors were likewise allowed to recover prejudgment interest from the surety company.

[2] Off-Site Bonds

On some private works of improvement, the public body will require the developer to bond the off-site improvements (streets, curbs, gutters, and sidewalks) under California Government Code §§ 66499 and 66499.10. If one's work relates to street improvements, one must be familiar with this type of bond.

The contractor has an election to pursue either the mechanic's lien or the bond.[67]

Suit on such a bond should be brought promptly. The bond will be released after the lien period has expired and after acceptance of the work by the public body unless a claimant has recorded a claim of lien and given notice in writing to the public body.[68]

§ 9.02 PROCEDURE FOR FILING MECHANICS' LIENS AND STOP NOTICES

[A] Twenty-Day Notice

California Civil Code § 3097 addresses the notice requirements for filing mechanics' liens and stop notices.

[1] In General

[a] When Given

Notice must be given within 20 days after the claimant first furnishes labor, services, equipment, or material to the jobsite.[69] If the claimant elected not to

---

[66] 48 Cal. 2d 793 (1957).
[67] Sukut-Coulson, Inc. v. Allied Canon Co., 85 Cal. App. 3d 648 (1978).
[68] Cal. Gov't Code § 66499.7.
[69] Cal. Civ. Code § 3097(d).

give the notice within the first 20 days, then the claimant may give it at any time during the job, in which event the claimant will be entitled to file a lien or stop notice for labor, service, equipment, or materials furnished 20 days prior to the notice and at any time thereafter.[70]

[b] By Whom Given

Notice must be filed by all persons claiming a lien or stop notice, except the original contractor and a person performing actual labor for wages. In other words, everyone except the prime contractor and wage claimants must give the notice.[71]

[c] To Whom Given

Everyone, except a subcontractor dealing directly with an owner/builder, must give the notice to the owner or reputed owner, the original contractor or reputed contractor, and the construction lender or reputed construction lender.[72]

When there is no prime contractor, a subcontractor, material supplier, or anyone else dealing directly with an owner/builder must give the notice to the construction lender or reputed construction lender.[73]

[d] Where Given

The notice may be served at the following locations of the person to be served if that person resides in California: the person's residence, the person's place of business, the address shown on the construction trust deed, or the address shown on the building permit.[74]

If the person to be notified resides outside California, then that person is served in the same manner as above, but if the person cannot be served as above, then the notice can be addressed to the construction lender or the original contractor.[75]

[e] How Given

Notice is served by personal service, registered mail, or certified mail. When the notice is sent by registered or certified mail, service is complete at the time of deposit. Therefore, this method of service is recommended.[76]

Service is proved by an affidavit showing compliance with the statute.[77]

---

[70] Id. § 3097(d).
[71] Cal. Civ. Code § 3097(a), (b).
[72] Cal. Civ. Code § 3097(a).
[73] Id. § 3097(b).
[74] Cal. Civ. Code § 3097(d).
[75] Id. § 3097(f)(2).
[76] Cal. Civ. Code § 3097(f).
[77] Id. § 3097.1. See also App. 5.

[f] Other Requirements

The preliminary notice statute also requires the following:

- Every written contract between a property owner and an original contractor shall provide space for the owner to enter its name, address of residence, and place of business, if any. The original contractor "shall" make available the owner's name and address of residence to any person seeking to serve a preliminary notice.[78]

- Every written contract between a property owner and an original contractor (except home improvement contracts and swimming pool contracts) shall provide a space for the owner to enter the name and address of the construction lender or lenders. The original contractor shall make available the name and address of the construction lender or lenders to any person seeking to serve a preliminary notice.[79]

- When one or more construction loans are obtained after commencement of construction, the property owner shall provide the name and address of the construction lender or lenders to each person who has given the property owner a preliminary notice.[80]

- Only one notice need be given for each job on which the claimant has furnished labor, service, equipment, or materials. If the claimant contracts with more than one subcontractor, then the claimant must give a notice for each subcontractor it deals with on the job.[81]

The 20-day notice preserves all of the claimant's remedies: it preserves the claimant's right to record a mechanic's lien, to serve a stop notice, and to bring an action on any payment bond on the project.

In summary, the following is an outline of the essential provisions of the preliminary notice requirements:

1. The 20-day notice preserves mechanic's lien, stop notice, and bond rights.

2. The preliminary notice must be served on the owner, contractor, and construction lender.

3. If the claimant fails to give the 20-day notice, the claimant may still pursue its claim on the payment bond (if any) by giving notice to the principal and surety on the payment bond within 15 days of the notice of completion (N.O.C.), or, if there is no N.O.C., within 75 days of completion.

---

[78] Id. § 3097(l).
[79] Id. § 3097(m).
[80] Id. § 3097(n).
[81] Id. § 3097(g).

[2] Contents of the Notice

The notice should contain:

1. General description of the labor, service, equipment, or materials furnished (any construction lender shall be furnished with an estimate of the total price of same)
2. Name and address of the person giving the notice
3. Name of the person who contracted for the purchase of the labor service, equipment, or materials
4. Description of the jobsite sufficient for identification
5. The following notice in bold type:

    NOTICE TO PROPERTY OWNER

    If bills are not paid in full for the labor, services, equipment, or materials furnished or to be furnished, a mechanic's lien leading to the loss, through court foreclosure proceedings, of all or part of your property being so improved may be placed against the property even though you have paid your contractor in full. You may wish to protect yourself against this consequence by (1) requiring your contractor to furnish a signed release by the person or firm giving you this notice before making payment to your contractor or (2) any other method or device which is appropriate under the circumstances.

6. If the notice is given by a subcontractor who is required pursuant to a collective bargaining agreement to pay supplemental fringe benefits into an express trust fund, the notice must contain the identity and address of such trust fund.[82]

The statute provides that if an invoice contains the above information, then it may be sent as required by the statute.[83]

An architect, registered engineer, or licensed land surveyor who has furnished services for the design of the work of improvement may give his notice within 20 days after the work of improvement has commenced.[84]

[3] Disciplinary Action

If a licensed contractor has a contract to supply more than $400 worth of labor, service, equipment, or materials, then failure to give the required notice is grounds for disciplinary action.[85] Also, if the subcontractor files a lien or stop notice but fails to include the name and address of trust funds he contributes to,

---

[82] Cal. Civ. Code § 3097(c).
[83] *Id.*
[84] *Id.*
[85] Cal. Civ. Code § 3097(h).

this also constitutes grounds for disciplinary action if the amount due the trust fund is not paid.[86] Finally, if a contractor or subcontractor becomes delinquent in the payment of fringe benefits, he must so notify the construction lender or reputed construction lender (setting forth the name of the owner and contractor, a description of the jobsite sufficient for identification, the identity and address of the trust fund, the total straight time and overtime hours on each delinquent job, and the amount then past due and owing) within five days of the date payment was due, or failure to do so will also constitute grounds for disciplinary action.[87]

[4] Filing Preliminary Notices with County Recorder

Effective July 1, 1988, California Civil Code § 3097 was amended to allow a claimant to file the preliminary notice with the county recorder. The law requires the county recorder, if a notice of completion or notice of cessation with respect to the work of improvement is recorded, to mail a notification to the claimant, endorsed with the date the notice of completion or cessation is recorded. The law also provides, however, that the failure of the county recorder to mail the notification or the failure of the person who filed the preliminary notice to receive the required notification does not affect the time within which a claim of lien must be recorded. As a result of this amendment, claimants should theoretically also file a copy of their preliminary notice with the county recorder. However, in practice this is rarely done, and claimants should be aware that the county recorder may fail to provide them with notice that a notice of completion has been recorded, and there is no remedy for such a failure to inform.

[B] Cases Interpreting Preliminary Notice

[1] *Romak v. Prudential*

In *Romak Iron Works v. Prudential Insurance Co. of America*,[88] a construction loan was recorded by Prudential on May 24, 1972. The record title to the property was held by MacDonald, Nelson & Heck, Inc., Marwin Garage Corporation, and Lician Sabella. On June 1, 1972, Romak Iron Works entered into a subcontract with MacDonald, Nelson & Heck, Inc., to furnish certain labor and materials for a contract price of $26,574. The subcontract identified Romak as the subcontractor, MacDonald, Nelson & Heck, Inc., Marwin Garage Corp., and Sabella as the owners of the project, and MacDonald Nelson & Heck, Inc., as the contractor. On June 20, 1972, Romak mailed a preliminary notice to MacDonald, Nelson & Heck, Inc., at their address in Oakland. In the space provided for the owner or reputed owner, Romak named MacDonald, Nelson & Heck, Inc., and Marwin Garage Corporation. In the space entitled "original contractor or reputed contractor," Romak identified MacDonald, Nelson & Heck, Inc. In the

---

[86] *Id.*
[87] *Id.*
[88] 104 Cal. App. 3d 767 (1980).

space entitled "construction lender or reputed construction lender," Romak wrote "not known." No preliminary notice was sent within 20 days after Romak first furnished its labor and material to Prudential, the lender.

Romak furnished labor and material from September 12, 1973, to November 27, 1973, for a total of $40,186.63. On January 18, 1974, Romak filed a stop notice with Prudential claiming that amount. Romak then brought a lawsuit seeking to recover on its contract with MacDonald, Nelson & Heck, Inc., seeking to foreclose a mechanic's lien and seeking to enforce its bonded stop notice served upon Prudential. Prudential filed a motion for *summary judgment,* a procedure whereby a judgment can be sought prior to trial if the court concludes that there are no issues of fact and that as a matter of law, Romak had no cause of action against the loan company, Prudential.

Prudential, in support of its motion, showed that it had recorded its construction trust deed on May 24, 1972. Romak showed in its opposition to the motion that the preliminary notice had been sent to MacDonald, the prime contractor. Romak further stated in its opposition that at the time the preliminary notice was sent out, Romak knew of no construction lender on the job. Romak further stated that prior to sending the notice, Romak spoke to MacDonald's controller and was told that it was then unknown whether there would be a construction lender. Romak further stated that they had dealt with MacDonald for years and never expected to be misinformed, and that at the time the preliminary notice was sent, no one at Romak knew of the existence of a construction lender. The prime contractor was adjudicated bankrupt, and Prudential foreclosed its deed of trust, thereby wiping out Romak's mechanic's lien. The only way Romak could recover in this case was if its stop notice were held to be effective.

Prudential's motion for summary judgment was granted, and Romak appealed, but the appellate court affirmed the judgment of the trial court. Romak's first argument on appeal was that because it had neither actual nor constructive knowledge that Prudential was a construction lender on the project, no preliminary notice to Prudential was necessary. The court, in response to this argument, stated that California Civil Code § 3097(b) (the preliminary notice section) provided that Romak had an absolute obligation to give a preliminary notice to Prudential as a necessary prerequisite to the validity of any stop notice that it might later file. The court further stated that the unqualified language of section 3097(b) makes no exception for the event that Romak had no actual knowledge that Prudential was the construction lender on the project. That event, stated the court, was covered by section 3097(f), which provided that a subcontractor could give the preliminary notice to any construction lender not known by mailing it to the jobsite, addressed merely to "Construction Lender." (It should be noted that this liberal method of service has now been eliminated from the statute.) The court said that Romak did not send the preliminary notice to the jobsite as was then permitted by the statute.

The court went on to say that section 3097(f) provides an array of alternative methods by which Romak might have given an effective preliminary notice to the construction lender. One method was by mail to the address shown by the building permit on file with the authority issuing the permit; another method was by mail to an address recorded pursuant to section 3097(j). Those alternative methods of service still remain in the preliminary notice section. Specifically, section 3097(f)(1) provides that service can be made at the address shown by the building permit on file with the authority issuing the permit, or at an address recorded pursuant to section 3097(j), which provides for the recording of the construction trust deed. The deed must have on its face the name and address of the lender, the name and address of the owner, a legal description of the property, and, if known, the street address of the property. Thus, when Romak sent out its preliminary notice, it could have ascertained the name and address of Prudential by examining the construction trust deed, but it failed to do so. In this regard, the court stated:

> The unmistakable thrust of subdivisions (i) and (j) is to impose on a prospective stop notice claimant the duty to examine either or both of the alternative sources of information they mention (the building permit and the specially indexed official records of the county) for the purpose of ascertaining the existence and identity of a "construction lender" to whom he must deliver a preliminary 20-day notice "as a necessary prerequisite to the validity" of a stop notice to be delivered later. (Civil Code § 3097(b).) If he fails to examine the two sources, subdivisions (i) and (j) operate to charge him with constructive notice of the information recorded in either.[89]

Romak had pointed out to the court that Prudential was not identified as the construction lender on the building permit, which Prudential had conceded. The court was unwilling to alter its opinion, however, because Prudential's name and address were in fact shown on the construction trust deed. Therefore, Romak was placed on "constructive notice" that Prudential was the construction lender on the project and of Prudential's address because the trust deed had been recorded before Romak sent out its preliminary notice.

Romak, struggling to win the case, argued that because it was not claiming an interest in the real property (the stop notice is against the construction funds only), the recording of the construction trust deed (which related to the real property) should not give Romak constructive notice of the contents of Prudential's construction trust deed. Again, the court disagreed, stating that Romak's argument overlooked the provisions of § 3097(i) and (j). Those subdivisions express the intent of the legislature to apply the doctrine of constructive notice to prospective stop notice claimants because they translate into actual knowledge that which a reasonable party could ascertain by reasonable inquiry of reliable sources, to wit, the building permit and the construction trust deed.

The court stated that Romak could have sent the preliminary notice to the jobsite. Failing to do so, Romak was charged with constructive notice of Prudential's identity as a construction lender and of its address as shown on the construction trust deed, which was a reliable record that Romak did not bother to consult. Romak's actual ignorance of the contents of the construction trust deed could not excuse Romak's failure to give Prudential a preliminary notice.

---

[89] 104 Cal. App. 3d at 775.

Romak still did not give up. It further contended that it complied with the 20-day notice requirement to Prudential by mailing its preliminary notice to MacDonald, the prime contractor. Romak argued that because Prudential was the construction lender and MacDonald was the original contractor when the preliminary notice was given, notice to the contractor was notice to the lender. The court rejected this argument as well.

Romak kept trying. It argued that the stop notice statute was to be liberally construed in favor of those persons in the construction industry whom the stop notice statute was designed to protect, to wit, subcontractors and material suppliers. The court agreed that the legislature did intend to protect subcontractors and others, but it also imposed the notice requirement for the valid purpose of alerting owners and lenders to the fact that the property or funds involved might be subject to claims arising from contracts to which they were not parties and would not otherwise have any knowledge. The court went on to state that the rule requiring liberal interpretation could not be applied to frustrate the legislature's manifested intent to exact strict compliance with the preliminary notice requirement.

Thus, notwithstanding Romak's valiant efforts, it lost the appeal. This case represents a strict construction of the preliminary notice requirement. Later cases have shown a more liberal interpretation of the statute. Claimants in the construction industry must use extreme diligence in ascertaining the correct name and address of the owner, contractor, and lender; if necessary, claimants should search the building permit and the construction trust deed to obtain that information.

[2] *Brown Co. v. Superior Court*

In *Brown Co. v. Superior Court*,[90] the appellate court had before it a case in which the claimant, Brown Company, doing business as Livingston-Graham, was advised by an employee of East Valley Development Company that the owner of the property was East Valley Development and that Crawford Investments was the construction lender. In reliance upon that information, Livingston-Graham forwarded its preliminary notice to East Valley Development as the owner and to Crawford Investments as the construction lender.

The information that had been given to Livingston-Graham was in error. In fact, the property was owned by Arthur W. Crawford and his wife, and East Valley Development was the prime contractor. Another factor in the case was that Livingston-Graham had sent its preliminary notice seven months before the first delivery of ready-mix concrete. Consequently, the owner contended that the preliminary notice was invalid for two reasons: (1) it was premature; and (2) it was not sent to the owner as required by California Civil Code § 3097.

The court held that the preliminary notice was not premature and that under the facts of this case, Livingston-Graham had given the notice to the "reputed owner." The court distinguished *Romak* on two major factors. First, the lien claimant in *Romak* was given no name and was told that there was no known construction lender. In *Brown,* Livingston-Graham was given the name and address

---

[90] 148 Cal. App. 3d 891 (1983).

of a specific owner said to be the owner by an employee of the general contractor. The court stated that there is a significant difference between being given the owner's name and address upon inquiry and being given no name and address for the construction lender.

The court also stated that section 3097(*l*) and (j), relating to notice to the lender or reputed lender suggest that the plaintiff in *Romak* should have inquired further about the construction lender's identity. In contrast, section 3097(*l*), relating to notice to the owner or reputed owner, does not contain a special provision to the effect that persons required to give preliminary notice under this section are not relieved of the duty by the contract's failure to set forth the owner's name and address.

It thus appears under these two cases, *Romak* and *Brown Co.*, that the issue for future cases will be whether the claimant had acted "reasonably" in finding out the names and addresses of the owner, contractor, and lender. It would appear that each case will have to be decided on its own facts. In obtaining information with regard to the service of the preliminary notice, it is suggested that a potential claimant make a notation of all inquiries made and information furnished in attempting to ascertain the correct names and addresses of the various parties to be notified.

[3] IGA v. Manufacturers Bank

In *IGA Aluminum Products, Inc. v. Manufacturers Bank*,[91] the claimant (a material supplier to a subcontractor) sent its preliminary notice by first-class mail. The owner, contractor, and lender received the notice. The claimant filed a lien and stop notice. The lender moved for summary judgment on the ground that the notice had been sent by first-class mail rather than certified or registered mail. The claimant lost.

[4] James v. Five Points Ranch

In *Harold L. James, Inc. v. Five Points Ranch, Inc.*,[92] the claimant (a subcontractor to a prime contractor) sent the owner and contractor a preliminary notice with the following provision: "If bills are not paid in full for labor, services, equipment, or materials furnished or to be furnished, the improved property may be subject to mechanics' liens." The court held that the lien was invalid because the notice did not have the precise language required by the California Civil Code. The court stated:

> By our holding today, we conclude that the transmittal methods and notice requirements must be strictly construed. However, the issue of minor errors in the body of the notice must be independently addressed on a case-by-case basis, if and when such a case is presented.[93]

---

[91] 130 Cal. App. 3d 699 (1982).
[92] 158 Cal. App. 3d 1 (1984).
[93] 158 Cal. App. 3d at 8.

[5] *Industrial Asphalt v. Garrett Corp.*

In *Industrial Asphalt, Inc. v. Garrett Corp.*,[94] the claimant was a subcontractor to a prime contractor. There was no construction lender. The subcontractor sent its preliminary notice to the owner only. The owner contended that the failure to give the notice to the prime contractor (as required by the statute) invalidated the preliminary notice and the mechanic's lien. The court held for the subcontractor, stating:

> To construe the statute strictly would require us to invalidate a lien against an owner who received notice because someone else, the original contractor, did not receive notice. That strict statutory construction would allow a party who received the required notice to be insulated from liability because another party did not receive notice. We do not believe that the statute's purpose should, or does, lead to this aridly formalistic result. We hold that the plaintiff's notice to the defendant satisfied the prerequisites for a valid lien against the defendant, and we reverse the trial court's judgment.
>
> (3a) Ancient authority enunciates the purpose of the mechanics' lien: to prevent unjust enrichment of a property owner at the expense of a laborer or material supplier. "The principle upon which liens are allowed in favor of mechanics and materialmen is that their labor and materials have given value to the buildings upon which they have been expended, and that it is inequitable that the owner of land, who has contracted with them for such improvement, or who has stood by and seen the improvement in progress without making objection, should have the benefit of their expenditures without making compensation therefor." (*Avery v. Clark,* 87 Cal. 619 [25 P 919] (1891).)
>
> * * *
>
> The laborer or material supplier has invested his labor, or added materials originally in his possession, to improve property of another and increase its value. They thus "have, at least in part, created the very property upon which the lien attached."
>
> * * *
>
> (3b) In interpreting statutes effecting the constitutional lien remedy, the courts have traditionally supported this historic preference of laborers and suppliers by applying a rule of liberal construction.
>
> * * *
>
> From *Roystone* to *Truestone*, courts have "uniformly classified the mechanics' lien laws as remedial legislation, to be liberally construed for the protection of laborers and materialmen."
>
> * * *
>
> Although Industrial Asphalt in the case at bench did not serve statutory notice upon Ken Jones, it complied with the notice requirement as to defendant Garrett Corporation. That notice, among other functions, affords the property owner the chance to protect his real property interest by requiring a bond, by naming both the original contractor and the subcontractor as payees on

---

[94] 180 Cal. App. 3d 1001 (1986).

checks written to pay for the work, or by requiring other proof that the original contractor has paid the subcontractor.

* * *

In the case at bench, the defendant and plaintiff had no contractual relationship. Nevertheless, the party who needed notice in fact received it, while the one who did not need notice (because he knew plaintiff's identity, and owned no real property against which plaintiff could file a lien) received none.

* * *

Where the purpose of the [relevant statute] is achieved and no one is prejudiced, technical requirements shall not stand in the way of achieving the purpose of the Mechanics' Lien Law. Here the defendant received proper notice and has made no showing of any prejudice arising from plaintiff's failure to serve notice on the original contractor.[95]

[6] *Kodiak Industries v. Ellis*

In *Kodiak Industries, Inc. v. Ellis*,[96] the plaintiff was a plumbing contractor. At the time the plaintiff first commenced work on the job, it had no actual knowledge of any lender, nor was any lender's identity then discoverable from the public records. Believing it had no further duty to investigate, Kodiak Industries did not serve any notice. Within 20 days after the plaintiff first started working on the job, the Bank of America became the actual construction lender and recorded its construction deed of trust. The trial court found that Kodiak Industries' failure to give the preliminary notice within 20 days of commencing its work on the jobsite precluded Kodiak Industries from enforcing a mechanic's lien. Kodiak Industries had a direct contractual relationship with an owner/builder, and therefore, under California Civil Code § 3097(b), Kodiak was required to give a preliminary notice to the construction lender only. Kodiak Industries appealed. The appellate court reversed the trial court, holding that under the above facts, Kodiak Industries was not obligated to give a preliminary notice because at the time that Kodiak Industries commenced work, there was, in fact, no construction loan of record.

The court did acknowledge and cite with approval *Romak Iron Works v. Prudential Insurance Co. of America*[97] for the proposition that if a lien claimant has no actual knowledge of the identity of the construction lender, he is nevertheless charged with constructive notice of the lender's identity when it is revealed, either by the recorded construction deed of trust or by the public building permit on file. In *Kodiak Industries,* the space for the name of the construction lender in the building permit was left blank.

It is interesting to note that Kodiak Industries' witnesses testified that it was their practice to serve notice on the construction lender if they learned of the construction lender's identity when the work commenced. They also testified that at

---

[95] 180 Cal. App. 3d at 1006–1009.
[96] 185 Cal. App. 3d 75 (1986).
[97] 104 Cal. App. 3d 767 (1980).

the commencement of the jobs that were the subject of this case, the owner/builder was asked if there was a construction lender for the job. The witnesses had no specific recollection of their conversations with the owner/builder's employees, but based upon their past practices, they assumed in retrospect from the fact that no notice had been sent to the Bank of America that they were told that no construction lender had been obtained. Kodiak Industries did not undertake any independent investigation of the owner/builder's representation that there was no construction lender. Kodiak Industries' principal shareholder testified that as he understood the mechanic's lien law, if there was no lender of record on the date the work commenced, no preliminary notice to any subsequent lender was required. He also testified that there was no reason to go back and check. He further testified he was not surprised that the owner/builder did not have a lender for the jobs in question because it was a large concern and might be self-funding.

The court again confirmed the general understanding of the mechanic's lien law that a subcontractor (such as Kodiak Industries in this case) with a direct contractual relationship with an owner/builder must give a preliminary notice to the construction lender. The court also confirmed that under those circumstances, no notice to the owner/builder with whom the subcontractor had a direct contractual relationship was necessary.

Bank of America contended on appeal that Kodiak Industries was given constructive notice of its identity as the actual lender when it recorded the construction loan deed of trust within the first 20 days after Kodiak Industries started work on the job. The court said that the statute was ambiguous as to when the notice needs to be given because it provides for a 20-day period but does not specify at what point in the period the claimant must send the notice. The statute states only that the notice shall be given not later than 20 days after the claimant first furnishes labor, service, equipment, or materials to the jobsite. Bank of America contended that such language required the lien claimant to check the records throughout the entire 20-day period. The bank further contended that a lien claimant cannot simply check once on the first day of construction and then assume thereafter that no construction lender will be involved. The court stated that, in essence, the bank's reading of the statute would require a lien claimant to recheck the record every day within the first 20 days, or wait until the 20th day, or else risk being assigned constructive knowledge of any subsequently filed documents. The court stated that to impose a duty on the claimant both to check and recheck the records, and then to serve the notice on the last day, seemed unduly onerous and fraught with unnecessary risks of lien forfeiture caused by unavoidable delays on the final day. The court stated that such a construction would be inconsistent with the policy of the law favoring protection of laborers and materialmen and the requirement that the lien law be liberally construed. The court said it would be manifestly unfair to a claimant who actually checked the records to penalize him by imputing to him knowledge of subsequent actions by the lender. The court held that a lien claimant need check the records only once, whether on the day it begins construction or at some other time before the end of the 20 days. Once a claimant has inspected the records during the notice period, there is no further duty of reinspection.

The court made a further interesting comment concerning *Brown Co. v. Superior Court.*[98] In *Brown Co.,* the court had suggested in dictum that a claimant's duty of inquiry might be different if it is told that there is no lender rather than being misinformed about the lender's identity. The court said that it rejected that dictum because it confuses inquiry duty with good faith. The court stated that it is only when the claimant mistakenly serves notice on the wrong party that the extent of the inquiry becomes critical to good-faith belief. According to the court, unless the lien claimant makes some reasonable inquiry, it normally cannot have a good-faith belief about the identity of the construction lender. The court went on to state that when the claimant is told by a responsible representative that there is no lender and does not serve any reputed lender, good-faith belief is not an issue, and the only duty of inquiry is to check those records that will impute constructive notice to him. In other words, if a claimant is told that there is no lender, then he has a minimum duty of checking the construction trust deed and the building permit. A "reputed" construction lender is a person or entity the claimant believes—reasonably and in good faith—to be the actual construction lender. The court stated that if the lien claimant has sufficient information to reasonably believe that a putative lender is the actual lender, it must either serve such lender or bear the risk that the putative lender is the actual lender. In this connection, if the claimant fails to give the preliminary notice to the reputed construction lender and that lender turns out not to be the actual lender, then the claimant's lien rights are not defeated. In that circumstance, the failure to give notice to the reputed lender cannot possibly prejudice the true construction lender. To hold the lien forfeited when the true lender would not have been served in any event would result in the defeat of the lien on meaningless technicalities. But, by failing to give notice to the reputed lender, the claimant bears the risk that the reputed lender is, in fact, the true lender. In that case, the failure to give the notice will defeat the lien.

The court stated that the test as to who was the reputed construction lender is an objective one, and a claimant must be deemed to possess sufficient information about a reputed lender when a reasonable person given the claimant's information would have been led to believe in good faith that the putative lender was the actual lender. The statute thus cuts both ways. A claimant is entitled to serve a reputed lender, and a reputed lender is entitled to the notice. The court states that the information on which a reasonable claimant should rely must be cloaked with sufficient indicia of reliability (such as statements from the owner, general contractor, or lender itself, or their agents) so as to distinguish this information from a mere guess or some ill-founded conjecture. Because possession of such presumptively reliable information would protect a lien claimant from a mistake, it is only fair to hold that possession of such knowledge imposes a concomitant duty to notify the lender thus indicated.

---

[98] 148 Cal. App. 3d 891, 901 (1983), *cited in* Kodiak Indus., Inc. v. Ellis, 185 Cal. App. 3d 75, 87 n.5.

[7] Additional Case Examples

Other cases addressing the preliminary notice requirement follow.

M. Arthur Gensler Jr. & Associates, Inc. v. Larry Barrett, Inc.[99] A claimant dealing with a lessee, when the owner has knowledge of the work and fails to post and record a notice of nonresponsibility, need not give preliminary notice because the lessee is considered an agent of the owner for notice purposes.[100]

Schrader Iron Works, Inc. v. Lee.[101] The claimant was not required to give preliminary notice to subsequent purchasers of the property on which labor and material were furnished.

Frank Pisano & Associates v. Taggart.[102] When the claimant contracts with an agent of the owner, no preliminary notice need be given to the owner; the claimant need only name the reputed owner in the mechanic's lien when the claimant does so in good faith. The lien is not lost if the claimant subsequently determines someone else is the actual owner. Filing a mechanic's lien claim in conjunction with a judicial proceeding to enforce it is privileged under California Civil Code § 47(2), and slander of title will not lie for a wrongful mechanic's lien.

Rich-Lee Equipment Rentals, Inc. v. Intermountain Construction Co.[103] A second-tier subcontractor who rented equipment and provided the operators to a subcontractor was not performing labor for wages and therefore had to give a 20-day notice.

Windsor Mills v. Richard B. Smith, Inc.[104] Oral preliminary notice was invalid.

Neptune Gunite Co. v. Monroe Enterprises.[105] Notice of rescission cannot serve as a preliminary notice.[106]

Fidelity Sound Systems, Inc. v. American Bonding Co.[107] The court in this case in the area of notices on public works applied a liberal standard. The case will probably be authority to which a claimant on private works could analogize.

---

[99] 7 Cal. 3d 695 (1972).
[100] *See also* Halspar, Inc. v. Barthe, 238 Cal. App. 2d 897 (1965); Benson Elec. Co. v. Hale Bros. Assocs., Inc., 246 Cal. App. 2d 686 (1966); Scott, Blake & Wynne v. Summit Ridge Estates, Inc., 251 Cal. App. 2d 347 (1967).
[101] 26 Cal. App. 3d 621 (1972).
[102] 29 Cal. App. 3d 1 (1972).
[103] 79 Cal. App. 3d 581 (1978).
[104] 272 Cal. App. 2d 336 (1969).
[105] 229 Cal. App. 2d 439 (1964).
[106] *See generally* Moss, *Preliminary Notices in Construction Industry Litigation—A Trap for the Unwary,* 51 L.A. B.J. 348 (1976).
[107] 85 Cal. App. 3d 13 (1978).

The facts showed that the plaintiff filed a stop notice but gave no preliminary notice whatsoever. The public body sent the prime contractor a copy of an internal letter that referred to the filing of the stop notice. This letter was sent by regular mail, and the prime contractor acknowledged that he received it. The plaintiff contended this was notice under California Civil Code § 3091, which required the claimant to send a written notice to the prime contractor by registered or certified mail within 90 days after the claimant's last work as a condition precedent to the right to recover on the bond. Although the plaintiff did not send the notice, the letter did not state the amount due the plaintiff, and the letter was not sent by certified or registered mail, it was deemed sufficient and in substantial compliance with the statute. This case could be analogized to a private works case in which there are preliminary notice problems.

Westfour Corp. v. California First Bank. [108] This case addressed the question of whether a preliminary notice was necessary. Blatte was the owner of certain real property. Westfour Corporation entered into a contract with U.S. Allied Development Corporation (Allied) to perform tenant improvement work on the project. That contract designated Allied as "owner" and Westfour as "contractor." Westfour believed that it had a direct contractual relationship with the owner and so gave no preliminary notice to the construction lender. On appeal, the construction lender contended that Westfour could not have a valid mechanic's lien because it did not have a direct contractual relationship with the owner; therefore, it had to give a preliminary notice under Civil Code § 3097. The court cited Civil Code § 3097, which requires all persons other than a person who is "under direct contract with the owner" and "the contractor" to give a preliminary notice to the construction lender. The court noted that the term *contractor* in section 3097 has been interpreted to mean "the general or prime contractor for the entire project," citing *Kodiak Industries, Inc. v. Ellis.*[109] The court stated that if a person is required to serve a preliminary notice but fails to do so, that person cannot enforce a mechanic's lien, citing Civil Code § 3114.

In this case, it was undisputed that Westfour did not serve a preliminary notice on the construction lender. Thus, unless Westfour was found to have been the general contractor under a direct contract with the owner, its failure to serve a 20-day notice was fatal to its ability to enforce a mechanic's lien. The appellate court looked at the trial court's findings of fact on this issue. The trial court had found that Westfour was a general contractor with respect to the work of improvement and that therefore no 20-day notice was required. Furthermore, the trial court found that in all respects, Westfour was dealing with and contracting directly with the owner of the property. The trial court found that U.S. Allied Development Corporation was substantially equivalent to Blatte for the purposes of Westfour's dealings with the owner. As a result, the trial court held that Westfour was under a direct contract with the owner, was a contractor as described in Civil Code § 3097(b), and was not required to give a preliminary notice.

---

[108] 3 Cal. App. 4th 1554 (1992).
[109] 185 Cal. App. 3d 75 (1986) (*see* § 9.02[B][6]).

The appellate court felt that there was substantial evidence to support the findings of the trial court, including evidence that the contract between Allied and Westfour referred to Allied as the "owner" and that Westfour treated Allied as the owner of the property. The finding was further supported by evidence that the construction contract referred to Westfour as "the contractor" (specifically contemplating that Westfour might hire subcontractors) and by testimony that Westfour always acted as a general contractor and never as a subcontractor. California First contended that Westfour could not possibly have been the general contractor for the entire project because its contract related only to a discrete portion of the total improvements, which included two years of demolition and renovation of the exterior before the appellant was even under contract to do the tenant improvements. The appellate court again stated that, based on the evidence presented, the trial court could properly conclude that Westfour had replaced Barnes as the general contractor on the project. Thus, the court concluded that Westfour was not required to give a preliminary notice to the lender.

Kim v. JF Enterprises.[110] In this case, a piece of real property owned by a trust was leased to JFE, which, in turn, subleased the premises to Cyns-Knabben, who operated a Jane Fonda Workout Exercise Gym on the premises. The trust, JFE, and Cyns-Knabben agreed to improve the property. JFE was to provide the financing to Cyns-Knabben, who would then be in charge of construction and responsible for paying for materials and laborers. All three were to benefit economically from the project.

Cyns-Knabben employed several persons to furnish labor, services, equipment, and materials for the improvement of the property. When they were not paid, they all recorded mechanics' liens, though none of the unpaid claimants had served a preliminary notice under Civil Code § 3097. Neither the trust, the lessee (JFE), nor the sublessee (Cyns-Knabben) had posted and recorded notices of nonresponsibility under Civil Code § 3094. The property owners demurred to the complaint to foreclose the mechanics' liens on the ground that the claimants had not served preliminary notices. The court of appeal held that the trial court had properly sustained the demurrer to the cause of action for foreclosure of mechanics' liens, without leave to amend, since the complaint did not allege that any of the plaintiffs had filed the preliminary notice required by Civil Code § 3097. Although section 3097(a) provides that notice is not required from a person under a direct contract with the owner, the court noted that the plaintiffs in this case had contracted with a sublessee, not the owner.

The court further stated that there was no authority or rationale for a rule that an owner's "constructive knowledge" exempts the claimant from serving a preliminary notice. The court noted that the purpose of the preliminary notice statute is to advise a noncontracting owner that construction is proceeding on its property so that the owner may notify the claimant, by posting and recording a notice of nonresponsibility under Civil Code § 3094, that the owner is not responsible for the work and that the claimant may not look to the property to secure

---

[110] 42 Cal. App. 4th 849 (1996).

payment for the construction. This purpose would be seriously undermined by equating knowledge of the lease provisions allowing construction with knowledge of actual construction. Permitting a lien claimant to avoid filing a preliminary notice and instead relying on the owner's "constructive knowledge" would also defeat the purpose of Civil Code § 3097, which requires express notice, and would result in repeated litigation over whether or not an owner's knowledge of lease terms or other facts was sufficient to put the owner on notice of actual construction. The court also noted that a lien claimant shall be entitled to enforce a lien only if the claimant has given the 20-day notice in accordance with the provisions of Civil Code § 3097 and has made proof of service in accordance with section 3097.1. In this case, the complaint did not allege that any of the plaintiffs had filed a preliminary notice.

The court stated that Civil Code § 3097(a) creates only two exceptions to the preliminary notice requirement—to wit, one under a direct contract with an owner and one performing actual labor for wages do not have to give notice. In rejecting the contention of the lien claimants in *Kim* that they had a direct contract with the owner, the court gave a detailed analysis of the rationale for that exception.

The rationale for excepting those under direct contract with the owner is that the owner is generally apprised of potential lien claims by those with whom the owner deals directly, whereas it is difficult for an owner to learn of potential lien claims by those with whom the owner does not have a direct contract. The *Kim* court also noted that Civil Code § 3129 provides that every work of improvement constructed with actual knowledge of the owner shall be deemed to have been constructed at the instance of the owner. That statute creates a presumption that, where the owner has actual knowledge of the construction work on its property, the work is done at the instance of the owner and hence is "under direct contract with the owner" within the meaning of Civil Code § 3097. Under such circumstances, even though the contract is with a lessee and not with the owner, no preliminary notice is required because of the owner's actual knowledge that work is being done on its property.

The *Kim* court pointed out that the cases relied upon by the claimants all antedated the preliminary notice statute by some 20 years and were based upon the notice of nonresponsibility statute. Those older cases held that where a noncontracting owner has even constructive knowledge of the improvements being made to its property, but fails to post and record a notice of nonresponsibility under Civil Code § 3094, the owner is estopped to deny that the work was done at its instance and request. Under those circumstances, a claimant could be deemed to have a direct contract with the owner for the purposes of Civil Code § 3097 and consequently is not required to give the notice prescribed therein.

The *Kim* court cited *Truestone, Inc. v. Simi West Industrial Park II*[111] and *Scott, Blake & Wynne v. Summit Ridge Estates, Inc.*[112] for the proposition that an

---

[111] 163 Cal. App. 3d 715 (1984).
[112] 251 Cal. App. 2d 347 (1967).

owner with actual knowledge of construction may be likewise estopped. The *Scott* case is cited for the proposition that the lessee is the agent, for preliminary notice purposes, of an owner with actual knowledge. The claimants in *Kim* tried to extend this principle to owners with "constructive knowledge" of the project, arising from the lease between JFE and the owners, which permitted JFE to make improvements that would benefit the owner. The *Kim* claimants argued that those lease terms also created a duty on the owner's part to inquire as to whether Cyns-Knabben had contracted for work on the project and that a reasonable inquiry would have revealed the construction work in question. The court concluded that this was still "constructive knowledge" and was insufficient to exempt the claimants from the 20-day notice requirement.

It is the authors' opinion that the preliminary notice section should receive a liberal interpretation.[113] As indicated above, the most recent cases on preliminary notices on private works have been strictly construed, as contrasted with the liberal interpretation given preliminary notices on public works.

§ 9.03 RECORDING THE LIEN

[A] Recording Requirements

A general contractor cannot record a lien until completing the contract.[114] All others cannot record their liens until they have ceased furnishing labor, services, equipment, or materials.[115] When a valid notice of completion has been recorded, the general contractor has 60 days to record, and all others have 30 days.[116] When there has been no notice of completion but actual completion has occurred, all persons have 90 days from actual completion to record.[117] If there is no notice of completion and no actual completion of the job, then the following are deemed equivalents of completion, and all persons have 90 days from the following events to record their liens:

- Occupation and use of the work of improvement by the owner or its agent, accompanied by a cessation of labor.[118]

- Acceptance of the work of improvement by the owner or its agent.[119] This acceptance must be communicated to the lien claimant.[120]

- A cessation of labor for a continuous period of 60 days.[121]

---

[113] *See* California Continuing Educ. of the Bar, California Mechanics' Liens and Other Remedies ch. 3, at 84, 85 (1972).
[114] Cal. Civ. Code § 3115.
[115] *Id.* § 3116.
[116] *Id.* §§ 3115, 3116.
[117] *Id.*
[118] *Id.* § 3086(a).
[119] *Id.* § 3086(b).
[120] Hammond Lumber Co. v. Yeager, 185 Cal. 355 (1921); Munger & Munger v. McBratney, 131 Cal. App. 2d Supp. 866 (1955).
[121] Cal. Civ. Code § 3086(c).

If the owner records a notice of cessation of labor (after a 30-day cessation of labor), this is the same as a notice of completion, and the prime contractor has 60 days and all others have 30 days to record their liens.[122] The lien must be recorded in the office of the county recorder where the real property is located.

California Civil Code § 3086 defines *completion* for the purposes of the mechanic's lien law. It states that completion in the case of any work of improvement other than a public work means actual completion of the work of improvement, and defines the equivalents of completion as set forth above.

There is a provision in section 3086 that is often overlooked. It says that "[i]f the work of improvement is subject to acceptance by any public entity, the completion of such work of improvement shall be deemed to be the date of such acceptance." This section applies to a very narrow area of mechanic's lien law, which has commonly been referred to as "off-site work" in the construction industry. *Off-site work* is defined in Civil Code § 3102, under the heading "Site Improvement," as the "demolition or removing of improvements, trees, or other vegetation located thereon; or drilling test holes; or the grading, filling, or otherwise improving of any lot or tract of land, or the street, highway, or sidewalk in front of or adjoining any lot or tract of land; or constructing or installing sewers or other public utilities; or constructing any areas, vaults, cellars, or rooms under said sidewalk, or making any improvements thereon." Essentially, the type of work that is subject to these Code sections is street work and the utilities in the street. As a result of this section in the mechanic's lien law, some unusual case results have been obtained.

In *Southwest Paving Co. v. Stone Hills*,[123] the plaintiff performed off-site work in October and November of 1959. The contractor's lien was not recorded until almost a year later, on September 16, 1960. The work done by the plaintiff was subject to acceptance by the city of Los Angeles, which at the time of filing of the plaintiff's complaint to foreclose its mechanics' lien had not as yet accepted the plaintiff's work. The owner interposed a demurrer, contending that a lien recorded almost a year after the contractor finished its work was not timely. The demurrer was sustained, the contractor appealed, and the appellate court reversed the judgment of the trial court. The court held that when the work is off-site work, the time for filing liens does not commence until the work has been accepted by the public body. Because the work that was the subject of the contractor's lien had not been accepted by the city of Los Angeles, the lien was timely.

In another case with slightly different facts, *Howard A. Deson & Co. v. Costa Tierra, Ltd.*,[124] various subcontractors who had performed work for an owner/builder on an apartment complex brought actions to foreclose their mechanics' liens. The liens had been recorded more than 60 days after the notice of completion. The trial court found that the entire project, including both the construction of the apartment buildings and the on-site and off-site lot improvements, was a single work of improvement subject to acceptance by the city of

---

[122] *Id.* §§ 3087(a), 3092, 3115, 3116.
[123] 206 Cal. App. 2d 548 (1962).
[124] 2 Cal. App. 3d 742 (1969).

Palm Springs and that the project was not completed for the purpose of timely liens filing until the city of Palm Springs accepted the project. The trial court found that the project was accepted by the city of Palm Springs on a date subsequent to the recording of the notice of completion; therefore, the mechanics' liens were timely.

The court of appeal reversed the judgment of the trial court except as to the lien for installation of streets, curbs, and gutters, and held that no acceptance by the city of Palm Springs was required for the other work that was performed. The appellate court sent the case back for a determination of whether the notice of completion filed by the owner was valid. In that connection, the appellate court pointed out that work was done by certain of the lien claimants after the notice of completion was filed. If that work were found to have been performed pursuant to their basic contracts, it would render the notice of completion void. The significant part of the opinion as it relates to the present issue is that the court made a clear holding that the installation of streets, curbs, and gutters was off-site work subject to acceptance by the public body and that the lien period had therefore not run until that work was in fact accepted. A notice of completion would not cut down the time for filing liens.

Another case to discuss this issue is *A.J. Raisch Paving Co. v. Mountain View Savings & Loan Ass'n.*[125] In that case, the lien claimant, through a subcontractor, did sewer work that was potentially beneficial to two contemplated subdivision tracts, both of which were subject to acceptance by a governmental agency. One tract was accepted, but the plaintiff failed to file a mechanic's lien against that tract within the statutory lien period. Subsequently, the plaintiff did file a claim of lien that included all sewer costs against the other tract, which had never been accepted by the public body, and was allowed a judgment of foreclosure by the trial court. The court of appeal modified the judgment to allocate the lien only to the cost of the sewer work for the benefit of the tract that had not been accepted. As so modified, the judgment was affirmed. The court noted that the contractor had, by failing to timely file its mechanic's lien with respect to the accepted tract, waived its lien rights against that tract. With respect to the unaccepted tract, the court rejected the owner's claim that the lien was untimely and held that the project came within the purview of the Code section providing that the date of acceptance is the date of completion. Because the tract had never been accepted, the claim was timely.

It is clear from the foregoing cases that the acceptance referenced in this section of the lien law is not the usual inspection and approval by a building department of work done on a construction project. Rather, it is acceptance of that portion of work that necessitates and requires separate governmental acceptance and approval, such as streets, curbs, gutters, sidewalks, sewers, and other off-site utilities.

---

[125] 28 Cal. App. 3d 832 (1972).

[1] Notice of Completion

As indicated above, the time for filing liens runs from the date of the recording of a notice of completion when the owner actually records a notice of completion. The requirements for a notice of completion are governed by California Civil Code § 3093, which provides that a notice of completion is a written notice signed and verified by the owner or the owner's agent, containing the following:

1. Date of completion. The recital of an erroneous date of completion does not affect the validity of the notice if the true date of completion is within 10 days preceding the date of recording of the notice of completion.

2. Name and address of the owner.

3. Nature of the interest or estate of the owner.

4. Description of the site sufficient for identification and containing the street address of the site, if any. If a sufficient legal description of the site is given, the validity of the notice of completion shall not be affected by the fact that the street address recited is erroneous or that the street address is omitted.

5. Name of the original contractor, if any, or, if the notice of completion is given after completion of a contract for a particular portion of the work, as provided in Civil Code § 3117, then the name of the original contractor on that contract and a general statement of the kind of work done or material furnished pursuant to the contract.

The notice of completion must be recorded in the office of the county recorder where the project is located within 10 days after completion. If there is more than one owner, a notice of completion signed by less than all of the co-owners must recite the names and addresses of all of the co-owners. Furthermore, a notice of completion signed by a successor in interest must recite the names and addresses of the transferor(s).

This section describes what constitutes completion. Many times a notice of completion will be invalid because it was prematurely recorded—in other words, completion work was done after recordation of the notice of completion. A notice of completion is sometimes challenged when it contains errors. Numerous cases have ruled on these types of questions.

In *Doherty v. Carruthers*,[126] the owners had a prime contract with Bried which provided that Bried was to file the notice of completion. Bried entered into contracts with subcontractors to provide labor and materials. The work was actually completed on August 9, 1955. On October 17, 1955, Bried filed a notice of completion that recited that the work had been completed on October 14, 1955.

---

[126] 171 Cal. App. 2d 214 (1959).

The mechanic's lien claimants filed their liens within 30 days of the date of the notice of completion, but 99 days after the actual date of completion. As noted, the notice of completion was recorded 69 days after the actual date of completion rather than within 10 days after completion as required by the statute. The trial court upheld the validity of the mechanics' liens. The owner appealed, contending that because the notice of completion had been recorded more than 10 days after completion, it was invalid, and because the lien claims were recorded 99 days after actual completion, they were too late. The appellate court held that the owner was estopped to deny the validity of the notice of completion and that the lien claimants were entitled to rely upon that notice of completion in filing their liens. The court stated that if one of two innocent parties is to be injured, the party to be injured should be the one who can be charged with having allowed the situation to develop. The lien claimants took advantage of the established way of doing things by recording their liens within 30 days after recordation of the notice of completion. The court further noted that, under the contract, the prime contractor Bried had express authority to file the notice of completion. The court concluded that the lien claimants were entitled to watch the filings of notices of completion in the Office of the County Recorder and rely upon them, and therefore it held the liens valid.

In *Howell v. Gunderson*,[127] a person by the name of Don Mowat deeded land to Mowat Associates, a corporation. Mowat Associates hired a general contractor to build a building. The lien claimant was a subcontractor. The job was completed, and Don Mowat filed a notice of completion, which he signed naming himself as the owner and as owner/builder but not naming the prime contractor, a Mr. Gunderson. The subcontractor recorded its lien within 90 days of completion of the job but more than 30 days after the notice of completion. The trial court held that the lien was not timely filed, and the lien claimant appealed. The appellate court reversed. The court held that the notice of completion was invalid because it was not, in fact, signed by the owner of the real property and the name and address of the true owner of the property was not inserted in the notice of completion. Mr. Mowat, who had signed the notice of completion in his individual capacity, was not, in fact, the owner/builder, and the name of the contractor had not been inserted. The court held that because of all these errors in the notice of completion, it was invalid.

In *Gary C. Tanko Well Drilling, Inc. v. Dodds*,[128] Tanko had a direct contractual relationship with the owner to drill a well and install a pump and pressure system. Tanko completed its work on April 1, 1977, and the owner (Dodds) recorded a notice of completion on April 8, 1977. In the portion of the notice of completion requiring the name of the contractor, Dodds inserted his own name. On April 11, 1977, three days after recording the notice of completion, Dodds transferred title to a new owner (Tinney). Tanko recorded its mechanic's lien 74 days after the notice of completion and 82 days after actual completion. The trial court denied Tanko's mechanic's lien on the grounds that it was not recorded

---

[127] 250 Cal. App. 2d 961 (1967).
[128] 117 Cal. App. 3d 588 (1981).

within 60 days after the notice of completion, as required by law. Tanko appealed, contending that under Civil Code § 3093(e), the notice of completion was invalid because the name of the original contractor was not inserted.

The judgment of the trial court was affirmed by the court of appeal, which stated that the purpose of the notice of completion is to protect the owner by shortening the time period in which mechanics' liens must be filed, as well as to give notice to interested parties in order that they might file their lien claims in due time. Therefore, the notice of completion should be upheld if it provides proper notice to lien claimants. The court of appeal also held that, where only two parties (that is, the owner and the contractor) are involved and are dealing face to face, the purposes of both the mechanic's lien law and the notice of completion law are realized when the owner inserts his own name as the original contractor. Thus, the notice of completion in this case was valid because it substantially complied with Civil Code § 3093.

In *Insul-Acoustics, Inc. v. Sung Soo Lee*,[129] the defendant, Sung Soo Lee, doing business as Buster Brown Shoes, leased store space in a shopping mall in Thousand Oaks owned by Village Associates. The general contractor for the shopping mall was Ernest W. Hahn, Inc. The defendant Sung Soo Lee contracted with Realty Design & Construction, Inc. for the fixturization of his store. Realty Design, in turn, subcontracted with Insul-Acoustics to do the drywall and acoustical ceiling. The fixturization at the store was completed on or before July 31, 1978, and the store was opened on July 31, 1978. No work was done after that date. Sung Soo Lee paid Realty Design, but Realty Design failed to pay Insul-Acoustics. Insul-Acoustics recorded its mechanic's lien on December 8, 1978, more than 90 days after the completion of the work at that particular store. The trial court granted a summary judgment in favor of Sung Soo Lee on the basis that the mechanic's lien was not recorded within 90 days after completion of the job, as required by Civil Code § 3116.

Insul-Acoustics appealed, contending that the mechanic's lien was timely because the work of improvement should be interpreted to include not only the work on the store in which Insul-Acoustics did its work but also work that was later performed at other stores in the mall by Insul-Acoustics, pursuant to separate subcontracts between Insul-Acoustics and Realty Design. The court held that because each job and each tenant shop was the subject of separate contracts between Insul-Acoustics and Realty Design, and between Realty Design and the individual tenant, each shop was a separate work of improvement, and therefore the time for filing liens ran from the completion of each such shop. Insul-Acoustics tried to rely upon Civil Code § 3106, which provides that the work of improvement means the entire structure or scheme of improvement as a whole. The court responded that the job on which Insul-Acoustics performed its work and each of the tenant shops were clearly separate works of improvement; there was no basis for holding that the separate tenant improvements could be an entire structure or scheme of improvement as a whole.

---

[129] 136 Cal. App. 3d 552 (1982).

Civil Code § 3086(c) provides that if the work of improvement is subject to acceptance by any public entity, the completion of such improvement shall be deemed to be the date of such acceptance, except as to contracts awarded under the State Contract Act; under those contracts, a cessation of labor on any public work for a continuous period of 30 days shall be a completion thereof. In *W.F. Hayward Co. v. Transamerica Insurance Co.*,[130] the court had occasion to interpret that particular provision of the lien law. In that case, the prime contractor contracted with the county of Los Angeles to construct the Lost Hills Sheriff's Station. W.F. Hayward was a subcontractor on the project. The job began, but a number of delays ensued. A dispute between the prime contractor and the county arose as to who was responsible for the damages caused by the delay. Both the prime contractor and the subcontractors sought additional payments from the county because of the delay. In a letter dated June 4, 1990, the county advised Transamerica Insurance Company, the surety on the prime contractor's performance and payment bonds, that the prime contractor had been suspended effective June 5, 1990, and that the Board of Supervisors would adopt a formal termination of the prime contractor on June 19, 1990. In a letter dated June 7, 1990, the prime contractor requested that the county approve a change order in the amount of $1,227,771.79, which reflected the claims of the prime contractor and its subcontractors for the delay. That sum included Hayward's claim of $243,650. The prime contractor and the subcontractors ceased work on the project, and the prime contractor never returned to the project.

On August 3, 1990, the prime contractor and the county entered into an agreement entitled "Termination for Convenience Agreement and Mutual Release," whereby the county released the prime contractor from any further obligation or liability regarding the contract to construct the sheriff's station. The agreement also purported to assign to the county all of the prime contractor's subcontracts, including the subcontract with Hayward. In a letter to the county dated August 15, 1990, Hayward proposed that the county and Hayward enter into a new contract. The county took the position that it had acquired a valid assignment of Hayward's contract with the prime contractor, and therefore Hayward's proposal was unacceptable. Using the former prime contractor's subcontractors, including Hayward, the county completed construction of the project on March 19, 1991. Thereafter, Hayward filed suit against the prime contractor, the county, and Transamerica (the surety on the prime contractor's payment bond).

The question in the case was whether Hayward's suit was timely. The court stated that under Civil Code § 3184, which is referenced in Civil Code § 3249, the stop notice on a public job must be served within 30 days of a notice of completion or, if there is no notice of completion, within 90 days of completion or cessation. The court noted that on this project there was no notice of completion, and therefore the question was whether there had been a completion or cessation. The court then turned to Civil Code § 3086, which provides that if the work of improvement is subject to acceptance by any public entity, the completion of such work of improvement shall be deemed to be the date of such acceptance;

---

[130] 16 Cal. App. 4th 1101 (1993).

however, a cessation of labor on any public work for a continuous period of 30 days shall be deemed a completion thereof. The court noted that the record in this case made it clear that there was a cessation of labor on the project. Hayward took the position that there was no cessation of labor because its subsequent performance was pursuant to the original contract between the prime contractor and the county. The court then turned to Civil Code § 3249, which provides that suit on the payment bond may be brought after the claimant has furnished the last of the labor or materials on the project under the contract between the original contractor and the owner. The court stated that any work a subcontractor performs after the original contractor's obligations have ended should not extend the surety's liability, because it is the conduct of the original contractor to which the bond relates. As noted earlier, the contract between the county and the original prime contractor was terminated. The court concluded, therefore, that although the subcontractor could look to the prime contractor's payment bond for satisfaction, suit against the bonding company must be brought within six months after the stop notice period had expired.

The court then had to analyze whether the stop notice period had expired. The question was whether the 90-day stop notice period commenced to run 30 days after the county suspended the original prime contractor's contract on June 5, 1990, or, because the job continued to go forward, that section did not apply. The court noted that if there was at the time of termination of the original prime contractor's contract on June 5, 1990, a 30-day cessation of labor within the meaning of Civil Code § 3086 then that work stoppage was a completion pursuant to Civil Code § 3086 and started the running of the stop notice period. The court found that, in this particular case, no work was done between June 5, 1990, when the original prime contractor was terminated, and August 3, 1990. The court stated that it was undisputed that no work relative to the contract between the original prime contractor and the subcontractor, Hayward, nor on the contract between the original prime contractor and the county, had occurred between June 5, 1990, and August 3, 1990. Therefore, there was a cessation of labor for more than 30 days, and the 90-day period for filing the stop notice had run. The court noted that because there was a dispute concerning payment for work on the project, and because work had ceased, Hayward, the subcontractor, could have served a stop notice on the county. That stop notice would have to have been served within 90 days after the 30-day cessation of labor. The 30-day cessation of labor expired on July 5, 1990, and therefore the stop notice period expired 90 days after that, on October 3, 1990. Civil Code § 3249 provides that suit must be brought on the payment bond within six months after the period in which stop notices may be filed. Six months from October 3, 1990, is on April 5, 1991. The court noted that the suit in this case was not filed until July 17, 1991, and therefore was untimely.

This case yields a harsh result for the subcontractor. The subcontractor did know that the original prime contractor had been terminated. However, that very subcontractor came back on to the job and completed the job on behalf of the county. Certainly, the subcontractor would not be sophisticated with regard to these very fine distinctions with regard to the time to file suit on the bond. The subcontractor was probably under the impression that its time to sue on the bond did not run until after the job as a whole had been completed. However, according to the *Hayward* court's interpretation of this unique section, Civil Code § 3086(c), a 30-day period of cessation of labor had expired, the 90-day stop notice period had expired, and the six-month period after expiration of the 90-day period had expired—and therefore the suit on the bond was untimely. Thus, the subcontractor was denied recovery on the payment bond.

As noted in *Doherty v. Carruthers*,[131] the owners recorded a notice of completion more than 10 days after completion—in fact, 69 days after the date of actual completion. The lien claimant recorded its lien within 30 days of the notice of completion, and the court held that the owner was estopped from denying the validity of its notice of completion. In a slightly different case, *Fontana Paving, Inc. v. Hedley Bros., Inc.*,[132] the lien claimant recorded its mechanic's lien more than 90 days after the completion of the job. After that lien had been recorded, the owner recorded a notice of completion reciting a completion date predating the claim of lien. The court, acknowledging *Doherty,* stated that, had the lien claimant relied upon the notice of completion, the lien would have been deemed valid. However, in *Fontana,* there was no evidence that the lien claimant had relied upon the notice of completion. In fact, the lien claimant had recorded its mechanic's lien before the notice of completion. Because the lien was recorded more than 90 days after actual completion, it was held to be untimely.

[B] Multiple Contracts

Under California Civil Code § 3117, if the work of improvement is made pursuant to two or more original contracts, each covering a particular portion of the work of improvement, the owner may record a separate notice of completion within 10 days after the completion of each separate contract. The lien period is then the same as that set forth in California Civil Code § 3117: 60 days for the original contractor and 30 days for any claimant under the original contractor.

[C] Contents of Mechanics' Liens

Mechanics' liens should include the following:

- Statement of the claimant's demand after deducting all just credits and offsets
- Name of the owner or reputed owner, if known
- Name of person by whom the claimant was employed or to whom the claimant furnished its materials
- General statement of the kind of labor, service, equipment, or materials furnished
- Description of the jobsite sufficient for identification.[133]

---

[131] 171 Cal. App. 2d 214 (1959).
[132] 38 Cal. App. 4th 146 (1995).
[133] *See* Cal. Civ. Code § 3084.

[1] Description of Site Sufficient for Identification

One of the contents of the mechanic's lien under California Civil Code § 3084 is a "description of the site sufficient for identification."[134] Numerous cases have discussed the issue of what happens when the description set forth in a lien contains some errors.

In *Howard A. Deson & Co. v. Costa Tierra, Ltd.*,[135] one of the lien claimants recorded a mechanic's lien that omitted the words "northwest corner" from the description but described the project as "commonly known as Palo Fierro Estates." The court stated that, as a general rule, the description of the property sought to be charged with a lien will be deemed sufficient if it enables a party familiar with the locality to identify the property with reasonable certainty to the exclusion of other properties (citing *Hollenbeck-Bush Planing Mill Co. v. Roman Catholic Bishop*,[136] *Union Lumber Co. v. Simon*,[137] and *Borello v. Eichler Homes, Inc.*[138] The court stated that errors in the description may be disregarded if the identification of the property is otherwise sufficient and if the recorded notice of lien is not fraudulent and does not mislead the owner or innocent third parties (citing 54 Cal. L. Rev. 179 (1966), *Borello v. Eichler Homes, Inc.*, and *American Transit Mix Co. v. Weber*[139]). The court held that use of the words "commonly known as Palo Fierro Estates" was a sufficient description, and therefore it upheld the mechanic's lien.

In *Borello v. Eichler Homes, Inc.*, the lien described the property as "Unit 3 and Unit 4 Terra Lina, San Rafael, California," whereas the correct description of the property was "Terra Linda Valley, Unit 3 and Unit 4." Civil Code § 3261 provides that no mistake or errors in the description of the property against which the lien is recorded shall invalidate the lien unless the court finds that such mistake or error was made with intent to defraud or that an innocent third party, without notice, has, since the claim of lien was recorded, directly or constructively become a bona fide owner of the property and that the notice of claim was so deficient that it did not in any manner put the party on further inquiry. As a result, the court held that the description of the property was sufficient.

[2] Name of Owner or Reputed Owner

Another requirement of the mechanic's lien is that the name of the owner or reputed owner, if known, be included. In an early case, *Allen v. Wilson*,[140] the claim of lien named no one as the owner or reputed owner. In fact, the claim of lien stated, in the blank for the name of the owner or reputed owner, the word

---

[134] *See* Cal. Civ. Code § 3084(a)(5).
[135] 2 Cal. App. 3d 742 (1969).
[136] 179 Cal. 229 (1918).
[137] 150 Cal. 751 (1907).
[138] 221 Cal. App. 2d 487 (1963).
[139] 106 Cal. App. 2d 74 (1951).
[140] 178 Cal. 674 (1918).

"unknown." The court held the lien valid. Specifically, the court stated that the lien claimant is only required to state the name of the owner if known. Therefore, if the name of the owner is not known, the claim of lien is sufficient if it is silent on the subject. The court went on to state that the name of the owner or reputed owner is not presumed to be within the knowledge of the claimant.

In *Frank Pisano & Associates v. Taggart*,[141] the court held that, in a claim of lien, it "is sufficient to give only the name of the reputed owner. Where an individual does so in good faith, he does not lose his lien if he subsequently determines that some other individual is the actual owner." Thus, if the claimant, in good faith, reasonably believes that a given individual is the owner of the property and names that person as reputed owner, and it later turns out that someone else is the actual owner, the lien will not be held invalid.

Compare, however, *H&L Supply, Inc. v. Ewing*.[142] In that case, the plaintiff was a material supplier to a subcontractor of a subcontractor. The president and sole stockholder of the material supply company prepared the preliminary notice and directed an employee to serve it upon the owner, Mr. Ewing. Although the opinion is unclear on the point, it appears that the preliminary notice named Mr. Ewing as the owner of the property. There was a conflict of testimony as to whether the notice was served upon him. In the mechanic's lien, the plaintiff described the property as 160 acres and also indicated that it was commonly known as "Ewing's Café." The space for the name of the owner was left blank. At the time the lien was recorded, the Ewings did operate a café, but it was known as "Ewing's Tam-O-Shanter Inn" and was located several miles away, on the opposite side of the city of Bakersfield from the place described in the lien. The court held that the lien was invalid, stating that it was rendering that holding based upon the "limited facts of the case" and that "we do not intend to say that a preliminary lien notice cannot serve to estop an owner from asserting a defect in the Notice of Claim of Mechanics' Lien." The lien claimant argued that, because it had served Mr. Ewing with the preliminary notice, the owner was estopped to assert the defect in the name of the owner in the lien. It is the opinion of the authors that this case appears to be in conflict with *Allen v. Wilson* and *West Coast Lumber Co. v. Newkirk*.[143] Evidently, the court in *H&L Supply, Inc.* based its decision on the idea that since the lien claimant did name an owner in the preliminary notice, he should also have done the same thing in the mechanic's lien.

[3] Name of Person by Whom Claimant Was Employed or to Whom Claimant Furnished Its Materials

Another requirement of the contents of the mechanics' lien is the name of the person by whom the claimant was employed or to whom the claimant furnished its labor, service, equipment, or materials. In *Wand Corp. v. San Gabriel Valley*

---

[141] 29 Cal. App. 3d 1 (1972).
[142] 253 Cal. App. 2d 283 (1967).
[143] 80 Cal. 275 (1889).

Lumber Co.,[144] the plaintiff furnished materials and labor to the general contractor. Sixteen days before recordation of the mechanic's lien, the lien claimant sent a preliminary notice to the owner, which properly contained the name of the person who contracted for the purchase of the plaintiff's labor and material. The lien claimant recorded its mechanic's lien and, in that section for the name of the person by whom the claimant was employed or to whom the claimant furnished its materials, the claimant put the word "claimant"—an obvious typographical error. The owner moved for a summary judgment on the ground that the lien did not comply with Civil Code § 3084. The trial court granted the motion, and the lien claimant appealed. The appellate court reversed, noting that "[i]f there is a single unifying thread which explains most though not all of the bewildering array of cases in this field, it is the principle that where the purpose of the requirements of Section 1193.1, Subdivision (J)(4) is achieved and no one is prejudiced, technical requirements shall not stand in the way of achieving the purpose of the Mechanics' Lien Law."[145] California Civil Procedure Code § 1193.1 is now Civil Code § 3086. The court held that, when the preliminary notice correctly designated the person to whom the claimant had furnished its labor or materials and who also should have been named in the mechanic's lien but was not, and when the mechanic's lien recorded is not fraudulent and does not mislead the owner or innocent third parties, the mistake is not fatal to enforcement of the lien. Thus, the court held that the lien was valid.

[4] Signature of Claimant

Moreover, the lien must be signed and verified by the claimant or his agent.[146]

[D] Time for Filing Suit to Foreclose

The suit to foreclose a mechanic's lien is subject to the following provisions:

- It must be commenced within 90 days after the lien is recorded in the county where the real property is located.[147] The time to sue may be extended up to one year from completion, provided a proper extension of time to foreclose is recorded within 90 days of recording the lien.[148]

- It should name as defendants all parties who have any interest in the property. The plaintiff should consult the lot book report, mechanic's lien guaranty, or preliminary title report.

- It is subject to discretionary dismissal if not brought to trial within two years.[149]

---

[144] 236 Cal. App. 2d 855 (1965).
[145] 236 Cal. App. 2d at 861.
[146] Cal. Civ. Code § 3084.
[147] Cal. Civ. Code § 3144.
[148] *Id.* § 3144.
[149] *Id.* § 3147.

§ 9.03[D]

- The plaintiff should record a lis pendens as soon as the suit is filed.[150]
- The plaintiff may consolidate mechanic's lien actions on the same property.[151]
- The plaintiff can get an attachment in a lien foreclosure action.[152]

On occasion, a lien claimant will record a mechanic's lien, let the 90-day period for foreclosure go by, and not bring a foreclosure action. The question then arises as to whether the lien claimant can record a second mechanic's lien on the same property for the same work. The answer is that the claimant may record a second lien, even though the first lien has not been foreclosed, if the time for filing liens has not expired.[153]

[1] Petition to Remove Lien Not Foreclosed

The mechanic's lien law contains within it a provision pursuant to which owners may remove as a cloud on their title mechanics' liens that have not been foreclosed within the 90-day period. That provision appears in section 3154 of the Civil Code, which provides that if the claimant does not foreclose the lien within 90 days of its recordation, the owner may give the lien claimant notice to remove the lien. If the lien claimant does not remove the lien, the owner may then file a petition in court to have a court order indicating that the lien is no longer valid, and may seek up to $500 in attorney's fees for pursuing that procedure. In one case, the lien claimant recorded a mechanic's lien on a project. The 90 days for foreclosure of that lien expired. The owner demanded that the lien claimant release the lien, but the lien claimant refused. The owner then brought a proceeding under Civil Code § 3154 and obtained a court order to the effect that the lien was null and void. The lien period was still open, and the lien claimant recorded a second mechanic's lien for the same work. The court held that, under those circumstances, when the owner had taken steps under Civil Code § 3154 to have the lien removed and obtained a court order to that effect, the second lien was invalid.[154]

In another case involving multiple liens, an architect was employed by an owner of real estate to perform certain work. In *Santa Clara Land Title Co. v. Nowack & Associates*,[155] the architect performed the work, was not paid, and recorded mechanic's lien 1. The architect brought a suit to foreclose that lien. After the suit was filed, the architect got paid and released mechanic's lien 1. The

---

[150] *Id.* § 3146.
[151] *Id.* § 3149.
[152] *Id.* § 3152.
[153] *See* Electric Supply Distribution Co. v. Imperial Hot Mineral Spa, 122 Cal. App. 3d 131 (1981); Schwartz & Gottlieb, Inc. v. Marcuse, 175 Cal. 401 (1917); *In re* Paul Potts Builders, Inc., 608 F.2d 1279 (9th Cir. 1979); Coast Cent. Credit Union v. Superior Court, 209 Cal. App. 3d 703 (1989); Solit v. The Tokai Bank, 68 Cal. App. 4th 35 (1999).
[154] *See* Maris Management Corp. v. Assured Drywall & Texture, 152 Cal. App. 3d 268 (1984).
[155] 226 Cal. App. 3d 1588 (1991).

architect then entered into a new contract with the owner to do more work, was not paid, and recorded mechanic's lien 2. A construction lender that was about to record a construction loan opened an escrow for that purpose. As a condition of recording its construction loan and as a condition of obtaining title insurance, a demand was made upon the architect to release mechanic's lien 2 in exchange for payment of the amount due the architect out of the escrow. The architect did release mechanic's lien 2 and got paid out of the escrow, and the construction loan was recorded. After that, the architect did still more work for the owner, was not paid, recorded mechanic's lien 3, and brought a suit to foreclose that lien. The construction lender foreclosed on its deed of trust. Ultimately, the issue before the court was whether the construction lender had priority over the architect's third mechanic's lien. The architect contended that its priority related back to the time when the architect did its first work on the project, which was long before the construction loan was recorded. The court concluded that because the architect had released mechanics' liens 1 and 2 in consideration for payment, the architect did not have priority; the construction lender had priority over the architect's lien. Some attorneys in California have read this case to mean that if a lien claimant records a lien and then releases it, the lien claimant cannot record another lien on the same project. The authors do not believe that this case stands for that proposition. Rather, it merely stands for the proposition that on the peculiar facts of this case, the construction lender had priority over the architect's third mechanic's lien.

In *Koudmani v. Ogle Enterprises, Inc.*,[156] the lien claimant, Ford Wholesale Company, recorded a mechanic's lien. Koudmani (the owner) then recorded a notice of completion. Within 30 days after the notice of completion, Ford recorded a second claim of lien for the same materials. The owner, through his attorney, sent Ford a letter demanding that Ford release its first claim of lien because no foreclosure action had been brought within 90 days. Ford complied with this demand and executed a release of lien stating that the claim of lien was "satisfied or otherwise released and discharged."

When the owner demanded that Ford release its second claim of lien as well, Ford refused. The owner filed a petition to release the property, under Civil Code § 3154, from Ford's second mechanic's lien. The trial court granted the owner's petition, ruling that the property had been released from both liens, and awarded attorneys' fees to the owner. Ford appealed.

In reversing the trial court, the court of appeal distinguished *Maris Management Corp.*, noting that in *Maris*, the owner had, in fact, obtained an order decreeing the original lien invalid under Civil Code § 3154 prior to recordation of the second lien. The court of appeal stated that the distinction drawn in *Maris* between the terms "lien" and "claim of lien" is inaccurate and misleading. The true distinction is that between the inchoate constitutional right of lien and the lien itself. The court noted that the California Constitution gives material suppliers an

---

[156] 47 Cal. App. 4th 1650, 55 Cal. Rptr. 2d 330 (1996).

inchoate right to a lien, which is meaningless unless it is exercised by recording a claim of lien. Unless and until a claim of lien is recorded, there is no lien because there is no charge imposed upon the property to secure the owner's act of payment to the claimant. Thus, said the court of appeal, it is more accurate to view the terms "lien" and "claim of lien" (as used in the mechanics' liens statutes) as synonymous rather than as having distinct meanings. The court concluded, following Coast Central Credit Union, that the inchoate right of lien is not terminated by operation of the 90-day statute, and a subsequent lien may be recorded for the same work, so long as the lien period is still open. Therefore, Ford's failure to commence a foreclosure action within 90 days of recording its first lien did not bar Ford from foreclosing on its second lien.

Koudmani had also argued that *Santa Clara Land Title Co.* likewise precluded Ford from recovering on its second lien. The court of appeal held that the principal issue in that case was whether the claimant's mechanic's lien had priority over an earlier recorded deed of trust. The court of appeal specifically found the decision in *Santa Clara* distinguishable by reason of the fact that, in *Koudmani,* the amount owing to Ford was never paid and Ford's release did not conclusively and unequivocally represent that the claim was "fully satisfied." Ford's release merely stated that the first lien was "hereby satisfied or otherwise released and discharged." The *Koudmani* court noted that a release that states a claim of lien is "fully satisfied" necessarily extinguishes the inchoate right upon which the claim of lien is based, because the underlying obligation has been paid according to the express terms of the release. On the other hand, a release that indicates that a specific claim of lien may not be "satisfied" (that is, paid), but is "otherwise released or discharged," does not extinguish the inchoate right to record a subsequent lien based upon the same work or material as in the released claim of lien. Such a release frees the property only of the particular claim of lien expressly identified.

The court of appeal noted that the essential purpose of the mechanic's lien statutes is to protect contractors, laborers, and material suppliers from nonpayment—which purpose would be frustrated by holding that a release of an unsatisfied claim bars the right to later record a timely claim of lien for the same work or materials. Accordingly, the *Koudmani* court held that the release executed by Ford as to its first claim of lien did not release Ford's inchoate right of lien. Therefore, the decision of the trial court was reversed, and the case was remanded to the trial court to determine the amount of costs and attorneys' fees to be awarded to Ford as the prevailing party.

[2] Lien Foreclosure Actions Stayed Pending Arbitration

Many times the general contract between the owner and the general contractor has an arbitration clause. Likewise, subcontracts between the general contractor and the subcontractors may have arbitration clauses. The mechanic's lien law, in contrast, requires an action to be brought to foreclose the lien within 90 days after its recordation. The contractor or subcontractor may not, in fact, be able to file its demand for arbitration and complete the arbitration within that 90-day period.

The mechanic's lien law itself does not specifically cover this situation. However, California Civil Procedure Code § 1281.5 sets forth the method pursuant to which the contractor is obligated to proceed. Section 1281.5 provides that the commencement of an action to foreclose a mechanic's lien shall not waive the lien claimant's right to arbitration if, in filing the action to foreclose the lien, the lien claimant at the same time presents to the court an application that the action be stayed pending the arbitration. When faced with this dilemma, the lien claimant should apply in the foreclosure action to stay the case pursuant to section § 1281.5. The manner in which the claimant must proceed is not set forth in the statute, which merely requires that the claimant, when filing an enforcement action, "at the same time present to the court an application that the action be stayed pending the arbitration." Trial courts have handled this problem in varying ways. The safest procedure for the claimant is to file, along with the complaint, a motion to stay the action pending arbitration. Most courts will not accept an *ex parte* application; they require a noticed motion. What that means is that the attorney will have to simultaneously file both the complaint and the motion, obtain a hearing date for the motion, and serve the owner with both the complaint and the motion at the same time. Section 1281.5(b) provides that if a defendant in the action to foreclose the lien fails to file a petition to stay the action and compel arbitration at or before the time he or she answers the complaint, this failure shall constitute a waiver of that party's right to compel arbitration. Thus, if an owner is served with a summons and complaint to foreclose a mechanic's lien, and the plaintiff has not taken steps to have the action stayed pursuant to section 1281.5, and if the owner desires to have arbitration, the owner should, pursuant to section 1281.5(b), file a petition to stay the action and compel arbitration.

In *R. Baker, Inc. v. Motel 6, Inc.*,[157] plaintiff Baker performed construction work for Motel 6 under an agreement containing an arbitration clause. On May 6, 1982, Baker filed a complaint to foreclose its mechanic's lien. Shortly after the action was filed, Baker's attorney and the attorney representing Motel 6 discussed resolving the dispute before the American Arbitration Association. Baker's attorney gave the attorney for Motel 6 an open extension of time to answer and, in the letter confirming that open extension, suggested that the matter be arbitrated. The complaint was amended several times. The attorney who had represented Motel 6 moved to withdraw as the attorney for the owners. He stated in his declaration that he had tentatively agreed with the attorney for Baker to resolve their dispute by way of arbitration, but his clients opposed binding arbitration. On November 30, 1984, Baker filed a demand for arbitration. The new attorneys representing Motel 6 answered the complaint and brought a motion to enjoin the arbitration. The court issued a preliminary injunction restraining Baker from proceeding with the arbitration because Baker had not complied with Civil Procedure Code § 1281.5, in that it had failed to request a stay when it filed its complaint. The court held that Baker had indeed waived its contractual right to

---

[157] 180 Cal. App. 3d 928 (1986).

arbitrate its claim under section § 1281.5 by bringing the action to foreclose the lien without "simultaneously" applying for a stay of the action pending arbitration. This case thus confirms that the claimant who desires to arbitrate should file a motion to stay the action pending arbitration at the same time it files its complaint.

[E] What Constitutes Completion

In the case of any work of improvement, other than a public work of improvement, *completion* means actual completion of the work of improvement.[158] From a factual standpoint, the courts have struggled to apply this definition.

Rockwell v. Light.[159] A second coat of paint, required by the contract, was applied to porch floors and steps within 90 days of the lien. The court held that this constituted completion.

Boscus v. Waldmann.[160] The lien claimant did two hours of painting work on a given date, and the court held that the lien period ran from that date. The court stated:

> The defendant, C.H. Waldmann, testified that the last work done on the building was by the painter on the twenty-sixth day of October 1912. On that day he did some work of painting on the house. It required him about two hours to do the work. This work was done, it is to be assumed, in pursuance of the building contract or the specifications attached thereto as a part thereof. The court was justified from this testimony in finding, as it did, that the building was completed on the twenty-sixth day of October. So long as the work performed was called for by the contract and essential to the completion of the building, the extent of the work or the length of time required to do it is wholly immaterial, for in such case the last stroke of the painter's brush marked the time of the completion. It may be true that if, when the lien was filed and it was sought to enforce it through a judicial decree, the two hours' work of the painter was still unfinished, and such work constituted all that was necessary to complete the building according to the terms and conditions of the contract and specifications, the omission to do that work might justly be treated as a "trivial imperfection" in the work of completion; but this argues nothing against the proposition above stated that the finishing of the work of painting marked the date of the actual completion of the building, said work being all that was necessary actually to complete it.[161]

Klan v. Hoffman.[162] The lien claimant finished his work, which consisted of two chimneys and a porch made of brick. He installed an ashpit door in one

---

[158] Cal. Civ. Code § 3086.
[159] 6 Cal. App. 563 (1907).
[160] 31 Cal. App. 245 (1916).
[161] 31 Cal. App. at 257.
[162] 176 Cal. 763 (1917).

of the chimneys that was 8″ × 10″ and cost 75 cents. It was stolen. He replaced it. The court held that the installation of the replacement door did not extend the time to file a lien.

Hogan Lumber Co. v. Boyk.[163] A "speculative house builder" built a house for sale. His practice was not to install the electrical fixtures or the furnace until the house was sold; in this way, the purchaser could select the fixtures to be installed. The owner/builder sold the house, and the buyers picked out the fixtures, which were then installed. The court held that under these facts, the installation of the electrical fixtures and furnace did not extend the lien period.

Hubbard v. Jurian.[164] Labor ceased on a construction project. After cessation, the architect advised the owner to paint window panels to protect them from cracking in the sun. This work was not called for in the original contract but was carried out by the owner on the advice of the architect, in order to protect his property from deterioration. The court held that this work did not extend the lien period.

Hammond Lumber Co. v. Yeager.[165] Labor ceased on a house, and the owner occupied the home. Later, the contractor discovered that certain stone work constituting the floor of the bathroom and the sinks in the kitchen was defective and persuaded the owner and the subcontractor to replace it at a cost of $35.50. The court held that replacing the defective work did not extend the lien period.

Hammond Lumber Co. v. Barth Investment Corp.[166] Electrical work valued at $260.61 was "necessary to correct and complete the work in accordance with the plans and specifications." The court held that the lien period did not commence to run until that work was completed. In so holding, the court specifically noted that the work was "[n]ot to correct imperfections but . . . necessary to complete the work as called for by the plans and specifications."[167] The owner argued that it had accepted the buildings and had communicated that to the contractor. The claimant in the case was a material supplier, and the court specifically stated: "Of course a secret acceptance by the owner from the contractor of an uncompleted building could not defeat the lien claimant."[168]

Nevada County Lumber Co. v. Janiss.[169] A sewer pipeline was installed defectively. (Cement obstructed the drainage.) This was called to the attention of the contractor. The old drainpipes were removed and replaced with a new sewer

---

[163] 77 Cal. 477 (1918).
[164] 47 Cal. App. 543 (1920).
[165] 185 Cal. 355 (1921).
[166] 202 Cal. 606 (1927).
[167] 202 Cal. at 608.
[168] *Id.* at 611.
[169] 25 Cal. App. 2d 579 (1938).

line. It took three men 11 hours. The court held that the lien period did not start to run until the new line was installed because the building, with its necessary sewer system, was not complete until the new sewer line was put in.

Munger & Munger v. McBratney.[170] After a notice of completion was recorded, certain ducts in the heating and ventilating system were covered and insulated, as required by the plans and specifications, which took 84 hours. The court held that the notice of completion was invalid. As in *Hammond Lumber Co. v. Barth Investment Corp.*,[171] the owner argued that she had accepted the building and the court repeated the supreme court's statement from *Hammond Lumber* that a secret acceptance by the owner not communicated to the claimant would not start the lien period.

Lewis v. Hopper.[172] The court held that the lien period did not start to run until four soap dispensers, which took four hours to install and were worth about $100, had been completed.

As a result of these cases, completion is a question of fact in each case. The rule seems to be that if the work that is done is work called for in the original plans and specifications, the job will not be considered complete until that work is, in fact, done.

[F] Other Mechanic's Lien Issues

The following case examples address a variety of issues related to mechanics' liens:

Petersen v. W.T. Grant Co.[173] The requirement of bringing action to foreclose a mechanic's lien in 90 days is a statute of limitations rather than a statute affecting a substantive right; therefore, it can be waived if not pleaded as an affirmative defense.

Vaughn Materials Co. v. Security Pacific National Bank.[174] The plaintiff furnished materials directly to an owner. The owner recorded a notice of completion. The material supplier filed its mechanic's lien 40 days after the notice of completion. The court held that a material supplier furnishing materials directly to an owner had to file its mechanic's lien within 30 days of the notice of completion.

Sawyer Nurseries v. Galardi.[175] When the owner files bankruptcy, the time to file suit to foreclose the lien is tolled. When the property is removed from the jurisdiction of the bankruptcy court, the time to sue on the mechanic's lien starts to run again.

---

[170] 131 Cal. App. 2d Supp. 866 (1955).
[171] 202 Cal. 606 (1927).
[172] 140 Cal. App. 2d 365 (1956).
[173] 41 Cal. App. 3d 217 (1974).
[174] 170 Cal. App. 3d 908 (1985).
[175] 181 Cal. App. 3d 663 (1986).

Monterey S.P. Partnership v. W.L. Bangam, Inc.[176] The plaintiff brought an action to foreclose a mechanic's lien and named as defendants the trustee and beneficiary of the deed of trust. Because there were 252 beneficiaries, the lien claimant served only the trustee and got a judgment foreclosing the mechanic's lien. The beneficiaries brought an action for declaratory relief in quiet title to declare that the mechanic's lien foreclosure judgment did not bind their interest. The California Supreme Court held that the mechanic's lien foreclosure judgment did not, in fact, bind the beneficiaries of the deed of trust because they were never served with the summons and complaint.

[G] Stop Notice

In general, all mechanic's lien requirements apply to stop notices with regard to the preliminary notice requirement and the time for filing. Anyone entitled to a mechanic's lien is also entitled to file a stop notice.[177] The stop notice is served on the owner, construction lender, or any other person holding construction funds.[178] If a payment bond has been recorded, then withholding by the owner or construction lender pursuant to the stop notice becomes optional.[179] In order to make withholding of funds mandatory by the construction lender, the stop notice must be accompanied by a bond one and a quarter times the amount of the claim.[180] If a payment bond has previously been recorded in the office of the county recorder where the site is located, in accordance with the mechanic's lien law, then the construction lender shall withhold funds pursuant to a bonded stop notice filed by an original contractor and may, at its option, withhold funds pursuant to a stop notice or bonded stop notice given by anyone other than an original contractor. The construction lender may elect not to withhold pursuant to a bonded stop notice by reason of a payment bond.

If, when giving the construction lender the stop notice or bonded stop notice, the claimant makes a written request for notice of the election, accompanied by a preaddressed, stamped envelope, the construction lender shall furnish the claimant a copy of the bond within 30 days after making the election. A lender shall not be liable for failure to furnish a copy of the bond if the failure was not intentional and resulted from a bona fide error, as long as the lender maintains reasonable procedures to avoid such an error, and if the error was corrected not later than 20 days from the date on which the violation was discovered. In the case of a stop notice or bonded stop notice filed by an original contractor or by a subcontractor contracting directly with an owner, the original contractor or subcontractor shall be entitled to recover on his stop notice or bonded stop notice only the net amount due the original contractor or subcontractor after deducting the stop notice claims of all subcontractors or material

---

[176] 49 Cal. 3d 454 (1989).
[177] Cal. Civ. Code §§ 3158, 3159.
[178] *Id.* §§ 3158, 3159.
[179] *Id.* §§ 3161, 3162.
[180] *Id.* § 3162.

§ 9.03[G]   CALIFORNIA CONSTRUCTION LAW

suppliers who have filed bonded stop notices on account of work done on behalf of the original contractor or the subcontractor.

The claimant should make sure that the bond is indeed one and a quarter times the amount of the stop notice claim. In *Manos v. Degen*,[181] the claimant filed a stop notice in the sum of $41,000 and also filed, along with the stop notice, a bond in the sum of $41,000. Thus, the bond was not one and a quarter times the amount of the claim. The lender did not hold back any money because the bond was not one and a quarter times the amount of the claim. When the stop notice claimant brought an action on the stop notice, the construction lender filed a motion for summary adjudication of issues, contending that it had no liability because the bond was not in the amount required by the statute. The construction lender was successful, and the court specifically held that California Civil Code § 3162 required the lender to withhold funds only if the stop notice were accompanied by a bond in a sum of one and a quarter times the amount of the stop notice claim. The stop notice claimant argued, unsuccessfully, that the construction lender had a duty to withhold at least $32,800 from the loan funds in light of the $41,000 bond. The court also rejected that contention.

The purpose of the bond is to protect the construction lender in the event of a wrongful stop notice. This was clearly illustrated in *Flintkote Co. v. Presley of Northern California*.[182] In that case, Presley was the owner and developer of a tract. Hyre was a painting subcontractor who had subcontracts with Presley to do painting and install wallboard. Flintkote was a material supplier supplying the wallboard to Hyre. Payments were made by joint check to Hyre and Flintkote. Flintkote filed a bonded stop notice with Presley and the construction lender, United California Bank. Flintkote also recorded a mechanic's lien. Flintkote brought an action to foreclose its lien and to enforce the stop notice. The action was tried, and the trial court entered a judgment enforcing the stop notice and the mechanic's lien. Presley appealed.

The court of appeal held that Flintkote was not entitled to a stop notice, so it reversed the judgment on the stop notice. Presley filed a cost bill for the appeal, and Flintkote paid it. For several months thereafter, counsel for Presley and counsel for the bonding company that had furnished the bond accompanying the stop notice negotiated with Flintkote regarding recovery of the attorney's fees for defending the stop notice claim. Presley demanded damages from the bonding company and stated that if it was not paid within 30 days, it would file a motion to recover attorney's fees. The bonding company demanded indemnification from Flintkote, and Flintkote rejected the demand. Presley filed a motion seeking to recover attorney's fees and interest from Flintkote and the surety on the stop notice bond. The trial court granted Presley's motion, and Flintkote and the bonding company appealed.

The judgment of the trial court was affirmed. The court stated that the stop notice is a powerful remedy in the hands of a claimant and that the counterpoise to this most powerful weapon is Civil Code § 3083, which imposes a penalty for

---

[181] 203 Cal. App. 3d 1237 (1988).
[182] 154 Cal. App. 3d 458 (1984).

improper use of the stop notice, which penalty includes "costs" and "all damages that such owner may sustain by reason of the equitable garnishment not exceeding the sum specified in the bond." The court stated that the inappropriateness of the filing of the stop notice had been determined by the appellate court. The court, citing *Systems Investment Corp. v. National Auto & Casualty Insurance Co.*,[183] held that attorney's fees were recoverable on the bond. The court held, therefore, that in being forced to defend against the improper imposition of the stop notice, the owner Presley had suffered "damages" within the contemplation of Civil Code § 3083. The trial court judgment was affirmed, and therefore Presley was entitled to recover its attorney's fees on appeal.

[1] Construction Funds Are Trust Funds

Generally, the funds used to finance a construction project are held by the construction lender and disbursed as the project progresses. Those funds are, of course, subject to the filing of bonded stop notices. Various statutes and cases would lead one to believe that funds made available for construction purposes are unique and have received special treatment in the state of California. This was confirmed, in part, in *Chang v. Redding Bank of Commerce*,[184] in which a property owner made a progress payment to its general contractor for the purpose of paying subcontractors on the project. The contractor deposited those funds in a bank account that it maintained for all of its business operations. The bank seized the money, by way of setoff, due to the contractor's default on a promissory note pursuant to which it owed the bank money. After the contractor filed for bankruptcy—and because the contractor had not paid its subcontractors (as a result of the offset made by the bank)—the subcontractors filed mechanics' liens against the owner's property, which the owner was obligated to satisfy. The owner then brought an action against the bank for unjust enrichment and the imposition of a constructive trust.

The court held that both the subcontractors and the owner, under the principles of equitable subrogation, had standing to impose a constructive trust on the funds held in the general contractor's bank account when those funds are considered to be trust funds and if the bank has knowledge of facts sufficient to put it on inquiry notice as to whether those funds were being held in trust by the depositor. In this particular case, the court held that there was sufficient evidence to put the bank on notice that the contractor was depositing progress payments into its general account for the specific purpose of paying its subcontractors. Thus, the court imposed a constructive trust on those funds in favor of the owner.

The *Chang* case may have far-reaching consequences in the construction industry. If a contractor, subcontractor, or material supplier has failed to perfect its mechanic's lien or stop notice rights on a construction project, it should investigate whether any funds have been deposited in any bank and then diverted for

---

[183] 25 Cal. App. 3d 1057 (1972) (property owner brought action against surety on bond posted in connection with stop notice).

[184] 29 Cal. App. 4th 672 (1994).

other purposes under circumstances where the bank knew or should have known that those funds were construction funds deposited for the express purpose of paying subcontractors or material suppliers on the project.

[2] Service of Stop Notice

The stop notice may be served by:

1. Personal service;
2. Registered mail; or
3. Certified mail.[185]

The stop notice must be served upon the manager or other responsible officer or person at the office of the construction lender at the branch where the construction funds are being administered.[186] If the job involves the use of a joint control agency, then a stop notice should be filed there as well because there will be funds at both locations.

[3] Contents of Stop Notice

The contents of the stop notice must:

- Be signed and verified by the claimant or its agent;
- Include the kind of labor, service, equipment, or materials furnished;
- Include the name of the person to whom labor, service, equipment, or materials were furnished; and
- Include the amount already furnished and the total amount agreed to be furnished.[187]

[4] Time for Filing Suit

Conditions of the suit include:

- It cannot be brought within the first 10 days after service but must be brought no later than 90 days after the expiration of the lien period.[188]
- Notice of suit must be given within five days after commencement.[189]
- The suit cannot be brought to trial until 90 days after the lien period expires.[190]

---

[185] Cal. Civ. Code § 3083.
[186] Id.
[187] Cal. Civ. Code § 3103.
[188] Cal. Civ. Code § 3172.
[189] Id.
[190] Id.

- Discretionary dismissal applies if the suit is not brought to trial in two years.[191]
- Stop notice suits may be consolidated, and the owner and the lender may compel claimants to interplead in one action.[192]

Effective January 1, 1992, California Civil Code §§ 3176 and 3176.5 were added to the stop notice statutes. Section 3176 provides that in any action against an owner or construction lender to enforce a stop notice, the prevailing party shall be entitled to attorneys' fees. Section 3176.5 provides that the prevailing party in an action on a stop notice shall be entitled to interest at the legal rate, calculated from the date the stop notice was served.

As noted above, suit must be brought on the stop notice within 90 days after the period of time for filing stop notices has expired. Thus, if there is a notice of completion, the time to file would be 120 days from the notice of completion or, in the event of no notice of completion, 180 days from the date of completion.

On a public works project, if a claimant pays a $2 fee to the public body when the claimant files its stop notice, and makes a request under Civil Code § 3185 to be notified when the job is completed, then the public body must, no later than 10 days after the filing of a notice of completion or after the cessation of labor has been deemed a completion, or after the acceptance of completion, whichever is later, give notice of the expiration of such period to each stop notice claimant by personal service or registered or certified mail. When the service is by registered or certified mail, the service is complete at the time of the deposit of the registered or certified mail in a United States Post Office, addressed to the claimant at the address shown on the claimant's stop notice claim.

In *J.A. Thompson Corp. v. D.C. Contractors*,[193] Thompson served a stop notice on the public body, served the request for the notice, and paid the $2 fee. The public body withheld money from the prime contractor. The stop notice period expired on July 7, 1984. On December 2, 1984, the public body released the money to the prime contractor. On January 15, 1986, Thompson brought an action to enforce the stop notice. The period for filing suit on the stop notice, at that point, had long since expired. The prime contractor and the public body asserted the 90-day statute of limitations as a defense, and the trial court ruled against the stop notice claimant. The appellate court reversed. The court stated that because the public body had accepted the $2 fee and had failed to give the required notice, it was therefore estopped from asserting the statute of limitations. The court held that equitable estoppel will be applied against a public body when justice and right require it. In this case, that was necessary.

When an action is brought to enforce a stop notice, Civil Code § 3172 requires that the claimant give a notice of commencement of any such action within five days after commencement of the suit to the same persons and in the

---

[191] *Id.* § 3173.
[192] *Id.* § 3175.
[193] 4 Cal. App. 4th 1355 (1992).

same manner as provided for the service of a stop notice or bonded stop notice. What this means is that if the action is to enforce a bonded stop notice served against a construction lender, a notice of commencement of the action must be given to the construction lender within five days after the suit is commenced. The five-day notice can be served either personally or by registered or certified mail.

Section 3172 was interpreted in *Sunlight Electric Supply v. McKee*.[194] In that case, the claimant supplied materials to a subcontractor on a public job. The material supplier filed a stop notice and then filed suit on March 7, 1962. The notice required by the Code was given on March 21, 1962, 14 days after commencement of the action. The court ruled in favor of the stop notice claimant, holding that because no detriment was caused by the claimant's failure to give the five-day notice, the requirement was not mandatory but merely directory and did not result in a loss of jurisdiction. In other words, the court held that unless someone is prejudiced by the failure to give the notice, it will not defeat recovery.

The court noted that, as to public works, there can be no mechanic's lien on the property belonging to the public body, as this would seriously affect the public interest. The court noted, therefore, that to give protection to material suppliers and labor claimants and still protect the public and public property, the legislature has, without providing lien rights on public works projects, provided the stop notice procedure for intercepting funds due from the public agency to the contractor. The underlying objective is the same for a lien right on private works or the stop notice right on public works. The mechanic's lien law, including the stop notice provisions, is an integrated and harmonious scheme, and the code sections must be construed together. These laws are remedial in character and should be liberally construed in their entirety with a view to effect their objectives and to promote justice. In other words, the rule of liberal construction of the mechanic's lien law is equally applicable to public works projects and the notice requirements regarding stop notices.

[5] Statute of Limitations on Filing Suit to Enforce Stop Notice Is Subject to Doctrine of Equitable Tolling

Under California Civil Code § 3210, suit to enforce a stop notice against the original contractor and the owner of a public work "shall be commenced not later than 90 days following the expiration of the period within which stop notices must be filed as provided in Section 3184." Until now, that time limit was strictly enforced and was not subject to any exception other than estoppel. In *Structural Steel Fabricators v. City of Orange*,[195] the court of appeal announced, for the first time, that the judicially created doctrine of equitable tolling[196] is applicable to toll the section 3210 period for bringing an action to enforce a stop notice on a public works project.

---

[194] 226 Cal. App. 2d 47 (1964).
[195] 40 Cal. App. 4th 459 (1995).
[196] *See* Addison v. State, 21 Cal. App. 3d 313 (1978).

As noted at the outset of the court's opinion, this is the second case of the same name involving the same parties and the same set of operative facts. In the earlier case,[197] the court of appeal reversed the trial court's grant of summary judgment in favor of the city pursuant to the limitations period in section 3210, on the ground that a triable issue of fact existed as to "whether Structural could rely on an estoppel theory to extend the time for filing its action." In a footnote, the court declined to reach Structural's alternative argument that the doctrine of equitable tolling is applicable, while noting that the position might have merit.

The court of appeal summarized the procedural history of the case, and its present holding:

> Following trial of the action, the court ruled Structural failed to prove estoppel, but found the action timely by virtue of equitable tolling of the statute of limitations while Structural pursued its claim against the general contractor and the bonding company. The city's appeal squarely presents the issue we had no need to decide in our prior decision. We affirm and publish our opinion in light of the dearth of law on the subject.[198]

Under the facts of the case, Structural was a structural steel fabrication subcontractor to IDC, which walked off the job in October 1988, leaving its surety (CSBC) to complete the project. Structural, which was not paid, served its stop notice on January 4, 1989. The city permitted CSBC to "bond around" the stop notice. On May 26, Structural filed suit against IDC and CSBC and obtained a default judgment in September. A month earlier, however, IDC filed a Chapter 7 bankruptcy action, and CSBC went out of business. While its suit was pending, Structural made demand on the city on June 21, 1989, and a corrected demand on June 27.

On November 6, 1989, Structural filed a separate lawsuit against the city. In its amended complaint, filed in June 1990, Structural "admitted the statute of limitations to enforce the stop notice had expired on July 4, 1989, however, it claimed the City's conduct had tolled the statute." As stated by the court of appeal:

> We need not concern ourselves with dates. All parties agree if tolling occurred, the action was timely; they also agree that without tolling, the section 3210 statute of limitations expired four months before the suit was filed.[199]

Quoting the supreme court's decision in *Addison v. State, supra,* the court of appeal set forth the following three elements of equitable tolling: "timely notice, and lack of prejudice, to the defendant, and reasonable and good faith conduct on the part of the plaintiff."[200] The court stated further: "The timely

---

[197] Structural Steel Fabricators v. City of Orange, 234 Cal. App. 3d 1206 (1991).
[198] 40 Cal. App. 4th at 461.
[199] *Id.* at 462.
[200] *Id.* at 463.

notice requirement essentially means that the first claim must have been filed within the statutory period" and "the filing of the first claim must alert the defendant in the second claim to the need to begin investigating the facts which form the basis for the second claim."[201] The second element, intended to protect defendants from "stale claims and deteriorated evidence . . . translates to a requirement that the facts of the two claims be identical or at least so similar that the defendant's investigation of the first claim will put him [or her] in position to appropriately defend the second claim."[202] The third element was more difficult for the court to define but rested on the delays not having been too great and the plaintiff's not having acted in any way to lull the defendant into thinking that the second claim would not be filed.

In this case, the first element of equitable tolling was met by the fact that Structural had twice made demand upon the city, in May and again in June, during the pendency of its first action. Moreover, the court noted that Structural had no choice, under Civil Procedure Code § 3264, but to sue CSBC once the public entity bonded around the stop notice. The court declined to consider the issue of prejudice to the city in light of the fact that the city failed to order a transcript of the lower court proceedings. Also, because the court had no evidence of prejudice in the record before it on appeal, the reviewing court must "presume all intendments to support the judgment." As for the third element, the court of appeal likewise deferred to the lower court's specific finding that Structural "in good faith, pursued a claim first against the general contractor and the bonding company."[203]

[H] Payment (Labor and Materials) Bonds

If the original contract is filed with the county recorder and a payment bond amounting to 50 percent of the contract price is recorded before a work of improvement is commenced, then the court must, when it would be equitable to do so, restrict recovery on mechanics' liens to the amount found due from the owner to the contractor and render judgment against the surety for the deficiency. The intent of the law is to limit the owner's liability to its contract price with the contractor.[204]

Labor and materials bonds can be filed on private works under California Civil Code §§ 3235 and 3236. In addition to these statutory payment bonds that may sometimes exist on a private work of improvement, there are payment bonds called common-law payment bonds, which are not provided pursuant to the Civil Code.

In *T&R Painting Construction, Inc. v. St. Paul Fire & Marine Insurance Co.*,[205] the subcontract between a subcontractor and a contractor provided that

---

[201] *Id.* at 464.
[202] *Id.* at 465.
[203] *Id.*
[204] Cal. Civ. Code §§ 3235, 3236.
[205] 23 Cal. App. 4th 738 (1994).

the prevailing party would be entitled to recover its attorney's fees. A dispute between the subcontractor and the contractor was submitted to arbitration. The arbitrator ruled in favor of the subcontractor and awarded the subcontractor a principal amount plus attorney's fees and costs as the prevailing party. A judgment was entered on the arbitration award, and the subcontractor then went to trial against the contractor's surety to determine whether the surety was liable on its payment bond for the attorney's fees incurred in prosecuting the subcontractor's claims, including the attorney's fees incurred in the arbitration proceedings. The trial court entered judgment for the surety, and the subcontractor appealed.

The court of appeal reversed, sending the case back to the trial court for a determination of the amount of the attorney's fees that should be awarded to the subcontractor. The court held that the surety was liable to the subcontractor for attorney's fees based upon the attorney's fees clause in the subcontract. The court stated that the subcontractor was, by law, a party for whose benefit the bond was given and could therefore enforce liability on the bond against both the principal and the surety.

In addition, the court noted that the bond stated on its face that it was "for the use and benefit of claimants," and a *claimant* was defined as including one having a direct contractual relationship with the principal. The principal on the bond was the contractor, with whom the subcontractor did have a direct contractual relationship. Civil Code § 2808, which provides that a surety's liability is commensurate with the liability of the principal, does not distinguish between subcontractors and owners, or between payment and performance bonds. Therefore, the subcontractor could recover from the surety the attorney's fees provided for in its subcontract, so long as the recovery against the surety, including attorney's fees, did not exceed the penal sum of the bond.

[1] Preliminary Notice

In order to recover on a payment bond, the claimant must have either served a preliminary notice under Civil Code § 3097 or served the surety and principal on the bond with a written notice within 15 days of notice of completion or 75 days after completion if there is no notice of completion.[206]

Civil Code §§ 3242 and 3252 make the 20-day preliminary notice under § 3097 (private works) and § 3098 (public works) applicable to payment bond claims as well as to mechanic's lien and stop notice claims.

If the 20-day private work preliminary notice was not given as provided in section 3097, a claimant may enforce a claim by giving written notice to the surety and the bond principal as provided in section 3242 within 15 days of recordation of a notice of completion. If no notice of completion has been recorded, the time for giving written notice to the surety is extended to 75 days after completion of the work of improvement.

---

[206] Cal. Civ. Code § 3242(a), (b).

A functionally identical provision was added with regard to public works.[207] Civil Code § 3252 provides that the preliminary notice to the surety and the bond principal must be given by personal delivery or certified or registered mail, addressed to the surety. The statutes provide various alternative addresses where the notice can be sent.

The 20-day preliminary notices on public projects relate back in the same way as they have on private projects. This means that if a subcontractor or material supplier who is required to give preliminary notice on a public project fails to give the preliminary notice within 20 days of first furnishing labor, services, equipment, or materials, it may do so later, and the notice will cover all work performed for 20 days preceding the notice through the end of the project.

The law also allows a subcontractor, as well as the owner or original contractor, to be a principal on a payment bond.

[2] Time for Filing Suit

Suit must be filed within six months after completion of the work of improvement.[208]

[3] Cases Interpreting Bond Claims

In *California Electric Supply Co. v. United Pacific Life Insurance Co.*,[209] a list of the subcontractor's creditors that was handed to a representative of the bonding company within the 90-day period was held to constitute compliance with California Government Code § 4209.[210]

The court in *General Electric Co. v. Central Surety & Insurance Corp.*[211] held a stop notice sent to the prime contractor to be in compliance with California Government Code § 4209.[212]

Because many "performance" bonds have been held to be "labor and material" or "payment" bonds, the parties must examine the contract and the bond closely.[213]

In 1985 the California Court of Appeal held in *General Insurance Co. of America v. Mammoth Vista Owners Ass'n, Inc.*[214] that an obligee under a surety

---

[207] *Id.* § 3252.
[208] Cal. Civ. Code §§ 3239, 3240.
[209] 227 Cal. App. 2d 138 (1964).
[210] Now Cal. Civ. Code § 3091.
[211] 232 Cal. App. 2d 590 (1965).
[212] Now Cal. Civ. Code § 3091.
[213] C.O. Sparks, Inc. v. Pacific Constr. Paving Co., 159 Cal. App. 2d 513 (1959); Woodhead Lumber Co. v. E.G. Niemann Invs., 99 Cal. App. 456 (1929); Sunset Lumber Co. v. Smith, 91 Cal. App. 746 (1928); Culbertson v. Cizek, 225 Cal. App. 2d 451 (1964); Gordy v. United Pac. Ins. Group, 243 Cal. App. 2d 445 (1966); Rexroth & Rexroth, Inc. v. General Cas. Co. of Am., 242 Cal. App. 2d 363 (1966); Continental Cas. Co. v. Hartford Accident & Indem., 243 Cal. App. 2d 565 (1966).
[214] 174 Cal. App. 3d 810 (1985).

bond could bring an action against the surety for engaging in unfair claims settlement practices in violation of California Insurance Code § 790.03(h) and seek punitive damages for breach of that provision. Although not directly overruled, that decision will be impacted by the case of *Parvaneh Moradi Shalal v. Fireman's Fund Insurance Co.*[215] In *Fireman's Fund,* the California Supreme Court held that a third party did not have a right of action under section 790.03(h) and therefore could not sue for violation of that provision and seek punitive damages. The supreme court did not, however, eliminate the doctrine of implied covenant of good faith and fair dealing as it exists between the parties to an insurance contract. It did not specifically discuss the issue in relationship to bonds.

There is a difference of opinion among practitioners as to whether *Mammoth Vista* is still good law. A principal could maintain an action based on the doctrine of breach of the implied covenant of good faith and fair dealing against the surety. It is also arguable that the named obligee of a bond could claim breach of the implied covenant of good faith and fair dealing. Furthermore, the intended beneficiaries of a labor and material bond (subcontractors and material suppliers) might likewise argue that there would be an implied covenant of good faith and fair dealing running from the surety company to those intended beneficiaries. Because this issue has not been decided by the courts, it must be left for future decisions.

The *Mammoth Vista* case was an action between the property owners and the surety on the performance bond relating to construction of recreational facilities within the common area of a planned unit development. The bond was required by the owner, in the penal amount of $42,350. The contractor did not finish the work and was financially unable to do so. The surety was placed on notice of a default by its principal (the contractor) twice in mid-1977. The surety offered $10,000 to settle the dispute, but the owner wanted the surety to pay the full penal obligation of the bond. In July 1978, the surety took the position that the claim was barred by the statute of limitations—but then offered $20,000 in settlement, which was also refused.

The surety then filed a declaratory relief action in which the owner filed a cross-complaint to recover the full penal amount of the bond, plus a cause of action for bad faith. The jury awarded the owner $120,500 in compensatory damages and $150,000 in punitive damages. The court of appeal upheld the decision, noting that, under common law, every insurer has an implied-in-law duty to act fairly and in good faith in handling the claim of an insured. The court further noted that tortious violation of that duty may occur in several ways, including the insurer's failure without proper cause to compensate its insured for a loss covered by the policy. The court focused on the Unfair Practices Act (Insurance Code § 790.038(h)), which sets forth standards of conduct substantially similar to those at common law, and held that "bad faith" may be found under those standards. Specifically, the Unfair Practices Act sets forth the following as improper conduct by an insurance company:

---

[215] 46 Cal. 3d 287 (1988).

(1) Misrepresenting to claimants pertinent facts or insurance policy provisions relating to any coverages at issue;

(2) Failing to acknowledge and act reasonably promptly upon communications with respect to claims arising under insurance policies;

(3) Failing to adopt and implement reasonable standards for prompt investigation and processing of claims arising under insurance policies;

(4) Failing to affirm or deny coverage of claims within a reasonable time after proof of loss requirements have been completed and submitted by the insured;

(5) Not attempting in good faith to effect prompt, fair and equitable settlements of claims in which liability has become reasonably clear;

(6) Compelling insureds to institute litigation to recover amounts due under an insurance policy by offering substantially less than the amounts ultimately recovered in actions brought by the insureds when the insureds have made claims for amounts reasonably similar to the amounts ultimately recovered;

(7) Attempting to settle a claim by an insured for less than the amount to which a reasonable man would have believed he was entitled by reference to written or printed advertising material accompanying or made a part of an application;

(8) Delaying investigation or payment of claims by requiring an insured claimant with a physician either to submit a preliminary claim report and then requiring the subsequent submission of formal proof of loss forms, both of which submissions contain substantially the same information;

(9) Failing to settle claims promptly where liability has become apparent under one portion of the insurance policy coverage in order to influence settlements under other portions of the insurance policy coverage; and

(10) Failing to provide promptly a reasonable explanation of the basis relied on in the insurance policy in relation to facts or applicable law for denial of the claim or for the offer of a compromise settlement.

As noted above, in *Parvaneh Moradi Shalal v. Fireman's Fund Insurance Co.*, the California Supreme Court held that no private right of action arises under the Unfair Practices Act in favor of third parties. It also held that common-law theories of breach of the implied covenant of good faith and fair dealing still were viable and available to the insured.

A case decided by the court of appeal, *Cates Construction, Inc. v. Talbot Partners*,[216] sheds further light on this controversy. This dispute centered around a performance bond issued by the surety on a condominium project as to which the contractor and the owner had a dispute regarding payment. The owner contended that it had overpaid the contractor, which abandoned the job in December 1990 because the owner did not pay additional sums that the contractor had demanded. The contractor recorded a $645,000 mechanic's lien and then went out of business.

---

[216] 53 Cal. App. 4th 1420 (1997).

In January 1991, the surety told the owner that it believed the owner had materially breached the contract by failing to pay the contractor and that the surety would therefore not arrange for performance of the remainder of the contract. In February 1991, the owner informed the surety that, as a result of delayed project completion, the construction lender was foreclosing on the property. In March 1991, the surety filed a lawsuit, in the name of the contractor, to foreclose the mechanic's lien; the owner cross-complained, seeking recovery on the performance bond and punitive damages for breach of the implied covenant of good faith and fair dealing. The construction lender foreclosed on the property in June 1991.

At trial, the owner was awarded $3.1 million in compensatory damages and $28 million in punitive damages, both against the surety. The court of appeal held that a surety is an insurer and that the obligee on a performance bond can assert a common-law tort cause of action for breach of the implied covenant of good faith and fair dealing. The surety had argued that it was "caught in the middle" of the dispute between the contractor and the owner and that it did not perform under the bond because it did not know whether the contractor or the owner was at fault. The owner argued that the surety had conducted a "sham investigation," had withheld information, and had delayed and deliberately avoided coming to a conclusion, despite the surety's knowledge of the owner's impending loss of the project through foreclosure, so that the surety could avoid its obligations under the bonds and could instead collect on the contractor's mechanic's lien. The court of appeal agreed with the trial court and held that bad faith did exist because the contractor's duties were clear and the contractor's nonperformance was easily ascertainable. Key evidence upon which the court relied included, but was not limited to, the following:

1. Testimony from the owner's accounting expert that a 30-minute review of the disbursements cash journal revealed that the contractor was not owed any additional payments when the contractor abandoned the job

2. Evidence that the surety never made a determination of how much the owner owed the contractor or even how much the contractor had been paid

3. Evidence that the surety filed the lawsuit to foreclose the lien without investigating the validity of the lien, and that the surety could not back up the amount of the lien and had just pulled the $645,000 figure out of thin air

4. Evidence that the surety had access to everything it needed to make a decision, but nonetheless engaged in dilatory fumbling by requesting additional documentation information, much of which had already been provided.

Thus, the surety was held liable for punitive damages for breach of the implied covenant of good faith and fair dealing, which the court of appeal reduced from $28 million to $15 million.

The California Supreme Court has now rendered a decision in the *Cates Construction* case, holding that suit cannot be brought for bad faith on a performance bond.[217]

As the *Mammoth Vista* and *Cates* cases both apply to performance bonds, a question remains as to whether such a bad-faith tort cause of action can also be maintained against a surety that issued a payment bond. In *United States ex rel. Ehmcke Sheet Metal Works v. Wausau Insurance Co.*,[218] a U.S. District Court, applying California law, held that a bad-faith cause of action could not be brought against a surety that had issued a Miller Act payment bond.

Obviously, with regard to a performance bond, the named obligee is the named beneficiary of the bond; its common-law tort cause of action for breach of the implied covenant of good faith and fair dealing was affirmed in *Moradi Shalal* and was further confirmed in the *Cates* decision. Subcontractors and material suppliers who are the beneficiaries of payment bonds are not named obligees on those bonds because, at the time the bond is executed, their names are unknown to the principal and surety on the bond. They are, however, direct beneficiaries of the bond, and clearly, under the *Cates* decision, the surety would owe a duty of good faith to the beneficiaries of the bond to promptly act upon any claims made on the bond. It remains to be seen how the appellate courts will resolve this issue. One author concluded that no published opinion of any California court has overruled the holding in *General Insurance Co. v. Mammoth Vista Owners Ass'n, Inc.* That author likewise concluded that the *Ehmcke* case is not binding on California courts. The final conclusion of the author was that the courts should recognize bond claimants' rights, as insureds or express beneficiaries, to sue sureties who respond to bond claims in bad faith.[219] In light of the holding in *Cates,* it would appear that a tort claim for breach of the implied covenant of good faith and fair dealing seeking punitive damages could not be maintained against the surety on a payment bond.

§ 9.04 SPECIAL PROBLEMS WITH MECHANICS' LIENS

[A] Priority

The general rule is that if the deed of trust is recorded before the construction commences, then it shall have priority over the mechanics' liens, but if work commences before the trust deed is recorded, then the mechanics' liens have priority and all mechanics' liens relate back to the commencement.[220]

Two exceptions to this rule should be noted. First, nonobligatory advances lose their priority.[221] Second, it is the writer's opinion that disbursements made

---

[217] Cates Constr., Inc. v. Talbot Partners, 21 Cal. 4th 28 (1999).
[218] 755 F. Supp. 906 (E.D. Cal. 1991).
[219] *See* Donald A. McIsaac, *Can Construction Bond Claimants Sue Sureties for Bad Faith?*, Cal. Constr. L. Rep., Sept. 1991, at 186, 191.
[220] *See generally* Cal. Civ. Code §§ 3134, 3135, 3136.
[221] *See id.* § 3136.

by a lender for nonconstruction purposes, ahead of the progress of the construction and in violation of its building loan agreement with the owner, are "optional" and therefore lose their priority over mechanics' liens.[222]

Off-site work done under a separate contract with the owner will not give persons doing site work priority even though the off-site work was commenced before the trust deed was recorded.[223]

The definition of an off-site improvement has been the subject of many problems and much controversy in the construction industry. The typical example involves an owner who hires a grading contractor to cut, fill, grade, and compact the land. In doing the work, the grading contractor may work not only on the streets in the subdivision but also on the building pad sites. The street work is clearly off-site, but what about the work on the building pads? If that is done under a separate contract with the owner, will the lien claimants be able to look to that work for priority purposes? There is no satisfactory answer to the question. However, it seems that such work should constitute commencement of the structures for priority purposes.[224]

Off-site work will have priority over the trust deed even if it is done after the trust deed is recorded, unless the lender places the funds for site improvement aside under a binding agreement to apply them solely to site claimants, with none to be distributed to the owner until there is satisfactory evidence that all claims have been paid.[225]

In *Schmitt v. Tri-Counties Bank*,[226] Tri Counties Bank made a $550,000 construction loan secured by a deed of trust on the property. The project ran over budget. The project ran over budget. A contractor who was unpaid for asphalt work recorded a mechanic's lien. The construction loan agreement provided that the bank's obligation to make advances was subject to the owner's obligation to provide mechanic's lien releases, but it did not provide, as required by Civil Code § 3137, that the proceeds of the loan be applied to the payment of mechanic's lien claims and that no portion of the proceeds would be paid to the owner in absence of satisfactory evidence that all mechanic's lien claims had been paid or the time for recording claims of lien had expired and no such lien claims had been recorded. Pay request number 1 contained a line item for payment to the owner marked "mobilization," and the owner used part of that money to pay off a land loan. The construction lender knew that part of the advance would be held to repay the land loan but did not ensure that all mechanic's lien claims had been paid or that the time for recording claims of lien had expired. The trial court held that the mechanic's lien took priority over the deed of trust securing the construction loan. The appellate court affirmed. The court held that the construction loan agreement did not comply with Civil Code § 3137. The court further noted

---

[222] Hunt, *California Mechanic's Lien Law: Need for Improvement,* 9 Santa Clara Law. 101–118 (1968); Hunt, *The Stop Notice Revisited,* 54 Cal. St. B.J. 24 (1979).
[223] Cal. Civ. Code § 3135.
[224] *See* Hopkins, *Selected Mechanic's Lien Priority Problems,* 16 Hastings L.J. 155 (1964).
[225] Cal. Civ. Code § 3137.
[226] 70 Cal. App. 4th 1234 (1999).

that the construction lender had introduced no evidence that it had determined that all mechanic's lien claims had been paid before it disbursed $228,108.68 to the owner. The court followed the normal admonition that the mechanic's lien laws are remedial legislation that must be construed for the protection of laborers and material suppliers. The court further noted that under Civil Code § 3137, the lender is required to control the loan proceeds until all liens have been paid. Otherwise, the trust deed securing the construction loan loses priority to mechanic's lien claims for site work. The court noted that loss of priority is the legislature's intended sanction for a lender's failure to comply with the requirements of Civil Code § 3137. The court noted that the construction lender, by foreclosing on its deed of trust, retained the value of the improvements installed by the lien claimant, and it was equitable for the mechanic's lien claim to take priority over the deed of trust.

In *Barr Lumber Co. v. Shaffer*,[227] the court held that a trust deed recorded prior to commencement of work has priority over a lien, a decision that was followed in *Rheem Manufacturing Co. v. United States*.[228] It should be noted, however, that *Rheem* seemed to indicate in dictum that alleging and proving bad faith or fraud and unjust enrichment might defeat priority. Specifically, the *Rheem* court stated:

> The judgment cannot be sustained on the theory of unjust enrichment. The only finding that relates to the matter, namely, that Rheem paid the fair market value of the property at the trustee's sale, tends to show that there was no unjust enrichment. No California case has been found which has denied priority to a trust deed in a situation similar to the one present here, and the cases relied on by respondents do not support their position. In *City Lumber Co. v. Brown*, the beneficiary under the deed of trust who was denied priority was a joint venturer with the owner and was to share in profits derived from improvements to be made on the land, whereas here the court found that Rheem had no such interest. The language in *Fickling v. Jackman*, indicating that the priority ordinarily given a recorded trust deed might be denied if there was bad faith or fraud, does not assist respondents because here the court found that Rheem acted in good faith throughout the transactions.[229]

In *Tracy Price Associates v. Hebard*,[230] the court held that off-site work done under a separate contract with the owner and performed before the trust deed was recorded did not give an architect's lien priority.[231]

---

[227] 108 Cal. App. 2d 14 (1951).
[228] 57 Cal. 2d 621 (1962).
[229] 57 Cal. 2d at 626–627 (citations omitted).
[230] 266 Cal. App. 2d 778 (1968).
[231] *See also* Walker v. Lytton Sav. & Loan Ass'n, 2 Cal. 3d 152 (1970); McDonald v. Filice, 252 Cal. App. 2d 613 (1967); Design Assocs., Inc. v. Welch, 224 Cal. App. 2d 165 (1964).

Generally, commencement for priority purposes means some work or labor done on the ground,[232] but at least one case has held that the delivery of building materials (to an adjacent lot, no less) was a commencement for priority purposes.[233]

Off-site work done under a separate contract with the owner will not give persons doing site work priority even though the offsite work was commenced before the trust deed was recorded. This is provided in Civil Code § 3135. The case of *Lambert Steel Co. v. Heller Financial Inc.*[234] interpreted this Code section. In *Lambert Steel,* the owner, between 1981 and 1984, had various site improvement activities conducted on the project, including building demolition; removal of trees, brush growth, and other debris; installation of a security fence; drilling of test holes, which were later filled with gravel; and preparation of surveys. The owner did not hire a general contractor for the work and in fact entered into separate contracts with each contractor to perform the work. On October 22, 1984, a deed of trust securing a construction loan was recorded. Two days later, on October 26, 1984, construction on the project began. Ultimately, persons who did work on the construction of the building (a motel and restaurant) did not get paid, recorded mechanics' liens, and brought actions to enforce those liens. The owner defaulted on the loan, and the construction lender foreclosed.

The question before the court was whether the mechanics' liens had priority over the construction loan. The court first of all cited section 3134 of the Civil Code, which provides that mechanics' liens are preferred to any deed of trust that attaches subsequent to the work of improvement. The court then turned to an analysis of Civil Code § 3135, which provides that if any site improvement is provided for in a separate contract, with respect to the erection of residential units or other structures, then the site improvement shall be considered a separate work of improvement, and the commencement thereof shall not constitute a commencement of the work of improvement consisting of the erection of any residential unit or other structure. In other words, if the site improvement work is done under a separate contract from the contract for erection of the residential units or other structures, then the persons who work on the residential units or other structures cannot rely on the work done under the site improvement contracts for priority purposes. The court specifically held that because the building demolition, the removal of the trees, brush, and other debris, the installation of the security fence, the drilling of the test holes, and the preparation of the surveys were all done under separate contracts with the owner, they constituted site improvements under Civil Code § 3135 and therefore did not constitute commencement of work with regard to the erection of structures on the property. As a result, the court held that the mechanics' liens of the claimants who had worked on construction of the structures on the property did not have priority over the construction loan. Their mechanics' liens were wiped out by foreclosure of the construction loan.

---

[232] English v. Olympic Auditorium, Inc., 217 Cal. 631 (1933).
[233] Bank of Italy v. MacGill, 93 Cal. App. 229 (1928).
[234] 16 Cal. App. 4th 1034 (1993).

In a case involving a dispute over priority between mechanic's lien claimants (a construction lender and a subordinated purchase money deed of trust), the court in *Santa Cruz Lumber Co. v. Bank of America*[235] worked out the priority of issues based upon the facts of the case and the reasonable expectations of the parties.

In that case, Western Real Estate Exchange (Western) sold land to Wittenburg Construction, Inc. (Wittenburg) and took back a purchase money deed of trust recorded in October 1978. At that time, the purchase money deed of trust was in a first-priority position. In 1979, various lien claimants furnished materials to Wittenburg, who was constructing condominiums on the project. At that point, the liens were subordinate to the purchase money deed of trust. In December 1979, Bank of America (Bank) made a construction loan secured by a deed of trust, and Western subordinated its purchase money deed of trust to the Bank's deed of trust. The Bank's deed of trust, at that point, was subordinate to the mechanics' liens. The question that obviously arose was whether Western had lost its original priority by subordinating to the construction loan. Wittenburg used one-half of the construction loan funds for materials that had not been supplied on that particular project. The lien claimants brought an action to foreclose their liens and named Wittenburg, the owner, Bank of America, and Western as defendants. The question presented to the trial court was one of priority among those parties. The trial court determined that, under these peculiar facts, the most equitable solution was to give first priority to the mechanic's lien claimants in equal rank; second trust deed priority to the Bank, but only to the amount of its loan that was actually expended on that specific project; third priority to the subordinated second trust deed held by Western; and fourth priority to the remainder of the Bank's loan.

The Bank appealed, and the appellate court affirmed the judgment. The court first noted that neither treatise authors on the subject in California nor the courts have formulated any rule by which the kind of circuity of lien priorities present in this case can be broken. The court stated that courts have considered the expectations, knowledge, and intentions of the parties to arrive at an equitable solution that best represents the benefits of each party's bargains (citing *Community Lumber Co. v. Chute*,[236] *Pike v. Tuttle*,[237] and *W.P. Fuller & Co. v. McClure*[238]). The court noted that when the Bank made its construction loan, it knew that work had already begun on the project and therefore was in the best position to protect itself from mechanics' liens that it knew would have priority over its deed of trust. The court also noted that when the holder of a deed of trust agrees to subordinate its deed of trust to a construction loan, it does so with the understanding that the construction loan will be used to improve the property. The subordinating party is benefited by the subordination agreement because the

---

[235] 160 Cal. App. 3d 858 (1984).
[236] 215 Cal. 268 (1932).
[237] 18 Cal. App. 3d 746 (1971).
[238] 48 Cal. App. 185 (1920).

property will increase in value by the improvements placed on it. Western, therefore, expected that all of the Bank's loan to Wittenburg would go to the improvement of the property. In fact, half of the loan proceeds were used by Wittenburg to pay debts to Santa Cruz Lumber that had been incurred on other projects. As a result, Western should be subordinated only to the amount of the bank's loan that was, in fact, actually used to improve that property. Western should not be penalized because the Bank failed to supervise Wittenburg's use of the loan funds. Western was a sophisticated seller and was therefore on notice that when it subordinated to the Bank's construction loan, there might be mechanics' liens attached to the property. This put Western in a position to protect itself against the liens by obtaining consideration for the subordination and seeking indemnification from Wittenburg against the liens, or something of that nature.

The court, citing *Connolly Development, Inc. v. Superior Court*,[239] noted that California has a long history of protecting the claims of laborers and material suppliers, and that the courts have liberally construed the laws to ensure that protection. The courts have also taken the position that subordinating sellers should be protected, because of their vulnerable position, and have imposed upon lenders the obligation of an implied agreement that the lender's priority extends only to the loan amount properly expended for construction purposes. As a result, the court stated that no hard-and-fast rule can be developed; consequently, the trial court had the discretion within its equitable powers to set the order of payment, which the appellate court confirmed.

In an unusual case, a mechanic's lien was held to have priority over a construction loan where the work was done by a separate contractor on the project. In *Westfour Corp. v. California First Bank*,[240] the owner of the property (Blatte) entered into a construction loan agreement with California First Bank. Blatte executed a deed of trust to secure the construction loan, but California First Bank did not record the deed of trust until July 17, 1984. Before the deed of trust was recorded, Blatte hired Joseph L. Barnes Construction Company (Barnes) as the contractor for the project and let excavation work on the property begin. Thus, the excavation work preceded the recordation of the deed of trust securing the construction lender's construction loan. Blatte's contract with Barnes did not include tenant improvements, although tenant improvements were included in the disbursement schedule to the construction loan agreement between Blatte and the lender. On September 16, 1985, Blatte executed a certificate of substantial completion, indicating that Barnes's work was substantially completed as of that date. Nine months later, on June 25, 1986, another company, U.S. Allied Development Corporation (Allied), entered into an agreement with Westfour Corporation (Westfour) to do the tenant improvement work. The opinion of the court does not specify the relationship between Allied and Blatte; however, the opinion does state as a conclusion that Allied was "acting on behalf of Blatte." The contract between Allied and Westfour named Allied as owner and Westfour as contractor.

---

[239] 17 Cal. 3d 803 (1976).
[240] 3 Cal. App. 4th 1554 (1992).

Westfour performed $450,587 worth of tenant improvement work and was paid $161,447 from Allied, which, in turn, obtained that money from California First with the bank's knowledge that those funds were being spent on tenant improvements. On April 2, 1987, Westfour recorded seven mechanics' liens on the property. On June 11, 1987, it filed a complaint to foreclose those mechanics' liens. The trial court held that Westfour's mechanics' liens had priority over California First's deed of trust by reason of the excavation work done by Barnes prior to recordation of that deed of trust. California First contended on appeal that if Westfour was a general contractor under a separate contract with the owner, then Westfour's mechanics' liens could only relate back to the date when Westfour began work, almost two years after California First's deed of trust was recorded, and therefore California First's deed of trust should retain priority.

The appellate court sustained the judgment of the trial court that Westfour's mechanics' liens had priority over California First's deed of trust. The court stated that the trial judge had found that Westfour's work was part of the same work of improvement commenced by Barnes prior to the recording of California First's deed of trust on July 17, 1984. The appellate court said that that finding was supported by substantial evidence, namely, California First's own witness's admission that tenant improvements were part of the original project and that California First did not consider the project complete until the tenant improvement work was finished. Under those circumstances, the appellate court stated that the trial court properly concluded that the tenant improvements performed by Westfour were part of the same work of improvement commenced by Barnes prior to the recording of California First's deed of trust, citing Civil Code § 3106, which defines a work of improvement as the entire structure or scheme of improvement as a whole.

[B] Separate Works of Improvement

California Civil Code § 3117 provides as follows:

> *Notice of Completion: Work completed pursuant to two or more original contracts.* Where the work of improvement is not made pursuant to one original contract for the work of improvement but is made in whole or in part pursuant to two or more original contracts, each covering a particular portion of the work of improvement, the owner may, within 10 days after completion of any such contract for a particular portion of the work of improvement, record a notice of completion. If such notice of completion be recorded, notwithstanding the provisions of Sections 3115 and 3116, the original contractor under the contract covered by such notice must, within 60 days after recording of such notice, and any claimant under such contract other than the original contractor must, within 30 days after the recording of such notice of completion, record his claim of lien. If such notice is not recorded, then the period for recording claims of lien shall be as provided for in Sections 3115 and 3116.[241]

---

[241] 1969 Laws ch. 1362 § 2; former § 3117 repealed 1963 Laws ch. 819 § 2.

This section was enacted to allow an owner/builder to record a notice of completion after each subcontractor finishes its work, in order to start the lien period running for those in that chain of contracting.

[C] Separate Residential Units

California Civil Code § 3131 provides as follows:

Commencement of running of lien on two or more separate residential units: Status of materials consumed on portion of entire work of improvement. If a work of improvement consists in the construction of two or more separate residential units, each such unit shall be considered a separate "work of improvement," and the time for filing claims of lien against each such residential unit shall commence to run upon the completion of each such residential unit. A separate residential unit means one residential structure, including a residential structure containing multiple condominium units, together with any common area or any garage or other improvements appurtenant thereto. The provisions of this qualification shall not impair any rights conferred under the provisions of Sections 3112 and 3130. Materials delivered to or upon any portion of such entire work of improvement or furnished to be used in such entire work of improvement and ultimately used or consumed in one of such separate residential units shall, for all the purposes of this title, be deemed to have been furnished to be used or consumed in the separate residential unit in which the same shall have been actually used or consumed; provided, however, that if the claimant is unable to segregate the amounts used on or consumed in such separate units, he shall be entitled to all the benefits of Section 3130.

This section of the Civil Code allows an owner of a tract of homes to record notices of completions as houses are completed.[242] If the owner does so, then the lien claimant must record its lien. If the owner does not, does the lien period run from completion of each house or from completion of the tract as a whole? No cases have yet answered this question. The statute, on its face, indicates that the lien period runs from the completion of each house.

[D] Segregation of Lien

California Civil Code § 3130 provides as follows:

Single claim filed against two or more buildings or works of improvement by claimant against sole owner or employer: Procedure. In every case in which one claim is filed against two or more buildings or other works of improvement owned or reputed to be owned by the same person or on which the claimant has been employed by the same person to do his work or furnish his materials, whether such works of improvement are owned by one or more owners, the person filing such claim must at the same time designate the amount due to him on each of such works of improvement; otherwise the lien

---

[242] Cal. Civ. Code § 3131.

§ 9.04[D]   CALIFORNIA CONSTRUCTION LAW

of such claim is postponed to other liens. If such claimant has been employed to furnish labor or materials under a contract providing for a lump sum to be paid to him for his work or materials on such works of improvement as a whole, and such contract does not segregate the amount due for the work done and materials furnished on such works of improvement separately, then such claimant, for the purposes of this section, may estimate an equitable distribution of the sum due him over all of such works of improvement based upon the proportionate amount of work done or materials furnished upon such respective works of improvement. The lien of such claimant does not extend beyond the amount designated as against other creditors having liens, by judgment, mortgage, or otherwise, upon either such works of improvement or upon the land upon which the same are situated. For all purposes of this section, if there is a single structure on more than one parcel of land owned by one or more different owners, it shall not be the duty of the claimant to segregate the proportion of material or labor entering into the structure on any one of such parcels, but upon the trial thereof the court may, where it deems it equitable so to do, distribute the lien equitably as between the several parcels involved.

If the lien claimant can segregate its claim, the claimant should do so, but if the claimant cannot, the claimant is not required to do so.[243]

Flintkote Co. v. Lisa Construction Co.[244] is an example of the liberal interpretation of the lien law. In that case, defendant Lisa Construction Company owned lots 1 through 10 of tract 21769. Lisa hired the Shiffman Company to install concrete foundations on each of the 10 lots. Shiffman purchased materials from the plaintiff for the job as a whole in the sum of $7,108.76. The plaintiff gave an appropriate pre-lien notice and filed its lien on all 10 lots for the full amount. The plaintiff did not segregate among the various lots. Suit was brought on April 16, 1964, naming only Lisa Construction as the owner of the 10 lots. The lis pendens was recorded on May 1, 1964. Lisa Construction had sold five of the lots to individual purchasers, who had recorded their deeds before the complaint was filed. None of these purchasers was named as defendants in the complaint, and the judgment excluded their lots from the lien. The trial court gave the plaintiff a lien for the full amount due and owing to the plaintiff on the remaining five lots. Those five lots had been sold by the defendant Lisa Construction after the complaint was filed, some of them being conveyed before the lis pendens was recorded. Lisa Construction appealed, making various contentions.

The first contention made by Lisa Construction was that, under California Civil Procedure Code §§ 1194.1 and 1195,[245] because the work of improvement consisted of 10 separate, single-family residences on 10 separate lots, the claim of the plaintiff should be equally apportioned among the 10 lots. In response to this contention, the court stated that the only penalty for failure to segregate under

---

[243] Cal. Civ. Code § 3130.
[244] 268 Cal. App. 2d 606 (1968).
[245] Now Cal. Civ. Code §§ 3130, 3131.

§ 1194.1 is that the lien claim is postponed to other liens, and when it is impossible to segregate, there is no responsibility upon the claimant to segregate.[246]

Lisa Construction next contended that the interest of the third parties who purchased the five lots upon which the foreclosure was ordered should be adjudged unaffected by the present action, pointing out that these five purchasers were not made parties to the action. It claimed that when the plaintiff's suit was dismissed as to fictitious defendants at the pretrial, this acted as a cancellation of the effect upon them of the notice of lis pendens. The defendant Lisa Construction pointed out that the deed to lot 1 was recorded before the lis pendens, but after the complaint, and that the deeds to lots 6, 8, 9, and 10 were recorded after the notice of lis pendens was recorded. The court stated that it is necessary to name and serve the purchaser of property who records his title after the lien attaches, but before suit is filed, citing *Packard Bell Electronics Corp. v. Theseus, Inc.*[247] The appellate court pointed out that the judgment does bind the five purchasers who purchased after the recording of the lis pendens but does not affect the purchaser who purchased the lot before the recording of the lis pendens and on that basis affirmed the judgment, stating that if that lot is sold, those owners can bring quiet title or other forms of relief.

The *Flintkote* court held:

- If a contractor furnishes labor or materials on a lump-sum basis among several different lots, the contractor does not have to segregate if it is impossible to do so. Furthermore, the lien can be on the lots that remain if some are sold.

- The plaintiff should name everyone with an interest in the property at the time the suit is filed; otherwise, those parties will be unaffected by the judgment.

[E] Condominiums

Condominiums may consist of many units in one building or groups of units in separate buildings. The Code was not designed to handle this modern construction phenomenon. What is the lien period? Can the owner record separate notices of completion as each unit or group of units is completed? *E.D. McGillicuddy Construction Co. v. Knoll Recreation Ass'n, Inc.*,[248] seems to say the lien period runs from completion of the entire project. This was a condominium project with separate buildings containing condominium units. The owner filed separate notices of completion after each unit was completed. The lien claimant (the prime contractor) filed its lien after the completion of the job as a whole, and the court held the lien timely, stating in effect that California Civil Code § 3131 applies to tract houses only.[249]

---

[246] Kritzer v. Tracy Eng'g Co., 16 Cal. App. 287 (1911); Vowels v. Witt, 149 Cal. App. 2d 257 (1957); Pugh v. Moxley, 164 Cal. 374 (1912); Hendrickson v. Bertelson, 1 Cal. 2d 430, 432 (1934).

[247] 244 Cal. App. 2d 355 (1966).

[248] 31 Cal. App. 3d 891, 897, 898 (1973).

[249] *See also* J. R. Kemper, Annotation, *Enforceability of Single Mechanic's Lien upon Several Parcels against Less Than the Entire Property Liened*, 68 A.L.R.3d 1300, 1308 (1976).

[F] Core Work versus Tenant Work

Many projects are now built in two phases. The core of the building is constructed first, and then as tenants are found, the tenant work is performed. Can the owner file a valid notice of completion after the core work? If the claimant's contract calls for core and tenant work, must he file two liens? There are no cases on this subject. It is the writer's opinion that if the lien claimant has one contract for core and tenant work, he need not file his lien until the work of improvement as a whole has been completed.[250] If, on the other hand, the core work and tenant work are done under separate contracts and the owner files separate notices of completion, then separate liens would have to be filed.

[G] Proof

The lien and stop notice claimant must prove that its labor and materials were furnished for use in the job and that its labor and material were used in the job. The burden is not difficult for a contractor who has itself installed the materials. The problem is difficult, however, for a material supplier whose customer (normally the subcontractor) has become insolvent and has disappeared. The material supplier then finds it difficult, if not impossible, to find witnesses who actually saw the material being installed in the job. Trial judges hold lien claimants to varying degrees of proof. Some hold the claimant to a strict burden of proof. Some will draw the inference that all the material was used in the job if it can be proved that most of it was actually used. Some judges, in the absence of any evidence that material was taken from the job or diverted from one job to another, will draw the inference that the material was used in the job if the claimant can produce sufficient evidence tracing the material from the claimant's business to the jobsite. The latter is the better rule. As a practical fact of the construction industry, once materials get to the jobsite, they are installed, and therefore, the logical inference to be drawn from the proved fact of delivery to the jobsite is incorporation into the job.

Consolidated Electrical Distributors, Inc. v. Kirkham, Chaon & Kirkham, Inc.[251] is an example of the plaintiff doing a very good job of proving its case. In that case, the plaintiff was a supplier of electrical material to an electrical subcontractor on a school job. The plaintiff sued on a stop notice and on a labor and material bond. It introduced into evidence all the subcontractor's invoices showing shipment of the electrical materials to the jobsite. The manager of the plaintiff's office had made a detailed physical inventory of the electrical fixtures after the school had been completed. A certified copy of the blueprints on the job was introduced into evidence. The plaintiff's manager testified that he had reviewed in detail the blueprints, including the electrical site plan and schedules and the

---

[250] *See* Cal. Civ. Code § 3106, which defines a *work of improvement* as the entire structure or scheme of improvement as a whole.

[251] 18 Cal. App. 3d 54 (1971).

fixture schedule, and had made two visits to the school to identify the fixtures on the jobsite. He made special reference to the blueprints, the subcontractor's purchase orders, and the plaintiff's invoices. The essence of his testimony was that the types of fixtures provided for in the blueprints were in the finished school and that they corresponded to the types and amount of fixtures ordered by the subcontractor from the plaintiff. He also testified that he was present on two different occasions when the merchandise was delivered and at various times he saw the subcontractor incorporating the materials into the jobsite. He did not physically inspect every room in the school but assumed from the blueprints that they were the same.

The trial court held for the material supplier, and the bonding company and the prime contractor appealed.

The appellate court said the materials not only must be furnished for or delivered to the site of the particular building but also must have actually been used in the construction. Furthermore, proof of delivery of the materials to the building site does not create a presumption of their use, but the material supplier must prove that the materials were actually used therein. The court held that the above evidence was sufficient to sustain that burden of proof. The court stated that the testimony of the plaintiff's manager was such that actual use could be inferred. Although this court noted that evidence of shipment and delivery of materials to the construction site by itself is not sufficient to support a finding that the materials supplied were actually used in the construction, most judges will infer use if the claimant gets the materials to the jobsite and there is no evidence of diversion.[252]

§ 9.05 DEFENSES TO MECHANICS' LIENS AND STOP NOTICES

[A] Application of Payments

A material supplier will often sell materials to a subcontractor who is working on several different construction projects at the same time. When the subcontractor makes payment to the material supplier with its own check, a question arises as to which of the construction projects should be credited with the payment. The owner has an interest in being certain that payments from its construction project are credited to that particular project by the material supplier.

The interest of the owner in ensuring that payment is credited to the construction project out of which it arises may be satisfied by written communication to the subcontractor and its material supplier indicating that a particular payment arose out of that construction project. However, the realities of the construction industry do not generally allow for such communication. Accordingly, the application of payments made by material suppliers will often be dependent upon any

---

[252] *See* San Pedro Lumber Co. v. Kreis, 111 Cal. App. 466 (1931); H.G. Fenton Material Co. v. Noble, 127 Cal. App. 338 (1932); Arthur v. New House Bldg. Corp., 217 Cal. App. 2d 526 (1963).

communication from the subcontractor to the material supplier regarding the construction project to which credit should be given when a payment is made.

To avoid the vulnerability arising from dependence on the communication from the subcontractor to the material supplier, the owner should issue joint checks to all persons or entities that have provided the owner with a preliminary notice on the project and should note the particular construction project out of which payment arose on the joint check.

If neither a written communication to the material supplier nor payment by joint check has been made, the owner may still insist upon application of payment to its particular construction project under certain circumstances. If the material supplier knows that the payment is from a source of funds connected with a particular construction project, it must apply the payment to that project.[253]

When payment is made to the material supplier directly by the owner, without notation of the particular construction project out of which payment arose, and the material supplier knows that the owner is financing more than one project, the material supplier cannot escape the duty of inquiring of the owner as to the application of the check so received.[254]

When a payment has been made on a lienable obligation, the application of payment as agreed by the subcontractor and material supplier cannot ordinarily later be changed by an agreement between them so as to injuriously affect the rights of the owner.[255] When a subcontractor owes several obligations to a material supplier and makes no designation as to the application of the payment, the material supplier may, in accordance with Civil Code § 1479, apply the payment to the extinction of the oldest obligation.[256]

[B] Releases

Civil Code § 3262 governs releases and waivers of mechanics' liens, stop notices, and bond rights. The section reads as follows:

> *Prohibition against contractual provision for waiver of claims or liens of other persons; Requirements and form of written waiver and release.*
>
> (a) Neither the owner nor original contractor by any term of their contract, or otherwise, shall waive, affect, or impair the claims and liens of other persons whether with or without notice except by their written consent, and any term of the contract to that effect shall be null and void. Any written consent given by any claimant pursuant to this subdivision shall be null, void, and unenforceable unless and until the claimant executes and delivers a waiver and release. Such a waiver and release shall be binding and effective to release the owner, construction lender and surety on a payment bond from claims and

---

[253] Ewing Irrigation Prods. v. Rohnert Park Golf Course, 20 Cal. App. 3d 862 (1973).
[254] Modesto Lumber Co. v. Wylde, 217 Cal. 421 (1933).
[255] Johnston v. Groom, 99 Cal. App. 462 (1929); Sunlight Elec. Supply Co. v. Pacific Homes Corp., 226 Cal. App. 2d 110 (1964).
[256] *See* Hollywood Wholesale Elec. Co. v. Jack Baskin, Inc. 121 Cal. App. 2d 415 (1953), and 146 Cal. App. 2d 399 (1956).

liens only if the waiver and release follows substantially one of the forms set forth in this section and is signed by the claimant or his authorized agent, and, in the case of a conditional release, there is evidence of payment to the claimant. Evidence of payment may be by the claimant's endorsement on a single or joint payee check which has been paid by the bank upon which it was drawn or by written acknowledgment of payment given by the claimant.

(b) No oral or written statement purporting to waive, release, impair or otherwise adversely affect a claim is enforceable or creates any estoppel or impairment of a claim unless (1) it is pursuant to a waiver and release prescribed herein or (2) the claimant had actually received payment in full for the claim.

(c) This section does not affect the enforceability of either an accord and satisfaction regarding a bona fide dispute or any agreement made in settlement of an action pending in any court provided the accord and satisfaction or agreement and settlement make specific reference to the mechanic's lien, stop notice, or bond claims.

(d) The waiver and release given by any claimant hereunder shall be null, void, and unenforceable unless it follows substantially the following forms in the following circumstances:

(1) Where the claimant is required to execute a waiver and release in exchange for, or in order to induce the payment of, a progress payment and the claimant is not, in fact, paid in exchange for the waiver and release or a single payee check or joint payee check is given in exchange for the waiver and release, the waiver and release shall follow substantially the following form:

CONDITIONAL WAIVER AND RELEASE UPON PROGRESS PAYMENT

Upon receipt by the undersigned of a check from _____ (Maker of Check) in the sum of $_____ (Amount of Check) payable to _____ (Payee or Payees of Check) and when the check has been properly endorsed and has been paid by the bank upon which it is drawn, this document shall become effective to release any mechanic's lien, stop notice, or bond right the undersigned has on the job of _____ (Owner) located at _____ (Job Description) to the following extent. This release covers a progress payment for labor, services, equipment or material furnished to _____ (Your Customer) through _____ (Date) only and does not cover any retentions retained before or after the release date for which payment has not been received; extras or items furnished after the release date. Rights based upon work performed or items furnished under a written change order which has been fully executed by the parties prior to the release date are covered by this release unless specifically reserved by the claimant in this release. This release of any mechanic's lien, stop notice, or bond right shall not otherwise affect the contract rights, including rights between parties to the contract based upon rescission, abandonment, or breach of the contract, or the right of the undersigned to recover compensation for furnished labor, services, equipment, or material covered by this release if that furnished labor, services, equipment or material was not compensated by the progress payment. Before any recipient of this document relies on it, said party should verify evidence of payment to the undersigned.

Dated: _____

_____ (Company Name)

By _____ (Title)

(2) Where the claimant is required to execute a waiver and release in exchange for, or in order to induce payment of, a progress payment and the claimant asserts in the waiver it has, in fact, been paid the progress payment, the waiver and release shall follow substantially the following form:

UNCONDITIONAL WAIVER AND RELEASE UPON
PROGRESS PAYMENT

The undersigned has been paid and has received a progress payment in the sum of $_____ for labor, services, equipment or material furnished to _____ (Your Customer) on the job of _____ (Owner) located at _____ (Job Description) and does hereby release any mechanic's lien, stop notice, or bond right that the undersigned has on the above referenced job to the following extent. This release covers a progress payment for labor, services, equipment or materials furnished to _____ (Your Customer) through _____ (Date) only and does not cover any retentions retained before or after the release date; extras furnished before the release date for which payment has not been received; extras or items furnished after the release date. Rights based upon work performed or items furnished under a written change order which has been fully executed by the parties prior to the release date are covered by this release unless specifically reserved by the claimant in this release. This release of any mechanic's lien, stop notice, or bond right shall not otherwise affect the contract rights, including rights between the parties to the contract based upon a rescission, abandonment, or breach of the contract, or the right of the undersigned to recover compensation for furnished labor, services, equipment or material covered by this release if that furnished labor, services, equipment or material was not compensated by the progress payment.

Dated: _____

_____ (Company Name)

By _____ (Title)

Each unconditional waiver in this provision shall contain the following language, in at least as large a type as the largest type otherwise on the document:

"NOTICE: THIS DOCUMENT WAIVES RIGHTS UNCONDITIONALLY AND STATES THAT YOU HAVE BEEN PAID FOR GIVING UP THOSE RIGHTS. THIS DOCUMENT IS ENFORCEABLE AGAINST YOU IF YOU SIGN IT, EVEN IF YOU HAVE NOT BEEN PAID. IF YOU HAVE NOT BEEN PAID, USE A CONDITIONAL RELEASE FORM".

(3) Where the claimant is required to execute a waiver and release form in exchange for, or in order to induce the payment of, a final payment and the claimant is not, in fact, paid in exchange for the wavier and release or a

single payee check is given in exchange for the waiver and release, the waiver and release shall follow substantially the following form:

CONDITIONAL WAIVER AND RELEASE UPON FINAL PAYMENT

Upon receipt by the undersigned of a check from _____ in the sum of $_____ payable to _____ and when the check has been properly endorsed and has been paid by the bank upon which it is drawn, this document shall become effective to release any mechanic's lien, stop notice, or bond right the undersigned has on the job of _____ (Owner) located at _____ (Job Description). This release covers the final payment to the undersigned for all labor, services, equipment or material furnished on the job, except for disputed claims for additional work in the amount of $_____. Before any recipient of this document relies on it, the party should verify evidence of payment to the undersigned.

Dated: _____

_____ (Company Name)

By _____ (Title)

(4) Where the claimant is required to execute a waiver and release in exchange for, or in order to induce payment of, a final payment and the claimant asserts in the waiver it has, in fact, been paid the final payment, the waiver and release shall follow substantially the following form:

UNCONDITIONAL WAIVER AND RELEASE UPON FINAL PAYMENT

The undersigned has been paid in full for all labor, services, equipment or material furnished to _____ (Your Customer) on the job of _____ (Owner) located at _____ (Job Description) and does hereby waive and release any right to a mechanic's lien, stop notice, or any right against a labor and material bond on the job, except for disputed claims for extra work in the amount of $_____.

Dated: _____

_____ (Company Name)

By _____ (Title)

Each unconditional waiver in this provision shall contain the following language, in at least as large a type as the largest type otherwise on the document:

"NOTICE: THIS DOCUMENT WAIVES RIGHTS UNCONDITIONALLY AND STATES THAT YOU HAVE BEEN PAID FOR GIVING UP THOSE RIGHTS, THIS DOCUMENT IS ENFORCEABLE AGAINST YOU IF YOU SIGN IT, EVEN IF YOU HAVE NOT BEEN PAID. IF YOU HAVE NOT BEEN PAID, USE A CONDITIONAL RELEASE FORM".

The purpose of Civil Code § 3262 is to provide the "rules of the game" with regard to the use of releases in the construction industry. The rules are as follows:

1. The owner and prime contractor cannot by any term of their contract waive the lien rights of others.

2. A lien claimant can waive his lien rights in writing.

3. If the claimant is not paid, he should execute only a conditional release in the form set out in the statute.

4. If the claimant has, in fact, been paid, he should sign an unconditional release.

These forms were enacted in an attempt to make it clear what is and what is not being released by the execution of the waiver and release forms set forth in the statute. The language sets forth the following rules with regard to what is and what is not released:

1. The release covers a progress payment for labor, services, equipment, or materials furnished through a certain date only.

2. The release does not cover any retentions retained before or after the release dates.

3. Extras furnished before the release date for which payment has not been received are not released.

4. Extras or items furnished after the release date are not being released.

5. Work performed on items furnished under a written change order, which has been fully executed by the parties prior to the release date, is covered by the release, unless specifically reserved by the claimant in the release form itself.

6. The release does not affect the contract rights, including rights between the parties to the contract, based upon rescission, abandonment, or breach of the contract.

7. The release does not affect the right of the claimant to recover compensation for furnished labor, services, equipment, or material covered by the release if that furnished labor, services, equipment, or material was not compensated by the progress payment.

Civil Code § 3262, in its present form, was the result of the lobbying efforts of construction trade associations in response to the decision in *Halbert's Lumber v. Lucky Stores, Inc.*[257] The resulting legislation, unfortunately, creates more problems than it solves. Before considering those problems, it is important to understand the historical context and the impact of the *Halbert's Lumber* decision.

[1] The Historical Context

More than a century ago, owners attempted to eliminate the constitutionally created mechanic's lien rights of subcontractors and material suppliers by inserting waiver clauses in construction contracts. The validity of these waiver

---

[257] 6 Cal. App. 4th 1233, 8 Cal. Rptr. 2d 298 (1992).

provisions was debated in California courts.[258] The 1885 legislature ended that debate by enacting former Civil Procedure Code § 1201 (now Civil Code § 3262), which permitted the waiver of mechanics' liens only upon the claimant's written consent. In response, some owners and contractors demanded written lien waivers from subcontractors and material suppliers before allowing them to do work.

In *Bentz Plumbing & Heating v. Favaloro*,[259] the court of appeal declared the use of lien waivers to induce payment null and void. In response, the 1984 legislature amended section 3262 to specify the forms of waivers and release set forth in the main text: (1) Conditional Waiver and Release Upon Progress Payment; (2) Unconditional Waiver and Release Upon Progress Payment; (3) Conditional Waiver and Release Upon Final Payment; and (4) Unconditional Waiver and Release Upon Final Payment.

The conditional version of the release allowed the subcontractor or material supplier to provide a release to the owner to induce payment, without being prevented from asserting its lien if payment was never made. The unconditional release protected the owner and general contractor from lien claims regardless of actual payment.

The language of the statutory releases raised serious practical problems. Many in the construction industry came to regard the statutory release as applying only to that labor, services, equipment, or material actually paid for by the progress payment in question. Literally, however, the progress payment release language appeared to release any claim for labor, services, equipment, or materials "furnished" to the project through the release date.

[2] The *Halbert's Lumber* Decision

The issue was settled when *Halbert's Lumber* ruled that by signing the statutory form of Conditional Waiver and Release Upon Progress Payment, a material supplier waived all of its lien rights for all materials "furnished" through the date specified in the release, including material delivered to the job but not included in the amount of the progress payment. The court's ruling made it clear that the term *furnished* meant delivered to the jobsite and that it included all labor, services, equipment, and materials provided to the project before the release date, whether or not the claimant had invoiced for the work.

Halbert's Lumber raised difficult questions. Did the exchange of a release for a progress payment also release claims for added labor, services, equipment, and materials attributable to change orders, extra work, delay, disruption, and acceleration? *Halbert's Lumber* could be interpreted to preclude such claims, if the extra work was furnished prior to the release date. In contrast, it was argued that the statutory forms did not release unpaid, pending claims.

Halbert's Lumber put material suppliers in jeopardy because they were often not aware of manufacturer shipments direct to the jobsite; according to *Halbert's Lumber,* suppliers' releases covered such direct shipments. Moreover, owners

---

[258] *See, e.g.,* Brown v. Aubrey, 22 Cal. 566 (1863). *Cf.* Whittier v. Wilber, 48 Cal. 175 (1874).
[259] 128 Cal. App. 3d 145, 180 Cal. Rptr. 223 (1982).

asserted that the statutory forms released claims for delay, disruption, and acceleration claims on the ground that these extras were furnished prior to the release date. The authors understand that several trial courts agreed with this interpretation and, based on *Halbert's Lumber,* denied claims for extra work.

In response to *Halbert's Lumber,* some lawyers recommended that their clients modify the statutory forms to exclude extras and materials "furnished but not invoiced"—but the Code required the release language to "follow substantially" that specified in the statute. No one was certain how a court would interpret a modified release form, and some owners, lenders, and general contractors refused to accept such modifications. The need for a legislative amendment was readily apparent.

[3] *Halbert's Lumber* Developments

It should come as no surprise, in the wake of the *Halbert's Lumber* decision and the 1994 amendments to Civil Code § 3262 in response thereto, that new litigation would occur seeking to sort out the relationship of the case to the new statutory provisions. The substance of the court of appeal's decision in *J.A. Jones Construction Co. v. Superior Court*[260] is foreshadowed in the second paragraph: "While the Legislature clearly was 'responding' to the decision, reports of *Halbert's Lumber's* death are somewhat exaggerated." The court went on to state that

> [n]othing in the actual changes in the statutory text makes them retroactive and the Legislature passed up the perfect opportunity to undo *Halbert's Lumber* regarding the old forms: at the same time the Legislature was considering changing the existing release forms, *another* bill was introduced declaring that the modifications were to "reinforce what the Legislature originally intended." That particular bill did not pass.
>
> In the face of the actual text of the recent changes—which by their terms operate only prospectively—and the lack of a clear statement from the Legislature that the recent modifications merely *clarified,* rather than *changed,* existing law, we conclude that *Halbert's Lumber* remains applicable to the old forms. Having made that determination, the instant writ petition, which depends on the demise of *Halbert's Lumber* as it pertains to those forms, must be denied.[261]

The court in *J.A. Jones* was considering two lien waivers signed by the general contractor after substantial completion of a $29 million hotel project. The releases, on the old forms, covered payments received through August 15, 1990, and September 15, 1990. There was no dispute that only $5,000 worth of labor and materials were furnished after the September 15 date; however, "Jones claimed that work changes meant it was owed $3 million more than the contract price and filed a mechanic's lien for an unpaid $1.876 million."

---

[260] 27 Cal. App. 4th 1568, 33 Cal. Rptr. 2d 206 (4th Dist. 1994).
[261] 27 Cal. App. 4th at 1571.

Following a trial, a judge determined that Jones was owed a net $803,396 from the owner and that Jones's mechanics' lien had priority over the bank's first trust deed because the bank had not recorded its deed until after Jones had commenced its work. The judge also concluded, however, that "the two lien waivers had released everything Waterfront owed Jones except for the $5,000 worth of work done after the September 15 release date."

Before judgment could be entered, Waterfront filed for Chapter 11 bankruptcy, and the bank requested the court to expunge the lis pendens on the property. The bank's motion was heard and granted one day after the legislature passed Senate Bill 934, amending the release provisions of Civil Code § 3262. Citing this legislation as new facts, Jones moved for reconsideration of the bank's motion. When reconsideration was denied, Jones sought a writ of mandate from the court of appeal.

Jones was clever enough not to argue merely that the new amendments were retroactive, which they clearly were not. Rather, Jones contended that the amendments were made to clarify what the legislature had always intended and thus constituted a legislative repudiation of the court's decision in *Halbert's Lumber*. After a lengthy discussion of the legislative history, the court concluded:

> The history of the legislation effecting the most recent changes to section 3262 of the Civil Code demonstrates the Legislature's "response" to *Halbert's Lumber* was *not* one of simply declaring that this court got it all wrong and that the *old* release forms must be interpreted as they were originally meant to be—to provide for release only to the extent of actual payment. *If the Legislature had wanted to do that, it would have passed AB 1845.* Rather, those who did not like the result in *Halbert's Lumber* acquiesced in a compromise, the end product of which was a set of *new* forms which, in the zero-sum game of mechanics' lien releases, tilt in favor of contractors rather than lenders or owners. It is the *compromise* that the Legislature wanted, not what one set of interests wanted but could not get.[262]

[4] Analysis of the New Release Forms

It is also readily apparent that the new forms contain many changes that create new problems.

[a] *Definitions*

Halbert's Lumber spilled much ink defining *furnished* and *pro tanto*. The term *pro tanto* has been removed from the Code, and the term *furnished* remains. *Halbert's Lumber* defined *furnished* as simple delivery. The court found that the term *pro tanto* did not restrict the release to items covered by the progress payment. Therefore, the new release "covers a progress payment for labor, services, equipment or materials furnished to [the customer] through [the release date]" and releases "any mechanic's lien, stop notice, or bond right to the following extent," that is, the extent not covered by the exceptions.

---

[262] *Id.* at 1582.

[b] Exceptions to the Release

Retentions, extras, and items. The new form excepts "any retentions retained before or after said date; extras furnished before the release date for which payment has not yet been received; extras or items furnished after the release date." The new language specifically indicates that retentions before and after the release date are not included. This seems to make sense, in that clearly no one intends to release claims for the retention by executing releases in exchange for progress payments.

Next, the new release excludes "extras furnished before the release date for which payment has not been received; extras or items furnished after the release date." The term *extras* is not defined in the statute. *Webster's New Riverside University Dictionary* (1984) defines the noun *extra* as "[s]omething more than the expected, usual, or necessary." Extras then must be something more than is necessary or contemplated under the original contract. They are excluded from the release. Thus, it would seem that claims for additional labor, services, equipment, and materials caused by delay, disruption, and acceleration are not released.

But what about the *Halbert's Lumber* problem? The materials supplied in *Halbert's* were not extras. They were part of the materials furnished under the original contract, "furnished before the release date." Such materials continue to be released under the new forms. This analysis is bolstered by the next phrase, excluding "extras or items furnished after the release date."

The fact that "items" are only excluded when furnished after the release date, while "extras" are excluded when furnished both before and after the release date, indicates that the *Halbert's Lumber* problem has not been addressed in this part of the release form.

Written change orders included, unless specifically reserved. "Rights based upon work performed or items furnished under a written change order which has been fully executed by the parties prior to the release date are covered by this release unless specifically reserved by the claimant in this release." This clause sets a trap for the unwary. Literally read, this language would release a fully performed, but not yet invoiced, written change order by the acceptance of a progress payment that does not include payment for the change order. It is incredible to the authors that lien and stop notice rights for work performed pursuant to a legitimate and approved written change order should be lost merely because they were not specifically reserved in a progress payment release. Yet, read literally, this is what the statute says. Contractors and subcontractors must reserve unpaid written change orders from the progress payment release form, if the change order work was performed before the release date.

Contract rights not released.

This release of any mechanic's lien, stop notice, or bond right shall not otherwise affect the contract rights, including rights between parties to the contract based upon a rescission, abandonment, or breach of the contract, or the right of the undersigned to recover compensation for furnished labor, services, equipment, or material covered by this release if that furnished labor, services, equipment or material was not compensated by the progress payment.

This convoluted, run-on sentence has two potential consequences. First, the clause clearly states that the claimant is not releasing contract rights or remedies against those with whom the claimant directly contracted. This is of great significance to the owner, general contractor, and subcontractor. No longer can an owner accept only a statutory release from the general contractor, nor can general contractors accept only a statutory release from subcontractors. They should additionally require releases of personal (i.e., contract) rights. An alternative would be to modify the statutory form to also exclude personal (contract) rights. Doing so, however, could invalidate the effectiveness of the release, as section 3262(b) requires that the release language "substantially follow" that set out in the statute.

Second, it is debatable whether this clause solves the concerns of material suppliers arising from the *Halbert's Lumber* decision. The first part of the clause is clear: "This release of any mechanic's lien, stop notice, or bond right shall not otherwise affect the contract rights." This is followed by a complex enumeration of what contract rights are not released:

> including rights between the parties to the contract based upon a rescission, abandonment, or breach of the contract, or the right of the undersigned to recover compensation for furnished labor, services, equipment, or material covered by this release if that furnished labor, services, equipment, or material was not compensated by the progress payment.

This murky enumeration invites speculation as to whether the form (1) does not release the right to recover compensation for labor, services, equipment, or materials under the contract, or (2) does not release mechanic's lien, stop notice, or bond rights by which the claimant may "recover compensation" for the labor, services, equipment, or materials not included in the progress payment. It will be argued that the right to recover compensation includes the right to enforce a mechanic's lien, stop notice, or bond. It will also be argued, however, that because the release, by its nature, releases mechanic's lien, stop notice, and bond rights, the only remedy reserved to recover compensation is the personal/contract right.

[c] *Inherent Ambiguity of the Unconditional Release Form*

The amendments to the Unconditional Release track the amendments to the Conditional Release, but one provision of the unconditional form enhances the ambiguity. The unconditional form provides: "THIS RELEASE IS ENFORCEABLE AGAINST YOU IF YOU SIGN IT EVEN IF YOU HAVE NOT BEEN PAID." But the preceding language of the form specifically states that some claims that are not compensated for by the progress payment are not released! Thus, the unconditional release form is inherently ambiguous—a fact evidently lost upon the lawmakers.

[5] Recommendations and Conclusions

In light of the preceding, the authors have the following recommendations:

1. The legislature should make another effort to address the ambiguities in the statutory language and to address *Halbert's Lumber.*

2. Anyone accepting a statutory release from a lower tier should also obtain a release of personal (contract) rights.

3. Contractors and subcontractors should expressly exclude from the progress payment release form any fully executed written change order not compensated by the progress payment.

4. Material suppliers should expressly state on the release form that the release is limited to materials invoiced through the release date.

The court of appeal has decided the issue of whether, under the *Halbert's Lumber* and *J.A. Jones* decisions, releases signed by a contractor also apply to retention amounts withheld from progress payments for which section 3262 releases have been provided. In *Western Landscape Construction v. Bank of America,*[263] the trial court relied upon dictum in the *Halbert's Lumber* decision to hold that a plaintiff contractor's section 3262 releases also served to waive its stop notice and mechanic's lien rights to the retention amounts withheld from progress payments as to which the releases had been provided.

The court of appeal reversed and remanded the matter to the trial court to be resolved in light of its holding that the discussion of retention in *Halbert's Lumber* was dictum (and therefore not controlling legal authority) and that the discussion of "work changes" in *J.A. Jones* did not concern the subject of retention. In its review of the *Halbert's Lumber* decision, the *Western Landscape* court noted that the section 3262 release includes a sentence that the form "does not cover retention or items furnished after said date" and concluded that the discussion of that sentence in *Halbert's Lumber* was not only "somewhat enigmatic" but also unnecessary to the decision in that case: "Retention payments were not involved in *Halbert's,* and whatever *Halbert's* has to offer on the subject of retention is [therefore] not binding precedent."[264]

In reaching its decision, the court in *Western Landscape* noted that an owner or lender could not confirm whether materials had been delivered to a project "except by an inspection of the site," whereas the amount of retention withheld "can be determined by mere examination of documents. Thus the concern of *Halbert's* that owners and lenders be able to rely on the release without inspecting the site is not implicated in the case of retention."[265] In the final analysis, the court of appeal declined to force upon a contractor a choice to "either forego all

---

[263] 58 Cal. App. 4th 57 (1997).
[264] *Id.*
[265] *Id.*

progress payments (thus converting the retention rate into an effective 100%), or forego its constitutional rights to security for the retained amounts."[266]

[C] Effect of Assignment for Benefit of Creditors

In a decision that promises to put California subcontractors on the horns of a dilemma when dealing with financially troubled general contractors, the court of appeal required a subcontractor to repay a general contractor's assignee for the benefit of creditors the funds received in exchange for the release of the subcontractor's mechanic's lien.[267] An assignment for the benefit of creditors is a state court procedure analogous to filing for reorganization under the Federal Bankruptcy Act.

The court of appeal held that the payment, made within 90 days prior to the general contractor's making of the assignment for the benefit of creditors, was an improper preference because "release of the mechanic's lien did not constitute 'new value' to the general contractor under the contemporaneous exchange exception to the rule allowing an assignee for benefit of creditors to recover preferential payments."[268]

Although recognizing the subcontractor's constitutional right to a mechanic's lien, the court held that the subcontractor had been "paid" and that its "express release" of its mechanic's lien remained binding. Thus, the subcontractor, though stripped of its payment, was left with no recourse against the owner or the subject property.

Remarkably, the court suggested that its decision might have been different if the subcontractor had been paid directly by the owner rather than by the general contractor. Besides elevating form over substance, this suggestion ignores the fact that the subcontractor has little control over the payment process between the owner and the general contractor. It is hard to imagine that a subcontractor could successfully decline payment from the party with whom it contracted and insist on receiving payment from a third party. It is also highly problematic whether a subcontractor could assert a mechanic's lien after having refused the general contractor's tender of payment in full, regardless of the general contractor's financial condition.

The opinion is rife with awareness of the injustice of its result upon the subcontractor but offers nothing better than the suggestion that it might be possible to protect the subcontractor "in other ways, through other legal instruments." Rather than suggesting what those other ways might be, the opinion concludes meekly that the legislature ought to change the law. Until that happens, what can a subcontractor do to protect itself when it suspects its payments are being made by someone in financial trouble? Obtaining early legal assistance may permit the

---

[266] *Id.*

[267] Angeles Elec. Co. v. Superior Court, 27 Cal. App. 4th 426, 32 Cal. Rptr. 2d 660 (2d Dist. 1994).

[268] 27 Cal. App. 4th at 428–429, 32 Cal. Rptr. 2d at 662.

[D] Restrictive Endorsements

payment transaction to be restructured to preserve the mechanic's lien or to defeat the claim of preference by receiving payments directly from the owner.

At the end of a construction project, there are often disputes between owners and contractors, contractors and subcontractors, and subcontractors and material suppliers over the "amount due" for various reasons. Sometimes the owner, contractor, or subcontractor will be claiming offsets for delay, defective workmanship, and other causes. It is not unusual for an owner, contractor, or subcontractor to tender a check to the creditor for an amount less than the claim of the creditor and to mark the check as "payment in full" or similar language.

In 1987, the legislature enacted California Civil Code § 1526, which provides that where a claim is disputed or unliquidated, and a check is tendered by the debtor in settlement of the claim and the words "payment in full" or other words of similar meaning are notated on the check, the acceptance of the check does not constitute an "accord and satisfaction" if the creditor totally protests against accepting the tender in full payment by striking out or otherwise deleting that notation or if the acceptance of the check was inadvertent or without the knowledge of a notation. In other words, if there was a dispute between the creditor and the debtor, and the debtor tendered a check with the notation "payment in full," all the creditor had to do was strike out those words and could go ahead and sue the debtor for the balance that was in dispute, notwithstanding the cashing of the check. This Code section was enforced in the courts. For example, in the case of *Red Alarm, Inc. v. Waycrosse, Inc.*,[269] a dealer sent a check to a manufacturer, stating that it was payment in full of the manufacturer's claim. The check was deposited in the manufacturer's lockbox, the manufacturer being unaware of the statement that the check was being tendered as payment in full. When the manufacturer determined that the check had such restrictive language, it promptly notified the dealer that it did not consent to the terms upon which the check was offered. The court held that the manufacturer was not bound by the release language. Furthermore, in the case of *In re Van Buren Plaza LLC*,[270] the bankruptcy court held that the debtor properly protected itself under this Code section by sending a letter to the maker of the check, indicating that it was accepting the check only in part payment of obligations between the parties, and therefore, cashing the check with the words "full payment" on it did not bind the creditor. Specifically, in that bankruptcy case, the court noted that the California statute authorized a creditor to retain and use the check while still preserving its rights against the maker of the check even when the check contains "payment in full" language because the statute was designed to prevent the debtor from making payment on the debt, which payment is clearly a minimal amount owed to the creditor, which forces the creditor into a compromise of an otherwise legal entitlement.

---

[269] 47 F.3d 999 (9th Cir. 1995).
[270] 200 B.R. 384 (Bankr. C.D. Cal. 1996).

In 1992, the Uniform Commercial Code was amended to provide in section 3311 that if a person against whom a claim is asserted proves that (1) that person in good faith tendered an instrument to the claimant as full satisfaction of the claim, (2) the amount of the claim was unliquidated or subject to a bona fide dispute, and (3) the claimant obtained payment of the instrument, the claim is discharged if the person against whom the claim is asserted proves that the instrument or accompanying written communication contained a conspicuous statement to the effect that the instrument was tendered as full satisfaction of a claim. This provision of the Uniform Commercial Code is in conflict with and contrary to Civil Code § 1526. This dilemma was presented to a court in 1998. In the case of *Directors Guild of America v. Harmony Pictures, Inc.*,[271] a check was tendered to the Directors Guild of America, which check was marked "full and final settlement for the audit period 6/1/90 to 5/31/94." The Directors Guild of America crossed out that language before cashing the check and mailed Harmony Pictures, Inc., a letter stating that the payment was not a full settlement. The suit by Directors Guild of America was for pension and health fund contributions plus interest, audit fees, and liquidated damages. The court looked at the familiar rule of statutory construction that states that where there is a conflict between statutes, they should be reconciled if reasonably possible. The court, in effect, recognized that these two statutes could not be reconciled. The court then turned to another rule of statutory interpretation to the effect that a statute enacted later in time should prevail. The court concluded therefore that because Uniform Commercial Code § 3311 was enacted later than Civil Code § 1526, Uniform Commercial Code § 3311 should prevail. Therefore, Directors Guild of America's cashing of the check marked "full and final settlement" constituted an "accord and satisfaction" and prevented Directors Guild of America from recovering the balance in dispute.

In light of the foregoing, the only safe practice that can now be followed where there is a dispute between a debtor and a creditor, and the debtor submits a check to the creditor marked "payment in full" or similar language, is for the creditor to return the check. If the creditor strikes out the "payment in full" language and cashes the check, it will be faced with the defense that there was an accord and satisfaction, and at least one court has held that Uniform Commercial § 3311 would preclude recovery. Until there is further clarification by the courts, the only safe practice would be to return the check.

[E] Economic Duress

Although the case did not involve an interpretation of California Civil Code § 3262, it has been held that under certain circumstances, a release may not be enforceable. In *Rich & Whillock, Inc. v. Ashton Development, Inc.*,[272] Bob Britain, Inc., the prime contractor on a project owned by Ashton Development, Inc., subcontracted the excavation work to Rich & Whillock, Inc. When Rich & Whillock

---

[271] 32 F. Supp. 2d 1184 (C.D. Cal. 1998).
[272] 157 Cal. App. 3d 1154 (1984).

submitted its final invoice for $72,286, Britain informed Rich & Whillock that there was no money left for the final payment. Rich & Whillock responded that as a new company, heavily indebted, it would go bankrupt if it did not receive the final payment. The president of Bob Britain, Inc., offered $50,000 in exchange for a full release and said, "I have a check for you and just take it or leave it. This is all you get. If you don't want this, you have got to sue me." Rich & Whillock accepted the check and signed the release, and four months later sued Bob Britain, Inc., for the difference. Confronted with the unequivocal release, Rich & Whillock argued that it was a product of economic duress and therefore unenforceable. The court stated it was reluctant to set aside any agreement but would not enforce an agreement that resulted from one party's coercive use of its superior bargaining position. The court noted that the doctrine of economic duress had been in existence for at least 50 years. Thus, the court held that the release was unenforceable and allowed Rich & Whillock to recover.

A similar result was obtained in *Centric Corp. v. Morrison Knutsen Co.*[273]

[F] Joint Checks

The general rule is that when there is no communication between the owner and material supplier regarding apportionment of the proceeds of a joint check prior to endorsement, the material supplier will be deemed paid the full face amount of joint checks endorsed, regardless of the amount it actually retains.[274]

The focus of a number of cases that have reiterated the general rule, however, has been on exceptions to the rule. In *Rodeffer Industries, Inc. v. Chambers Estates, Inc.*,[275] the court stated that the issuance of a joint check payable to the material supplier and subcontractor in payment for labor and materials for specifically designated lots does not require that the material supplier credit the entire proceeds of such check to lots other than those designated. Accordingly, the material supplier was not deemed paid the entire proceeds of joint checks endorsed when he retained less than those proceeds; he himself had paid in full for materials delivered to the lots designated on the joint check and allowed the subcontractor to retain the balance of the proceeds.

In *Bohannan Bros. v. Lo Jean Development Co.*,[276] the joint checks had attached stubs designating the specific work of improvement for which each check was issued. Each check was attached to the subcontractor's billing showing the work of improvement for which the check constituted payment. The subcontractor, however, removed the stubs and his billings from the joint checks before presenting them to the material supplier for endorsement. Instead, he presented the material supplier with his own check together with the material supplier's invoices to the subcontractor that he was paying by means of his own check. The

---

[273] 731 P.2d 411 (Okla. 1986).
[274] Post Bros. Constr. Co. v. Yoder, 20 Cal. 3d 1 (1977).
[275] 263 Cal. App. 2d 116 (1968).
[276] 3 Cal. App. 3d 200 (1969).

material supplier then endorsed the joint check of the owner over to the subcontractor in exchange for the subcontractor's check, which of course was always less than the owner's check. The court ruled that the material supplier was bound to follow the owner's designation and therefore held that the material supplier was deemed paid in connection with the specifically designated work of improvement.

In *Post Bros. Construction Co. v. Yoder,*[277] the supreme court stated that checks jointly payable to three or more persons constitute payment of the full face amount of the check to each of the payees in the absence of a communication between the issuer of the check and the payees. The court of appeals had held that when there were three or more payees, such payees need only make application of payment for those sums actually received out of the proceeds of the joint check. The court of appeals reasoned that when there are multiple payees not in privity of contract with one another, there is no way any one of them is going to know what is owed to the other by the common debtor; accordingly, the only practical procedure is for them all to endorse the joint check and turn it back to the common debtor, leaving it to him to pay them what is owed. The supreme court, however, held that the number of payees had no impact on the general rule pertaining to joint checks.

Owners may easily communicate to material suppliers the amount of materials being paid by a joint check by designating the subcontractor invoice or describing the work being paid on the joint check itself. Here, owners face a dilemma: they generally do not wish the material supplier to be deemed paid only a portion of the joint check if that is all that is being paid for materials by that joint check because the general rule would protect owners to a greater extent in the absence of a designation. However, absence of such a designation will often lead to expensive litigation. Accordingly, owners are well advised to designate the particular lots for which payment is made or the subcontractor's invoice that induced payment and to recognize that only a portion of the proceeds of the joint check will be retained by the material supplier. In addition, the owner should hold sufficient funds from the subcontractor to be sure that the balance of the material supplier's claim will be paid out of future joint checks.[278]

[G] Unclean Hands: Fraud by the Lien Claimant

In *Burton v. Sosinsky,*[279] Burton was a paving subcontractor to Katotakis, who in turn had a contract to build a car dealership for Sosinsky, the owner. Burton would invoice Katotakis for his work, and Katotakis would then submit a

---

[277] 20 Cal. 3d 1 (1977).

[278] For other cases on payment and joint checks, see *Michel & Pfeffer v. Oceanside Properties, Inc.,* 61 Cal. App. 3d 433 (1976); *Phillips Construction Co. v. Argonaut Insurance Co.,* 77 Cal. App. 3d 575 (1978); *Westwood Building Materials Co. v. Valdez,* 158 Cal. App. 2d 107 (1958); *Hollywood Wholesale Electric Co. v. John Baskin, Inc.,* 121 Cal. App. 2d 415 (1953); *Re-Bar Contractors, Inc. v. City of Los Angeles,* 219 Cal. App. 2d 134 (1963); *Petaluma Building Materials, Inc. v. Foremost Properties, Inc.,* 180 Cal. App. 2d 83 (1960); *J.S. Schirm Co. v. Rollingwood Homes Co.,* 56 Cal. App. 2d 789 (1961).

[279] 203 Cal. App. 3d 569 (1988).

marked-up billing to Sosinsky, the owner. The owner would then pay by joint check to Burton and Katotakis. The joint checks always exceeded the amount due Burton because Katotakis had marked up Burton's billings. Burton would deposit the checks and refund the excess to Katotakis.

On one occasion, Burton received a check for $45,000 from Katotakis; it was made payable to Burton and was to replace a joint check to M.J. Rudy & Sons. Sosinsky's project was known as Fireside Dodge, but someone had written the word "Dodge" over the word "Toyota" on the check. The job next to the name "Yosemite Asphalt and Paving" (Burton's fictitious name) had also been altered. Burton was told by Katotakis to credit a Toyota project with this $45,000 instead of the Fireside Dodge paving job owned by Sosinsky. After contacting Mr. Sosinsky and informing him of those events, Burton applied the $45,000 to the Toyota job. Burton then brought an action to foreclose his mechanic's lien. The owner defended on the basis of *unclean hands,* claiming that Burton and Katotakis had entered into a conspiracy to defraud the owner.

Katotakis testified that he had altered the invoices from Burton without Burton's knowledge because in the contract with Sosinsky the line item for paving was more than Katotakis's contract with Burton. When Sosinsky provided joint checks in excess of Burton's invoice amounts, Burton was required to return the excess to Katotakis. When Sosinsky and Burton talked about the matter on the phone, Burton explained, "I know it sounds illegal, but it is a very common practice in the construction business." On the basis of this evidence, the trial court determined that Burton had conspired with Katotakis to defraud Sosinsky. The trial court stated that although there was no evidence specifically showing that Burton directly benefited from the conversion of Sosinsky's money, it could reasonably be inferred that a long-range benefit to Burton was expected by reason of his cooperation with Katotakis in that there was a continued business relationship between Katotakis and Burton that financially benefited Burton. Burton appealed, contending that an equitable defense such as unclean hands had no place in the mechanic's lien law. The court held just the opposite—that Burton had forfeited his lien rights by engaging in the aforesaid conduct with Katotakis.

§ 9.06 MISCELLANEOUS POINTS

[A] Removal of Materials: Right to Lien

Generally, removal of materials after incorporation into the structure will not eliminate lien rights.[280]

Destruction by fire will not destroy a lien right.[281] On the other hand, if the lien claimant himself removes his equipment, he loses his lien right.[282]

---

[280] Pacific Sash & Door Co. v. Bumillier, 162 Cal. 664 (1912); Johnson v. Smith, 97 Cal. App. 752 (1929).
[281] Butler v. Ng Chung, 160 Cal. 435 (1911); Humboldt Lumber Mill Co. v. Crisp, 146 Cal. 686 (1905).
[282] Cornell v. Sennes, 18 Cal. App. 3d 126 (1971). In the following cases, the courts addressed the issue of loss of lien rights: *Fraters Glass & Paint Co. v. Southwestern Construction Co.,* 200

[B] Voidable Preferences

The circuit court in *Keenan Pipe & Supply Co. v. Shields*[283] held that payments made to a material supplier within four months of bankruptcy are not a voidable preference because waiver of job rights affords present consideration for the payment.[284]

Keenan Pipe & Supply v. Shields is cited for the proposition that payments made within four months of bankruptcy, in exchange for waiver of job rights, does not constitute a voidable preference. The cases cited in footnote 284 are to the same effect. Another question that often arises when a bankruptcy ensues is whether claimants can file liens and/or stop notices in light of the bankruptcy proceeding and whether the mechanic's lien or the stop notice is a voidable preference.

In *In re KDR Building Specialties Inc.*,[285] the debtor, KDR Building Specialties Incorporated, was a subcontractor on a construction project. Squires-Belt supplied materials to KDR. Squires-Belt was KDR's largest creditor, with a claim of $129,000. KDR filed a Chapter 11 proceeding on April 24, 1987. After the filing of the Chapter 11 proceeding, Squires-Belt filed and served stop notices on construction lenders and owners of property upon which construction was being performed by KDR. Squires-Belt also recorded preliminary notices of mechanics' liens against owners on all jobs in which KDR was delinquent in paying Squires-Belt. KDR sought an order holding the material supplier in contempt for violation of the automatic stay. The court stated that the issue in the case was whether a post-petition filing of a stop notice pursuant to California Civil Code § 3156 violated the automatic stay. The court noted that under California law, subcontractors and material suppliers may serve a stop notice upon the owner or a construction lender and that, upon receipt of the stop notice, the owner or construction lender must withhold from the general contractor sufficient money to pay the stop notice claimant. If the owner or construction lender fails to withhold sufficient funds, as required by the stop notice, it becomes personally liable to the stop notice claimant, notwithstanding the absence of privity of contract. The stop notice operates as a lien against the unexpended balance of the loan and survives foreclosure of the deed of trust securing the loan. The court also noted that California law provides a separate and distinct remedy for subcontractors and

---

Cal. 688 (1927); *Santa Cruz P.C. Co. v. Snow Mountain Water & Power Co.*, 96 Cal. App. 615 (1929); *Ferry v. Ohio Farmers Insurance Co.*, 211 Cal. App. 2d 651 (1963); *Powers Regulator Co. v. Seaboard Surety Co.*, 204 Cal. App. 2d 338 (1962); *Bonded Products. Co. v. R.C. Gallyon Construction Co.*, 228 Cal. App. 2d 186 (1964). *But see* Card Constr. Co. v. Ledbetter, 16 Cal. App. 3d 472 (1971) (prime contractor who paid subcontractor in reliance on material supplier's release was entitled to sue material supplier for indemnity).

[283] 241 F.2d 486 (9th Cir. 1956).

[284] Arnold Braid Elec. Co., 908 F.2d 52 (6th Cir. 1990); O'Rourke v. Seaboard Sur. Co., 887 F.2d 955 (9th Cir. 1989); E.R. Fegert, Inc. v. Coral Constr., Inc., 88 B.R. 258 (Bankr. 9th Cir. 1988), *aff'd sub nom.* O'Rourke v. Seaboard Sur. Co., 887 F.2d 955 (9th Cir. 1989); *In re* Flooring Concepts, Inc., 37 B.R. 957 (Bankr. 9th Cir. 1984); Bel Marin Drywall v. Groves, 470 F.2d 934 (9th Cir. 1972).

[285] 76 B.R. 778 (Bankr. S.D. Cal. 1987).

material suppliers in the form of a mechanic's lien, which constitutes a direct lien on the improvement and the real property to the extent of the interest of the owner or person who caused the improvement to be made.

Squires-Belt argued that construction funds in the hands of the owner or the construction lender do not become property of the subcontractor's estate, and therefore the automatic stay does not apply. In *In re Flooring Concepts, Inc.*,[286] Shaw Industries supplied carpeting to the debtor for installation in an apartment complex. The debtor failed to pay Shaw, and Shaw filed a preliminary 20-day notice under Civil Code § 3097. The general contractor paid Shaw directly within 90 days of the debtor's filing of a Chapter 11 petition. The bankruptcy court held that those payments were preferences, but on appeal the bankruptcy appellate panel reversed, relying upon *Keenan Pipe & Supply v. Shields*. The court cited Civil Code § 3167(b), which provides, in effect, that if the monies withheld pursuant to stop notices are insufficient to pay all claims in full, then the claimants share on a pro rata basis. The court stated that because Civil Code § 3167(b) allows stop notice claimants to share in the distribution of funds from a construction lender without regard to the order in which the stop notices were filed, it therefore appeared that the California legislature intended the stop notice to relate back to the time when the work was first performed.

The *KDR* court concluded that, given the strong legislative desire, as illustrated by the California mechanic's lien law and as noted by the California Supreme Court in *Connolly Development, Inc. v. Superior Court*,[287] the California stop notice scheme is intended to relate back to the date that services are first performed. Therefore, a stop notice may be perfected under § 546(b) of the Bankruptcy Code without violating the automatic stay provisions of § 362(a) of the Bankruptcy Code.

The voidable preference rule was also analyzed in the case of *Angeles Electric Co. v. Superior Court*.[288] That action arose in a proceeding by an assignee for the benefit of creditors of a general contractor to set aside a voidable preference under the provisions of the state law governing assignments for the benefit of creditors. A payment had been made by the general contractor to a subcontractor that met all of the elements of voidable preference under Civil Procedure Code § 1800. The subcontractor, which had provided a typical conditional waiver and release upon final payment pursuant to Civil Code § 3262, defended the action on the basis that the payment by the contractor to the subcontractor constituted "new value" by reason of the fact that the subcontractor had given up its mechanic's lien rights in consideration of the payment. The trial court granted the assignee's motion for summary adjudication of that issue in its favor, and the subcontractor sought review by a prerogative writ. The court of appeal denied the subcontractor's application for a peremptory writ of mandate, holding that the payments made by the contractor to the subcontractor were preferences recoverable by the assignee for the benefit of creditors and that the subcontractor's

---

[286] 37 B.R. 957 (Bankr. 9th Cir. 1984).
[287] 17 Cal. 3d 803 (1976).
[288] 27 Cal. App. 4th 426 (1994).

surrender of its mechanic's lien rights against the owner's property on account of that payment did not result in "new value" to the contractor.

Subsequent to this decision, the Civil Procedure Code section relating to assignments for the benefit of creditors was amended so as to specifically provide, in § 1800(c)(7), that such a payment is not a voidable preference:

> (c) The assignee may not recover under this section a transfer: . . .
>
> (7) That is payment to a claimant as defined in Section 3085 of the Civil Code, in exchange for the claimant's waiver or release of any potential or asserted claim of lien, stop notice or right to recover on a payment bond or any combination thereof.

Thus, under the assignment for benefit of creditors law in the state of California, a payment in exchange for a claimant's waiver or release of any potential or asserted claim of lien, stop notice, or right to recover on a payment bond, or any combination thereof, is no longer a voidable preference. The corresponding statute under the Federal Bankruptcy Act has not been so amended.

The *Angeles Electric* court did distinguish other cases that would not constitute a voidable preference. Specifically, the *Angeles Electric* case held that the case of *Keenan Pipe & Supply Co. v. Shields*,[289] continues to be good law. In the *Keenan Pipe* case, a subcontractor arranged to have its debt to a supplier satisfied by payments from the general contractor directly to the supplier. That was accomplished by joint checks issued by the general contractor, payable to the subcontractor and the supplier, which the subcontractor endorsed over to the supplier. The subcontractor then filed bankruptcy, and the trustee sought to have the payments set aside as voidable preferences under the Bankruptcy Act. The court held there was no voidable preference because the general contractor had dedicated a specific sum to pay the supplier, and those dedicated funds could not be reached by anyone else. The *Keenan* court also held that the money used by the general contractor to pay the subcontractor never became a part of the subcontractor's estate because it was a joint check payable to the subcontractor and the material supplier. Also, the payment was made as the result of a new agreement supported by new consideration (the material supplier's forbearance) or a tripartite agreement in which new consideration was furnished.

The *Angeles Electric* court also commented on the decision of the Ninth Circuit Bankruptcy Appellate Panel in *In re Flooring Concepts, Inc.*,[290] in which a subcontractor failed to pay a supplier on a construction project. An arrangement was made pursuant to which the general contractor would pay the supplier, who would refrain from enforcing its mechanic's lien rights. After payments were made on that basis, the subcontractor filed bankruptcy. The court held that the payments were not voidable preferences, by reason of the fact that no money was ever paid by the bankrupt subcontractor. Relying upon the *Keenan Pipe* case, the court in *Flooring Concepts* concluded that payments made by a contract debtor

---

[289] 241 F.2d 486 (9th Cir. 1956).
[290] 37 B.R. 957 (Bankr. 9th Cir. 1984).

of a bankrupt to a creditor of the bankrupt do not become a part of the bankruptcy estate where there is an independent obligation on the part of the debtor to pay the creditor.

The *Angeles Electric* court then commented upon the case of *In re Anderson Plumbing Co.*,[291] which involved a plumbing contractor who had a contract to do plumbing work on a public school. A supplier that sold materials to the plumbing contractor for use on the job perfected its stop notice and bond claims against the project. The general contractor made a payment to the supplier during the preference period, and the plumbing contractor then filed bankruptcy. When the trustee in bankruptcy sought to recover the payment as a voidable preference, the plumbing supplier conceded that the payment was a voidable preference, and the only issue was whether there was a contemporaneous exchange of "new value." The court held that the payment was not a voidable preference, concluding that, just as a direct payment by the owner to a supplier of a contractor does not result in a voidable preference as to the contractor, neither should a direct payment by the contractor to a supplier result in such a preference. In each case, the contractor's creditor surrenders lien rights against third parties (the owner and owner's retention fund or a surety), resulting in a contemporaneous exchange for new value.

In reaching its decision, the *Anderson Plumbing* court had relied on *In re Dick Henley, Inc.*[292] and *In re Advanced Contractors*.[293] However, the *Angeles Electric* court chose not to follow *Anderson Plumbing, Dick Henley,* or *Advanced Contractors*.

Another case commented upon by the *Angeles Electric* court was *In re E.R. Fegert, Inc.*,[294] which involved payments by a contractor to subcontractors who, in turn, released unsecured claims against the contractor and against the surety on the contractor's payment bond. The *Fegert* court had reasoned that, as the holder of a contingent claim, the surety was a creditor of the debtor and held an equitable lien on proceeds due the contractor on a Miller Act project. The debtor paid the subcontractors under a tripartite arrangement, by which the subcontractors released their claims against the surety in exchange for direct payments by the debtor/contractor. The court observed that, if the surety had paid the subcontractors directly, there would have been no preference, and concluded that the result should not be different because they were paid by the debtor instead. The *Angeles Electric* court noted that while the *Fegert* decision might be explained as one of the tripartite agreement cases, which do not result in a voidable preference, the fact that the debtor made the payments and the court's statement that "new value" was given in a substantially contemporaneous exchange obviate the basis of that distinction. However the *Angeles Electric* court stated that the case was one of the few that hold, in a situation analogous to *Angeles Electric*, that there is no voidable preference.

---

[291] 71 B.R. 19 (Bankr. E.D. Cal. 1986).
[292] 38 B.R. 210 (Bankr. M.D. La. 1984).
[293] 44 B.R. 239 (Bankr. M.D. Fla. 1984).
[294] 88 B.R. 258 (Bankr. 9th Cir. 1988).

Angeles Electric has, in effect, been overruled with regard to assignments for the benefit of creditors by the amendment to Civil Procedure Code § 1800(c)(7), as noted above. It remains to be seen whether the federal courts, in a voidable preference action by a trustee in bankruptcy, will follow the *Angeles Electric* case or the contrary cases, referenced above, that the *Angeles Electric* court declined to follow. At the federal level, there is a split of authority, and, as noted above, there are cases that hold that payment made by a debtor to a creditor in exchange for a waiver of job rights would not constitute a voidable preference. It also remains to be seen whether federal courts in the Ninth Circuit will follow the *Angeles Electric* case or the cases that *Angeles Electric* reviewed and rejected.

Another defense to a claim of voidable preference is that the payment was made "in the regular course of business" of the relationships between the parties. A case illustrating that defense is *National Lumber & Supply, Inc. v. Orange Commercial Credit*,[295] in which National Lumber & Supply, Inc. (the debtor) was in the retail hardware and home improvement business and hired installers who installed garage doors for National Lumber's customers. Orange Commercial Credit was in the business of purchasing or factoring accounts receivable. The installers factored or assigned to Orange Commercial Credit the receivables owing from National Lumber. Within 90 days prior to filing a Chapter 11 bankruptcy case, National Lumber gave Orange Commercial Credit 11 checks totaling $33,533. During that 90-day period, the debtor had also received services from the installers with a total value of $21,935. The trustee in bankruptcy brought an action against Orange Commercial Credit to recover the payments as voidable preferences. Orange Commercial Credit, in response to a motion for summary judgment brought by the trustee in bankruptcy, raised two defenses under the Bankruptcy Code: first, that the payments were made in the ordinary course of business and therefore were an exception under Bankruptcy Code § 547(c)(2); and second, that there was new value given, falling within the exception contained in Bankruptcy Code § 547(c)(4).

The bankruptcy court determined that the transfers made within 45 days of invoicing were within the ordinary course of business and therefore not voidable preferences under § 547(c)(2), leaving only $3,981 in liability. The bankruptcy court also concluded that Orange Commercial Credit had given new value in the amount of $1,939, pursuant to § 547(c)(4) and the other subsequent advances of $17,493, having been repaid by unavoidable transfers. As a result, judgment was rendered in favor of the trustee for $2,042, representing the difference between the $3,981 in liability and the $1,939 of new value. The court first considered the defense that the payments were made in the ordinary course of business and therefore fell within the exemption of § 547(c)(2). The United States Bankruptcy Appellate Panel for the Ninth Circuit stated that to qualify for the ordinary course of business exception, a creditor must prove that (1) the debt and its payment are ordinary in relation to past practices between the debtor and that particular

---

[295] 184 B.R. 74 (Bankr. 9th Cir. 1995).

creditor; and (2) the payment was ordinary in relation to prevailing business standards, citing *In re Grand Chevrolet, Inc.*[296] and *Foo Catering & Housing, Inc.*[297] The trustee contended that Orange Commercial Credit had failed to present any admissible evidence as to the course of payment between the debtor and Orange, or the applicable industry standard. Since the bankruptcy court had found that all the transfers made on account of invoices dated within 45 days of the payment were within the ordinary course of business, the Bankruptcy Appellate Panel looked at the evidence on that issue. The court noted that the only evidence as to the course of payment between the debtor and Orange was the testimony of David Harden, a garage door installer who, in his declaration, stated that as long as he did business for National Lumber, it was in the ordinary course of his business relationship that National Lumber would make payment on an average within 45 days from the date of the invoice. With regard to the industry standard, Mr. Harden testified that, during the same time period in which he performed garage door installation services for National Lumber, he also performed the same kind of services for Lumber City, Builders Emporium, Builders Square, and Builders Discount; and with those other businesses, he followed the same pattern and was paid by them on undisputed invoices on an average of within 45 days from the date of the invoice. He further testified that Builders Square and Builders Discount often did not pay until 60 days from the date of the invoice and that it was still a common practice in the industry to follow the payment pattern set forth above.

The court also looked at the new value defense under § 587(c)(4). The court referred to the Ninth Circuit case of *In re IRFM, Inc.*,[298] which had ruled on the issue of whether new value that has been repaid by the debtor may be used as a preference defense under § 547(c)(4). The court noted that the Ninth Circuit held that interpreting § 547(c)(4) as requiring creditors' payments to remain unpaid is an incomplete description of the statute. Rather, a creditor who raises the new value defense must prove that the new value has not been repaid with an otherwise unavoidable transfer. The Bankruptcy Appellate Panel noted that Orange Commercial Credit could include as new value all payments it made to the debtor within 90 days of the bankruptcy, to the extent that those payments did not exceed the amount of prior preferential transfers from the debtor. Under the construction of the new value exception as provided for in the *IRFM* case, Orange Commercial Credit could use $19,432 of new value it gave to the debtor as an offset against the preferential payments of $33,533 received from the debtor, leaving Orange liable for a total of $14,101 in preferential payments. The court concluded that, because Orange Commercial Credit had not provided any "admissible evidence" in support of its ordinary course of business defense, the summary judgment should have been granted to the trustee on that issue.[299]

---

[296] 25 F.3d 728 (9th Cir. 1994).
[297] 971 F.2d 396 (9th Cir. 1992).
[298] 52 F.3d 228 (9th Cir. 1995).
[299] *See also* Hunt, *The Ordinary Course of Business Defense to Voidable Preference Claims in the Construction Industry*, 8 Cal. Constr. L. Rep. No. 2, at 33–39 (Feb. 1998).

[C] Are Wall-to-Wall Carpets Lienable?

[1] In General

The following cases address the issue of wall-to-wall carpeting:[300]

Shelley v. Kofka.[301] The plaintiff furnished and installed wall-to-wall carpeting. The carpeting was secured to the floor with tackless stripping and was nailed around the floor casing. When the plaintiff filed a mechanic's lien and went to trial, this evidence was introduced. The defendant/owner moved for a nonsuit, taking the position that the plaintiff was not entitled to a lien for wall-to-wall carpeting affixed by the tackless strip method of installation. The trial court agreed with the defendant owner and granted the nonsuit.

When the carpet company appealed, the court of appeals said that it was error for the trial court to grant the nonsuit, that the evidence was sufficient to require the trial court to determine the controlling fact: Was the carpet affixed to and was it a part of the real property or was it removable personal property? The case was sent back to the trial court to rule on that question. Thus, the court did not make a firm holding one way or the other as to whether wall-to-wall carpeting was lienable.

Plough v. Petersen.[302] Around the perimeter of concrete floors, wooden strips were affixed from which tacks were sticking up and wall-to-wall carpeting was placed thereon. This was obviously the tackless strip method of installation. The carpets could be removed by pulling them up. The carpets were purchased on a conditional sales contract, and a chattel mortgage relating to the carpets had been recorded prior to the recordation of the first deed of trust. (This case arose before the Uniform Commercial Code came into effect. Today, in order to have a security interest in the carpets, one would need a security agreement together with a financing statement filed in Sacramento.)

The plaintiff in this case was the beneficiary of the first deed of trust. He had not discussed nor seen the carpeting before the trust deed was recorded. The defendant was the owner of the property, not the carpet installer. He testified that he intended to have the carpets remain in the premises for the period of their usefulness.

The defendant/owner defaulted on the first trust deed, and the plaintiff/beneficiary of the first trust deed foreclosed. The defendant owner removed the wall-to-wall carpeting and transferred it to another house. The plaintiff sued to recover it, contending that it was a fixture and part of the realty and therefore part of the security for his first deed of trust. The plaintiff lost in the trial court and appealed.

---

[300] *See also* Dean Vincent, Inc. v. Redisco, Inc., 373 P.2d 995 (Or. 1962); Fell v. Masseroff, 145 So. 2d 238 (Fla. Dist. Ct. App. 1962); Merchants & Mechanics Fed. Sav. & Loan v. Harold, 201 N.E.2d 237 (Ohio Ct. App. 1964). *Dean Vincent* is the only case of the group that held carpeting was not a fixture but instead was removable and therefore not lienable.

[301] 107 Cal. App. 2d 827 (1951).

[302] 140 Cal. App. 2d 595 (1956).

§ 9.06[C]   CALIFORNIA CONSTRUCTION LAW

The appellate court affirmed the trial court judgment, saying in effect that the carpets were not fixtures and not security for the loan of the plaintiff, that they were personalty, and that the owner had the right to remove them. In arriving at this decision, the appellate court made some interesting points. It stated that whether an object in a house is a fixture or personalty is a question of fact to be determined by considering several factors: the manner of its annexation, its adaptability to the purpose for which the realty is used, and the intention of the party making the annexation.

The court went on to state that whether the rugs were fixtures or personalty under the facts of this case is subject to reasonable difference of opinion. Therefore, because the judgment of the trial court that the carpeting was personalty and therefore removable and not a fixture was a factual question and subject to reasonable difference of opinion, it was binding upon the appellate court.

Finally, the court stated that in determining the intent of the owner in this case, there was a conflict of testimony. On the one hand, the owner testified that he intended the carpets to remain in place during their usefulness. Contrary to this, they were sold on a conditional sales contract and could easily be removed, and the owner had granted a chattel mortgage on them.

Finley Gordon Carpet Co. v. Bay Shore Homes, Inc.[303] The plaintiff carpet company installed the carpets in the defendant's apartments by the tackless strip method. The apartments were designed so that the carpets were interchangeable between rooms in an apartment and between apartments. There was no molding above the carpeting. The carpets could be removed without damaging the carpets, the pads, or the floor.

The plaintiff, who did not have a specialty contractor's license to install the carpeting, sued for the price of the carpeting. The defendant was defending on the grounds that the plaintiff was unlicensed and therefore not entitled to recover. The plaintiff won in the trial court, and the defendant appealed. On appeal, the judgment was affirmed. The court cited *Shelley v. Kofka*[304] and then stated that because the evidence showed that the carpets could be easily removed without damaging the apartments, this supported the trial court's finding that they did not become a fixed part of the structure. Therefore, under California Business and Professions Code § 7045, the plaintiff was exempt from licensing.

At the time the case was decided, section 7045 provided that the licensing law did not apply to the sale or installation of any finished products that did not become a fixed part of the structure. That section has since been amended and now reads as follows:

> This chapter does not apply to the sale or installation of any finished products, materials or articles of merchandise, which do not become a fixed part of the structure, nor shall it apply to a materialman or manufacturer furnishing finished products, materials, or articles of merchandise who does not

---
[303] 247 Cal. App. 2d 131 (1966).
[304] 107 Cal. App. 2d 827 (1951).

install or contract for the installation of such items. The term "finished products" shall not include installed carpets.

Larkin v. Cowert.[305] The plaintiff sold wall-to-wall carpets and drapes to a joint venture developing an apartment building. The original owners of the property had sold it to the joint venture, taken back a second mortgage, and subordinated it to the construction loan. The carpets were installed by the tackless strip method, and the drapes were installed by fitting hangers to the walls at the windows and then hanging the drapes on a rod that could be lifted off the hangers. When the joint venture defaulted on the second mortgage, the prior owners of the property foreclosed and bought it at the foreclosure sale. The plaintiff sued for conversion of the carpets and drapes and lost in the trial court.

The court cited with approval *Dean Vincent, Inc. v. Redisco, Inc.*[306] and distinguished *Plough v. Petersen.*[307] The court stated that the primary question was: What was the intent of the parties when the carpets were installed? The court held that because they were installed in an apartment building in order to make the apartments more rentable and were intended to stay there during their useful life, the carpets were therefore fixtures and passed to the owners' second deed of trust when they foreclosed, and there was no chattel mortgage or conditional sales contract on the carpets and drapes.

People v. Custom Craft Carpets, Inc.[308] In this case, the attorney general and the city attorney brought an action against a carpet seller, alleging unfair competition and false and misleading statements concerning the selling and installation of wall-to-wall carpeting. The carpet company had secured the amount owing to it by taking trust deeds on the buyers' real property. The attorney general sought to have the trust deeds rescinded and to enjoin the carpet company from using them in the future. The court held that carpets installed by the tackless strip method are not attached to real property and, therefore, would not be eligible to be the subject of a deed of trust.

[2] Conclusion

In light of this discussion, it is clear that the question as to whether wall-to-wall carpets can be the subject of a mechanic's lien is still an unsettled question in the state of California. It is the opinion of the authors that if the wall-to-wall carpets constitute the finished floor and are intended to remain in the building until worn out or remodeled, then they should be the subject of a mechanic's lien.

---

[305] 263 Cal. App. 2d 27 (1968).
[306] 373 P.2d 995 (Or. 1962).
[307] 140 Cal. App. 2d 595 (1956).
[308] 159 Cal. App. 2d 676 (1984).

[D] Unreasonable Delay in Payment/Pay-If-Paid Clauses Unenforceable

Many contracts contain a clause stating that a subcontractor will not be paid until the contractor is paid by the owner. Unless it is clearly spelled out that receipt of payment by the contractor from the owner is a condition precedent to payment, such a clause may be unenforceable.[309] The California courts have now held that "pay if paid" clauses in subcontracts are unenforceable.[310]

[E] Diverting Funds

It is a crime for a contractor to fail to pay for labor or materials willfully when he has received money from the owner to do so.[311]

In *People v. Stark*,[312] a contractor received progress draws from the owners on a construction project. The contractor diverted the funds from that job to pay bills that he had incurred on other jobs. Ultimately, the owners suffered a loss of approximately $46,000 as a result of the contractor's diversion of funds.

At trial, the contractor admitted that he had used the money from the project to defray costs on some other jobs but said he always intended to pay the money back. The contractor contended that, to be convicted of a crime under California Penal Code § 484(b), he had to have a specific intent, at the time he diverted the funds, not to pay the subcontractors and material suppliers.

The contractor was convicted of violating Penal Code § 484(b), which provides that any person who receives money for the purpose of paying for services, labor, materials, or equipment, and willfully fails to apply such money for such purpose, by either willfully failing to complete the improvements for which the funds were provided, or by willfully failing to pay for services, labor, materials, or equipment provided incident to the construction project, and wrongfully diverts the funds to a use other than that for which the funds were received, is in violation of Penal Code § 484(b). The court specifically held that Penal Code § 484(b) was a general intent criminal statute and that, to violate the statute, all that is required is the wrongful diversion of funds, which means not applying the funds for the purpose for which they were disbursed. The diversion could be the cause of at least one of the described failures—that is, failure to complete or failure to pay subcontractors and suppliers. The court stated that the violation of Penal Code § 484(b) is complete if the wrongful diversion of the funds was the cause of a failure either to complete the improvement or to pay for services, labor, materials, or equipment. The court noted that it was immaterial whether

---

[309] Yamanishi v. Bleily & Collishaw, Inc., 29 Cal. App. 3d 457 (1972).

[310] *See* Wm. R. Clarke Corp. v. Safeco Ins. Co. of Am., 15 Cal. App. 4th 882 (1997); Capital Steel Fabricators, Inc. v. Mega Constr. Co., 58 Cal. App. 4th 1049 (1997).

[311] Cal. Penal Code § 484(b), held to be constitutional in People v. Howard, 70 Cal. 2d 618 (1969).

[312] 26 Cal. App. 4th 1179 (1994).

the contractor intended that there be a failure either to complete the project or to pay the subcontractors and material suppliers.

[F] Lien Law Liberally Construed

The mechanic's lien right of a material supplier in the state of California is a right guaranteed by the California constitution, and the legislature has been given a constitutional mandate to provide for the speedy and efficient enforcement of said lien right. The purpose of both the constitutional provision and the statute setting up the procedure for the enforcement of the lien right is to ensure that those persons whose material has contributed to a work of improvement are paid the reasonable value thereof. To hold otherwise would allow the owner to obtain the benefit of the material without compensation and result in unjust enrichment of the owner at the expense of the lien claimant. The legislature is obligated to devise a reasonable scheme for the enforcement of the constitutional right of lien, and the scheme should not be so cumbersome or ultra-technical as to impair or unduly hamper the exercise of that constitutionally guaranteed right.[313]

The following quotations amply support the conclusion that the mechanic's lien law must be liberally construed to effectuate its purpose:[314]

Sunlight Electric Supply Co. v. McKee.[315]

The mechanics' lien law, including the stop notice provisions, is an integrated and harmonious scheme, and applicable code sections must be construed together. It "is remedial in character, and should be liberally construed in its entirety with a view to effect its objects and to promote justice." (*Hendrickson v. Bertelson,* 1 Cal. 2d 430, 432 [35 P.2d 318]; *Nolte v. Smith,* 189 Cal. App. 2d 140, 144 [11 Cal. Rptr. 261, 87 A.L.R.2d 996].)

Bentz Plumbing & Heating v. Favaloro.[316]

[T]he courts have uniformly classified the mechanics' lien law as remedial legislation, to be liberally construed for the protection of laborers and materialmen. (Fn. omitted.) (*Connolly Development, Inc. v. Superior Court,* 17 Cal. 3d 803, 826–827 (1976) [132 Cal. Rptr. 477, 553 P.2d 637]; *Hendrickson v. Bertelson,* 1 Cal. 2d 430, 432 (1934) 35 P.2d 318].)

---

[313] Hammond Lumber Co. v. Moore, 104 Cal. App. 528 (1930); English v. Olympic Auditorium, Inc., 217 Cal. 631 (1933); Jones v. Great S. Fireproof Hotel Co., 86 F. 370 (1898); Martin v. Becker, 169 Cal. 301 (1915); Hammond Lumber Co. v. Barth Inv. Corp., 202 Cal. 606 (1927); Hendrickson v. Bertelson, 1 Cal. 2d 430 (1934); Bay Lumber Co. v. Pickering, 120 Cal. App. 163 (1932); E.D. McGillicuddy Constr. Co. v. Knoll Recreation Ass'n, Inc., 31 Cal. App. 3d 891 (1973); Hunt, *California Mechanic's Lien Law; Need for Improvement,* 9 Santa Clara Law. 101 (1968).

[314] *See also* Connolly Dev., Inc. v. Superior Court, 17 Cal. 3d 803, 807–808, 825–829 (1976); Industrial Asphalt, Inc. v. Garrett Corp., 180 Cal. App. 3d 1001, 1006–1009 (1986).

[315] 226 Cal. App. 2d 47, 50 (1964).

[316] 128 Cal. App. 3d 145, 148, 149 (1982).

Brown Co. v. Superior Court.[317]

The statute is a remedial statute, adopted in obedience to the requirements of the constitution (art. XX, sec. 15), and is to be liberally construed in furtherance of the purposes for which it was authorized. The persons for whose benefit the statute is enacted are not presumed to be versed in the niceties of pleading, and the notices, which under its provisions they are authorized to give, have regard to substance rather than form. The terms of the section clearly indicate that it was not the intention of the legislature that in the claim of lien which he files for record the claimant shall state the name of the real owner, at the risk of losing his lien if it shall turn out that he was in error. The provisions therein that the claimant shall give the "name of the owner or reputed owner, if known," implies that, if he does not know the name of the owner, he may state this fact, and perfect his lien without naming an owner, and, also, that *if, in good faith he gives the name of a reputed owner, he shall not lose his lien if he shall afterward ascertain that some other person was the owner.* . . . In either case [whether the person is being designated as owner or reputed owner], *it is only the opinion of the claimant upon matters that are not presumptively within his knowledge, but which he has formed from external information;* and in that respect the notice which he is to file differs from a pleading in which a fact essential to a recovery must be definitely averred. (*Corbett v. Chambers, supra,* 109 Cal. 178, 184–185; emphasis added.)

Truestone, Inc. v. Simi West Industrial Park II.[318] Holders of mechanics' liens are protected by constitutional mandate.

The mechanic's lien derives from the California Constitution itself; the Constitution of 1879 mandated the Legislature to grant laborers and materialmen a lien upon the property which they have improved; no other creditors' remedy stems from constitutional command. (See *Martin v. Becker* (1915) 169 Cal. 301, 316 [146 P. 665].) Indeed this state, from the earliest days, and consistently thereafter has asserted its interest in protecting the claims of laborers and materialmen. In 1850 the first session of the California Legislature enacted a mechanics' lien law. (Stats. 1850, ch. 87, §§ 1–14, at pp. 211–213.) Moreover, the courts have uniformly classified the mechanics' lien laws as remedial legislation, to be liberally construed for the protection of laborers and materialmen. [Fns. omitted.] (*Connolly Development, Inc. v. Superior Court,* 17 Cal. 3d 803, 826–827 (1976) [132 Cal. Rptr. 477, 553 P.2d 637].)

Harold L. James, Inc. v. Five Points Ranch, Inc.[319]

If there is a single unifying thread which explains most, though not all, of the bewildering array of cases in this field, it is the principle that where the purpose of the requirement of [the relevant statute] is achieved and no one is prejudiced, technical requirements shall not stand in the way of achieving the

---

[317] 148 Cal. App. 3d 891, 901 (1983) (citations omitted).
[318] 163 Cal. App. 3d 715, 723 (1984).
[319] 158 Cal. App. 3d 1 (1984).

purpose of the Mechanics Lien Law. Professor Bottomley, in a recent article felt justified in saying this: "The decisions dealing with defective claims of lien seem generally to be in accord with the objectives of [the relevant statutes]. In the absence of a showing of intent to defraud (extremely difficult to prove) the courts have almost uniformly upheld the claim unless the defect is one which would not charge the owner or, more importantly, a new owner with constructive notice of the claims." (*Wand Corp. v. San Gabriel Lumber Co.,* 236 Cal. App. 2d 861, 862, citation and footnote omitted.)

[G] Oral Agreements for Extras

As a general rule, the courts will allow a contractor to recover for extras even when they are orally agreed upon and the contract provides that the agreement for extras must be in writing.[320] The *Hickey* and *Healy* line of reasoning was followed in *Weeshoff Construction Co. v. Los Angeles County Flood Control District.*[321]

In the *Weeshoff* case, the prime contractor sought to recover from the flood control district the cost of laying temporary pavement as an extra. Under the terms of the contract, the use of temporary resurfacing was prohibited. The prime contractor sought to comply with that contractual requirement by placing hard sand to keep the traffic lanes open. The flood control district was dissatisfied with the use of hard sand and on its own restored the traffic lanes with temporary pavement. Having observed the flood control district using temporary pavement, the contractor used temporary pavement on the traffic lanes and sought to recover the extra expense incurred. The contractor never obtained a written change order, and the contractor did not submit daily expenditure reports while performing the extra work, even though both were required by contract to recover for extra work. The trial court awarded the contractor damages for the extra work, and the court of appeal affirmed. The court stated the following rule of law when there is a written contract stating that extras must be authorized in writing:

> As the district claims, California courts generally have upheld the necessity of compliance with contractual provisions regarding written "change orders." *Acoustics, Inc. v. Trepte Constr. Co.* 14 Cal. App. 3d 887, 912 [92 Cal. Rptr. 723]; *A. Teichert & Son, Inc. v. State of California* 238 Cal. App. 2d 736, 758 [48 Cal. Rptr. 225]; overruled on other grounds in *E.H. Morrill Co. v. State of California,* 65 Cal. 2d 787, 792 [56 Cal. Rptr. 479, 423 P.2d 551]. However, California decisions have also established that particular circumstances may provide waivers of written "change order" requirements. If the parties, by their conduct, clearly assent to a change or addition to the contractor's required performance, a written "change order" requirement may be waived. (*C.F. Bolster Co. v. J.C. Boespflug Etc. Co.,* 167 Cal. App. 2d 143, 150–151

---

[320] Frank T. Hickey, Inc. v. L.A.J.C. Council, 128 Cal. App. 2d 676 (1954); Healy v. Brewster, 59 Cal. 2d 455 (1963), *remanded,* 251 Cal. App. 2d 541 (1967).

[321] 88 Cal. App. 3d 579 (1979).

[334 P.2d 247]; *Frank T. Hickey, Inc. v. L.A.J.C. Council,* 128 Cal. App. 2d 676, 682–683 [276 P.2d 52].)[322]

The *Weeshoff* court distinguished the *Acoustics, Inc.* case by saying that, in the current case, the contractor was given instructions by the flood control district to use temporary pavement and the contractor could rely upon those instructions. In a later case, a lien claimant was allowed "the reasonable value" of extras in its mechanic's lien even though no amount had been agreed upon for those extras.[323]

The *Acoustics, Inc.* case involved a state contract on which the state inspectors had orally ordered work to be done by the contractor. The court held that compliance with the contractual provision requiring written orders for changed work was indispensable in order for the contractor to recover for the alleged extra work. The reason for this decision, as stated by the court, was that subordinate field personnel cannot waive the mandatory contract requirements that ordered changes or additions or extras must be approved in writing by the state architect.

[H] Amount of Lien

Effective January 1, 1991, California Civil Code § 3123 was amended. It previously provided that a mechanic's lien was limited to the lesser of the agreed contract price or the reasonable value of the labor, services, equipment, or materials furnished. The newly amended section 3123 does not preclude the mechanics' lien claimant from including in the lien any amount due for labor, services, equipment, or materials furnished based on a written modification of the contract, or as a result of the rescission, abandonment, or breach of the contract. In the event of rescission, abandonment, or breach, the amount of the lien is limited only by the reasonable value of the labor, services, equipment, and materials furnished by the claimant.

One question that often arises is whether a lien claimant can include attorney's fees in its mechanic's lien. In *Abbett Electric Corp. v. California Federal Savings & Loan Ass'n,*[324] the court held that attorney's fees could not be included in a mechanic's lien.

In *Abbett,* an electrical contractor entered into a contract that provided for attorney's fees to the prevailing party. The contractor filed a suit for breach of contract and to foreclose its mechanic's lien. The complaint sought $314,652 plus attorney's fees of $100,000. When the case finally came to trial, there were two deeds of trust against the property. The trial court awarded $113,827.36 plus interest and costs of suit in excess of $27,000 and attorney's fees of $230,000. The attorney's fees were included in the mechanic's lien. The court noted that there was absolutely no dispute that the property owner was personally liable for

---

[322] 88 Cal. App. 3d at 589.
[323] *See* Western Sierra, Inc. v. Ramos, 97 Cal. App. 3d 482 (1979).
[324] 230 Cal. App. 3d 355 (1991).

the attorney's fees and that they were properly included in the judgment against the owner.

The issue before the court was whether the attorney's fees were properly includable in the mechanic's lien. The lien claimant cited four cases in which attorney's fees had been recovered in mechanic's lien foreclosure actions: *Vitek, Inc. v. Alvarado Ice Palace, Inc.*,[325] *Robinson v. Diller*,[326] *Distefano v. Hall*,[327] and *California Viking Sprinkler Co. v. Cheney*.[328] The court stated that none of those opinions indicates whether the attorney's fees were included in the mechanic's lien, and the point was never raised on appeal. The court opined that Civil Code § 3123(a) limits the amount of a mechanic's lien to the reasonable value of the labor, services, equipment, or materials furnished by the claimant or the price agreed upon between the parties, whichever is less. The court noted that the Code section makes no provision for attorney's fees. Furthermore, California Constitution article 14, § 3, provides for mechanic's liens "for the value of the labor done and material furnished"—it does not provide for attorney's fees. The court cited *Lambert v. Superior Court*,[329] (discussed further hereinafter) for the propositions that delay damages could not be included in a mechanic's lien and that the function of a mechanic's lien is to secure reimbursement for services and materials actually contributing to a construction project, not to facilitate recovery of consequential damages. Accordingly, the court held that, like delay damages, attorney's fees are beyond the contemplation of the mechanic's lien remedy. The court rejected an argument that attorney's fees should have been included in the mechanic's lien claim under Civil Code § 3150, which provides that "[i]n addition to any other costs allowed by law, the court in an action to foreclose a mechanic's lien must also allow as costs the money paid for verifying and recording the lien, such costs to be allowed each claimant whose lien is established whether he be plaintiff or defendant." Without acknowledging that the costs provided for in Civil Code § 3150 are to be included in the lien, the court was not persuaded that the costs referenced in that Code section would include attorney's fees. Thus, it would appear under the *Abbett* case that attorney's fees cannot be included in a mechanic's lien.

It should be noted, however, that even though *Abbett* was decided in 1991, it did not mention the amendment to Civil Code § 3123, effective January 1, 1991, which provides that a claimant may include in a claim of lien an amount due for labor, services, equipment, or materials furnished based upon a written modification to the contract or as a result of the rescission, abandonment, or breach of the contract. The court also placed emphasis on the *Lambert* case, which was likewise decided in 1991, after the amendment to Civil Code § 3123, but also did not mention that amendment. Notwithstanding the foregoing, it would appear that, under the *Abbett* case, attorney's fees cannot be included in a

---

[325] 34 Cal. App. 3d 586 (1973).
[326] 274 Cal. App. 2d 813 (1969).
[327] 263 Cal. App. 2d 380 (1968).
[328] 182 Cal. App. 2d 564 (1960).
[329] 228 Cal. App. 3d 383 (1991).

judgment of foreclosure of a mechanic's lien. Labor and material improve the property of an owner, but the services of an attorney for the contractor do not improve the property.

In *Royster Construction Co. v. Urban West Communities*,[330] the court held that just as attorneys' fees are not recoverable by a lien claimant for foreclosure of a mechanic's lien, they are not recoverable in an action against a mechanic's lien release bond. In that case, the trial court had entered a judgment in favor of the contractor (Royster) awarding damages, prejudgment interest, and attorneys' fees on each cause of action. Pursuant to Civil Code § 3143, the owner had posted bonds to release the land from Royster's mechanic's lien. The court of appeal modified the judgment to delete the award of attorneys' fees to the contractor on its mechanic's lien claims, holding that attorney's fees are never recoverable on a mechanic's lien claim, even when the recovery is sought on a Civil Code § 3143 release bond. The court held that an action to foreclose a mechanic's lien and an action against a surety on a release bond are one and the same, and the bond does not change the relation or rights of the parties other than substituting the bond obligation for the property subject to the lien. The court further noted that it was not within the legislative purpose, in permitting the substitution of the bond for the lien, to deteriorate the mechanic's lien claimant's rights. Thus, the contractor's remedies remain the same, including the bar against recovery of attorneys' fees. The court also stated that, even if the release bonds were subject to the bond and undertaking law (Civil Procedure Code §§ 995.010 *et seq.*), the only mention in the bond law of attorneys' fees pertains to procedures not employed in a mechanic's lien case. Finally, the court stated that since the owner, as principal on the bond, had no legal obligation to pay the contractor's attorneys' fees as a part of the mechanic's lien claim, the surety on the bond could raise that defense in its own behalf.

Lambert v. Superior Court[331] was another case decided after January 1, 1991, but involving a mechanic's lien recorded in 1990, wherein a California court held that a mechanic's lien could not include interest and delay damages. In that case, the general contractor, dealing directly with the owner, included in its mechanic's lien an amount for "interest" and for "delay." The owner made a motion to determine the lien invalid on the theory that Civil Code § 3123 does not allow a lien claim to include interest or delay damages. The appellate court held that such a motion was procedurally proper and further held that a mechanic's lien could not include interest or delay damages.

It is the opinion of the authors that if the issue with regard to delay damages were to go before the appellate courts on a mechanic's lien recorded after January 1, 1991, pursuant to the amendments to Civil Code § 3123 (effective January 1, 1991), a lien claimant could include and recover delay damages in its mechanic's lien on either a rescission, abandonment, or breach of contract theory, so long as such amounts were for labor, services, equipment, or materials

---

[330] 40 Cal. App. 4th 1158, 47 Cal. Rptr. 2d 684 (1995).
[331] 228 Cal. App. 3d 383 (1991).

furnished to the project. With regard to the issue of interest, the authors suggest to lien claimants that they not include interest in the mechanic's lien but pray for recovery of interest in the foreclosure action and allow the court to decide whether the lien claimant is entitled to interest on the mechanic's lien.

In the case of *Corech v. Hornwood*,[332] a subcontractor had a written contract with the prime contractor to provide electrical work. That contract had an attorney fee clause in it. During the progress of the job, the prime contractor walked off the job, and the owner met with the electrical subcontractor to discuss finishing the job. The electrical subcontractor completed the job but was not paid, so recorded a mechanic's lien. The electrical subcontractor brought an action to foreclose its mechanic's lien and also sued the owner for breach of contract and unjust enrichment, contending the owner adopted the contract and ratified it when the owner took over the project. The trial court allowed the subcontractor to foreclose its mechanic's lien but ruled against the subcontractor on its claims of breach of contract and for quantum meruit against the owner. Because the subcontractor had sought to impose the attorney fee clause against the owner, the trial court held that because the owner prevailed in defending the contract cause of action, the owner was the prevailing party, and the court awarded the owner its attorney's fees. The subcontractor appealed, and the appellate court affirmed the judgment on appeal. The court quoted Abraham Lincoln as follows:

> "In law it is a good policy never to plead what you need not, lest you oblige yourself to prove what you cannot." (Abraham Lincoln, Letter to Usher F. Linder (Feb. 20, 1848) in
>
> The Quotable Lawyer (Shrager & Frost edits. 1986 ¶ 96.6, p. 241.)

The appellate court held that the trial court was correct in its decision. The court noted that the subcontractor and its attorneys would have known by appropriate research that they could not recover attorney's fees on the mechanic's lien under the *Abbett* case referenced above. The subcontractor pursued the contract claim in order to take advantage of the attorney fee clause that it had in its contract with the prime contractor. Since the trial court properly determined, based upon the facts, that there was no contract between the subcontractor and the owner, the owner therefore properly prevailed on that contract cause of action and was therefore entitled to its attorney's fees under Civil Code § 1717.

The question whether a lien claimant can include in its mechanic's lien sums over and above the contract balance due was addressed in the case of *Basic Modular Facilities, Inc. v. Ehsanipour*.[333] In that case, the contract plus two change orders for improvements on property owned by Ehsanipour totaled $154,884. Ehsanipour had paid the contractor $137,092, leaving a balance due of $17,792. The contractor recorded a mechanic's lien for $18,192 but continued to work on

---

[332] 58 Cal. App. 4th 1412 (1997).
[333] 83 Cal. Rptr. 2d 462 (Ct. App. 1999).

the property for another month. The contractor sent a demand letter to the owner, breading down its claim as follows:

1.	Past Due Billings	$18,192.00
2.	Interest	709.00
3.	Extended Overhead	$18,891.00
4.	Additional Work Required by Owner	$24,319.00
5.	Billings Incurred from Delays/Interference from the Owner (including 15% profit and 10% overhead)	$183,188.00
6.	Total	$245,299.00

The contractor than recorded a second mechanic's lien for the foregoing amounts in the sum of $245,299. Ehsanipour moved the court to have the lien removed on the ground that it was overstated. The contractor conceded that the claim for extended overhead was impermissible and voluntarily recorded a partial release of lien in the sum of $53,237. The trial court ordered the removal of the mechanic's lien claim. The appellate court reversed. Citing Civil Code § 3123, the court stated that mechanics' liens must be for the reasonable value of the work or materials furnished or the price agreed upon by the parties, whichever is less, but that said section does not preclude a claimant from including amounts due under written change orders or as a result of rescission, abandonment, or breach of the contract. The court held that Civil Code § 3123(b) authorizes a mechanic's lien claim based on written change orders or as a result of rescission, abandonment, or breach of the contract. The use of the disjunctive "or" permits a claimant to include in the lien an amount for the reasonable value of work or materials furnished based upon the owner's breach of oral modifications to the contract. The contractor alleged that Ehsanipour authorized numerous changes to the work and then refused to pay for those changes. The court noted that if the foregoing allegations were true, this would have constituted a breach of contract and the contractor would have been authorized to include in its mechanic's lien the reasonable value of the work or materials furnished as a result of that breach. The appellate court further noted that the trial court has the authority to reduce an excessive lien to its proper amount. In addition, the court noted that there is no authority for the proposition that a lien holder who includes a few dollars for impermissible items loses the entire lien. The court indicated that the claimant has the burden of proof to prove the reasonable value of the work and materials and the amount due on the lien, which presents a question of fact for the trial court. The appellate court remanded the case with directions to restore the lien in the amount of $192,062 and determine the actual amount due.

Thus, it is clear that under Civil Code § 3123(b), mechanic's lien claimants *can include in their mechanics' liens amounts due under written change orders or as a result of rescission, abandonment, or breach of the contract.*

[I] Ninety-Day Foreclosure Limit

In *Petersen v. W. T. Grant Co.*,[334] suit was filed 94 days after the lien was recorded. California Civil Code § 3144 provides that suit must be brought in 90 days. The defendants did not raise an affirmative defense that suit was not brought within 90 days. At trial, the defendants orally moved to dismiss by raising the 90-day statute by amendment. The judge heard the evidence and ruled against the plaintiff, who later appealed, contending that the statute should have been raised as an affirmative defense. The court held that because the statute setting forth the 90-day period is a statute of limitations rather than a statute affecting a substantial right, the trial court erred in treating the matter as a substantive right and granting judgment on the pleadings. With regard to the defendant's motion to amend to raise the statute of limitations at trial, the court acknowledged that was proper procedure, but because the trial court did not rule on the motion, the case was sent back for the trial court to balance the equities and rule on the motion.

Consequently, owners should always raise this affirmative defense if suit is brought more than 90 days after the lien is recorded.

In *Robinson v. S&S Development*,[335] the court ruled that the 90-day limitations period had been tolled against the trustee in bankruptcy when the claimant filed bankruptcy.

[J] Interest

In *Southwest Concrete Products v. Gosh Construction Corp.*,[336] the California Supreme Court held that a material supplier's charge of 18 percent interest per annum for late payments is not usurious. The court held that the California usury statutes do not apply to a credit sale between merchants.

The plaintiff in *Southwest Concrete Products* was a concrete pipe supplier. Its contract with the contractor provided that overdue payments would bear interest at the rate of 1.5 percent per month (18 percent per year). This amount is more than the allowed interest under California's usury laws. *Crestwood Lumber Co. v. Citizen's Savings & Loan Ass'n*[337] and *Mark McDowell Corp. v. LSM*[338] had held that interest charges of 18 percent on overdue commercial accounts were subject to the usury law.

Southwest Concrete Products expressly overrules *Crestwood Lumber* and *Mark McDowell* and permits late charges of 18 percent in commercial contracts. The supreme court reasoned that late charges provided in a contract do not constitute a "loan" or "forbearance of money" so as to bring the usury law into play.

---

[334] 41 Cal. App. 3d 217 (1974).
[335] 256 Cal. App. 2d 13 (1967).
[336] 51 Cal. 3d 701 (1990).
[337] 83 Cal. App. 3d 819 (1978).
[338] 214 Cal. App. 3d 1427 (1989).

In *Associated Wholesale Electric v. S.H. Kress & Co.*,[339] prejudgment interest on a lien was not allowed. However, it was allowed in *Vowels v. Witt*,[340] to a prime contractor dealing directly with the owner. Prejudgment interest was also allowed on stop notices in *A-1 Door & Materials Co. v. Fresno Guarantee Savings & Loan Ass'n.*[341]

[K] Jurisdiction

The county where the real property is located is the proper county in which to bring a lien foreclosure action.[342] If the action is brought in the wrong county (that is, a county other than the county where the real property is located), then the lien action is improper.[343]

In *L.A.M. Construction, Inc. v. Kriz*,[344] the plaintiff filed an action to foreclose a $23,000 mechanic's lien in the superior court. The actual jurisdiction for claims of less than $25,000 is the municipal court. The real property was evidently located in the city of Los Angeles, and therefore the Los Angeles Judicial District of the Los Angeles Municipal Court was the proper place for the action to be brought. More than 90 days after the action was brought in the superior court, at a status conference, the judge transferred the case to the municipal court because the lien fell below the jurisdictional minimum of the superior court. The defendants then moved for summary judgment, arguing that the action had not been transferred to "a proper court" within 90 days after the lien was recorded, relying upon Civil Code § 3144, which provides that the lien does not bind the property for more than 90 days unless, within that time, an action to foreclose that lien is commenced "in a proper court." The trial court dismissed the action, and the lien claimant appealed. The appellate court reversed.

The appellate court held that the county of Los Angeles had subject matter jurisdiction because the real property was located in the county of Los Angeles. Specifically, the court held that a proper court in the context of lien foreclosures against real property means either the superior court or the municipal court in the county of location of the real property. The court reversed the judgment of the trial court, and thus the lien claimant prevailed.

The decision of the court of appeals in *L.A.M. Construction, Inc. v. Kriz* is contrary to another district's prior decision in *Halbert's Lumber, Inc. v. Burdett.*[345] In *Halbert's Lumber,* the lien claimant filed its foreclosure action in the wrong judicial district of the municipal court, and the court held that the action was improper. As a result, there is now a conflict of decisions in the appellate courts in the state of California. The only safe practice for a lien claimant and its attorney to follow, with regard to a lien that falls within the jurisdiction of the

---

[339] 11 Cal. App. 2d 592 (1936).
[340] 149 Cal. App. 2d 257 (1957).
[341] 61 Cal. 2d 728 (1964).
[342] Division of Labor Law Enforcement v. Egnew Inv., Inc., 247 Cal. App. 2d 863 (1966).
[343] Douglas v. Donner Pines, 73 Cal. App. 3d 268 (1977).
[344] 14 Cal. App. 4th Supp. 1 (1993).
[345] 202 Cal. App. 3d 14 (1988).

municipal court, is to file the action in the particular judicial district of the municipal court where the real property is located.

The court of appeal has made it clear that California's stop notice remedy is limited to construction projects built within California. In *Mechanical Wholesale Corp. v. Fuji Bank, Ltd.*,[346] a California subcontractor that worked on a project in Hawaii brought an action to enforce a bonded stop notice against a California bank that funded the project. The plaintiff sought recovery from a construction loan fund established by an agreement between the general contractor, the project owner, and the defendant bank (Fuji). The trial court sustained Fuji's demurrer, without leave to amend, and dismissed the complaint, on the ground that California's statutory stop notice remedy is limited to works of improvement constructed within California.

The court of appeal affirmed. When read as a whole, the court stated, California's mechanic's lien scheme, of which the stop notice remedy is a part, was clearly intended to apply only to California projects. In addition, public policy considerations compel the same conclusion. Not only does California have no real interest in ensuring payment to contractors on projects in other states, but also lenders and owners in the other state would effectively be deprived of the provisions of California's mechanic's lien laws that protect them (that is, securing and recording a payment bond from the general contractor). In addition, because mechanic's liens are a creature of statute that are not a part of common law, an out-of-state lender would have no notice or reasonable expectation that it could be subject to such a remedy based upon another state's laws. This, the court of appeal noted, would impose a burden on California lenders that "might well result in an increase in California borrowing costs." The court therefore declined to adopt a position that it felt would be detrimental to Hawaii's economy with "no apparent benefit to California's economy or vindication of any significant California interest."[347]

The court of appeal also awarded Fuji its attorneys' fees as the prevailing party in an "action against a construction lender to enforce payment of a claim stated in a bonded stop notice"[348] notwithstanding the fact that the action was held not to be valid under the California statutes. This required the court to interpret the phrase "from the party held liable by the court for payment of the claim" as applicable to an unsuccessful stop notice claimant. The court's logic was that the subcontractor was "liable for payment of the claim in the sense that the amount of the claim becomes a loss which must be absorbed." Thus, perhaps the most lasting impact of this case will be its holding that "a construction lender who successfully defends a suit on an invalid stop notice can legitimately claim to have recovered 'a greater relief' and is a prevailing party entitled to receive its attorney fees."[349]

---

[346] 42 Cal. App. 4th 1647 (1996).
[347] 42 Cal. App. 4th at 1660.
[348] Cal. Civ. Code § 3176.
[349] 42 Cal. App. 4th at 1661.

[L] Forged Endorsement of Joint Check

A joint payee whose endorsement has been forged may recover against the payor bank;[350] the drawee bank can collect from the collecting bank.[351]

[M] Lien Forfeiture: Willful Overstatement

A lien can be forfeited if it is willfully overstated.[352]

[N] Slander of Title

In *Frank Pisano & Associates v. Taggart*,[353] the court deemed a mechanic's lien absolutely privileged; the lien claimant cannot be sued for slander of title even if the lien is invalid.

[O] Lien Contents

In one case, although the lien named no one as owner, it was upheld by the court.[354] The court held in another case that if the claimant names a reputed owner in good faith and the owner turns out to be someone else, the lien is still valid.[355]

[1] Description of Property

If the property's description is sufficient to enable a party familiar with the locality to identify the property, then even though the description is erroneous, it will be upheld.[356]

[2] Amount of Liens

Effective January 1, 1991, California Civil Code § 3123 was amended so that the liens provided for by the chapter shall be for the reasonable value of the labor, services, equipment, or materials furnished, or for the price agreed upon by the claimant and the person with whom he or she contracted, whichever is less. The statute goes on to provide that the lien shall not be limited in amount by the price stated in the contract as defined in § 3088, except as provided in § 3123(c).

---

[350] Harry H. White Lumber Co. v. Crocker-Citizens Nat'l Bank, 253 Cal. App. 2d 368 (1967); Indiana Plumbing-Supply Co. v. Bank of Am., 255 Cal. App. 2d 910 (1967).

[351] Feldman Constr. Co. v. Union Bank, 28 Cal. App. 3d 731 (1972).

[352] B&J Constr. Co. v. Spacious Homes, Inc., 204 Cal. App. 2d 216 (1962); Callahan v. Chatsworth Park, Inc., 204 Cal. App. 2d 597 (1962).

[353] 29 Cal. App. 3d 1 (1972).

[354] Allen v. Wilson, 178 Cal. 674 (1918).

[355] Frank Pisano & Assocs. v. Taggart, 29 Cal. App. 3d 1 (1972). *See also* West Coast Lumber v. Newkirk, 80 Cal. 275 (1889).

[356] Borello v. Eichler Homes, Inc., 221 Cal. App. 2d 487 (1963).

California Civil Code § 3123(b) does not preclude the claimant from including in the lien any amount due for labor, services, equipment, or materials furnished based on a written modification to the contract or as a result of the rescission, abandonment, or breach of the contract. However, in the event of rescission, abandonment, or breach of the contract, the amount of the lien may not exceed the reasonable value of the labor, services, equipment, and materials furnished by the claimant. This is a new provision added to the lien law. It has not been interpreted by the courts but would appear to mean that in addition to the amount due on the contract, the lien claimant could also include any damages being claimed as a result of rescission of the contract, abandonment of the contract, or breach of the contract. In no event, however, may that amount exceed the total reasonable value of the labor, services, equipment, and materials furnished.

New § 3123(c) provides that the owner shall notify the prime contractor and construction lender of any changes in the contract if the change has the effect of increasing the price stated in the contract by 5 percent or more.

For further discussion of the right of the lien claimant to include in his or her mechanic's lien amounts due as a result of written change orders or as the result of rescission, abandonment, or breach of contract, see the discussion of the case of *Basic Modular Facilities, Inc. v. Ehsanipour*[357] in § 9.06[H], *supra*.

[P] Implied Warranty of Design

In *Pollard v. Saxe & Yolles Development Co.*,[358] the court held that sellers of new construction impliedly warrant that the completed structure was designed and constructed in a reasonably workmanlike manner.

[Q] Personal Liability of Owner

Owners who have no contractual relationship with the lien claimant are not personally liable to the lien claimant.[359]

[R] Parties

A contractor is not a necessary party to a lien foreclosure action.[360] On the other hand, all parties who have an interest in the property are necessary parties[361] to a lien foreclosure action.

---

[357] 83 Cal. Rptr. 2d 462 (Ct. App. 1999).
[358] 12 Cal. 3d 374 (1974).
[359] Sinnock v. Young, 61 Cal. App. 2d 130 (1943); R.D. Reeder Lathing Co. v. Allen, 66 Cal. 2d 373 (1967). *See also* 57 C.J.S. *Mechanic's Liens* § 250; 36 Am. Jur. *Mechanic's Liens* § 225; 18 Ruling Case Law Mechanic's Lien § 114; 1911 Cal. Stat.§ 14, ch. 681 (liens direct and independent of any indebtedness between owner and contractor).
[360] Green v. Clifford, 94 Cal. 49 (1892); Hazard Gould & Co. v. Rosenberg, 177 Cal. 295 (1918).
[361] Packard Bell Elecs. Corp. v. Theseus, Inc., 244 Cal. App. 2d 355 (1966).

One of the primary parties to be named as a defendant in the foreclosure action is the owner of the property. In *Barr Lumber Co., Inc. v. Old Ivy Home Builders, Inc.,*[362] the lien claimant named the general partner of the limited partnership as a defendant in its foreclosure action. The limited partnership itself was not named as a defendant and was never served in the action.

The court of appeal held that because the plaintiff failed to name the proper party in the foreclosure action within 90 days of the filing of the mechanic's lien, the limited partnership could not be bound by the judgment, and its property could not be used to satisfy the lien claim. The court held further that notice to the general partner was not sufficient to constitute notice to the limited partnership and that the general partner could not itself be deemed the owner of the property.

The court cited and relied upon *Grinell Fire Protection Systems Co. v. American Savings & Loan Ass'n,*[363] wherein the court concluded that all parties to be bound by the judgment must be joined in the foreclosure action. It should be noted, however, that under *Grinell,* if the mechanic's lien claimant had brought the limited partnership in as a Doe defendant, that action would have been proper and the judgment would have been sustained.

[S] Release Bond

When a release bond has been recorded, the owner can get a summary judgment on the mechanic's lien.[364] California Civil Code § 3143 states that real property subject to a mechanic's lien is released from the lien upon the filing of a lien release bond; consequently, the owner can obtain a summary judgment on the mechanic's lien. Section 3144.5 of California's Civil Code requires that notice of the recording of the release bond be given to the lien holder by mailing a copy of the bond to the lien holder at the address appearing on the lien. It further provides that service of the notice shall be by certified or registered mail, return receipt requested, and that failure to give the notice shall not affect the validity of the lien release bond, but the statute of limitations on any action on the bond shall be tolled until the notice is given. Finally, any action on the lien release bond must be commenced by the claimant within six months after the claimant is given notice of the recording of the lien release bond.

In *Hutnick v. United States Fidelity & Guaranty Co.,*[365] Hutnick recorded a mechanic's lien and filed a suit to foreclose that mechanic's lien. After the suit had been filed, the owners obtained from United States Fidelity & Guaranty a mechanic's lien release bond filed according to Civil Code § 3143. The bond was recorded on July 26, 1985 (more than a year after the original complaint had been filed), and Hutnick was given notice of recording on July 30, 1985. On March 18, 1986, Hutnick served United States Fidelity & Guaranty as a "Doe Defen-

---

[362] 34 Cal. App. 4th 1 (1995).
[363] 183 Cal. App. 3d 352 (1986).
[364] Frank Curran Lumber Co. v. Eleven Co., 271 Cal. App. 2d 175 (1969).
[365] 199 Cal. App. 3d 49 (1988).

dant." The district court of appeal held that the action against United States Fidelity & Guaranty was barred by reason of the six-month statute of limitations set forth in California Civil Code § 3144.5, because the lien claimant, Hutnick, had failed to bring an action on the bond within six months after he had been given notice of the recording of the bond. The court stated that Hutnick should have filed a supplemental complaint stating a separate cause of action on the release bond.

The California Supreme Court later reversed and held that when a mechanic's lien foreclosure action is commenced and a release bond is thereafter recorded, the lien claimant may obtain recovery against the bond's surety in the pending action without having to plead a new cause of action or comply with an additional limitations period.[366]

In *Grade-Way Construction Co. v. Golden Eagle Insurance Co.,*[367] a subcontractor recorded mechanics' liens and served stop notices on a private work of improvement. The subcontractor then brought suit to foreclose the liens and enforce the stop notices. After the suit had been filed, Golden Eagle Insurance issued mechanic's lien release bonds and stop notice release bonds. When the subcontractor brought Golden Eagle into the lawsuit, Golden Eagle filed an answer raising certain affirmative defenses. The subcontractor and the general contractor stipulated to the entry of a judgment by the subcontractor against the general contractor. The subcontractor then made a motion for summary enforcement of Golden Eagle's liability on the release bonds pursuant to Civil Procedure Code § 996.440, which provides for enforcement of claims against bonds "given in an action or proceeding . . . on motion made in the court without the necessity of an independent action." The subcontractor contended that the release bonds had been given in an action or proceeding and that therefore, once the subcontractor had judgment against the prime contractor, the subcontractor could seek judgment on the release bonds by way of motion. The trial court granted the subcontractor's motion, and Golden Eagle appealed. The court of appeal held that the mechanic's lien release bonds and the stop notice release bonds could be enforced pursuant to section 996.440, that is, by way of a noticed motion. The importance of this case is that it holds that release bonds can be enforced by virtue of a noticed motion without a full-blown trial. The surety does have the right to contest the motion. In this particular case, the surety filed no affidavit setting forth any facts pursuant to which it would not have liability.

[T] Contractual Waivers of Mechanic's Lien Remedies

Civil Code § 3262 deals with the waiver of lien rights during the progress of the job through the execution of conditional and unconditional waivers for progress payments and final payments. *See* § 9.05[B]. An issue that is not addressed is whether a lien claimant may waive its lien rights by contract.

---

[366] Hutnick v. United States Fidelity & Guar. Co., 47 Cal. 3d 456 (1988).
[367] 13 Cal. App. 4th 826 (1993).

It is arguable that under Civil Code § 3262(b), such a contractual waiver is unenforceable. Civil Code § 3262(b) states that no oral or written statement purporting to waive, release, impair, or otherwise adversely affect a claim is enforceable or creates any estoppel or impairment of the claim unless it is pursuant to a waiver and release prescribed under Civil Code § 3262 or unless the claimant has actually received payment in full for the claim. In contrast, it is arguable that Civil Code § 3262 applies only to progress payments and final payments, not to a knowing and intelligent waiver of lien rights by contract. Under those circumstances, it is arguable that the person executing the contract has, for valuable consideration, by contract, given up its right to enforce its lien remedies. Assuming that Civil Code § 3262 does not prohibit these contractual waivers, the courts ordinarily will uphold and enforce such clauses.

For example, *Aetna Casualty & Surety Co. v. United States*[368] applied California law and held that such a clause is enforceable. Again, it should be noted that this decision preceded the enactment of Civil Code § 3262.[369]

The *Aetna* court stated that certain principles have evolved over the years concerning contracts in general and waivers of mechanics' liens in particular. A mechanic's lien is a valuable and substantial right. It takes precedence over virtually all other liens, whether filed before or after. In the absence of clear evidence to the contrary, it is presumed that one will not preclude oneself from exercising a right granted by statute. The court stated that the doubt about a provision in the contract is to be resolved against the person for whose benefit the language was intended and that the court will not deprive subcontractors and material suppliers of their liens unless they expressly waive their lien rights or expressly or by clear implication agree to be bound by the lien waiver of another.

Another set of cases applying this rule has interpreted prime contract clauses requiring a general contractor to deliver the project free of liens. Most of those cases hold that that language does not waive the general contractor's right to pursue a mechanic's lien remedy for nonpayment by the owner.[370] The Indiana Supreme Court has written contradictory decisions interpreting such language in a construction contract, while the Illinois Appellate Court has held that such language is effective to waive mechanic's lien remedies.[371]

---

[368] 655 F.2d 1047 (9th Cir. 1981).

[369] *See also* Fuller Co. v. Brown Minneapolis Tank & Fabricating Co., 678 F. Supp. 506 (E.D. Pa. 1987). The Virginia Supreme Court, in *McMerit Construction Co. v. Knightsbridge Development Co.,* 235 Va. 368, 367 S.E.2d 512 (1988), stated that it would strictly construe any lien waiver clause against the party seeking to enforce it.

[370] *See* Plunkett v. Winchester, 98 Ark. 160 (1911); First Atl. Bldg. Corp. v. Neubar Constr. Co., 352 So. 2d 103 (Fla. Dist. Ct. App. 1977); Porier v. Desmond, 177 Mass. 201 (1900); Kertscher & Co. v. Green, 205 N.Y. 522 (1911); Nelson v. Cohen, 160 Or. 336 (1938); Roberts v. Burleson, 284 S.W. 632 (Tex. Civ. App. 1926); Gray v. Hickey, 94 Wash. 370 (1970); Davis v. LaCrosse Hosp. Ass'n, 121 Wis. 579 (1904).

[371] *Compare* Kokomo F&W Traction Co. v. Kokomo Trust Co., 193 Ind. 219 (1923), *with* Peter & Burghard Stone v. Marion Nat'l Bank, 198 Ind. 581 (1926), *and* Moulding Brownell Corp. v. E.C. Defosse Constr. Co., 9 N.E.2d 459 (Ill. App. Ct. 1937).

The rule of strict construction of lien waiver clauses has also been applied when the lien waiver was granted to a lender. Under the rule, these have been construed as subordination agreements rather than true waivers of mechanics' liens.[372] Lien waiver clauses have also been held not to impair the general contractor's right to file a lien for extra work performed at the owner's request.[373]

In California, Civil Code § 3262 provides, in subdivision A, that neither the owner nor the original contractor may, by any term of the contract, waive or impair the lien claims of any subcontractors or suppliers.

The question whether California courts will allow a waiver of mechanic's lien rights only by the methods set forth in Civil Code § 3262 appears to have been answered in the affirmative by the California Supreme Court in *Wm. R. Clarke Corp. v. Safeco Insurance Co.*[374] As noted in § 1.09[C] of this book, the supreme court held that, under Civil Code § 3262, a subcontractor cannot waive its mechanic's lien rights except under certain specified circumstances. Civil Code § 3262(d) provides that a waiver and release of mechanic's lien rights shall be null and void and unenforceable unless it follows substantially the forms set forth in the statute. The California Supreme Court noted:

> Thus, under our mechanic's lien law, waiver and release of mechanic's lien rights is permitted only in conjunction with payment, or a promise of payment, and a conditional release is effective only if the claimant is actually paid. (See Cal. Mechanics' Liens and Other Remedies, (Cont. Ed. Bar 1988) § 4.21, at p. 200.)[375]

> A pay if paid provision in a construction agreement does not take the form of a waiver of mechanic's lien rights. Yet "[t]he law respects form less than substance" (Civ. Code § 3528), and a pay if paid provision is in substance a waiver of mechanic's lien rights because it has the same practical effect as an express waiver of those rights . . . We may agree with Safeco that a pay if paid provision is not precisely a waiver of mechanic's lien rights and yet conclude that a pay if paid provision is void because it violates the public policy that underlies the anti-waiver provisions of the mechanic's lien laws. The Legislature's carefully articulated anti-waiver scheme would amount to little if parties to construction contracts could circumvent it by means of pay if paid provisions having effects indistinguishable from waivers prohibited under Civil Code section 3262.[376]

The Supreme Court concluded:

> By closely and carefully circumscribing subcontractors' freedom to waive mechanic's lien rights, and by forbidding waivers not accompanied by payment, or a promise of payment, the Legislature has already determined that

---

[372] *See* Dawson v. Eldridge, 84 Idaho 331 (1962).

[373] *See* G.R. Smonagle & Sons, Inc. v. McKnight Constr. Co., 304 A.2d 339 (Del. 1973); Sturgis Sav. & Loan Ass'n v. Italian Village, Inc., 265 N.W.2d 755 (Mich. Ct. App. 1978).

[374] 15 Cal. 4th 882 (1997).

[375] 15 Cal. 4th at 889.

[376] *Id.* at 890.

there are policy considerations here that override the value of freedom of contract. We merely recognize and enforce that legislative policy determination.[377]

The court of appeal, in *Capitol Steel Fabricators, Inc. v. Mega Construction Co.*,[378] on rehearing, held that the *Clarke* decision applies to a public works project where there is no pending action against the government entity. The case involved construction of a school gymnasium for the Long Beach Unified School District. The plaintiff subcontractor, Capitol, brought causes of action for breach of contract, enforcement of its public works stop notice, and a statutory payment bond claim.

Mega, the defendant general contractor, had settled its own action against the school district but had not yet been paid the amount due under the settlement agreement. Mega had also made a partial payment to Capitol, in exchange for a release of the stop notice. The rest of Capitol's claims then proceeded to a trial on stipulated facts, and the trial court found in favor of Mega, based upon a pay-if-paid clause in the Capitol subcontract that included "condition precedent" language.

On appeal, Capitol contended that the pay-if-paid clause was unenforceable in light of the supreme court's decision in *Clarke*. Mega contended that *Clarke* should be restricted to its facts and should not be applicable to public works projects, as to which contractors have no mechanic's lien rights due to the principle of sovereign immunity. The court of appeal rejected Mega's argument, finding that the anti-waiver provisions of California Civil Code § 3262 apply to stop notice and bond rights as well as to mechanic's lien rights; furthermore, there are no public policy reasons for giving subcontractors different protections from forfeiture of their rights to be paid by general contractors on public projects than the *Clarke* decision mandated on private projects. Thus, Mega was liable to Capitol for the work it performed on the project as was the surety, Fidelity. The court of appeal therefore reversed the trial court and remanded the matter "for further proceedings consistent with this opinion."

---

[377] *Id.* at 891–892.
[378] 58 Cal. App. 4th 1049 (1997).

CHAPTER 10
STOP NOTICES AND BONDS ON PUBLIC WORKS

§ 10.01 No Rights to Mechanics' Liens

§ 10.02 Stop Notice
    [A] Preliminary 20-Day Notice
        [1] Who Gives Notice?
        [2] When Is Notice Given?
        [3] To Whom Is Notice Given?
        [4] How Is Notice Given?
        [5] Where Is Notice Given?
        [6] When Is Service Complete?
        [7] Contents of Notice
        [8] Disciplinary Action
    [B] Stop Notice Requirements
        [1] Contents of Stop Notice
        [2] Who May Serve Stop Notice?
        [3] How Is Stop Notice Served?
        [4] Upon Whom Is Stop Notice Served?
        [5] Time for Service
    [C] Release of Stop Notice
    [D] Suit on Stop Notice

§ 10.03 Labor and Material Bond
    [A] Preliminary 20-Day Notice
        [1] Who Gives 20-Day Notice?
        [2] When Is 20-Day Notice Given?
    [B] Time for Filing Suit on Bond
    [C] Other Requirements

§ 10.04 Case Examples
    [A] Decisions Regarding Stop Notice and Bond Remedies
    [B] Attorney's Fees on Payment Bond on Public Works
    [C] Surety's Right to Raise Its Principal's Defenses

§ 10.01 NO RIGHTS TO MECHANICS' LIENS

There is no right to a mechanics' lien on public works of improvement.[1] Therefore, the state legislature has provided the alternative remedies of a stop notice and a bond.

§ 10.02 STOP NOTICE

[A] Preliminary 20-Day Notice

In order to preserve a claimant's right to file a stop notice, a preliminary notice is necessary under California Civil Code § 3098. The following is an outline of the procedural steps required in order to preserve stop notice rights in public works of improvement.

The first step in preserving the claimant's right to serve stop notices is the 20-Day Preliminary Notice required under Civil Code § 3098. "The preliminary notice must be served upon the public agency and the original contractor." If the claimant's contract is directly with the contractor, no notice is required. The foregoing is applicable only to public works projects done by the state of California or any of its political subdivisions, such as cities, counties, school districts, irrigation districts, and the like. There has been no change to the Miller Act, which is applicable to federal construction projects and which is discussed in Chapter 11.

Under a procedure similar to that for private works of improvement, if the claimant failed to give the 20-day notice, the claimant may still pursue its claim against the payment bond, if it gives a notice to the principal and surety on the payment bond within 15 days after the notice of completion, or within 75 days after actual completion. For further details, *see* § 10.03[A].

[1] Who Gives Notice?

All persons who do not have a direct contractual relationship with the prime contractor, except a person who performed actual labor for wages or an express trust fund described in California Civil Code § 3111, must give notice.[2]

[2] When Is Notice Given?

Notice must be given within 20 days after the claimant has first furnished labor, services, equipment, or materials to the jobsite.[3] Similar to the right of a claimant on a private works project, if the claimant failed to give the notice within the first 20 days of furnishing labor, service, equipment, or material, under Civil Code § 3098(d), the claimant may serve the notice at any time and may file

---

[1] Cooley v. Freeman, 204 Cal. 59 (1928); Pneucrete Co. v. U.S. Fidelity & Guar. Corp., 7 Cal. App. 2d 733 (1935).
[2] Cal. Civ. Code § 3098(a).
[3] Cal. Civ. Code § 3098(a).

a stop notice for any labor, service, equipment, or materials furnished within 20 days of the service of the notice and thereafter.

[3] To Whom Is Notice Given?

Notice is given to the contractor and the public agency.[4]

[4] How Is Notice Given?

Notice is given by first-class mail, registered or certified mail, or personal service.[5] This requirement is different from that for private works, for which notice by first-class mail is not allowed.

[5] Where Is Notice Given?

The notice is given to the contractor at any place the contractor maintains an office or conducts its business, or at the contractor's residence. In case of any public works constructed by the Department of Public Works or the Department of General Services of the state, the notice should be sent to the office of the disbursing officer of the department constructing the work.[6]

[6] When Is Service Complete?

If service is by personal service, it is completed at the time of service. If service is by registered or certified mail, service is complete at the time of deposit of mail. The Code is silent as to when service is completed if done by first-class mail, but it can be inferred that service is not complete until the notice is received by the contractor and the public body. Therefore, service by registered or certified mail is recommended.[7]

[7] Contents of Notice

The notice should contain a general description of labor, service, equipment, or materials furnished or to be furnished and the name of the party to whom they were furnished.[8]

[8] Disciplinary Action

When the contract price to be paid to any subcontractor exceeds $400, the failure to give the notice constitutes grounds for disciplinary action.[9]

---

[4] Cal. Civ. Code § 3098(a).
[5] Cal. Civ. Code § 3098(a).
[6] Cal. Civ. Code § 3098(a).
[7] Cal. Civ. Code § 3098(a).
[8] Cal. Civ. Code § 3098(a).
[9] Cal. Civ. Code § 3098(b).

[B] Stop Notice Requirements

The next step in perfecting a stop notice is the filing of the stop notice itself. The following is an outline of the necessary procedural steps.

[1] Contents of Stop Notice

The written notice, signed and verified by the claimant, should state the following:

1. Kind of labor, service, equipment, or materials furnished or agreed to be furnished by the claimant
2. Name of the person to or for whom the same was done or furnished
3. Amount in value, as near as may be, of that already done or furnished and of the whole agreed to be done or furnished.[10]

Although the statute does not require it, most stop notice forms also contain the amount paid and the balance due and the date from which interest is claimed.

[2] Who May Serve Stop Notice?

All persons furnishing labor, service, equipment, or materials to the job (except the original contractor), and persons furnishing provisions, provender, or other supplies may serve a stop notice.[11]

[3] How Is Stop Notice Served?

The notice is served by personal service or by registered or certified mail.[12]

[4] Upon Whom Is Stop Notice Served?

In the case of any public work for the state, the notice is filed with the director of the department that let the contract. In the case of any other public work, the notice is filed with the office of the controller, auditor, or other public disbursing officer whose duty it is to make payments under the provisions of the contract, or with the commissioners, managers, trustees, officers, board of supervisors, board of trustees, common council, or other body by whom the contract was awarded.[13]

---

[10] Cal. Civ. Code § 3103.
[11] Cal. Civ. Code § 3181.
[12] Cal. Civ. Code § 3103.
[13] Cal. Civ. Code § 3103.

[5] Time for Service

If a notice of completion (many times referred to as a "notice of acceptance" on public works) or notice of cessation is recorded, the stop notice must be filed within 30 days. If there is no notice of completion or cessation recorded, it must be filed within 90 days of completion.[14] If the work of improvement is subject to the acceptance of any public entity, the completion shall be deemed the date of acceptance. However, regarding contracts awarded under the California Contract Act, Chapter 3 (commencing with § 14250), Part 5, Division 3, Title 2 of the California Government Code, a cessation of labor on any public work for a continuous period of 30 days shall be a completion.[15]

[C] Release of Stop Notice

There are two ways the effect of a stop notice may be released—by bond or by summary proceedings.

If the original contractor or subcontractor disputes the correctness or validity or enforceability of the stop notice, the public entity may, in its discretion, permit the original contractor to file with the public entity a corporate surety bond in an amount of 125 percent of the stop notice claim, whereupon the withholding of money pursuant to the stop notice shall cease.[16] If the contractor or subcontractor can provide the bond, the public body will usually allow the release bond to be filed. There are some public agencies that have refused release bonds. In trial courts, contractors have been successful in getting a court order compelling the public agency to accept the bond.

To request release of the funds through summary proceedings, the contractor must assert:

1. That the claim upon which the stop notice is based is not included within the types or classifications of claims referred to in the California Civil Code;
2. That the claimant is not one of the persons named in Civil Code § 3181;
3. That the amount of the claim is excessive; or
4. That there is no basis in law for the claim.

The contractor can then serve upon the public entity an affidavit (along with a copy) alleging its legal grounds and demanding release of the funds.[17] The affidavit must set forth the address of the contractor where the contractor may be served by mail.[18]

---

[14] Cal. Civ. Code § 3184(a), (b).
[15] *Id.* § 3086.
[16] Cal. Civ. Code § 3196.
[17] *Id.* §§ 3197, 3198.
[18] *Id.* § 3198.

The public body then serves the claimant with a copy of the affidavit, along with notice that unless the claimant files a counter-affidavit in not less than 10 days nor more than 20 days after service, the public body will release the funds.[19] If the claimant contests the prime contractor's affidavit, the claimant files a counter-affidavit (with proof of service) alleging the details of the claimant's claim and the basis for contesting or rebutting the original contractor's allegations, a copy of which is served on the contractor.[20] If the counter-affidavit is not filed, the money is released.[21] If the counter-affidavit is filed, then either the contractor or the claimant can file an action for declaratory relief and move for a hearing within 15 days.[22] At the hearing, the contractor has the burden of proof. The affidavit and counter-affidavit are the pleadings and are received into evidence.[23] The court rules on the basis of the affidavits or may allow additional evidence. The court then determines whether the funds shall be released.[24] A jury trial is available on this hearing.[25] The determination at this hearing is not res judicata as to the claimant's right of action on the labor and material bond or the claimant's contract action.[26]

In *Breda Construzioni Ferroviarie S.P.A. v. Los Angeles County Metropolitan Transit Authority*,[27] the court of appeal has ruled that interest earned on funds withheld in response to a bonded stop notice must be paid to the contractor claimant and cannot be withheld by the project owner, in this case a public entity. One of Breda's subcontractors served a $4,255,000 stop notice, pursuant to which the MTA withheld funds, which were put into an interest-bearing account. By the time Breda, the prime contractor, bonded around the subcontractor's stop notice, the account had earned $616,743 in interest. The MTA released the principal amount to Breda, pursuant to its stop notice release bond, but kept the earned interest for itself. Breda brought an action and recovered the interest that had been actually earned but kept by the MTA; however, the trial court denied Breda's request to be awarded interest at the legal rate of 10 percent.

The court of appeal affirmed the judgment in favor of Breda, holding that the "takings clause" of the Fifth Amendment of the United States Constitution, which applies to the states under the Fourteenth Amendment, as well as California common law, bars a public entity from retaining for its own account the interest earned on funds withheld from a prime contractor by reason of a subcontractor's stop notice. The public entity acts solely as a stakeholder and has no right to reap a windfall in the form of interest earned on funds belonging to the prime contractor. Also, no provision of the stop notice law authorizes the withholding entity to "defray expenses of administering the stop notice procedure."

---

[19] *Id.* § 3199.
[20] *Id.* § 3200.
[21] *Id.*
[22] Cal. Civ. Code § 3201.
[23] *Id.* § 3202
[24] *Id.*
[25] *Id.* § 3204.
[26] *Id.* § 3205.
[27] 56 Cal. App. 4th 1433 (1997).

The court of appeal would not, however, award the higher interest rate (10 percent) under California Public Contract Code § 20104.50, because the funds withheld pursuant to a stop notice are not "undisputed" as that term is used in California's prompt payment statutes.

[D] Suit on Stop Notice

In order to perfect a stop notice, a suit must be brought. Suit on the stop notice is brought no sooner than 10 days after its service and no later than 90 days after the expiration of the time for filing stop notices.[28] Notice of suit must be given to the public body within five days after commencement.[29] The court has the discretionary right to dismiss the suit if not brought to trial within two years.[30] All claimants may join in one action. Separate stop notice suits may be consolidated. The public body may require interpleader.[31]

§ 10.03 LABOR AND MATERIAL BOND

[A] Preliminary 20-Day Notice

In addition to the stop notice remedy, all public works contracts in excess of $25,000 must be bonded with a labor and material bond by the prime contractor.[32] The following is a procedural outline of what must be done to perfect a claim on that bond.[33]

[1] Who Gives 20-Day Notice?

All claimants except laborers, trust funds, and claimants with a direct contractual relationship with the original contractor must serve the 20-day notice.[34]

[2] When Is 20-Day Notice Given?

The notice must be served within 20 days from the date on which the claimant first furnished labor, services, equipment, or materials to the jobsite.[35] Like the stop notice, if the claimant failed to serve the 20-day notice within the first 20 days, the claimant may serve the notice at any time and make a claim on the bond for labor, service, equipment, or materials furnished within 20 days of the notice and thereafter under California Civil Code § 3098(D). In addition, if the claimant failed to serve a 20-day notice, the claimant may still seek to recover

---

[28] Cal. Civ. Code § 3210.
[29] Id. § 3211.
[30] Id. § 3212.
[31] Id. § 3214.
[32] Cal. Civ. Code § 3247.
[33] Id. § 3098.
[34] Cal. Civ. Code § 3098.
[35] Cal. Civ. Code § 3098.

on the bond if the claimant serves a notice on the principal and surety on the payment bond within 15 days of a notice of completion or within 75 days of completion if there is no notice of completion pursuant to Civil Code § 3252(B). Such notice may be served by personal service, certified mail, or registered mail.[36] The notice must contain (1) the kind of labor, service, equipment, or materials furnished, (2) the name of the person to or for whom the labor, service, equipment, or materials were furnished, and (3) the amount in value, as near as may be determined, of any labor, service, equipment, or materials already furnished or to be furnished.[37]

[B] Time For Filing Suit On Bond

A claimant must file suit on the labor and material bond no later than six months after the period for filing stop notices expires. Specifically, California Civil Code § 3249 provides that suit against the surety on the payment bond must be brought by the claimant after the claimant has furnished the last of its labor or materials and no later than six months after the period in which stop notices may be filed, as provided in section 3184 of the Civil Code. Section 3184 of the Civil Code, in turn, provides that the time for filing stop notices is 30 days after the recording of a notice of completion or notice of cessation of labor or 90 days after completion or cessation of the work.

Civil Code § 3086 defines the term *completion*. With regard to public work, Civil Code § 3086 provides that, if the work of improvement is subject to acceptance by any public entity, the completion of such work of improvement shall be deemed the date of such acceptance; provided, however, that, except as to contracts awarded under the State Contract Act, a cessation of labor on any public work for a continuous period of 30 days shall be a completion thereof.

The meaning of this section was interpreted in *W.F. Hayward Co. v. Transamerica Insurance Co.*[38] Specifically, the court interpreted the portion of Civil Code § 3086 which provides that completion, except as to contracts awarded under the State Contract Act, occurs when there is a cessation of labor on any public work for a period of 30 days.

In *Hayward,* the contractor contracted with the county of Los Angeles to construct a sheriff's station. Hayward was a subcontractor. The project began, but a number of delays ensued. A dispute arose between the contractor and the county concerning responsibility for the delays. The contractor and its subcontractors, including Hayward, sought additional payments from the county for the delay. The county suspended the contractor on June 5, 1990. The prime contractor requested that the county approve a change order for $1,227,771.79, which included a claim from Hayward in the sum of $243,650. The prime contractor ceased work on the project effective June 5, 1990, and never returned to the project.

---

[36] *Id.* § 3227(a).
[37] *Id* § 3227(b).
[38] 16 Cal. App. 4th 1101 (1993).

In August 1990, after labor had been stopped on the project for more than 30 days, the prime contractor and the county entered into an agreement pursuant to which the county relieved the prime contractor of any further obligation or liability regarding construction of the sheriff's station, and pursuant to which the prime contractor assigned to the county all its subcontracts, including the subcontract with Hayward. The county then finished the job using the original prime contractor's subcontractors, including Hayward. After the job was finally completed, Hayward brought an action on the original prime contractor's payment bond for recovery of the $243,650 that was owed by the original prime contractor. The question in the case was whether the suit had been timely commenced.

The court, first of all, examined Civil Code § 3249, which provides that suit against the surety on the payment bond must be commenced before the expiration of six months after the period in which stop notices may be filed, as provided in Civil Code § 3184. The court then examined Civil Code § 3184, which provides that stop notices must be served before the expiration of 30 days after the recording of a notice of completion or notice of cessation or, if no notice of completion or notice of cessation is recorded, then within 90 days after completion or cessation. The court noted that, in this particular case, neither a notice of completion nor a notice of cessation had been filed or recorded. Thus, the question presented was: When was there a completion or cessation?

The court then turned to Civil Code § 3086, which provides that if a work of improvement is subject to acceptance by any public entity, the completion of such work of improvement shall be deemed to be the date of such acceptance; provided, however, that a cessation of labor on any public work for a continuous period of 30 days shall be deemed a completion. Looking at the facts of this case, the court noted that labor ceased on June 5, 1990. A 30-day cessation would, therefore, take place between June 5, 1990, and July 5, 1990. That would be deemed a completion. Thus, the stop notices would have had to have been served within 90 days of July 5, 1990—that is, on or before October 3, 1990. Suit on the bond would have to have been brought within six months of October 3, 1990—that is, no later than April 5, 1991. The suit in this particular case was filed on July 17, 1991, and therefore the court ruled that the suit was not timely brought.

This case seems to be a very strict reading and interpretation of the Code sections in light of the realities of the construction industry. It is assumed that the original subcontractor, Hayward, and certainly any material suppliers that it might have had on the project, would have considered that their time to sue on the bond did not run until completion of the job as a whole. It was only by virtue of fortuitous circumstance—that the original prime contractor was terminated and that there was a cessation of labor for 30 days—that the time for bringing suit on the original prime contractor's bond expired before the job was finished. This case seems to yield a very strict and harsh result. Certainly, subcontractors and material suppliers would be misled under these circumstances, and it seems to the authors that this case's strict interpretation of the statute is contrary to many of the cases that call for liberal interpretation of the statutes relating to claims on public works.

[C] Other Requirements

A payment bond is required *only* on public jobs in excess of $25,000.[39] The amount of the bond is calculated as follows:

1. Contract from $25,000 to $5 million: 100 percent of the contract amount
2. Contract from $5 million to $10 million: 50 percent of the contract amount
3. Contract in excess of $10 million: 25 percent of the contract amount.[40]

The bond must provide for attorneys' fees, and the prevailing party is entitled to reasonable attorneys' fees.[41]

Filing a stop notice is not necessary to sue on a labor and material bond.[42]

California Civil Code § 3267 on its face seems to indicate that only subcontractors and their subcontractors or suppliers can recover on the bond. This statute was enacted to prevent someone not in the chain of authority under the prime contractor from recovering on the labor and material bond, but the language of the section is more restrictive than that. The section states:

> *Work done for principal on bond; claim.* Nothing contained in this title shall be construed to give to any person any right of action on any original contractor's private or public work payment bond described in Chapter 6 (commencing with Section 3235) or 7 (commencing with Section 3247) of this title, unless the work forming the basis for his claim was performed by such person for the principal on such payment bond, or one of his subcontractors, pursuant to the contract between the original contractor and the owner.
>
> Nothing in this section shall affect the stop notice rights of, and relative priorities among, architects, registered engineers or licensed land surveyors and holders of secured interests on the land.

Although Civil Code § 3267 seems to indicate that only subcontractors and their subcontractors or suppliers can recover on the bond, the case *Union Asphalt, Inc. v. Planet Insurance Co.*[43] holds that anyone in the chain of construction can recover on the bond.

In *Union Asphalt,* the owner was developing a subdivision, and under the Subdivision Map Act (California Government Code §§ 66499.1 and 66994.2), the public body required performance and payment bonds. The developer entered into a contract with Spiess Construction to construct the work. Spiess subcontracted 100 percent of the work to Swift Techtonics. Swift, in turn, as a second-tier subcontractor, subcontracted portions of the work to Union Asphalt and

---

[39] Cal. Civic Code § 3247(a)
[40] *Id.* § 3248(a)(1), (2), (3).
[41] *Id.* §§ 3248(b), 3250.
[42] *Id.* § 3250.
[43] 18 Cal. App. 4th 633 (1993).

purchased materials from Southern Pacific Milling and Westburn. Union Asphalt, Southern Pacific, and Westburn brought actions on the payment bond. The surety defended on the basis of Civil Code § 3267, contending that said section limits claims on payment bonds to first-tier material suppliers and second-tier subcontractors; because Union Asphalt, Southern Pacific, and Westburn were third-tier subcontractors after Spiess and Swift Techtonics, they were not covered by the bond.

The court noted that no previous case had interpreted Civil Code § 3267. The court also noted that there appears to be no good reason why the legislature would cut off suppliers and subcontractors at the second tier for payment bond, but give suppliers and subcontractors of any tier the right to a mechanic's lien. The court stated that it would be unreasonable to draw such an arbitrary line. Finally, the court concluded—as the authors have concluded—that the purpose of the statute is to exclude architects, engineers, and land surveyors, who perform work prior to or otherwise outside the scope of the construction contract and have no claim on the contractor's payment bond.

§ 10.04   CASE EXAMPLES

[A]   Decisions Regarding Stop Notice and Bond Remedies

There are numerous cases interpreting the stop notice and bond statutes on public works. Following are some of the leading decisions:

- The public body may be liable for negligence if it fails to require the prime contractor to put up a labor and material bond.[44]

- A bonding company on a public works bond cannot use a release of lien as a defense.[45] Compare, however, *Card Construction Co. v. Ledbetter,*[46] which held that the prime contractor who had paid out money to a subcontractor in reliance upon a release given by the material supplier could seek indemnity against that material supplier.

- The notice requirement on public work is liberally construed. For example, a stop notice was construed to comply with the old 90-day notice requirement[47] (which has been replaced by the 20-day notice as set forth in § 10.03[A] above); giving a list of creditors, including the name and amount due the claimant, to a bonding company representative at a creditors' meeting was held to be sufficient notice,[48] and proof of receipt of notice, whether or not it complied strictly with the statute, was held to be sufficient.[49]

---

[44] C.A. Magistretti Co. v. Merced Irrigation Dist., 27 Cal. App. 3d 270 (1972).
[45] Powers Regulator Co. v. Seaboard Sur. Co., 204 Cal. App. 2d 338 (1962); Bonded Prods. Co. v. R.C. Gallyon Constr. Co., 228 Cal. App. 2d 186 (1964).
[46] 16 Cal. App. 3d 472 (1971).
[47] General Elec. Co. v. Central Sur. & Ins. Corp., 232 Cal. App. 2d 590 (1965).
[48] California Elec. Supply Co. v. United Pac. Life Ins. Co., 227 Cal. App. 2d 138 (1964).
[49] Fidelity Sound Sys., Inc. v. American Bonding Co., 85 Cal. App. 3d 13 (1978).

- A stop notice on public works affects only funds actually due the contractor. If the contractor defaults, the public body may use the contract funds to complete the job in disregard of the stop notice.[50] This is contrary to the rules on private works. Once the loan company receives a valid stop notice on private works, the loan company must honor it and withhold.[51]

- When a contractor posts a release bond, the claimant's cause of action is on the bond, not on the original stop notice.[52]

- Notice of suit that was required to be given within five days under California Civil Code § 3211 was in fact given 14 days after suit. The court held that because the public body was not prejudiced, the plaintiff was entitled to recover on the stop notice.[53]

- A stop notice may be filed on public work even though the money may not be owed to the subcontractor under the terms of the contract with the prime contractor.[54]

- Tender by the prime contractor to the subcontractor of the principal amount due on the subcontract after the subcontractor has brought suit on the labor and material bond may be rejected by the subcontractor, who may then also recover interest and attorneys' fees.[55]

- When the claimant on a labor and material bond loses its lawsuit, the bonding company is the prevailing party and entitled to attorneys' fees.[56]

- A premature payment to the contractor by the public body confers no rights on the stop notice claimant.[57]

- The public body is not required to anticipate that stop notices may be filed and therefore may make final payment to the contractor even though the time for filing stop notices has not expired.[58]

- A stop notice claimant can claim only that portion of the subcontract price that has actually been performed when the stop notice is filed.[59]

---

[50] Harsco Corp. v. Department of Pub. Works, 21 Cal. App. 3d 272 (1971).

[51] Calhoun v. Huntington Park First Sav. & Loan Ass'n, 186 Cal. App. 2d 451 (1960); Rossman Mill & Lumber Co. v. Fullerton Sav. & Loan Ass'n, 221 Cal. App. 2d 705 (1963); A-1 Door & Materials Co. v. Fresno Guar. Sav. & Loan Ass'n, 61 Cal. 2d 728 (1964); Idaco Lumber Co. v. Northwestern Sav. & Loan Ass'n, 265 Cal. App. 2d 490 (1968); H.O. Bragg Roofing, Inc. v. First Fed. Sav. & Loan Ass'n, 226 Cal. App. 2d 24 (1964).

[52] Cal-Pacific Materials Co. v. Redondo Beach City Sch. Dist., 94 Cal. App. 3d 652 (1979).

[53] Sunlight Elec. Supply v. McKee, 226 Cal. App. 2d 47 (1964).

[54] Central Indus. Eng'g Co. v. Strauss Constr. Co., 98 Cal. App. 3d 460 (1979).

[55] Id.

[56] Kalico, Inc. v. Dooley, 198 Cal. App. 2d 379 (1961); Winick Corp. v. Safeco Ins. Co., 187 Cal. App. 3d 1502 (1986). See also § 10.04 [B].

[57] Pacific Employers Ins. Co. v. State, 3 Cal. 3d 573 (1970).

[58] American Fidelity Fire Ins. Co. v. United States, 385 F. Supp. 1075 (N.D. Cal. 1974).

[59] University Casework Sys., Inc. v. Superior Court, 41 Cal. App. 3d 263 (1974).

- A city was held liable on a stop notice when it allowed the prime contractor to use the same surety on the stop notice release bond that had furnished the labor and material bond. To hold otherwise would defeat the purpose of the law requiring the two sureties to be jointly and severally liable.[60]

- A public body is entitled to attorney's fees when it files interpleader of stop notice claims.[61] This case also stands for the proposition that, if the surety on the labor and material bond pays off the stop notice claims, it is entitled to the funds being held by the public body. Also, the public body may not offset its costs and fees in handling the stop notice claims. The public body's remedy is interpleader. If the surety on the labor and material bond has to sue to recover the funds in the hands of the public body after the stop notice claimants have been paid by the surety, the surety can recover interest and attorneys' fees against the public body for wrongfully refusing to pay over the stop notice funds.

- When the state claimed unemployment insurance contributions against the labor and material bond, the court held that the state was not required to give a preliminary notice.[62]

- The bonding company's liability on the labor and material bond is determined by the contract price between the claimant and the bonding company's principal and is not limited to reasonable value, as is a mechanic's lien claim. Also, in this case, the plaintiff loaned equipment to the prime contractor. The lease had a minimum lease period. Labor ceased while the prime contractor negotiated with the city, during which time the equipment sat idle on the job. The plaintiff was allowed to recover for equipment rental against the bonding company.[63]

- No notice of completion was recorded on a public works project. Ninety-one days after labor ceased, the Department of Industrial Relations filed a stop notice because the contractor had not paid workers the prevailing wage rates. When the contractor agreed to pay the workers a "bonus" to compensate them for the lower wages, that money was released. The public body continued to hold a $10,000 "penalty" asserted by the Department of Industrial Relations in the stop notice. As noted under California Civil Code § 3184, if there is no notice of completion on a public works project, the stop notice must be filed within 90 days of completion. On the other hand, California Labor Code § 1727 states that no sum shall be withheld from the contractor's final payment without a full investigation by either the public agency or the Division of Labor Law Enforcement. The contractor filed a suit to cause the money to be released. The court held that Civil Code § 3184 did not apply to the Department of Industrial Relations because it was not a "person" under § 3184,

---

[60] Azusa W., Inc. v. West Covina, 45 Cal. App. 3d 259 (1975).
[61] Leatherby Ins. Co. v. Tustin, 76 Cal. App. 3d 678 (1977).
[62] State *ex rel.* Department of Employment v. General Ins. Co. of Am., 13 Cal. App. 3d 853 (1970).
[63] John A. Artukovich Sons, Inc. v. American Fidelity Fire Ins. Co., 72 Cal. App. 3d 940 (1977).

and furthermore, Labor Code § 1727 did apply as to the penalties claimed, and therefore, the Department could not give notice to withhold money until the matter had been investigated.[64]

- A growing practice in the construction industry is for certain firms to obtain the workers, and pay the wages of the contractor's and subcontractor's workers, in exchange for a fee. Those firms have relied upon the lien, stop notice, and bond rights for recovery in the event that their clients failed to pay. They usually perform their services for subcontractors and contractors who are not capable of performing those functions on their own. The issue often arises as to whether such companies are proper claimants on the bond.[65]

---

[64] Kruger Bros. Builders, Inc. v. San Francisco Hous. Auth., 198 Cal. App. 3d 1 (1988).

[65] In *Primo Team, Inc. v. Blake Construction Co.*, 3 Cal. App. 4th 801 (1992), such a company filed a claim on a prime contractor's payment bond on a state public works project. The company had assembled the subcontractor's workforce and administered the paperwork relating to payment of the subcontractor's workers. The workers were directly employed by the subcontractor, but the claimant, Primo Team, Inc., obtained the workers, paid their wages, paid the withholding taxes, handled other matters, and even had jobsite safety meetings. Primo advanced the wages, the employer tax contributions, and the workers' compensation insurance premiums, and was repaid 135.7 percent of the costs expended.

When the subcontractor failed to reimburse Primo, Primo filed claims on the prime contractor's payment bond. The issue was whether Primo was a proper claimant on the bond. Civil Code § 3110 extends mechanic's lien rights, and therefore public bond protection, to "all persons . . . performing labor upon or bestowing skill or other necessary services on . . . a work of improvement." The appellate court acknowledged that bond protection was not limited to persons who physically labor on the jobsite but can include entities that furnish laborers or equipment to the site. The court held, however, that Primo did not furnish laborers to the site; instead, it furnished them to the subcontractor, who in fact employed the laborers and supervised their work. The court observed that the bulk of Primo's claim rested on its request for reimbursement of the funds advanced to the subcontractor in the form of wages, benefits, withholdings, and insurance premiums. In this, Primo acted like a lender, and lenders are not entitled to coverage under payment bonds. The court observed that Primo did not provide its services to the job but instead to the subcontractor. Even though Primo's services were beneficial and necessary to the subcontractor's operation as a business, the services were not bestowed upon the work of improvement.

The *Primo Team* case was distinguished in *Contractors Labor Pool, Inc. v. Westway Contractors, Inc.*, 53 Cal. App. 4th 152 (1997), in which Contractors Labor Pool (CLP) entered into a contract, with a subcontractor, entitled "Contractors Labor Pool Agreement." It provided that (1) the subcontractor would pay CLP an hourly rate for each hour a CLP employee performed services for the subcontractor on a public works project; (2) CLP would be responsible for the employees' payrolls, taxes, worker's compensation, insurance premiums, and federal and state withholding taxes and fringe benefits; (3) the subcontractor would control the activities of the employees while on the job; (4) the subcontractor would indemnify CLP for damages to persons or property arising out of work performed or not performed by the employees; and (5) the subcontractor would provide CLP with all information necessary for CLP to maintain its mechanic's lien rights.

In allowing CLP to recover on the payment bond posted by the surety on the public work of improvement, the court of appeal distinguished the *Primo Team* case, by stating that the workers in the *Contractors Labor Pool* case were employed directly by CLP; therefore, CLP was a person performing labor upon or bestowing skill or other necessary services on the work of improvement, and thus was within the definition of persons entitled to a mechanic's lien under Civil Code § 3110. In contrast, the workers furnished by Primo Team in fact became employees of the subcontractor.

§ 10.04[A] CALIFORNIA CONSTRUCTION LAW

- A private stop notice claimant may be entitled to interest on the unpaid balance due on its contract, including prejudgment interest.[66] The court also held that the stop notice claimant had the right to question liquidated damages.

- Any attorney's fees incurred by a subcontractor in enforcing its claim on the payment bond are probably chargeable to the surety on the payment bond if the subcontractor prevails.[67]

- Public works payment bonds cover the obligation to pay prevailing wages, and the 90-day limitations period for actions under section 1775 of the Labor Code, regarding the action the DLSE (the Division of Labor Standards Enforcement of the State Department of Industrial Relations) is authorized to bring against the contractor for prevailing wage deficiencies, does not apply to an action for payment of prevailing wages against the surety on a payment bond. Instead, the limitations period for an action on a payment bond prescribed by Civil Code § 3249 applies (6 months plus 30 days after recording the notice of completion). Thus, if the DLSE misses the 90-day deadline for filing suit against the contractor, it will still have time to bring suit against the payment bond surety within the time limits of Civil Code § 3249.[68]

The *Contractors Labor Pool* case also stands for the proposition that a person furnishing employees for a work of improvement, who does not actually supervise their work on the jobsite, is not a contractor and therefore need not be licensed under California Business and Professions Code § 7031. Finally, the case stands for the proposition that, with regard to the award of attorney's fees to CLP (as the prevailing party on the action on the payment bond), CLP was entitled to an award of "reasonable" attorney's fee under Civil Code § 3250. Therefore, the trial court erred in applying a Los Angeles Superior Court local rule used to determine attorney's fees. In determining what constitutes a reasonable attorney's fee when a statute provides for such an award, courts should consider the nature of the litigation; its difficulty; the amount involved; the skill required; the success of the attorney's efforts; his or her learning, age, and experience in the particular type of work demanded; the intricacies and importance of the litigation; the labor and necessity for skilled legal training; the attorney's ability in trying the cause; and the time consumed.

[66] Westinghouse Electric Corp. v. County of Los Angeles, 129 Cal. App. 3d 771 (1982).
See also Washington International Insurance Co. v. Superior Court, 62 Cal. App. 4th 981 (1998); the prime contractor received progress payments from the public body which it failed to pay to its subcontractor. The subcontractor brought an action on the public works payment bond and sought the statutory penalty of 2% per month under Public Contract Code §10262.5, (the Prompt Pay Statute which was applicable to this particular public works project). The court allowed recovery of the 2% per month penalty on the theory that the public works payment bond secures payment of "claims" and the 2% per month penalty was part of the subcontractor's "claim" and therefore recoverable under Public Contract Code §10233.

[67] Litton Constr. Gen. Eng'g Contractor, Inc. v. United Pac. Ins.,16 Cal. App. 4th 577, 20 Cal. Rptr. 2d 200 (1993). *See also* § 10.04[B].

[68] Department of Indus. Relations v. Seaboard Sur., 50 Cal. App. 4th 1501 (1996). *See also* Department of Indus. Relations v. Fidelity Roof Co., 57 Cal. App. 4th 445, (1997). In that case, a roofing company contracted with a school district to reroof a high school. The DLSE filed suit against the contractor to recover unpaid wages and penalties under the prevailing wage law (Cal. Labor Code §§ 1720 *et seq.*). The DLSE also sued the surety on the contractor's payment bond. The trial court entered judgment in favor of the DLSE against the contractor for the unpaid wages and also entered judgment in favor of the surety and against the DLSE on the ground that the payment bond inured to the benefit of material suppliers and therefore was not surety for unpaid

- A subcontractor sued the original contractor and the surety on the payment bond on a public works project. The subcontract had an arbitration clause. The court held that the action against both the contractor and the surety should be stayed until the arbitration proceedings determined how much was due under the subcontract.[69]

[B] Attorney's Fees on Payment Bond on Public Works

One of the few areas in construction law where attorney's fees are provided by statute is for claims made on payment bonds on public works in the state of California. Specifically, California Civil Code § 3248(b) provides that any payment bond issued on a public work must provide for reasonable attorney's fees, to be fixed by the court in case any suit is brought upon the bond. In addition, Civil Code § 3250 provides that, in any action on the payment bond, the court shall award to the prevailing party a reasonable attorney's fee to be taxed as costs. Thus, it is clear that any claimant who brings an action on a payment bond is entitled to attorney's fees if the claimant prevails. For example, in *Kalico, Inc. v. Dooley*,[70] the court held that if the surety company prevails in the action on the bond, the surety company is entitled to attorney's fees.

In *Litton Construction General Engineering Contractor Inc. v. United Pacific Insurance*,[71] the California courts again reemphasized the right of a claimant on a payment bond on public works to recover attorney's fees. In this case, the state withheld payment from the prime contractor as liquidated damages for delay. The prime contractor, in turn, withheld liquidated damages from its subcontractor. The subcontract provided for arbitration and that each party would bear its own legal fees in connection with any arbitration.

---

wages. The appellate court reversed both the judgments of the trial court. The appellate court held that the DLSE's complaint against the contractor for wages and penalties was barred by Labor Code § 1775 since the complaint was not filed within 90 days after the public works project was accepted as completed by the school district. The court noted that under Labor Code § 1775, the 90-day period begins to run upon the occurrence of the latter of two events—to wit, the acceptance of the project or the recording of a valid notice of completion. Even though a valid notice of completion was never recorded, the statute did not require that both events occur. With respect to the DLSE's complaint against the surety, the court held that the DLSE was a proper claimant against the surety when acting on behalf of aggrieved workers to recover their unpaid prevailing wages from the surety on the payment bond. The contract between the contractor and the school district specified that the contractor would pay prevailing wages, and the aggrieved workers were third-party beneficiaries of that contract. Although Civil Code § 3248(c) provides that the payment bond shall inure to the benefit of those who provide labor or materials or their assigns and although there was no evidence that the DLSE received an assignment from the aggrieved workers to recover the unpaid wages, Labor Code § 96.7 permits the labor commissioner to collect unpaid wages and benefits without assignment. The court held that that specific statute took precedence over the more general statute touching on the same subject, and therefore, under Labor Code § 96.7, which is specifically directed to the DLSE's ability to recover unpaid prevailing wages, the commissioner could collect the unpaid wages and benefits without a formal assignment.

[69] Federal Ins. Co. v. Superior Court, 60 Cal. App. 4th 1370, (1998).
[70] 198 Cal. App. 2d 379 (1961).
[71] 16 Cal. App. 4th 577, 20 Cal. Rptr. 2d 200 (1993).

The subcontractor filed suit against the contractor and the payment bond surety, but the suit prayed for recovery of attorney's fees against the surety company only. The prime contractor and the surety filed motions to stay the litigation and to compel the subcontractor to arbitrate its claims against the contractor. The motion was granted, and an arbitration took place between the prime contractor and the subcontractor. The arbitrator ruled in favor of the subcontractor and awarded the subcontractor all of the liquidated damages that had been withheld ($46,480), plus interest. The arbitrator, following the contract, did not award the subcontractor attorney's fees. The award was then reduced to judgment and was paid by the contractor.

After the stay of litigation was lifted, because the arbitration had concluded, the subcontractor renewed its cause of action against the surety for payment of attorney's fees. The trial court entered summary judgment in the subcontractor's favor in the amount of $93,029, and the surety company appealed. The court held that the surety was liable to the subcontractor for the attorney's fees incurred by the subcontractor in the arbitration against the prime contractor. The court held that the post-award suit by the subcontractor against the surety was an action on the bond; thus, the Code section previously cited entitled the subcontractor to payment of attorney's fees as the prevailing party. The court noted that the subcontractor had been compelled by the prime contractor and the surety to go through the arbitration proceeding before the subcontractor was given an opportunity to present its claim on the payment bond. Under those circumstances, the court reasoned that the arbitration fees were incurred by the subcontractor as part of its action "upon the bond."

In *Granite Construction Co. v. American Motorists Insurance Co.*,[72] Longview Corporation contracted with the city of Roseville, in the county of Sacramento, to construct a subdivision. Those contracts required the construction of public improvements, for which Longview Corporation was required to provide performance and payment bonds. American Motorists Insurance Company was the surety on the bonds. Longview Corporation entered into a subcontract with Site Contractors to do site improvements. Site Contractors, in turn, entered into a sub-subcontract with Granite Construction Company to lay asphalt concrete street paving. That sub-subcontract provided for unit prices and interest in the amount of $1\frac{1}{2}$ percent per month on any balance unpaid when due. Likewise, the sub-subcontract provided that because Site Contractors was preparing the subgrade, Granite Construction Company would not be responsible for any drainage problems or subgrade failures. Site Contractors prepared the underlying base material for the asphalt concrete. On January 20, 1989, Granite Construction completed the initial laying of asphalt and billed Site Contractors $45,585.45. The billing went out in January 1989. The underlying base material failed, damaging the asphalt, and the county and city inspectors refused to grant final approval of the street work. Granite Construction agreed to apply a seal coat to the asphalt concrete and did so on May 26, 1989. Granite absorbed the $2,555 cost and made no billing for that work.

---

[72] 29 Cal. App. 4th 658 (1994).

On June 20, 1989, Granite Construction served a 90-day notice on Longview Corporation. (This case arose when a 90-day notice was required for an action on a payment bond. As noted in § 10.03[A] above, the 90-day notice has been replaced by the 20-day notice.) This notice was sent more than 90 days after January 20, 1989, but within 90 days of May 26, 1989. Longview Corporation and Site Contractors filed for bankruptcy, and Granite Construction brought an action against American Motorists to recover on the bond. American Motorists contended that the notice given by Granite Construction on June 20, 1989, was given more than 90 days after Granite completed its work on January 20, 1989. Granite Construction contended that because the notice was sent within 90 days of May 26, 1989 (the time when it put on the seal coat), the notice was proper.

The court discussed what it called the "90-day gap rule," which stems from interpretations of the Miller Act, in the cases of *United States v. Peter Reiss Construction Co.*[73] and *United States v. Sovereign Construction Co.*[74] In those two cases, after material suppliers had commenced deliveries on the jobsite, there was a 90-day period in which no deliveries were made, after which additional deliveries were resumed. The material suppliers' 90-day notice was not given until after resumption of the deliveries, and both cases held that, as to deliveries before the 90-day gap, the material suppliers could not recover. The court distinguished those cases and held that in this particular case, Granite Construction performed its work for Site Contractors pursuant to a single contract. Thus, until the work was completed on May 26, 1989, by the application of the seal coat, the period of time for sending the 90-day notice did not expire. The court specifically held that the notice was timely.

American Motorists also contended that the application of the seal coat was not a part of Granite's "claim" and therefore could not constitute labor, service, equipment, or materials for which a claim was made, because Granite did not charge for the work. The court, citing *United States ex rel. P.A. Bourquin & Co. v. Chester Construction Co.*,[75] *United States v. U.S. Fidelity & Guaranty Co.*,[76] and *United States v. Gunnar I. Johnson & Son, Inc.*,[77] held that the seal coat was a part of Granite's claim. The court noted that the seal coat was the last labor furnished on the project for which Granite sought payment under the bond; the fact that Granite did not pass the charge along to Site Contractors was of no significance. Granite Construction was also seeking attorney's fees and interest at the rate of $1\frac{1}{2}$ percent per month, as provided for in the contract. American Motorists contended that Granite should only recover attorney's fees incurred in the suit against the bond, and that the bond did not provide for interest at $1\frac{1}{2}$ percent per month, and therefore, the interest should be 10 percent. The court rejected these arguments as well. The court held that Granite was entitled to all attorney's fees that it incurred in pursuing its claim—not only the attorneys' fees incurred in

---

[73] 273 F.2d 880 (2d Cir. 1959).
[74] 338 F. Supp. 659 (S.D.N.Y. 1972).
[75] 104 F.2d 648 (2d Cir. 1939).
[76] 656 F.2d 993 (5th Cir. 1981).
[77] 310 F.2d 899 (8th Cir. 1962).

the action against American Motorists' bond but also the attorney's fees that Granite incurred in attempting to collect from Longview Corporation and Site Contractors, before they filed bankruptcy.

In addition, the court allowed the interest rate at $1^1/_2$ percent per month, under the case of *Southwest Concrete Products v. Gosh Construction Corp.*,[78] from the date that the amount became due to Granite Construction (to wit, 30 days after Granite rendered its billing to Site Contractors). Thus, this case stands for the proposition that a sub-subcontractor on a public works payment bond can recover interest, at the rate set forth in its contract, from the date the obligation became due, and likewise attorney's fees. It should be noted in passing that the payment bond in this case did provide as follows:

> NOW THEREFORE, said principal [Longview Corporation] and the undersigned as corporate surety [American Motorists Insurance Company], are held firmly bound unto the County of Sacramento and all contractors, subcontractors, laborers, materialmen and other persons employed in the performance of the aforesaid agreement and referred to in the aforesaid Code of Civil Procedure, in the sum of [$76,750], for materials furnished or labor thereon of any kind, or for amounts due under the Unemployment Insurance Act with respect to such work or labor, that said surety will pay the same in an amount not exceeding the amount hereinabove set forth, and also in case suit is brought upon this bond, will pay, in addition to the face amount thereof, costs and reasonable expenses and fees, including reasonable attorney's fees, incurred by County in successfully enforcing such obligation, to be awarded and fixed by the court, and to be taxed as costs and to be included in the judgment therein rendered.
>
> It is hereby expressly stipulated and agreed that this bond shall inure to the benefit of any and all persons, companies and corporations entitled to file claims under Title 15 (commencing with Section 3082) of Part 4 of Division 3 of the Civil Code, so as to give a right of action to them or their assigns in any suit brought upon this bond.

The court specifically held that the foregoing language included the right to recover attorney's fees in enforcing the "obligation," as Granite was enforcing the obligation to pay under the bond.

[C] Surety's Right to Raise Its Principal's Defenses

In *Kalfountzos v. Hartford Fire Insurance Co.*,[79] a subcontractor on a public works project brought an action against the general contractor and the surety on the general contractor's payment bond and stop notice release bonds. By the time the case came to trial, the general contractor (the principal on the payment and stop notice release bonds), because it was a corporation whose authorization to do business had been suspended due to nonpayment of franchise tax fees, had

---

[78] 51 Cal. 3d 701 (1990)
[79] 37 Cal. App. 4th 1655 (1995).

no legal capacity to defend itself in court. When the surety sought to raise the defenses and setoffs that its principal (the general contractor) had, the plaintiff subcontractor objected.

The trial court allowed the surety to assert the defenses and setoffs to the underlying obligation of its principal, even though its principal, who had obtained the bond, was legally disabled. As a result, the trial court determined that there was nothing due and owing to the subcontractor and rendered judgment for the surety.

The subcontractor appealed, and the court of appeal affirmed the decision of the trial court, stating that the general rule is that a surety may raise all defenses that would be available to the principal (citing *Flickinger v. Swedlow Engineering Co.*[80]). The surety must pay on a bond only if the claimant establishes, without reference to the bond, a legal obligation on the part of the principal to pay.[81] Under Civil Code § 2809, the obligation of the surety must be neither larger in amount nor, in any other respect, more burdensome than that of its principal. As a result, if the principal is not liable on the obligation, neither is the surety.[82]

With those general principles of suretyship in mind, the court of appeal held that the obligation of the principal on the bond was the obligation of the general contractor to the subcontractor. Because the surety was not liable to the subcontractor unless liability existed under the underlying obligation (that is, the liability of the general contractor to the subcontractor), the surety could raise any defenses or setoffs with respect to its principal's obligation to reduce or to eliminate the amount due on the bond, even though the principal (the general contractor) was precluded from raising those defenses and setoffs itself by reason of a legal disability. The court noted that any other conclusion would provide the creditor (i.e., the subcontractor) with an unjustified windfall.

---

[80] 45 Cal. App. 2d 388 (1995).
[81] Lewis & Queen v. N.M. Ball Sons, 48 Cal. 2d 141 (1957).
[82] United States Leasing Corp. v. Dupont, 69 Cal. App. 2d 275 (1968).

CHAPTER 11
MILLER ACT

§ 11.01   Labor and Material Bond
§ 11.02   Ninety-Day Notice
§ 11.03   Suit on the Bond
§ 11.04   Persons Entitled to Recover on the Bond
§ 11.05   Proof Required
§ 11.06   Practice Notes
§ 11.07   Illustrative Cases
    [A]   Claim Against Person Certifying Sufficiency of the Surety
    [B]   Pay-If-Paid Clause Did Not Waive Miller Act Bond Rights
    [C]   General Contractor Not Indispensable Party in an Action on the Miller Act Payment Bond
    [D]   Recovery of Money Due under Savings Clause Allowed
    [E]   Equitable Lien Not Allowed

MILLER ACT § 11.02

§ 11.01 LABOR AND MATERIAL BOND

The Miller Act[1] is the statutory procedure on federal construction projects that is equivalent to the bond provisions for state public jobs. *See* Chapter 10.

Section 270a of the Act provides that in contracts exceeding $100,000 for the construction, alteration, or repair of any public building or public work of the United States, the contractor will furnish a labor and material bond.

If the contract is less than $1 million, the bond is one-half the contract price. When the contract is between $1 million and $5 million, the bond is 40 percent of the contract price. When the contract is in excess of $5 million, the bond amount is $2 million.

There are two procedural steps to enforcing a claim under the Miller Act bond: (1) preliminary notice; and (2) suit on the bond.

§ 11.02 NINETY-DAY NOTICE

If the claimant does not have a direct contractual relationship with the prime contractor, the claimant must, as a condition precedent to the right to sue on the bond, send a notice in writing to the prime contractor by registered mail within 90 days after the claimant last furnished labor or material to the job. Conversely, if the claimant has a direct contractual relationship with the prime contractor, no preliminary notice is necessary. The courts have held that the registered-mail requirement is not mandatory as long as it can be shown that written notice was actually given and received by the general contractor.[2]

The cases have been very liberal in interpreting this notice requirement, but the claimant must, at a minimum, convey in writing to the prime contractor that there is a claim and that the claimant is looking to the contractor for payment.[3]

In *United States ex rel. Consolidated Electrical Distributors, Inc. v. Altech, Inc.*,[4] the claimant was a material supplier to a subcontractor. The material supplier met with the contractor and testified that nonpayment was discussed at the meeting. After the meeting, the contractor sent a follow-up letter which confirmed that three invoices provided by the supplier at the meeting appeared to cover work performed at the project. The trial court entered judgment for the supplier under the Miller Act, and the contractor appealed, alleging that there was no written notice from the material supplier to the contractor. The Fifth Circuit ruled that the supplier's oral demand for payment at the meeting, coupled with the contractor's confirming letter, resulted in substantial compliance with the 90-day notice requirement. At the meeting, the material supplier presented and

---

[1] 40 U.S.C. §§ 270a–270e (1988).

[2] Fleisher Eng'g & Constr. Co. v. United States *ex rel.* Hallenbeck, 311 U.S. 15 (1940); United States *ex rel.* American Radiator & Standard Sanitary Corp. v. Northwestern Eng'g Co., 122 F.2d 600 (8th Cir. 1941).

[3] Bowden v. United States *ex rel.* Malloy, 239 F.2d 572 (9th Cir. 1956); United States *ex rel.* Carter-Schneider-Nelson, Inc. v. Campbell, 293 F.2d 816 (9th Cir. 1961); United States *ex rel.* J.A. Edwards & Co. v. Thompson Constr. Corp., 273 F.2d 873 (2d Cir. 1959).

[4] 929 F.2d 1089 (5th Cir. 1991).

discussed its invoices. The court stated that the only reasonable inference was that the presentation of the invoices was intended to constitute written notice of the claim to the contractor.

The liberality of the *Altech* decision is in stark contrast to the hypertechnical strictness of the decision in *Pepper Burns Insulation, Inc. v. Artco Corp.*,[5] wherein the court of appeals effectively placed the Miller Act rights of sub-subcontractors and material suppliers at the mercy of the United States Postal Service. After paying lip service to the policy that "the Miller Act should receive a liberal construction to effectuate its protective purposes," the court of appeals ruled that the statutory phrasing of 40 U.S.C. § 270b(a)—"giving written notice to said contractor"—requires actual delivery of the 90-day notice to the general contractor.[6] This means that the mailing of written notice, even by registered mail, is insufficient if the general contractor does not actually receive the notice within the statutory period.

On this ground, the court of appeals in *Pepper Burns* not only reversed the district court's granting of summary judgment in favor of the plaintiff sub-subcontractor but also ordered that the case against the defendant general contractor be dismissed! The court went on to state that only the remedial provisions of the Miller Act are to be liberally construed, while the "notice provisions of section 2(a) should be strictly enforced to carry out the design of the statute."[7]

In the opinion of the authors, the *Pepper Burns* decision is not well reasoned, especially when it relegates to a dismissive footnote the last sentence of 270b(a), which provides: "Such notice shall be served by mailing the same by registered mail, postage prepaid, in an envelope addressed to the contractor at any place he maintains an office or conducts his business, or his residence."[8] Second-tier subcontractors and suppliers should be forewarned that extraordinary measures may be required to assure actual receipt of their 90-day notices in order to avoid the risk that postal inefficiency will deprive them of their Miller Act remedies.

§ 11.03 SUIT ON THE BOND

If the claimant is not paid, suit on the Miller Act bond must be brought in the federal district court where the job is located within one year after the last furnishing of labor and material.[9]

A question often arises in a Miller Act project as to when the one-year time for suit begins to run. In *United States ex rel. Pippin v. J.R. Youngdale Construction Co.*,[10] the Ninth Circuit Court of Appeals ruled that the one-year statute began to run on the last day that earth-moving equipment was available for use

---

[5] 970 F.2d 1340 (4th Cir. 1992).

[6] 970 F.2d at 1343.

[7] *Id.*

[8] *Id.* at 1343 n.6.

[9] United States *ex rel.* Austin v. Western Elec. Co., 337 F.2d 568 (9th Cir. 1964); United States *ex rel.* McGrath v. Travelers Indem. Co., 253 F. Supp. 330 (D. Ariz. 1966).

[10] 923 F.2d 146 (9th Cir. 1991).

on the jobsite, rather than on the last day the equipment was actually used for grading operations. Youngdale was the prime contractor on a job being performed for the Department of the Navy. Youngdale subcontracted excavation and grading work to Bishop. Bishop orally hired earth-moving equipment from Pippen. The last day that Bishop physically used the equipment on the site was February 8, 1985. On that date, Bishop stopped work because of a controversy with Youngdale about changes. Bishop negotiated with Youngdale for a written change order until February 18, 1985, on which date Youngdale terminated Bishop's contract and mailed a notice to Bishop that Bishop was replaced as of that date. Until that date, Bishop had intended to return to the project upon receiving a written change order, and Youngdale had remained hopeful that Bishop would return to the job. Pippen's equipment remained on the site until at least February 18, 1985. On February 12, 1986, Pippen, Youngdale, and the surety on the payment bond entered into a "freeze agreement" under which the parties waived all defenses based on the statute of limitations except for claims that were barred as of February 12, 1986.

The court held that Pippen was entitled to judgment against the surety on the Miller Act payment bond. The court noted that suit must be filed within one year from "the day on which the last of the labor was performed or material was supplied," citing 40 U.S.C. § 270b(b). The court stated that when a claimant provides equipment under a lease or rental, the claimant must allow the equipment to remain on the jobsite for the time specified under the lease. In this case, the equipment remained available for Bishop's use until Youngdale terminated its contract with Bishop. As a result, the statute of limitations began to run on the last day the equipment was available for use on the jobsite, rather than on the last day it was actually used.

Frequently, both the prime contract documents and the subcontract documents provide for arbitration or have disputes clauses requiring the prime contractor and the subcontractors to go through certain administrative procedures under the contract documents before bringing a lawsuit. In a Miller Act case, the court held that neither an arbitration clause in the subcontract nor the standard disputes clause in the government prime contract could prevent the unpaid subcontractor from pursuing a claim against the Miller Act payment bond. In *United States ex rel. N.U., Inc. v. Gulf Insurance Co.*,[11] N.U., Inc. was a subcontractor on a federal job. The subcontract called for arbitration and also had a clause stating, "Contractor shall have the same rights and privileges against the subcontractor herein as the owner in the general contract has against contractor." The prime contractor failed to pay the subcontractor and the subcontractor brought an action on the payment bond under the Miller Act. The surety, Gulf Insurance Company, argued that the payment bond claim should be stayed pending the arbitration. The district court rejected that argument, stating that the bonding company was not a party to the subcontract and therefore could not enforce the arbitration clause. Gulf Insurance argued, in the alternative, that the subcontract

---

[11] 650 F. Supp. 557 (S.D. Fla. 1986).

flow-down clause bound the subcontractor to the disputes clause found in the prime contract between the prime contractor and the government. Under that clause, the prime contractor would submit the subcontractor's claim to the government's contracting officer and pass any recovery through to the subcontractor. The surety company argued that if the subcontractor were not bound by the disputes clause, then the surety company faced the risk of inconsistent results, that is, having to pay the subcontractor money that the surety or the prime contractor would be unable to collect from the government. The court rejected this argument as well. The court stated that compelling the subcontractor to comply with the disputes provision in the prime contract documents would put the subcontractor in a vulnerable position because the subcontractor would have to rely upon the prime contractor to present the subcontractor's claim. The court stated that a subcontractor generally cannot be bound by a disputes clause that is incorporated into the subcontract by reference to the general contract. The court stated that the subcontractor will be bound by such a clause only if the subcontract contains a provision making the disputes clause expressly applicable and waiving the Miller Act remedy.

§ 11.04 PERSONS ENTITLED TO RECOVER ON THE BOND

It is clear, under the cases interpreting the Miller Act, that claimants beyond the third tier on the job cannot recover under the Miller Act:[12]

> Prime Contractor—First Tier
>
> Subcontractor—Second Tier
>
> Subcontractor or Material Supplier—Third Tier (covered)
>
> Subcontractor or Material Supplier—Fourth Tier (cannot recover)

Anyone relying on the Miller Act bond in extending credit on a federal construction project must take care that the customer is either the prime contractor or a person with a contractual relationship with the prime contractor.

In *United States ex rel. Metal Manufacturing, Inc. v. Federal Insurance Co.,*[13] McCarthy Western Constructors was the prime contractor on a federal prison project. McCarthy hired a first-tier subcontractor, who in turn hired a second-tier subcontractor, who in turn purchased flashing material from Metal Manufacturing. When the second-tier subcontractor failed to pay Metal Manufacturing, the first-tier subcontractor agreed to issue future checks payable jointly to the second-tier subcontractor and Metal Manufacturing. When Metal Manufacturing was not paid, it brought an action on the Miller Act bond. The general contractor and the surety on the payment bond moved to dismiss the claim. The

---

[12] Clifford F. MacEvoy Co. v. United States *ex rel.* Calvin Tomfine Co., 322 U.S. 102 (1944); United States *ex rel.* Whitmore Oxygen Co. v. Idaho Crane & Rigging Co., 193 F. Supp. 802 (D. Idaho 1961).

[13] 656 F. Supp. 1194 (D. Ariz. 1987).

court stated that the Miller Act requires a claimant to have a direct contractual relationship with either the prime contractor or a subcontractor who, in turn, has a contractual relationship with the prime contractor. Because Metal Manufacturing was dealing with a second-tier subcontractor, it was not entitled to recover under the Miller Act. Metal Manufacturing argued that the joint-check agreement established a contractual relationship between it and McCarthy's first-tier subcontractor. The district court disagreed. The court stated that the first-tier subcontractor's agreement to issue joint checks was simply an inducement to Metal Manufacturing to continue delivering materials. The first-tier subcontractor neither placed an order with Metal Manufacturing nor accepted sole responsibility for paying Metal Manufacturing. The court held that this was not the type of direct contractual relationship required by the Miller Act. Specifically, the court stated:

> Although the act is remedial and liberally construed to effect that purpose, the requirement of direct contractual relationship with the general contractor or a first tier subcontractor has long been strictly construed. Thus, a general contractor cannot be expected to answer the claims of every lower level supplier.

The Ninth Circuit Court of Appeals, in *Wright v. United States Postal Service*,[14] ruled that unpaid subcontractors on a facility constructed by the United States Postal Service could bring an action directly against the Postal Service. The decision was based upon the fact that the Postal Reorganization Act provides that the Postal Service has the power "to sue and be sued in its official name," and that this provision therefore constitutes a waiver of the Postal Service's sovereign immunity.

Although the opinion is not entirely clear on the subject, it appears that there probably was a Miller Act bond on that particular project, even though the subcontractors who were bringing the equitable lien claim had not brought an action on the Miller Act bond. Notwithstanding the foregoing, the Ninth Circuit court held that the subcontractors could assert any claim against the Postal Service for which they had an independent jurisdictional basis, including their equitable lien claim to the contract funds. The court further held that the subcontractors' suit was not preempted by the Contract Disputes Act. Finally, the court held that, because the Miller Act does not prevent subcontractors from bringing equitable lien actions against the Postal Service, the subcontractors' litigation was properly brought in the United States District Court.

§ 11.05 PROOF REQUIRED

One very interesting aspect of the Miller Act is that the federal courts have held that material suppliers need not prove that their material was actually used in the job. They need prove only that they furnished the material in good faith, reasonably believing that it would be used in the job.

---

[14] 29 F.3d 1426 (9th Cir. 1994).

§ 11.05 CALIFORNIA CONSTRUCTION LAW

In *United States ex rel. Krupp Steel Products, Inc. v. Aetna Insurance Co.*,[15] the court confirmed the decision in *United States ex rel. Westinghouse Electric Supply Co. v. EndebrockWhite Co.*[16] and held specifically that in order to permit recovery against a Miller Act surety, the following four elements must be proved:

1. That materials were supplied for work in the particular contract at issue;
2. That the supplier is unpaid;
3. That the supplier had a good-faith belief that the materials were for the specified work; and
4. That the jurisdictional requisites are met.

In *Krupp*, even though the steel company was unable to prove that its steel was used in the job, it was still allowed to recover. Furthermore, the steel company was a supplier to a subcontractor, and the steel supplier's contract with the subcontractor provided for attorneys' fees. The court held that the steel supplier could not recover attorneys' fees on the Miller Act bond.

The Ninth Circuit Court of Appeals held that a subcontractor could recover delay damages against the prime contractor's Miller Act payment bond and against a common-law payment bond. In *Mai Steel Service, Inc. v. Blake Construction Co.*,[17] the U.S. Navy hired Blake Construction Company to build a naval hospital in San Diego. The Miller Act payment bond was issued by Aetna Casualty Insurance. Blake subcontracted with Mai Steel Services to fabricate and install the steel framework. Mai Steel furnished a payment bond from Commercial Union Insurance. Blake also subcontracted with Molnick's to install steel decking. The steel decking was to be furnished no more than six months after the framework was in place. Mai Steel entered into a sub-subcontract with Molnick's to install the steel framework. Molnick's was given five months to install the steel framework, but Molnick's conditioned its performance on Mai Steel's ability to supply it with steel. Mai Steel never lived up to its part of the sub-subcontract; its steel shipments were routinely late and often included defective materials. Molnick's incurred sizable cost overruns and failed to complete the framework on time. Mai Steel's inability to deliver materials also prevented Molnick's from installing the steel decking on schedule. Molnick's sued Mai Steel for breach of the framework contract and for tortious interference with the decking contract. Molnick's sought to recover the unpaid balance of the framework contract plus additional labor and material costs it incurred in completing both the framework and the decking contracts. Molnick's also sued Blake for breach of contract and was awarded more than $450,000 in damages on that claim, which judgment was satisfied. Because Molnick's was unable to complete the framework contract on time, Mai Steel suspended payments.

---

[15] 831 F.2d 978 (11th Cir. 1987).
[16] 275 F.2d 57 (4th Cir. 1960).
[17] 1992 U.S. App. LEXIS 14719, 92 Daily Journal D.A.R. 9116 (9th Cir. 1992).

At trial, the jury found in favor of Molnick's on its breach of contract claim against Mai Steel and awarded Molnick's damages for the unpaid balance of the framework contract plus $1,154,008 in additional labor and material costs incurred in completing the framework contract. The jury also found that Mai Steel had tortiously interfered with Molnick's performance of its decking contract with Blake and awarded Molnick's an additional $758,818 for increased labor and material costs that Molnick's was forced to expend in completing the decking contract. Mai Steel had filed bankruptcy. At a posttrial hearing, the district court judge concluded that neither Aetna, the surety on the prime contractor's bond, nor Commercial Union, the surety on the subcontractor's bond, was liable for the increased costs of labor and material that Molnick's was awarded against Mai Steel. Molnick's appealed.

The Ninth Circuit court ruled, with regard to the liability of Aetna (the surety on the prime contractor's payment bond), that Molnick's was entitled to recover against Aetna in its capacity as a subcontractor and as a sub-subcontractor, provided the costs that it sought to recover fell within the scope of the Miller Act. The court stated that the appeal raised two issues of first impression in the Ninth Circuit: (1) whether a subcontractor or supplier can recover against a Miller Act surety for the subcontractor's increased labor and material costs caused by construction delays; and (2) if so, whether such recovery is limited to situations in which the general contractor caused the delays. The court answered both of those questions in the affirmative. The Ninth Circuit held that it agreed with three other circuits that a subcontractor could recover delay damages from the prime contractor's Miller Act payment bond surety. The court followed *United States ex rel. Lochridge-Priest, Inc. v. Kahn-Real Support Group, Inc.*,[18] *United States ex rel. T.M.S. Mechanical Contractors, Inc. v. Miller's Mutual Fire Insurance Co.*,[19] *United States ex rel. Pertun Construction Co. v. Harvesters Group, Inc.*,[20] and *United States ex rel. Heller Electric Co. v. William F. Klingensmith, Inc.*,[21] The court noted that in each of those cases, the surety was held liable for the subcontractor's increased labor and material costs but not for the subcontractor's lost profits caused by delay. The court looked at the language of the Miller Act and concluded that the Miller Act requires the payment bond "for the protection of all persons supplying labor and material in the prosecution of the work." The court stated that the additional out-of-pocket costs that Molnick's incurred directly contributed to the completion of the steel framework and decking subcontracts. Thus, by definition, those expenses were incurred in the prosecution of the work.

Specifically, the court held that increased out-of-pocket costs caused by construction delays fall within the intended coverage in the Miller Act, and therefore a subcontractor may recover those costs from the Miller Act payment bond surety. However, the subcontractor may not recover from the surety any lost profits

---

[18] 950 F.2d 284 (5th Cir. 1992).
[19] 942 F.2d 946 (5th Cir. 1991).
[20] 918 F.2d 915 (11th Cir. 1990).
[21] 670 F.2d 1227 (D.C. Cir. 1982).

caused by the delay. The court went on to state that the subcontractor may recover even when the general contractor is not at fault.

The court followed *United States ex rel. Superior Insulation Co. v. Robert E. McKee, Inc.*,[22] a district court opinion that held a Miller Act surety liable for delay costs. In the *McKee* case, the general contractor hired Kinney as a subcontractor. Kinney, in turn, hired Superior Insulation Company, the sub-subcontractor, to assist in the work. Because of Kinney's improper scheduling, Superior had to use additional labor and materials to complete the contract. Kinney filed for bankruptcy protection and Superior sued McKee and the Miller Act surety to recover its increased costs. The district court concluded that Superior could recover its increased costs from McKee's Miller Act surety even though McKee had not been responsible for the delays. The district court in the *McKee* case reasoned that if the sub-subcontractor were not allowed to recover against the general contractor and the surety, then the sub-subcontractor would be left with only a cause of action against the subcontractor, who was bankrupt — and that was the precise situation that the Miller Act was intended to avoid. In *Mai Steel*, the Ninth Circuit court agreed with the *McKee* case and specifically stated that it had previously allowed a sub-subcontractor to recover against a Miller Act surety for labor and materials furnished to a subcontractor, even though the general contractor was blameless, citing *United States ex rel. Martin Steel Constructors, Inc. v. Avanti Constructors, Inc.*,[23] *Harvis Construction, Inc. v. United States ex rel. Martin Constructors, Inc.*,[24] and *United States ex rel. Hillsdale Rock Co. v. Cortelyou & Cole, Inc.*[25] The court concluded that when a sub-subcontractor's claim falls within the Miller Act, the sub-subcontractor may recover from the general contractor's Miller Act surety all of its increased labor and material costs resulting from construction delays for which it is not responsible, even if those delays were caused by someone other than the general contractor.

The court in *Mai Steel* then determined the liability of Commercial Union, the payment surety of the subcontractor Mai Steel. The Ninth Circuit court stated that claims arising under a common-law payment bond are governed by state rather than federal law. Specifically, the court stated that, under California law, common law surety bonds are construed in the same manner as other contracts. The court then looked to the terms of Commercial Union's payment bond to determine the extent of Commercial Union's liability. Commercial Union's payment bond provided that if Mai Steel promptly made payments to claimants, as defined in the bond, for all labor and material used or reasonably required for use in the performance of the framework contract, then the obligation would be null and void, otherwise to remain in full force and effect. The court found that the bond therefore required Commercial Union to pay for "all labor and materials used in the performance of the contract." Hence, the increased labor and material costs that Molnick's incurred in completing the hospital's framework

---

[22] 702 F. Supp. 1298 (N.D. Tex. 1988).
[23] 750 F.2d 759 (9th Cir. 1984).
[24] 474 U.S. 817 (1985).
[25] 581 F.2d 239 (9th Cir. 1978).

subcontract, including costs attributable to construction delays, fell within the scope of the payment bond.

Finally, the court turned to the question of Commercial Union's liability to Aetna. The court noted that it had held that Aetna was liable under the Miller Act for the increased costs that Molnick's incurred in completing both the decking subcontract and the framework sub-subcontract. Aetna had cross-complained against Mai Steel and Commercial Union for indemnification. Mai Steel's performance bond required Commercial Union to indemnify Blake (and therefore Aetna) from any liability incurred by Mai Steel's failure to "promptly and faithfully perform" its contractual obligations. The court noted that a subcontractor does not perform its contractual obligations for the purposes of a performance bond until it pays for all the labor and materials used in completing its work, citing *Glenn Falls Indemnity Co. v. United States ex rel. Westinghouse Electric Supply Co.*,[26] and *Continental Casualty Co. v. Hartford Accident & Indemnity Co.*[27] The court stated that Mai Steel failed to "promptly and faithfully perform" its contract with Blake by not paying Molnick's for constructing the hospital's framework. Citing *Pacific States Electric Co. v. United States Fidelity & Guaranty Co.*,[28] the court concluded that, as the surety for Mai Steel's performance bond, Commercial Union was required to indemnify Aetna for any liability arising out of Mai Steel's breach, including any amount Aetna was ordered to pay in satisfaction of Molnick's breach of contract judgment against Mai Steel. Thus, the court held that both Aetna and Commercial Union were jointly liable for Molnick's increased labor and material costs incurred in completing the framework sub-subcontract, while Aetna alone was liable for Molnick's increased costs incurred in completing the decking subcontract. Under the terms of Mai Steel's performance bond, Commercial Union was required to indemnify Aetna for any amount it paid in satisfaction of Molnick's Miller Act claims. Thus, the Ninth Circuit reversed the district court's judgment and remanded the case to the district court to enter judgments in favor of Molnick's and against Aetna and Commercial Union consistent with the opinion. The court also noted that, to the extent the judgments included any recovery for lost profits, they should be reduced to exclude that recovery.

The Ninth Circuit Court of Appeals again confirmed that material suppliers are required to prove only four elements to recover under the Miller Act, to wit:

1. That the materials were supplied in prosecution of the work provided for in the contract;
2. The supplier has not been paid;
3. The supplier had a good-faith belief that the materials were intended for the specific work; and
4. The jurisdictional requisites were met.

---

[26] 229 F.2d 270 (9th Cir. 1955).
[27] 243 Cal. App. 2d 565 (1966).
[28] 109 Cal. App. 691 (1930).

In *United States ex rel. Hawaiian Rock Products v. A.E. Lopez Enterprises, Ltd.*,[29] the claimants were material suppliers to the general contractor. Their proof consisted of invoices and delivery tickets that indicated that they supplied the materials to Lopez for use in an Air Force base project, that the general contractor failed to pay, that the suppliers had a good-faith belief that the materials were for the project, and that they met the jurisdictional requirements of the Miller Act by timely notice and filing. Accordingly, the Ninth Circuit affirmed a summary judgment in favor of the material suppliers, holding that the claimants had established a prima facie case for relief and that the surety had failed to produce any evidence or to identify missing documents that would undermine the prima facie case. The Ninth Circuit likewise awarded interest to the material suppliers at the rate of 1 percent per month, which was the amount set forth on the invoices.

§ 11.06 PRACTICE NOTES

Parties to a federal construction contract must consider the following points:

1. The Miller Act applies to federal construction projects.
2. If there is no direct contractual relationship with the prime contractor, the claimant must give written notice of the claim by registered mail to the prime contractor within 90 days of last furnishing labor or material to the job.
3. Suit on the bond must be brought in federal court within one year after last furnishing labor or material.

Occasionally, there will not be a payment bond on a federal project. Recent cases have held that when the government fails to require a bond, the claimant has no right of recovery against the government. One court commented, "The hapless contractor is left holding the bag."[30] Therefore, anyone working on a federal project should always check with the contracting officer on the job to confirm that the prime contractor has, in fact, put up a payment bond.

§ 11.07 ILLUSTRATIVE CASES

[A] Claim Against Person Certifying Sufficiency of the Surety

Under the Miller Act, the sureties on the payment bond may be personal sureties. In *Loral Terracom v. Valley National Bank*,[31] a subcontractor brought an action against the prime contractor and the individual surety under the Miller Act. The subcontractor obtained a judgment against both defendants but was unable

---

[29] 74 F.3d 972 (9th Cir. 1996).
[30] Arvanis v. Noslo Eng'g Consultants, Inc., 739 F.2d 1287 (7th Cir. 1984). *cert. denied*, 469 U.S. 1911, 105 S. Ct. 964, 83 L. Ed. 2d 969 (1985). *See also* Baudier Marine Elecs. v. United States, 3 FPD 41 (Cl. Ct. 1984); 4-Star Constr. Corp. v. United States, 3 FPD 46 (Cl. Ct. 1984).
[31] 49 F.3d 555 (9th Cir. 1995).

to collect any portion of that judgment from either defendant. The subcontractor then filed a separate lawsuit against a bank, contending that the bank was liable to the subcontractor for negligence because it failed to properly investigate the surety's financial strength before signing a certificate required under the Miller Act (known as a Form SF28 certificate), which the bank had executed for the individual surety. The form contains a sworn "affidavit of individual surety" wherein the surety lists his or her assets and liabilities and declares his or her intent to become a surety on a government construction project. On the reverse side is a "certificate of sufficiency," providing that the surety named in the affidavit of individual surety is personally known to the persons signing the certificate of sufficiency and that "to the best of the signatory's knowledge," the facts stated in the affidavit are true. The certificate for the individual surety was signed by an official of a Kentucky bank.

The court ruled against the subcontractor. The subcontractor maintained that, because the contracting officer makes use of the certificate in approving the surety, persons who execute those certificates have a duty to investigate independently the truth of the financial information set forth in the surety's affidavit. The court held that argument to be without merit because the subcontractor failed to explain how a contracting officer's possible reliance upon that certificate can itself establish a duty of care.

The subcontractor also argued that the certificate itself created a duty of care. The court noted that the certificate uses the phrase "to the best of my knowledge," which indicates that the certifier is attesting merely for the surety on the basis of the certifier's previous personal knowledge about the surety and an examination of any documents provided with the affidavit. Because the language limits the representation to what the certifier personally knows, it puts the recipient of the certification on notice that further investigation may be needed. In effect, the court held that the person who certifies the sufficiency of the individual surety's affidavit cannot be held liable should that information turn out to be incorrect, and that the claimant, therefore, is unable to recover against the individual surety. In light of this decision, any person furnishing work on a federal project subject to the Miller Act should ascertain who the surety is on the payment bond and, if that surety is an individual surety, the claimant will have to look solely to that individual surety for ultimate recovery.

[B] Pay-If-Paid Clause Did Not Waive Miller Act Bond Rights

In *United States ex rel. DDC Interiors, Inc. v. Dawson Construction Co.*,[32] a subcontractor brought an action on the prime contractor's Miller Act payment bond. The subcontractor sought $600,000 for "alleged" approved change orders and extra work. The general contractor moved to stay the action pending completion of its dispute resolution process with the government. The district court judge stated that the issue was whether and under what circumstances a government subcontractor can surrender its rights to seek prompt payment under the

---

[32] 895 F. Supp. 270 (D. Colo. 1995).

Miller Act. The judge concluded that under the circumstances in that case, the subcontractor did not "clearly and expressly waive its Miller Act rights" and therefore denied the motion to stay the proceedings.

The court stated that the purpose of the Miller Act was to provide security for those who furnished labor and material in the performance of government contracts. The Miller Act is highly remedial in nature and was enacted for the protection of subcontractors and suppliers on government projects, and its terms should be liberally construed. The risk of nonpayment for goods and services provided on a government project should be borne by the surety, not the laborers and material suppliers. The purpose of the payment bond is to shift the ultimate risk of nonpayment from workers and suppliers to the surety company. Under the Miller Act, a subcontractor may sue 90 days after the last day on which it supplied labor and material. The right to sue under the Miller Act may be waived by "clear and express provisions in the contract between the prime contractor and the subcontractor."[33] However, the courts do not favor finding that a subcontractor has contractually abandoned its rights under the Miller Act. Indeed, such a drastic curtailment of a subcontractor's rights will not be read into a general agreement by the subcontractor to be bound by the terms of the prime contract.

The court then determined that the subcontract excluded "skim coating" that was done as an extra. The contract clearly provided that the subcontractor would perform all extra work and that no changes were to be made except upon written order from the contractor. The subcontract also provided that the prime contractor had the same rights against the subcontractor that the government had against the prime contractor. The subcontract specifically incorporated by reference the plans, specifications, and amendments to the prime contract documents. The disputes clause was not specifically referenced anywhere in the subcontract. As a result, the court held that the incorporation by reference of the prime contract's dispute clause is general rather than specific. The court stated that there were two questions: first, does the subcontract's general incorporation language waive the subcontractor's Miller Act rights by binding the subcontractor to wait for payment until resolution of the disputes process between the general contractor and the owner; and second, does Article 15 of the subcontract waive the subcontractor's Miller Act rights? Article 15 of the subcontract was what the district court judge referred to as a "pay upon payment" clause. The clause was not specifically quoted in the opinion, and therefore, its exact language is unknown.

In answering the first question, the court looked to the case of *Fanderlik-Locke Co. v. United States ex rel. M.B. Morgan*,[34] wherein the prime contract contained an appeal procedure for disputed claims. The subcontract contained a general incorporation by reference clause that stated that the subcontractor would be bound to the contractor by the terms of the agreement, general conditions, drawings, and specifications and would assume toward the general contractor all of the obligations and responsibilities that the general contractor (by those contract documents) assumed toward the owner. The prime contractor argued that

---

[33] United States *ex rel.* B's Co. v. Cleveland Elec. Co., 373 F.2d 585, 588 (4th Cir. 1967).
[34] 285 F.2d 939 (10th Cir. 1960).

this language in the subcontract incorporated the disputes procedure contained in the prime contract and therefore, the subcontractor was required to comply with the disputes procedures before it could bring a Miller Act action. The Tenth Circuit court held that the incorporation language was insufficient to waive the protections of the Miller Act. The court noted that the language in the subcontract did not refer to the settlement of disputes or the subcontractor's right to sue under the Miller Act. As a result, the court emphasized that there was no procedure by which the claim of a subcontractor could be presented to the United States except as it may become a claim of the prime contractor. The court pointed out that ordinarily the fact that the prime contractor has a claim for the same amounts pending under the disputes clause of the prime contract does not affect Miller Act cases. The court noted that, in this particular case, the subcontract did not mention the Miller Act, and the prime contractor's disputes clause was not specifically referenced anywhere in the subcontract. As a result, the district court concluded that the incorporation language in the subcontract did not effect a waiver of the subcontractor's Miller Act rights.

In commenting upon the second issue, the court noted that Article 15 of the subcontract was a "pay upon payment" clause. The court stated that at best, by implication, Article 15 may be said to give rise to a waiver of a subcontractor's Miller Act rights. The court stated that circumvention of Miller Act rights cannot rest upon mere implication, and the waiver of Miller Act rights must be "clear and express." The court then concluded:

> At a minimum, an effective waiver of Miller Act rights must include mention of the Miller Act and unambiguously express intention to waive the rights provided by it. No such language is found in the contract documents before me.[35]

As a result, the court denied the prime contractor's motion to stay the proceedings.

[C] General Contractor Not Indispensable Party in an Action on the Miller Act Payment Bond

In *United States ex rel. Henderson v. Nucon Construction Corp.*,[36] a subcontractor brought an action on a Miller Act payment bond. The subcontractor named the general contractor as a defendant, but the federal district court found that the general contractor had not been properly served and therefore was not a party to the action. The district court also found that, because the general contractor was an indispensable party to the lawsuit, the subcontractor could not maintain its action solely against the payment bond surety. The time had expired for the subcontractor to serve the general contractor for a second time. The Ninth Circuit court stated that the only issue on appeal was whether the general contractor was an indispensable party to a Miller Act suit. The court held, in accordance

---

[35] *Dawson Constr.*, 895 F. Supp. at 274.
[36] 49 F.3d 1421 (9th Cir. 1995).

[D] Recovery of Money Due under Savings Clause Allowed

There is no question that subcontractors or material suppliers can recover the balance due and owing to them on a Miller Act project against the Miller Act surety. In the case of *Taylor Construction, Inc. v. Service Corp., Inc.*[37] a question arose as to whether a subcontractor could recover amounts due under a savings clause in the subcontract. In that case, the subcontract provided that the subcontractor would bill for its cost of labor, service, and equipment and that the amount to be paid would not exceed $150,000. The subcontract further provided that "any savings realized in this work shall be divided evenly between the contractor and the subcontractor." During the progress of the job, the subcontractor submitted charge sheets on a daily basis, itemizing material, labor, and equipment used each day. That totaled $42,819. The contractor paid the $42,819 but did not pay the savings in the sum of $41,405.68. The subcontractor brought an action against the contractor and the surety on the Miller Act payment bond, and the appellate court held that the subcontractor was entitled to a judgment for the savings. The court noted that under the Miller Act, before a contract for the construction of a federal public work is awarded, the contractor must secure a payment bond for the protection of all persons supplying labor and material in the prosecution of the work. The Act further provides that any person who has furnished labor or material in the prosecution of his work and who has not been paid in full for it shall have the right to sue on the payment bond for the sums justly due him. The policy behind the Act is to provide a surety who, by force of the Act, must make good the obligations of a defaulting contractor to his suppliers of labor and material. The Act is entitled to a liberal construction and application in order to protect those whose labor and materials go into public projects. The court noted that the plain language of the Act provides the quickest answer to the question posed in the case at bar because it is clear from the face of the Act itself, that the subcontractor can recover from the surety. The court noted that the subcontractor was indisputably entitled to bring suit to recover under the payment bond because the subcontractor furnished labor or material in the prosecution of the work. What is disputed is not who can recover under the Act but whether the subcontractor could recover the amount due under the savings clause. The court noted that the "who" referred to in the Act is limited to those persons supplying "labor or material." "What" is due is not limited and is described simply as "sums justly due." The court noted that "sums justly due" refers back to the term "paid in full" contained in an earlier part of the same sentence. As a result, the court noted that a provider of labor and materials is entitled to all sums justly due, meaning that the provider is entitled to be paid in full under the subcontract. The court cited long-standing precedent for the proposition that "sums justly due" means the sum

---

[37] 68 Cal. App. 4th 70 (1998).

due the party under the bonded contract. The court thus affirmed a judgment in favor of the subcontractor for the amount due the subcontractor pursuant to the savings clause.

[E] Equitable Lien Not Allowed

In the case of *Department of the Army v. Blue Flocks, Inc.*,[38] an insolvent prime contractor failed to pay a subcontractor for work that the subcontractor performed on a construction project for the Department of the Army. The Department of the Army failed to require the prime contractor to obtain a payment bond on the project. When the subcontractor was not paid, the subcontractor notified the Department of the Army that it had not been paid, but notwithstanding said notices, the Army disbursed $86,132.33 to the prime contractor. The prime contractor had not paid the subcontractor $46,586.14. The subcontractor brought an action seeking an "equitable lien" on the funds that the Army had paid out to the contractor notwithstanding the subcontractor's notice and for any further funds that might become due on the contract. The Department of the Army terminated the contractor and used the balance of the contract funds to pay a follow-on contractor.

The United States Supreme Court held that the subcontractor could not sue the United States government by reason of sovereign immunity. The Supreme Court reasoned that the equitable lien sought by the subcontractor constituted a claim for "money damages" and therefore sought to seize or attach money in the hands of the government as compensation for the loss resulting from the default of the prime contractor. The Court further held that the equitable lien was being used as a substitute for money damages, and that the subcontractor's claim did not seek specific relief and therefore was not a legitimate claim under section 702 of the Administrative Procedures Act (APA). Specifically, section 702 of the APA provides that a person suffering a legal wrong because of agency action is entitled to judicial review. An action in a court of the United States seeking relief "other than money damages" and stating a claim that an agency acted or failed to act in an official capacity shall not be dismissed nor relief denied on the ground that the action is against the United States. In other words, section 702 waives sovereign immunity against the United States government in an action seeking relief "other than money damages." In the case at bar, the subcontractor was seeking damages and was merely using the "equitable lien" as a substitute for damages. Accordingly, the Supreme Court held that where the government fails to provide a Miller Act payment bond, the unpaid subcontractor may not seek to obtain money damages against the United States government by alleging that it is seeking a "equitable lien" on the funds owed by the government to the contractor.

---

[38] 119 S. Ct. 687 (1999).

Chapter 12

BANKRUPTCY IN THE CONSTRUCTION INDUSTRY

§ 12.01	Introduction
§ 12.02	Bankruptcy Process
§ 12.03	Performance Issues
§ 12.04	Assumption or Rejection of Executory Contract
§ 12.05	Time
§ 12.06	Cure of Default and Adequate Assurance
§ 12.07	Setoff
§ 12.08	Mechanic's Lien Claims
§ 12.09	Evaluating Potential Claims
§ 12.10	Pay Only If Protected
§ 12.11	Is the Debtor Bonded?
§ 12.12	Completing the Work

§ 12.01 INTRODUCTION

Bankruptcy filings at times reach epidemic proportions in the construction industry. All parties involved in a construction project are vulnerable to the events that bring about bankruptcy. During a recessionary period, subcontractors may be particularly vulnerable because they are the first to provide services and the last to receive payment under the trickle-down scheme by which most construction projects are financed. Therefore, it is imperative for parties to a construction contract and their attorneys to understand how to deal with another party's bankruptcy.

§ 12.02 BANKRUPTCY PROCESS

The most frequently encountered forms of bankruptcy are Chapter 7 (liquidation) and Chapter 11 (reorganization). In both instances, the automatic stay provisions of 11 U.S.C. § 362 go into effect immediately upon the filing of the debtor's petition. The stay prohibits creditors from taking any action that might adversely impact the debtor or improve their position over other creditors. The purpose of the stay is to maintain the *status quo,* prevent a further deterioration of the estate's value, and provide the debtor with some breathing room. A creditor who takes action against the debtor without prior court approval is subject to sanctions.

In a Chapter 7 proceeding, a trustee is appointed to oversee the orderly liquidation of the estate. Once the assets are liquidated and the relative priorities of the creditors determined, the trustee distributes the proceeds. Debtors who are individuals may be discharged from debts that arose before the filing. Discharge is not available to partnerships or corporations.

In a Chapter 11 bankruptcy, the debtor usually retains control over the business and is required to propose a plan for reorganization to the bankruptcy court. The plan must demonstrate how the debtor intends to operate or liquidate the business and how the creditors will be treated. If the plan is not in the "best interests of creditors," that is, does not provide more to the various classes of creditors than they would receive under a liquidation, the court may, upon the objection of a creditor, convert the proceeding to a Chapter 7 and liquidate. If the plan is approved, the debtor must carry it out in accordance with court supervision. Within this context, the parties must deal with myriad issues, including completion of the project, payment to parties still involved in the project, recovery of possible damages caused by the debtor, and other issues relating to the relationship between the debtor and the remaining parties.

§ 12.03 PERFORMANCE ISSUES

When an owner files for bankruptcy, the critical question is whether the project will go forward or cease. If the owner files for Chapter 11 reorganization, there is a chance the project will continue. If, however, the owner files for Chapter 7 liquidation, the project is most likely to cease. The situation is further complicated for general contractors and subcontractors by the fact that the stay prohibits the recording or foreclosing of mechanics' liens.

If the general contractor is the party filing for bankruptcy, the owner must decide whether to retain the contractor on the project or find another party to complete the project. The owner can also expect subcontractors to record mechanics' liens in an effort to protect their interests. Subcontractors are faced with the possibility that the general contractor will be unable to pay for work already performed. As the debtor is allowed to reject or assume executory contracts, the issue of continuing to perform the project does not initially rest with the subcontractor. If the owner has required a performance and payment bond from the general contractor, the issues may not be as critical.

Because Murphy's Law governs in the construction industry, a subcontractor inevitably will choose the most critical time in the performance of its work to file bankruptcy. This presents a number of practical and legal problems for the general contractor. First and foremost, the general contractor must decide whether to retain the subcontractor to complete performance of the contract. Changing horses in midstream is always a tricky proposition. On the other hand, a bankruptcy filing badly shakes the general contractor's confidence that the subcontractor will complete the work.

§ 12.04 ASSUMPTION OR REJECTION OF EXECUTORY CONTRACT

Although executory contracts are not specifically defined by the Bankruptcy Code, cases generally define an *executory contract* as one where the material obligations of both parties have not yet been completely performed.[1] As an example, when a subcontractor is in mid-performance of its work and the general contractor owes the subcontractor payment for the performance, the contract would be considered executory for the purpose of the Bankruptcy Code.

Under the Code, the trustee or the debtor has the power to assume or reject executory contracts.[2] This power allows the trustee to avoid contracts that impose burdensome liabilities upon the estate and to take advantage of favorable contracts, thereby assisting the reorganization or increasing the estate's net value through performance.

§ 12.05 TIME

One problem for the parties involved in a construction project is that under ordinary circumstances in a Chapter 7 proceeding, the debtor has 60 days to decide to assume or reject the executory contract.[3] In Chapter 11 cases, the debtor in possession may assume or reject an executory contract at any time prior to the confirmation of the plan of reorganization unless the bankruptcy court specifies

---

[1] *See, e.g., In re* Wegner, 839 F.2d 533 (9th Cir. 1988).
[2] *See* 11 U.S.C. § 365 (1994).
[3] *See* 11 U.S.C. § 365(d) (1994).

a shorter period.[4] These time periods may simply be too long for an owner, general contractor, or subcontractor to wait during the course of a project. The trustee has a reasonable time within which to assume or reject a contract, and the determination of reasonableness is left to the court's discretion after considering the facts and circumstances of each particular case.[5] However, a party to an executory contract may ask the court to set a specific date by which the executory contract will be assumed or rejected. This is accomplished by a motion to compel assumption or rejection of the contract by the trustee, filed in bankruptcy court. In a time-critical situation, an owner, general contractor, or subcontractor may well wish to file such a motion to obtain an expedited determination of whether a contract will be assumed or rejected. Time is always a major concern for parties in the midst of a project. A showing of such time-related factors as liquidated damages, delay claims, disruption, and coordination problems by the general contractor, owner, or subcontractor will demonstrate to the bankruptcy court the need to expedite the decision. Parties must remember that unless a motion by the contractor results in the trustee's rejection of the contract, the contract remains in effect. Hence, proceeding unilaterally to obtain substitute performance or to deal with sub-subcontractors and/or material suppliers may constitute both a breach of the contract and a violation of the automatic stay.

Like the owner, the primary problem faced by the subcontractor or supplier is that assumption or rejection of the contract is within the exclusive control of the prime contractor. If the contract is rejected, the only remedy for the subcontractor will be a claim for breach of contract, which becomes a general unsecured claim against the bankrupt contractor's estate.[6]

§ 12.06 CURE OF DEFAULT AND ADEQUATE ASSURANCE

One way for the parties to avoid the entanglements of bankruptcy court is to terminate the contract before a party files for bankruptcy relief. If the contract is terminated in compliance with the contract terms and pursuant to law prior to the bankruptcy filing, the contract is not part of the estate, and no further action is necessary in the bankruptcy court. The termination process, however, must be completed with no possibility that the contract can be reinstated.[7]

If the debtor has defaulted in its performance but has not been finally terminated, the Bankruptcy Code provides that the trustee may assume the contract but must cure the default or provide adequate assurance that the default will be cured in a timely manner.[8] For instance, if a subcontractor is behind schedule and has not provided adequate manpower to the project, the trustee is required either to compensate the general contractor or to provide adequate assurance of a cure

---

[4] *See id.* at § 352(d)(2).
[5] *See In re* Attorney's Office Management, Inc., 29 B.R. 96 (Bankr. C.D. Cal. 1983).
[6] *See* 11 U.S.C. § 365(g).
[7] *See In re* Burke, 76 F. Supp. 5 (S.D. Cal. 1948).
[8] *See* 11 U.S.C. § 365(b)(1)(A) (1994).

of the default as a condition of continued performance of the contract. The filing of a petition in bankruptcy, however, is not grounds for termination.[9] The term "adequate assurance" is purposely vague so that the court can fashion appropriate assurance based upon the facts of each particular situation.

The debtor must also compensate the other party for actual losses resulting from default or provide adequate assurance that compensation will be paid.[10] For example, the debtor must reimburse the general contractor for actual losses or must fashion some other means of protecting the general contractor so that continued performance will not diminish the general contractor's ability to be paid for its losses.

At mid-project, a general contractor's or owner's greatest concern is that the debtor provide adequate assurance of continued performance.[11] Obviously, this adequate assurance requirement is an invitation to the court and the parties to fashion creative means of assuring the completed performance of the contract. Completion bonds, personal guarantees, or increased priority of a claim might be sufficient to qualify as adequate assurance of future performance. It should be noted, however, that if the underlying contract is not in default, no adequate assurance is required in order for the court to permit an assumption of the contract.

As a practical matter, it may be difficult for the debtor to cure its default or provide adequate assurance of its performance, especially if the other parties have suffered or will suffer as a result of long delays, liquidated damages, lien claims, and disgruntled employees, unions, suppliers, and sub-subcontractors of the debtor. To force the rejection of the contract, a party to the contract must show that the debtor has defaulted on its performance and that adequate assurance is not possible under the circumstances. A complete showing of the problems (including project schedules, notices of default, mechanics' liens and/or stop notices from the debtor's suppliers or subcontractors, and affidavits of project managers concerning the debtor's performance) is essential to show that the debtor cannot cure the default and will be unable to provide adequate assurance. This evidence must be assembled in an effort to convince the bankruptcy court that assumption of the executory contract is not in the interests of the estate.

The court weighs the evidence of default against any potential benefit the continued performance of the contract might give to the estate. The court then renders a judgment based on the facts determining whether the debtor should assume or reject the contract. Obviously, the stronger the showing of default and that a cure would create an unproductive burden on the estate, the better the chances the court will determine that a contract should be rejected. In gathering the evidence, there is no substitute for a well-documented file verifying the facts of default and breach and demonstrating that contractual standards of performance have not been achieved.

---

[9] *Id.* § 365(e).
[10] *See id.* § 365(b)(1)(B).
[11] *See id.* § 365(b)(1)(C).

Once the court has ordered the contract rejected, the general contractor or owner is free to engage the services of a substitute contractor and may pursue its claims for compensation against the debtor through the bankruptcy court. In most cases, the general contractor and owner will become general unsecured creditors to the estate of the debtor, sharing in the assets of the estate after all secured, administrative, and priority claims have been paid. That does not mean, however, that the parties give up their right to setoff.

§ 12.07 SETOFF

The general contractor or owner may be tempted to use any funds owed in the debtor's contract to complete the work and pay lien claimants. Unless matters are properly handled, however, the contractor or owner may end up paying twice and may be exposed to sanctions for violating the automatic stay.

11 U.S.C. § 553 recognizes the equitable doctrine of setoff but adds certain restrictions. The elements necessary to justify a setoff are as follows:

1. There must be a debt owed by the creditor to the debtor that arose prior to commencement of the bankruptcy case;

2. There must be a claim of the creditor against the debtor that arose prior to commencement of the bankruptcy case; and

3. The debt and claim must be mutual obligations.[12]

The mutuality requirement simply means that something must be owed by both sides.

Marshaling of debts and claims among related corporate entities is not allowed. A setoff may not be acquired from a third party within 90 days of the filing and cannot be exercised within 90 days of the debtor's filing. Because the setoff may affect an asset of the bankrupt estate, a creditor seeking to exercise a post-petition setoff must first move for relief from the automatic stay. A creditor who attempts to set off mutual debts on a post-petition basis may be in violation of the automatic stay. In order to avoid waiving one's right to claim a setoff later, it is prudent, but not required, to file a claim against the debtor. This is true even if the claim is not liquidated. Claims for costs incurred in completing subcontracts breached prior to the bankruptcy petition are normally pre-petition claims even though the costs may be incurred post-petition. Thus, setoffs may be the only viable means to recover amounts that otherwise would constitute an unsecured claim.

§ 12.08 MECHANIC'S LIEN CLAIMS

Obviously, lack of money is a problem for bankrupt contractors and subcontractors. Usually they have not paid their subcontractors, sub-subcontractors, and suppliers. As a result, a flurry of mechanics' liens are recorded and stop

---

[12] *See In re* Fulghum Constr., 23 B.R. 147 (Bankr. M.D. Tenn. 1982).

notices served by unpaid creditors of the debtor. In most cases, the general contractor has a contractual obligation and a statutory obligation to keep the owner's property free of such claims.[13] However, if the general contractor blithely pays the debtor's sub-subcontractors and suppliers in discharge of its obligations, the debtor may be able to force the general contractor to pay it as well, claiming that the money was "owed" to the debtor or that the bankrupt subcontractor had legitimate backcharges against the sub-subcontractors and suppliers. Consequently, the general contractor may end up paying substantially more than the original contract specified.

When an owner files for bankruptcy, two issues are raised in regard to mechanics' liens. First, how does the contractor, subcontractor, or supplier preserve its lien rights in spite of the automatic stay provisions of the Bankruptcy Code? Second, how can abandonment be used to help satisfy lien claims? Preservation of mechanic's lien rights may present a problem for contractors, subcontractors, and suppliers because of the automatic stay provisions of the Bankruptcy Code. If the mechanic's lien is not properly perfected prior to the filing of the bankruptcy petition, the holder of such lien might be treated only as an unsecured creditor. In order to preserve lien rights, a contractor, subcontractor, or supplier will need to file a Notice of Perfection with the bankruptcy court (if the time for filing the lien has not lapsed under the lien laws and the lien holder has not already recorded the lien.)[14] If the lien is already recorded at the time of the bankruptcy filing, the lien holder will need to file a Notice of Continued Perfection with the bankruptcy court.[15] This notice has the effect of extending the time to foreclose on the mechanic's lien, since the automatic stay prevents filing such action. There is some authority that this step is unnecessary.[16] Furthermore, the general rule seems to be that a mechanic's lien may be recorded as long as it is not foreclosed.[17]

Actions to foreclose mechanics' liens are automatically stayed following the filing of a bankruptcy petition.[18] There may be situations in which the lien holder can seek to have the automatic stay lifted. This will of course require permission from the bankruptcy court. Reasons for the court to grant relief from the automatic stay include lack of adequate protection and lack of equity in the property.

---

[13] *See* Cal. Civ. Code § 3153.

[14] 11 U.S.C. § 362(b)(3).

[15] *See, e.g., In re* Baldwin Builders, 232 B.R. 406 (B.A.P., 9th Cir. 1999).

[16] *In re* Hunters Run Ltd. Partnership (Hand v. Hunters Run Ltd. Partnership), 875 F.2d 1425, 1428 (9th Cir. 1989) (under Washington law, notice of continued perfection is unnecessary since the time for filing action to foreclose mechanic's lien is tolled by 11 U.S.C. § 108(c)); *In re* Paul Potts Builders, Inc. (Grover v. County of Marin), 608 F.2d 1279, 1283 (9th Cir. 1979) (if a mechanic's lien is recorded before owner's bankruptcy filing, such lien is valid despite failure to file timely action under Civil Code § 3144).

[17] *In re* Cocolat, Inc. (Cocolat, Inc. v. Fisher Dev., Inc.) 176 B.R. 540, 550 (Bankr. N.D. Cal. 1995) ("recordation of a lien is excepted from the bar of the automatic stay"); *In re* Miller Lee Corp., 70 B.R. 780 (Bankr. S.D.N.Y. 1987).

[18] 11 U.S.C. § 362(d).

Contractors and suppliers need to be aware of the owner's option to abandon the property. Debtors can abandon assets that are a burden to the bankruptcy estate or that are of inconsequential value.[19] A partially completed construction project is likely to fit this profile. A showing that the owner has little or no equity in the property—i.e., the value of mechanics' liens on the property exceeds the value of the property—makes the case for abandonment stronger. The trustee or debtor in possession can agree to abandon, or creditors can move the court to compel the trustee or debtor in possession to abandon the property. Once the property is abandoned, mechanic's lien holders can move forward with the foreclosure action.

The subcontractor may not be in as bad a position in the case of a general contractor's for bankruptcy as when an owner files for bankruptcy. In the situation of a general contractor's bankruptcy, the subcontractor and supplier are still able to record and foreclose mechanics' liens. This is because the owner is a third party, and the Bankruptcy Code does not stay actions against third parties.

§ 12.09 EVALUATING POTENTIAL CLAIMS

Upon learning of the bankruptcy, the parties should suspend payments to the debtor pending a review of the potential claims. The first order of business if you are an owner or a general contractor is to determine which creditors of the debtor have valid lien rights. This entails reviewing preliminary notices, releases, joint checks, and project progress reports. The potential claimants should be contacted and documentation supporting their claims gathered. This information should be analyzed from a legal perspective to determine if the statutory prerequisites for perfection of the lien have been fulfilled. The data should also be evaluated from a practical standpoint to determine if the work claimed or materials supplied were in fact used or consumed in the construction of the project. This analysis will require review by an individual familiar with the project and the materials used by the debtor.

The reason for a painstaking analysis is that establishing the validity of any claim is essential to determinate whether a payment made to the claimant is a preference or an unauthorized setoff against the debtor. The general contractor or owner may pay the debtor's creditors only when there is a direct legal obligation to pay that claimant because of a mechanic's lien, stop notice, or payment bond claim.[20] If the claim is invalid in any respect, then payment cannot be made without risking possible liability to the debtor.[21]

If an owner or general contractor is the debtor, the subcontractor will want to evaluate the amounts owed to the subcontractor on the project and any setoffs that may be due. The primary problem faced by subcontractors is the fact that the assumption or rejection of the contract is wholly in the hands of the debtor. The subcontractor can object to the assumption of the contract on the basis that

---

[19] *Id.* § 554.
[20] *See In re* Flooring Concepts, Inc., 37 B.R. 957 (Bankr. 9th Cir. 1984).
[21] *See* Tucson House Constr. Co. v. Fulfor, 378 F.2d 734, 737 (9th Cir. 1967).

the debtor is unable to provide adequate assurance that the subcontractor will be paid in the future. If the contract is assumed, the subcontractor will receive a higher priority if the contract is later breached or rejected.[22]

§ 12.10 PAY ONLY IF PROTECTED

Once the validity of the claim is determined, the general contractor or owner can proceed in one of several ways. First, the general contractor or owner should attempt to arrange for a written stipulation between the debtor, its trustee in bankruptcy, and the lien claimant that the proposed payment to the lien claimant is valid, that no claims will be made by the estate to the contrary, and that the payment constitutes a valid setoff of any funds owed to the debtor. After giving notice to creditors, the bankruptcy court may approve the payment. In fact, such approval is required to obtain the optimum protection from claims that the payments are preferential or do not constitute a valid setoff. This time-consuming and expensive process is clearly the best way to protect the general contractor or owner from claims of preference and a possible double payment.

Another option for removing liens and paying claimants of the debtor is to interplead the money in the bankruptcy court in exchange for releases of the liens. As a requirement of the interpleader, the bankruptcy court may require the claimants to release the liens, transferring claims to the interpled funds. The court then decides the relative validity of each competing lien claim before ordering the disbursement of the funds. As an element of the interpleader, the general contractor or owner should receive the court's protection from multiple claims for the same money.

Another, although riskier, alternative for resolving the dilemma is to pay the lien claimant by a joint check. The debtor and the lien claimant must then resolve any dispute concerning the validity of the claim in order to negotiate the check. The disadvantage is that the lien may not be released by the claimant until its portion is paid.

Another option is to have the lien claimant indemnify and defend the general contractor or owner from any claim that the payment is in any way invalid, preferential, or illegal. Obviously, before obtaining any such indemnification, the general contractor or owner must consider the stability, size, and financial capacity of the indemnitor. If the claimant is substantial and a determination can be made that it has the ability to honor its indemnification, then payment can be made directly to the lien claimant in exchange for releases and its indemnification of the general contractor.

The larger problem for contractors or subcontractors will be Chapter 7 trustee's attempts to recover payments made from the bankrupt owner or contractor to the recipient as preferences. A trustee or debtor in possession generally may avoid (i.e., set aside and recover back into the bankruptcy estate) a prepetition payment to a non-insider creditor under 11 U.S.C. § 547(b) if:

---

[22] 11 U.S.C. § 365.

- The debt was incurred before payment was made;
- The debtor was insolvent at the time of the payment;
- The payment was made on or within 90 days of filing the bankruptcy; and
- The creditor received more than it would have received had the payment not been made, had the case been brought under Chapter 7, and had the creditor received payment as permitted by the Bankruptcy Code.

The debtor is presumed to have been insolvent during the 90-day period immediately preceding the filing.[23]

§ 12.11 IS THE DEBTOR BONDED?

The general contractor or owner can avoid many problems caused by the subcontractor's or general contractor's bankruptcy by ensuring that the subcontractor or general contractor provides performance and payment bonds prior to commencing work. Immediately upon the determination that the subcontractor or contractor is in default and certainly upon notification that the subcontractor or contractor has declared bankruptcy, the general contractor or owner should notify the surety and, if appropriate, make written demand for performance and/or payment. It is incorrect to assume that simply because the subcontractor or contractor is bankrupt, the surety must take over. The general contractor or owner must still demonstrate to the surety that the debtor has defaulted on its obligations. The surety should be kept informed of all efforts to deal with the debtor. A cooperative effort between the party making a claim against the bond and the surety in many instances results in a quicker substitute performance and less overall delay to the project. The act of contacting the surety and making claims on the performance or payment bond does not constitute a violation of the automatic stay, inasmuch as the bonds do not constitute property of the estate. Indeed, in some instances of default, the surety may be instrumental in making sure that the project is completed and may in fact assist the debtor in providing adequate assurance of performance, curing defaults, and performing the work. In any case, the general contractor or owner must involve the surety in order to preserve its rights, should problems arise later.

Subcontractors also benefit if a general contractor has a payment bond on a project. The surety is not the debtor; therefore, a claim by a subcontractor against a contractor's payment bond does not violate the automatic stay provisions of the Bankruptcy Code.

The protection provided by a payment and performance bond cannot be underestimated. Often, the surety will be the fastest and most responsive party to the general contractor's or owner's needs when a bankruptcy occurs. On the other hand, if the surety does not respond, legal action may need to be taken, and the general contractor stands a better chance, in most cases, of recovering from the surety for its damages than of recovering from the bankruptcy estate.

---

[23] 11 U.S.C. § 547(f).

§ 12.12 COMPLETING THE WORK

It is prudent for the general contractor or owner to contact all subcontractors and suppliers of the debtor to determine whether they have been paid for work, labor, and materials supplied to date on the project, and whether they are willing to continue to perform the work, either for the debtor or for another party in the event the debtor cannot perform. It is important to obtain input from the project management staff directly involved in evaluating past performance of the debtor, its subcontractors, and its suppliers. Certainly, in order to avoid liens and to determine whether any claim made by the subcontractor or supplier is valid, the general contractor or owner should obtain copies of all agreements, invoices, releases, and records of payment from the prospective claimants on the project. Often unpaid fringe benefit trust funds and workers who are unpaid as a result of the bankruptcy are overlooked.

The general contractor or owner should also obtain estimates for completion of the work from other competent and solvent subcontractors or contractors before deciding whether to attempt to assume or reject the contract with the debtor. The general contractor and owner should not, however, contract with any other party until the appropriate order has been obtained from the bankruptcy court.

The owner's main concern when a contractor files for bankruptcy will be the decision to have the contractor continue the work or substitute in a completion contractor. If the contractor had a performance bond on the project, completion of the project might be easier. Remember, however, that because most construction contracts are likely to be viewed as executory contracts, the option to reject or assume the contract lies with the contractor or the trustee. The owner will also need to consider recovery of damages for delay in the project.

Chapter 13
HOME IMPROVEMENT CONTRACTS

§ 13.01   Statutory Requirements
§ 13.02   Home Improvement Salespersons
§ 13.03   Home Improvement Contracts
§ 13.04   Swimming Pools
§ 13.05   Rescission of Home Solicitation Contracts
§ 13.06   Arbitration Provisions in Home Improvement Contracts

§ 13.01 STATUTORY REQUIREMENTS

In response to abuses of the law by certain contractors and home solicitation salespersons, the legislature has developed very specific requirements with regard to home improvement contracts. More importantly, the legislature has imposed certain misdemeanor penalties and fines on contractors who fail to comply with requirements of the law. Therefore, it is important that a contractor engaged in the home improvement business be aware of these laws and comply with them so as to avoid any penalties.

California Business and Professions Code § 7151 defines *home improvements* to include repairing, remodeling, altering, converting, modernizing, or adding to residential property. It includes, but is not limited to, the construction, erection, replacement, or improvement of driveways, swimming pools, terraces, patios, landscaping, fences, garages, porches, fallout shelters, basements, and other improvements of the structures or land adjacent to a dwelling house. The law also applies to the installation of home improvement goods or the furnishing of home improvement services. Home improvement goods or services include, but are not limited to, carpeting, texture coating, fencing, air-conditioning or heating equipment, and termite extermination. Home improvement goods include goods to be affixed to the real property whether or not they can be removed therefrom.[1] Furthermore, there are specific rules governing the swimming pool industry.

§ 13.02 HOME IMPROVEMENT SALESPERSONS

California Business and Professions Code § 7152 requires home improvement salespersons to be registered with the registrar of contractors. A *home improvement salesperson* is defined as any person employed by a contractor to solicit, sell, negotiate, or execute home improvement contracts under which home improvement may be performed by the contractor. A home improvement salesperson also includes any person selling home improvement goods or services. However, persons possessing a contractor's license, officers of licensed construction corporations, showroom salespersons, persons who contact prospective buyers only for scheduling appointments with home improvement salespersons, and those whose services are confined to repairs under the employment of a licensed contractor need not register as a salesperson under this law.[2]

Any person acting as a home improvement salesperson who fails to register is subject to possible misdemeanor liability.[3]

The prospective home improvement salesperson need only submit to the registrar an application and fee in order to become a registered home improvement salesperson. The registrar may refuse to register the applicant under the grounds specified in Business and Professions Code § 480, i.e., if the applicant

---

[1] *See* Cal. Bus. & Prof. Code § 7151.
[2] *See* Cal. Bus. & Prof. Code § 7152.
[3] *See id.* § 7153.

has been convicted of a crime, has done any act involving dishonesty, fraud, or deceit for personal gain, or has done any act that would call for the license's suspension or revocation. However, note that the registrar may not deny the license unless the crime or act in question is substantially related to the duties of home improvement sales. In addition, the registrar may deny the license if the applicant knowingly made a false statement of fact.[4]

§ 13.03 HOME IMPROVEMENT CONTRACTS

All home improvement contracts in excess of $500 must be evidenced by a writing and must be signed by all parties to the contract. In addition, any changes to the contract must be evidenced by a writing and signed by the parties to the contract. Furthermore, California Business and Professions Code § 7159 sets forth the following requirements for the written contract:

1. The name, address, and license number of the contractor, together with the name and registration number of the registered salesperson who solicited or negotiated the contract, must appear in the written contract.[5]

2. The approximate dates when the work will begin and will be completed must be included.[6]

3. The work to be done, materials to be supplied, equipment to be used or installed, and the agreed price for the work must be described.[7]

4. If a down payment is provided for, that down payment cannot exceed $1,000 or 10 percent of the total contract price, whichever is less.[8]

5. A schedule of payments must be included showing the amounts of each payment as a sum in dollars and cents (at no time may the contractor receive more than 100 percent of the value of the work performed on the project).[9]

6. The contractor must substantially commence work within 20 days of the approximate date specified in the contract for work to begin; otherwise, the contractor's next succeeding payment will be postponed by the amount of time between when substantial commencement was scheduled to occur and when it did occur.[10]

7. A statement that, upon payment of consideration for a portion of the work performed, the contractor shall provide the buyer with a full

---

[4] *See id.* Code § 7153.1. *See also id.* § 480.
[5] *See* Cal. Bus. & Prof. Code § 7159(a).
[6] *See id.* § 7159(b).
[7] *See id.* § 7159(c).
[8] *See id.* § 7159(d).
[9] *See id.* § 7159(e).
[10] *See id.*

release of any claim or mechanic's lien under section 3114 of the Civil Code must be included in the contract. This release must be granted prior to any further payment on the work.[11]

8. Contractors providing a payment, lien, or completion bond or its equivalent, or joint control, need not set forth in the contract a schedule for payments and may require a larger down payment.[12]

9. A statement in 10-point type must be included, near where the parties sign the contract, that the owner or tenant has the right to require the contractor to obtain a performance and/or payment bond.[13]

10. No extra or change order work may be required without prior written approval of the individual contracting for the improvement or swimming pool. Change orders will be enforceable only if they clearly set out the scope and price of the additional work. Change orders will be considered incorporated into the contract for improvement.[14]

11. If the contract sets forth a commission for the salesperson of the work, the payment shall be pro rata in proportion to the schedule of payments made by the party paying for the work.[15]

12. The language required by California Business and Professions Code § 7018.5, i.e., the notice to owner form, must be included.[16]

13. What constitutes substantial commencement of work pursuant to the contract must be defined;[17]

14. A notice must be included stating that if the contractor fails to substantially commence the work within 20 days of the approximate date specified in the contract, this will constitute a violation of the Contractors State License Law.[18]

In one case,[19] a contractor had an oral home improvement contract. He sued to recover. The defendants asserted that because the contract was not in compliance with section 7159 of the Business and Professions Code, the contractor was barred from recovery. The California Supreme Court held that the rule promulgated by section 7159 is "not an inflexible one." In cases where unjust enrichment would be avoided by enforcing the contract, the court may do so. In this case, because the consumer was relatively sophisticated when it came to home

---

[11] *See* Cal. Bus. & Prof. Code § 7159(f).
[12] *See id.* § 7159(g).
[13] *See id.*
[14] *See id.* § 7159(h).
[15] *See id.* § 7159(i).
[16] *See id.* § 7159(j).
[17] *See* Cal. Bus. & Prof. Code § 7159(k).
[18] *See id.* § 7159(*l*).
[19] Asdourian v. Araj, 38 Cal. 3d 276 (1985).

improvement agreements, and because the agreement in question was not particularly unjust, the court decided to enforce the contract.

The contract should include any other matters agreed upon by the parties to the contract. The provisions of this law are not exclusive; compliance with these provisions does not relieve the contractor from compliance with all other applicable provisions of law.

Anyone who falsely accepts a completion certificate while knowingly violating any of the above contract requirements may be punished, along with the licensed salesperson, for a misdemeanor violation and fined not less than $500 and not more than $5,000, or may be imprisoned in the county jail for no more than one year, or receive both a fine and imprisonment.[20]

If the requirements are violated "as part of a plan or scheme to defraud an owner of a . . . structure . . . in connection with the offer or performance of repairs" due to a natural disaster, the violator shall be forced to make restitution to the fraud victim and additionally may be forced to pay not less than $500 or more than $25,000 in fines, as well as receiving a prison sentence between one month and one year.[21]

In *Gonzales v. Covered Gardens Mobile Home Park*,[22] the contractor did not comply with the notice requirements of the license law.[23] He sued the owner to recover on his contract, and the owner raised failure to give notice as a defense. The court held that such failure by the contractor to give the notice did not preclude recovery.

In *Davenport & Co. v. Spieker*,[24] the homeowners entered into a written contract with the contractor to build a house on a time and materials basis plus a 15 percent fee not to exceed $130,000. Ultimately, the owners paid the contractor a total of $178,623 for work performed pursuant to the contract and for extras. The contractor claimed an additional $37,331 for extras and sued to recover for those extras. The extras were all oral and not in writing, contrary to California Business and Professions Code § 7159, which requires that both the initial contract and any extras be in writing. The homeowner in this case was a general partner of Trammell Crow Company, a real estate investment and development firm, and had been involved in the construction business for 20 years. The court, under these circumstances, held that the "oral extras" were recoverable.

If contractors or their registered salespersons represent to a prospective home improvement purchaser that certain trademark or brand-name products are to be used in the construction of the home improvement, a description of the product must be set forth in writing in the contract or specifications. The contractor's failure to install the goods or materials described in the contract or specifications

---

[20] *See* Cal. Bus. & Prof. Code § 7158(a).
[21] *See id.* § 7158(b).
[22] 90 Cal. App. 3d 871 (1979).
[23] Cal. Bus. & Prof. Code §§ 7018, 7019 (forerunners of Cal. Bus. & Prof. Code § 7018.5).
[24] 197 Cal. App. 3d 566 (1988) (depublished opinion).

constitutes a cause for disciplinary action under the Contractors State License Law.[25]

Two provisions in the Business and Professions Code[26] require any contract for the sale of home improvement goods or services that is secured by a lien on real property and is offered by door-to-door sale to include a notice, set forth in the Code, regarding the use of a home as security. Any such contract that does not include the notice is void and unenforceable, with certain specified exceptions. The required notice must be in 18-point bold type and must read as follows: "WARNING TO BUYER: IF YOU SIGN THE CONTRACT WHICH ACCOMPANIES THIS NOTICE, YOU WILL BE PUTTING UP YOUR HOME AS A SECURITY. THIS MEANS THAT YOUR HOME COULD BE SOLD WITHOUT YOUR PERMISSION AND WITHOUT ANY COURT ACTION IF YOU MISS ANY PAYMENT REQUIRED BY THIS CONTRACT."[27]

Business and Professions Code § 7159.2 provides that no contract for home improvement goods or services valued at $5,000 or less shall provide for a security interest in real property, except for mechanics' liens or other interests in property arising under operation of law; and any lien in violation of this subdivision is void and unenforceable. Additionally, when the proceeds of a loan secured by a mortgage on real property are used to fund goods or services pursuant to a home improvement goods or services contract of more than $5,000, the entity making the loan shall pay the contractor only (1) by an instrument payable to the borrower, or jointly to the borrower and contractor, or (2) through a third-party escrow agent under a written agreement signed by the borrower, the lender, and the contractor prior to the disbursement.[28]

§ 13.04 SWIMMING POOLS

As a result of many problems relating to the sale and construction of swimming pools, the legislature has enacted special provisions[29] regarding contracts for the construction of swimming pools. Moreover, the legislature has imposed severe penalties upon swimming pool contractors failing to comply with this law.

In addition to the requirements set forth for general home improvement contracts, the swimming pool contractor must include a plan and scale drawing of the shape, size, dimensions, construction, and equipment specifications of the pool.[30] In addition, the contractor can receive a down payment of only $200 or 2 percent of the contract price, whichever is less.[31] No additional work can be performed without the prior written authorization of the party contracting for the construction of the swimming pool. Also, the swimming pool contractor must

---

[25] *See* Cal. Bus. & Prof. Code § 7162.
[26] *See id.* §§ 7159.1, 7159.2.
[27] *Id.* § 7159.1.
[28] *See id.* § 7159.2.
[29] *See* Cal. Bus. & Prof. Code §§ 7159(b), (c), (d), (e), (f), (h), 7165–7168.
[30] *See id.* § 7159(c).
[31] *See id.* § 7159(d).

furnish a copy of the written agreement, signed by the contractor, to the owner prior to commencing any work on the pool.[32]

It should be noted that these provisions relate only to single-family residences. This law does not apply if the swimming pool is built for the use and enjoyment of other than a single-family unit or if the swimming pool is built by the original developer who constructed the single-family residence.[33]

If a lawsuit is initiated by any party to a swimming pool construction contract, the law allows the prevailing party to recover reasonable attorneys' fees.[34] The effects of the contractor's noncompliance with any of the regulations relating to swimming pools subject the contractor to both civil and criminal penalties. If the contractor does not comply with this law, the contract for the construction of the swimming pool is void and unenforceable by the contractor;[35] that is, the contractor cannot maintain a lawsuit to recover for the work, labor, services, goods, or materials supplied in the construction of the swimming pool. Moreover, failure to comply with the law may subject contractors, their agents, or salespersons to misdemeanor punishment: a fine of not less than $100 nor more than $5,000; imprisonment in the county jail not exceeding one year; or both fine and imprisonment.[36]

§ 13.05 RESCISSION OF HOME SOLICITATION CONTRACTS

Contractors engaged in the home improvement business should be aware that their offers and contracts to sell labor and/or materials to the homeowner are subject to rescission. In response to high-pressure sales techniques used by some home improvement contractors and other home solicitation salespersons, the legislature has afforded the consumer a three-day period during which the consumer may cancel the transaction.[37] The law allows the consumer to cancel the home solicitation contract at any time until midnight of the third business day after the day on which the consumer signs the agreement. In addition, under Civil Code § 1689.6(c), a buyer has the right to cancel a home solicitation contract or offer for the repair or restoration of residential premises damaged by a disaster until midnight of the seventh business day after the buyer signs and dates the contract. The consumer must give the contractor written notice of cancellation at the address required to be included on all contracts. The notice of cancellation, if given by mail, is effective when deposited in the mail, properly addressed with postage prepaid.

The form required by the legislature for the notice of cancellation is as follows:[38]

---

[32] See id. § 7165.
[33] See id. §§ 7159, 7166.
[34] See id. § 7168.
[35] See Cal. Bus. & Prof. Code § 7167.
[36] See id. § 7158(a).
[37] See Cal. Civ. Code § 1689.6(a).
[38] Id. § 1689.7.

Notice of Cancellation

(Date of the Transaction)

You may cancel this transaction, without any penalty or obligation, within three business days from the above date.

If you cancel, any property traded in, any payments made by you under the contract or sale, and any negotiable instrument executed by you will be returned within 10 days following receipt by the seller of your cancellation notice, and any security interest arising out of the transaction will be canceled.

If you cancel, you must make available to the seller at your residence, in substantially as good condition as when received, any goods delivered to you under this contract or sale, or you may, if you wish, comply with the instructions of the seller regarding the return shipment of the goods at the seller's expense and risk.

If you do make the goods available to the seller and the seller does not pick them up within 20 days of the date of your notice of cancellation, you may retain or dispose of the goods without any further obligation. If you fail to make the goods available to the seller, or if you agree to return the goods to the seller and fail to do so, then you remain liable for performance of all obligations under the contract.

To cancel this transaction, mail or deliver a signed and dated copy of this cancellation notice, or any other written notice, or send a telegram to _____ (name of seller), at _____ (address of seller's place of business) not later than midnight of _____ (Date).

I hereby cancel this transaction.

_____
(Date)

_____
(Buyer's signature)

It should be noted that if a consumer or his agent or insurance representative requests emergency repairs or services for the immediate protection of persons or property, the consumer does not have the right to cancel the transaction.[39] However, it is necessary for the contractor to obtain from the consumer a separate dated and signed personal statement describing the situation requiring the immediate repairs or services and expressly acknowledging and waiving the right to cancel the transaction within three or seven days, whichever is applicable.[40]

---

[39] See id. § 1689.13.
[40] See id.

If contractors wish to secure mechanic's lien rights on the property to which they are supplying goods and services for home improvement, they must comply with the Retail Sales Installment Act. The Retail Sales Installment Act includes very specific provisions regarding any financing arrangements by the consumer and certain requirements with regard to disclosure of any finance charges and the total cash price to be paid to the consumer.

One case has held that the Truth-in-Lending Act does not apply to a material supplier extending credit to a contractor who was dealing with the owner.[41] However, a different court reached the exact opposite result, holding invalid the lien of a material supplier who had failed to give the owner a Truth-in-Lending Act notice.[42]

One contract fell within the home solicitation law but did not contain the notice required by California Civil Code § 1689.7 (the notice of cancellation form).[43] When the contractor substantially completed the work, the homeowners canceled the contract. The contractor sued, and the homeowner defended on the basis that the contractor could not recover because he had failed to comply with the law and the owner had canceled the contract. However, the court held that the contractor could recover the reasonable value of what he had furnished.

In *Louis Luskin & Sons, Inc. v. Samovitz*,[44] a plumber's contract did not contain the cancellation notice required by California Civil Code §§ 1689.7 and 1689.8. Because the plumber started work on the job within the three-day cancellation period, the court denied recovery against the homeowner by the plumbing contractor.

§ 13.06 ARBITRATION PROVISIONS IN HOME IMPROVEMENT CONTRACTS

If a contract for a work of improvement on a residential property with four or fewer units contains a provision for arbitration, the provision must be clearly titled "ARBITRATION OF DISPUTES." If a provision for arbitration is included in a printed contract, it shall be set out in at least 10-point Roman bold type or in contrasting red print in at least 8-point Roman bold type; if the provision is included in a typed contract, it shall be set out in capital letters. Immediately before the line or space provided for the parties to indicate their assent or nonassent to the arbitration provision, and immediately following that arbitration provision, the following shall appear:

> NOTICE: BY INITIALING THE SPACE BELOW YOU ARE AGREEING TO HAVE ANY DISPUTE ARISING OUT OF THE MATTERS INCLUDED IN THE "ARBITRATION OF DISPUTES" PROVISION DECIDED BY

---

[41] Mid Am. Homes, Inc. v. Eldon Horn, 377 N.E.2d 657 (Ind. Ct. App. 1978), *vacated on other grounds*, 396 N.E.2d 879 (Ind. 1979).
[42] Hobbs Lumber Co. v. Shidell, 326 N.E.2d 706 (Ohio C.P. 1974).
[43] *See* Beley v. Municipal Court, 100 Cal. App. 3d 5 (1979).
[44] 166 Cal. App. 3d 533 (1985).

NEUTRAL ARBITRATION AS PROVIDED BY CALIFORNIA LAW AND YOU ARE GIVING UP ANY RIGHTS YOU MIGHT POSSESS TO HAVE THE DISPUTE LITIGATED IN A COURT OR JURY TRIAL. BY INITIALING IN THE SPACE BELOW YOU ARE GIVING UP YOUR JUDICIAL RIGHTS TO DISCOVERY AND APPEAL, UNLESS THOSE RIGHTS ARE SPECIFICALLY INCLUDED IN THE "ARBITRATION OF DISPUTES" PROVISION. IF YOU REFUSE TO SUBMIT TO ARBITRATION AFTER AGREEING TO THIS PROVISION, YOU MAY BE COMPELLED TO ARBITRATE UNDER THE AUTHORITY OF THE BUSINESS AND PROFESSIONS CODE OR OTHER APPLICABLE LAWS. YOUR AGREEMENT TO THIS ARBITRATION PROVISION IS VOLUNTARY. WE HAVE READ AND UNDERSTAND THE FOREGOING AND AGREE TO SUBMIT DISPUTES ARISING OUT OF THE MATTERS INCLUDED IN THE "ARBITRATION OF DISPUTES" PROVISION TO NEUTRAL ARBITRATION.[45]

Any provision for arbitration in a contract for work on a residential property with four or fewer units that does not comply with the foregoing requirements may not be enforceable against any person other than the licensee.[46]

---

[45] Cal. Bus. & Prof. Code § 7191(b).
[46] *See id.* § 7191(c).

CHAPTER 14
ALTERNATIVE DISPUTE RESOLUTION

§ 14.01 Overview of Alternative Dispute Resolution

§ 14.02 Arbitration of Construction Disputes
    [A] Advantages of Arbitration
    [B] Disadvantages of Arbitration
    [C] Compelling Arbitration
    [D] Consolidation of Arbitration Proceedings
    [E] Waiver of Right to Arbitrate
    [F] Resisting Arbitration
    [G] Binding Effect of Arbitration Awards
    [H] Disclosure Requirements for Arbitrators

§ 14.03 Mediation
    [A] The Mediation Process
    [B] Advantages of Mediation
    [C] Disadvantages of Mediation
    [D] Enforcement of Mediation Agreements

§ 14.04 Other Available Types of ADR
    [A] Contractually Mandated ADR
    [B] Court-Ordered and Statutorily Required ADR
        [1] Court-Ordered ADR
        [2] Statutorily Required ADR
            [a] Judicial Arbitration under Civil Procedure Code §§ 1141.10 *et seq.*
            [b] Judicial Mediation under Civil Procedure Code §§ 1775 *et seq.*
            [c] Arbitration Provisions in the Public Contract Code
    [C] "Private Trials" and "Mini-Trials"
    [D] Arbitration Clauses in Residential Construction Contracts

§ 14.05 Enforcement of ADR Settlement Agreements

§ 14.01 OVERVIEW OF ALTERNATIVE DISPUTE RESOLUTION

In recent years, many people have come to regard the investment of time, effort and money necessary to prosecute a case in the courts as so oppressive that almost any alternative way of enforcing their legal rights seems likely to be better than litigation. With the increased caseload clogging federal and state courts, a party pursuing a construction claim in court often must wait years for resolution of disputes. In the meantime, the party can lose money on the claim from lost operating funds, loss of use of money, costs from legal fees, and inflation. The efforts of the courts to streamline their procedures, such as California's Trial Delay Reduction Act, have speeded up the process considerably but have not solved these basic problems. In some instances, the expedited procedures merely mean that the costs of litigation are incurred faster, while the promised benefits of a "speedy trial" may prove illusory when the early trial date arrives and there are no courtrooms available.

In response to this reality, what has come to be called "alternative dispute resolution" (ADR) has developed and continues to grow. There are a variety of processes and procedures that come under the broad heading of ADR. By far the most established ADR process is arbitration, which is the subject of a considerable body of judicial opinions, only some of which are discussed below in §§ 14.02[B] through 14.02[D]. Another widely used ADR mechanism is mediation. Mediation is discussed below in §§ 14.03 through 14.03[D]. In addition, there are mixtures of these processes, as well as "private trials" and "mini-trials," which are discussed in §§ 14.04 through 14.04[D] below.

§ 14.02 ARBITRATION OF CONSTRUCTION DISPUTES

Arbitration of construction disputes and claims, a well-established process, has become ever more popular as owners, design professionals, subcontractors, and contractors look for less expensive methods of pursuing claims and handling disputes. A great many, if not the majority, of the construction claims proceeding to arbitration are heard under the auspices of the American Arbitration Association (AAA). Within the last several years, the number of claims administered by the AAA has increased dramatically, a reflection of disenchantment with the judicial system and a growing interest in arbitration. All parties to a construction contract should be aware that arbitration is an available alternative to traditional litigation of claims. At the same time, however, it should be noted that arbitration may not be the ideal solution to all claim situations; party may desire to litigate rather than arbitrate certain disputes.

[A] Advantages of Arbitration

Arbitration offers a number of advantages in resolving construction claim disputes. First, arbitration is usually a less expensive method of resolving disputes. Unlike litigation, it does not require complex pleadings, legal memoranda, and other such costly trappings of a formal lawsuit. Most significantly, arbitration allows much less discovery than accompanies litigation. In arbitration there is little, if any, discovery unless the parties or the arbitrator agrees to allow it. Even then, the ability of the parties to structure their own discovery process means that the scope of discovery is likely to be far less extensive. Therefore, unless discovery is essential to the examination of the adversary's case prior to the dispute's being heard, arbitration should be considered over litigation.

Arbitration proceedings are likely to be conducted within a shorter period of time than is typical in the traditional court system. The legal rules of evidence and procedure do not apply in arbitration, enabling the parties to streamline hearings significantly. This can also be a disadvantage, as discussed in § 14.02[B] below.

Another advantage of arbitration is the greater likelihood that a decision will reflect an understanding of the construction industry as well as the costs associated with construction delay, both of which are primary causes of disputes. The arbitrators of construction disputes are usually more experienced than most superior court judges in the specific problems of construction projects. Throughout the United States, the AAA maintains a national panel of arbitrators to hear construction disputes. This panel consists of experts in many construction-related trades and professions. By choosing arbitration under the AAA's Construction Industry Arbitration Rules, an arbitrator may be selected who is a practicing architect, engineer, contractor, construction lawyer, or professor of construction engineering. Generally, it is more advantageous for all parties to have such members of the construction industry hear disputes, rather than a judge or jury with no professional experience or specific knowledge of the construction industry.

Another distinct advantage of arbitration is that it offers the contractor a speedier resolution of the dispute. Crowded court calendars preclude most construction disputes from being heard until several years after suit is filed. This is especially true of disputes that require lengthy trials, such as those arising from large, complex projects and involving numerous parties. Construction defect litigation, which has been on the rise in recent years, typically involves multiple parties and issues. However, this kind of case often requires extensive discovery, including destructive testing and other expert analysis that may make arbitration inappropriate unless the parties can agree on the needed discovery program. During the extended period of time typical to traditional litigation, witnesses may disappear, memories tend to fade, and the contractor often continues to lose money and incur legal and expert costs. Nevertheless, even arbitration can be a time-consuming process, particularly when an adversary opposes the arbitration and attempts to block the proceedings by going to court. In such a case, the party expecting arbitration to provide a quick resolution to the dispute may find that months are spent in court, merely to obtain a decision to allow the arbitration to proceed.

In arbitration, a party is more likely to prevail on the "equities" than when a judge is making the decisions. The authority of an arbitrator to decide a case is far greater than that of a trial judge, who is bound to follow the letter of legal rules of procedure, evidence, and burdens of proof. For example, a party with a valid claim who has failed to comply with the notice requirements of the contract is much more likely to lose in court before a judge who is inured to harsh results flowing from strict enforcement of "technical" legal requirements. In arbitration, the party may still be able to recover damages because the arbitrator is much more likely to seek a "fair" result based on equitable considerations.

Finally, there is much greater flexibility in arbitration in terms of the scheduling hearings and the introduction of evidence and witnesses. Court proceedings, particularly in a lengthy dispute, may require days and days of hearings, evidence, and testimony, during which time all the parties must continually appear in court. Typically, both attorneys and expert witnesses charge hourly or daily rates during trials that are significantly higher than their rates for trial preparations. Arbitration hearings, however, are often scheduled with the needs of the parties and the arbitrator in mind so that their daily professional lives can continue while the arbitration hearings are being conducted. In fact, arbitration may enable parties to resolve their dispute without disrupting progress on the construction project.

In summary, the primary advantages of arbitration are its flexibility, its ability to provide a relatively quick and inexpensive means of resolving construction disputes, and the use of arbitrators who are familiar with the construction industry. Arbitration eliminates many of the prerequisites to a lawsuit, such as discovery and pleadings, which are costly to all parties in terms of both time and money.

[B] Disadvantages of Arbitration

In deciding whether to arbitrate or litigate, the advantages of arbitration must be carefully weighed against its disadvantages. For example, a party who has a good claim but needs the information obtained through the discovery process to develop its case may be better off utilizing the federal or state courts, in which the discovery process is an essential element of the system. It is often more advantageous for a party with a good technical defense to present its case in a courtroom rather than in an arbitrator's office. In addition, a party with a "home-court advantage" may feel more comfortable pursuing the claim in a local court rather than a potentially distant arbitration.

Other reasons for choosing the court system over arbitration include a desire to defuse the opponent's equitable arguments, which may be more appealing to a panel of arbitrators than to an experienced trial judge. In addition, it may be to a particular party's advantage not to seek quick resolution of a given dispute. If such is the case, that party would not want to proceed with arbitration, where the dispute would likely be heard more quickly than in the court system. Finally, proceeding in a courtroom, particularly with a trial by judge, allows a greater opportunity to avoid a compromise verdict. A panel of arbitrators, especially one faced with a difficult case in which a great amount of contradictory evidence is presented, may reach a decision by means of a compromise in which neither party really wins or loses. Juries often do the same thing, but a single trial judge is more likely to avoid the possibility of a compromise verdict and give both parties a greater chance of a big win.

Contractors who are inclined toward choosing arbitration, however, should be aware of certain pitfalls that could defeat the right to arbitrate. For instance, one of the potential pitfalls is the problem of waiver, which is discussed more fully in § 14.02[E] below. Some cases have held that filing suit, answering a complaint, or recording and enforcing a mechanic's lien waives a party's right to arbitrate. In California, however, through a 1977 statute, a contractor seeking foreclosure of a lien does not lose any right to arbitrate if, at the time the foreclosure action is filed, a written application is made to the court to stay the foreclosure action pending arbitration.[1] Furthermore, due to the strong public policy favoring arbitration, the waiver argument is rarely successful. Finally, under Rule 47 of the AAA Construction Industry Arbitration Rules, contracts that refer to the AAA's Construction Industry Arbitration Rules are exempt, by contract, from the normal rule that an action taken in court by the parties may constitute a waiver of the party's right to arbitrate.

A second pitfall the contractor may encounter is fighting on two fronts. Once the contractor decides to pursue a claim through arbitration, pending arbitration the contractor should be certain to oppose any judicial proceedings that are begun by the opposing party. Otherwise, the contractor will find itself arbitrating and litigating the same dispute. This issue is discussed in greater detail under the topic of "Resisting Arbitration" in § 14.02[F] below.

Any contractual preconditions to arbitration can be another problem area. For example, the construction contract may require the contractor to submit a dispute to the architect before seeking arbitration. Generally, the contractor will be reluctant to do so if it believes that a favorable decision will not be received from the architect, who, in any event, may be slow to act. If the contractor proceeds to invoke arbitration, the owner may point to the contractor's failure to submit the dispute to the architect as a reason why arbitration should not go forward. One way out of this dilemma is to give the architect notice of the dispute and a reasonable amount of time within which to act. If the architect does nothing, then the contractor can establish an attempt to comply with this precondition.

Subcontractors have a special pitfall to avoid when the prime contract, the terms of which are incorporated into the subcontract, contains an arbitration provision that conflicts with the arbitration provision in the subcontract. For example, the prime contract's arbitration provision may be a broad scope provision that allows the arbitration of virtually any dispute arising on the project. The subcontract arbitration provision, however, may refer only to disputes between the prime contractor and the subcontractor. The prime contractor will argue against arbitration by claiming that the dispute in question really is with the owner and is beyond the scope of the subcontractor's arbitration provision. The subcontractor, who would prefer to go to arbitration, points to the broader provision in the prime contract as justifying arbitration.

Perhaps the most significant disadvantage of arbitration is the fact that arbitrators are not bound to follow the law. In *Moncharsh v. Heily & Blase,*[2] the California Supreme Court held that, with limited exceptions found in the arbitration

---

[1] Cal. Civ. Proc. Code § 1281.5.
[2] 3 Cal. 4th 1 (1992).

statute, arbitrators' awards are not reviewable by the courts for errors of fact or law. Essentially, under this case, a disappointed party in an arbitration proceeding may not overturn the ruling of the arbitrator because the arbitrator made an error of fact or law. As a result of this decision, if the parties by contract choose arbitration, then the grounds for appeal will be extremely limited. Specifically, pursuant to Civil Procedure Code § 1286.2, the grounds for vacating an arbitrator's award are limited to the following:

1. The award was procured by corruption, fraud, or other undue means;
2. There was corruption in any of the arbitrators;
3. The rights of the party were substantially prejudiced by misconduct of a neutral arbitrator;
4. The arbitrators exceeded their powers and the award cannot be corrected without affecting the merits of the decision upon the controversy submitted;
5. The rights of the party were substantially prejudiced by the refusal of the arbitrators to postpone the hearing upon sufficient cause being shown therefor[,] or by the refusal of the arbitrators to hear evidence material to the controversy[,] or by other conduct of the arbitrators contrary to the provisions of this title.

The foregoing must be carefully considered by any party when the contract documents are being negotiated, to determine whether to include an arbitration provision in the contract documents.

It should be noted that Civil Procedure Code § 1296 specifically authorizes the parties to a construction contract with a public agency to include a provision that "the arbitrator's award shall be supported by law and substantial evidence." The statute also provides that a court reviewing an award under a contract with such an express provision shall "vacate the award if after review of the award it determines either that the award is not supported by substantial evidence or that it is based on an error of law." One seeking relief under section 1296 must follow the procedures stated in Civil Procedure Code §§ 1285 *et seq.* for vacating an arbitration award.

Another of the most frequently stated disadvantages of arbitration is that it is often not possible to require the participation of all of the parties needed to fully resolve a particular dispute. This issue is addressed in detail in the next section below.

[C] Compelling Arbitration

While any two or more parties may voluntarily agree to submit their disputes to arbitration, parties may only be compelled to arbitrate if either (1) there is a statute providing for arbitration that covers the dispute; or (2) there is an arbitration provision in the parties' written contract. When the parties agree that a dispute is subject to arbitration, they may proceed directly to arbitration without first filing a lawsuit. However, when one of the parties refuses to arbitrate the dispute, the party seeking to arbitrate must file a superior court action and petition the court for an arbitration order under Civil Procedure Code § 1281.2. If the party refusing to arbitrate itself commences a court action, instead of answering the complaint, the other party may respond with a petition to compel arbitration under § 1281.2.[3] If the petition is denied, the party then has an additional 15 days after the denial to respond to the complaint.

The right to binding arbitration is favored judicially and by public policy to promote judicial efficiency and to settle disputes quickly and fairly.[4] For this reason, written agreements (including construction contracts) are to be liberally interpreted in favor of arbitration of disputes that arise thereunder.[5] As stated by the United States Supreme Court: "[I]t has been established that where the contract contains an arbitration clause, there is a presumption of arbitrability in the sense that '[a]n order to arbitrate the particular grievance should not be denied unless it may be said with positive assurance that the arbitration clause is not susceptible of an interpretation that covers the asserted dispute.'" Doubts should be resolved in favor of coverage.[6]

However, even when their agreement contains an arbitration provision, the parties still cannot be compelled to arbitrate specific controversies that they did not agree to arbitrate and as to which there is no statute compelling arbitration.[7] Therefore, if the court determines that an agreement to arbitrate does exist, it will not order contracting parties to arbitrate a controversy, if either (1) grounds exist for revocation of the agreement or the arbitration clause; or (2) the right to compel arbitration has been waived.

The issue of waiver is discussed in § 14.02[E] below. The issue of whether a particular dispute is subject to arbitration is usually to be decided by the court, not by the arbitrator. Under Civil Procedure Code § 1281.2, the court is required to make an advance determination of whether arbitration is required under the express terms of a contract between the parties.[8] Moreover, both the U. S. Supreme Court and the California Supreme Court have ruled that an arbitrator is only empowered to decide the issue of arbitrability if the contract between the parties expressly grants such power to the arbitrator.[9] Furthermore, when only some of the issues are subject to an arbitration provision, the courts have broad discretion to order all issues arising from the subject of the agreement to be

---

[3] Cal. Civ. Proc. Code § 1281.7.

[4] *See, e.g.,* Charles J. Rounds Co. v. Joint Council of Teamsters No. 42, 4 Cal. 3d 888 (1971); Keating v. Superior Court, 31 Cal. 3d 584 (1982).

[5] AT&T Tech., Inc. v. Communication Workers of Am., 475 U.S. 643, 106 S. Ct. 1415, 89 L. Ed. 2d 648 (1986); United Steelworkers of Am. v. Warrior & Gulf Navigation Co., 363 U.S. 574, 80 S. Ct. 1347, 4 L. Ed. 2d 1403 (1960); Service Employee Int'l Union, Local 347 v. City of L.A. Dep't of Transp., 24 Cal. App. 4th 136, 143 (1994).

[6] AT&T Tech., 475 U.S. at 650 (quoting *United Steelworkers,* 363 U.S. at 582–583).

[7] Freeman v. State Farm Mut. Auto. Ins. Co., 14 Cal. 3d 473, 481 (1975); United Public Employees v. San Francisco, 53 Cal. App. 4th 1021, 1031 (1997).

[8] *See also, e.g.,* Lawrence v. Walzer & Gabrielson, 207 Cal. App. 3d 1501, 1505 (1989).

[9] First Options of Chicago, Inc. v. Kaplan, 514 U.S. 938, 943, 115 S. Ct. 1920, 1923, 131 L. Ed. 2d 985, 993–994 (1995); Freeman v. State Farm, 14 Cal. 3d at 480.

consolidated either for arbitration or for trial in a court of law.[10] Thus, in *AT&T Technologies, Inc. v. Communications Workers of America*,[11] the United States Supreme Court held that "unless the parties clearly and unmistakably provide otherwise," it is the Court, not the arbitrator, that decides the issue of arbitrability, but that the Court should not at that point decide the merits of the underlying case.[12]

As noted above, and in § 3.02[B], if the contract documents so provide, then either party may be required by the other to submit to arbitration any claims arising from a construction project. Chapter 3 also makes the point that, for the prime contractor, it is crucial that the arbitration provisions in its contract with the owner be consistent with those in its contracts with subcontractors; otherwise, the prime contractor could find itself fighting a two-front war, with the possibility of inconsistent results and substantially increased attorney's fees and costs.

A case illustrating the importance of this principle is *Slaught v. Bencorno Roofing Co.*,[13] in which a dispute arose between the general contractor (Bencorno) and a property owner (Slaught) concerning a construction project. When the matter was submitted for arbitration, the general contractor demanded that certain subcontractors join in the arbitration, but they declined to participate. The general contractor then filed a petition to compel arbitration and to consolidate the arbitration proceedings, but the trial court denied the general contractor's petition.

The court of appeal reversed and directed the trial court to enter an order compelling the subcontractors into arbitration. The court held that the subcontractors were subject to arbitration because their subcontracts incorporated by reference the terms of the construction contract between the owner and the general contractor, and that contract, in turn, contained an arbitration clause. The court further ruled that the subcontractors had to participate in the same arbitration proceeding as the owner and the general contractor, even though the subcontracts contained their own arbitration clauses. The court held that when the construction contract was incorporated by reference into the subcontracts, the arbitration terms of the construction contract were likewise incorporated.

An earlier case, *Boys Club v. Fidelity & Deposit Co.*, applied essentially the same rule to sureties, holding that a surety on a bond that incorporates by reference a construction contract containing an arbitration clause can be required to participate in an arbitration.[14]

In the *Boys Club* case, the court of appeal reversed the trial court, which had denied the petition of the plaintiff (the Club) to compel a surety to take part in (and thus be bound by) an arbitration between the Club and a construction company. After completion of a recreation facility, the Club brought a construction defect action against the contractor and filed an arbitration demand pursuant

---

[10] *See* Cal. Civ. Proc. Code § 1281.2(c); Parker v. Twentieth Century-Fox Film Corp., 118 Cal. App. 3d 895, 905–906 (1981).
[11] 475 U.S. 643, 106 S. Ct. 1415, 89 L. Ed. 2d 648 (1986).
[12] 475 U.S. at 649.
[13] 25 Cal. App. 4th 744 (1994).
[14] 6 Cal. App. 4th 1266 (1992).

to an express provision in the construction contract. Three years later, the Club amended its arbitration demand, seeking to add as a defendant party the surety on the contractor's performance bond. When the surety resisted, the Club petitioned the superior court for an order requiring the surety to participate in the arbitration.

The trial court upheld the surety's opposition to the Club's arbitration petition on two grounds: (1) that the surety was not a party to the construction contract containing the arbitration provision; and (2) that the Club was time-barred from adding the surety as a party. In reversing the trial court, the court of appeal held that by including language in the performance bond that incorporated the construction contract, the surety intended and agreed to be bound by the arbitration clause, in spite of the fact that it was not a party to the contract. The appellate court also held that the Club had not waived its right to demand arbitration by its three-year delay in filing the amended demand, because of the surety's inability to show that it had been materially prejudiced by the delay.

In reaching its decision in *Boys Club,* the court of appeal cited with approval a recent series of federal district court and federal circuit court cases compelling sureties to arbitrate in similar cases.[15] The court of appeal chose not to follow the precedent of two New York cases.[16] Instead of following the decision in the *Fidelity* case, which had interpreted and extended the logic of the *Transamerica* decision, the *Boys Club* court agreed with and adopted the logic of the dissenting opinion in the *Fidelity* case. It also distinguished another federal case[17] on the ground that the surety bond in that case did *not* expressly incorporate the construction contract by reference.

Thus, it is now established law in California that if the express language of its bond incorporates the terms of an underlying construction contract, a surety will be bound to those terms even though it is not a party to the agreement. Furthermore, it is incumbent upon the surety to demonstrate actual prejudice to itself if it seeks to avoid such arbitration on the grounds of waiver. The ability to compel the participation of a surety will depend upon the express terms of the bond, but under the right circumstances this could be the deciding factor in favor of choosing arbitration.

The importance—for better or for worse—of being able to compel a responsible party to participate in binding arbitration is amply illustrated by another case, *Creative Plastering, Inc. v. Hedley Builders, Inc.,*[18] which cites and builds upon *Moncharsch v. Heily & Blase*[19] to even further buttress the mounting legal authority for the inviolability of an arbitrator's decision, even if based upon an error of law or fact.

---

[15] 6 Cal. App. 4th at 1272.
[16] Transamerica Ins. Co. v. Yonkers Contracting Co., 49 Misc. 2d 512, 267 N.Y.S.2d 669 (1966); Fidelity v. Parsons & Whittemore Constr., 397 N.E.2d 380, 421 N.Y.S.2d 869 (1979).
[17] Windowmaster Corp. v. B.G. Danis Co., 511 F. Supp. 157 (S.D. Ohio 1981).
[18] 19 Cal. App. 4th 1662, 24 Cal. Rptr. 2d 216 (1st Dist. 1994).
[19] 3 Cal. 4th 1 (1992).

In *Creative Plastering,* the superior court granted a petition to vacate or modify an arbitrator's award and struck the award of attorney's fees by choosing to take issue with the arbitrator's determination that the plaintiff subcontractor (Creative) was the prevailing party in a payment dispute with a developer (Hedley). Both parties stipulated that $13,099.60 was due Creative under the subcontract, but Hedley claimed a backcharge of $22,176 for alleged delay damages. The arbitrator awarded Creative a net recovery of $11,251, plus $1,549 in interest and $11,178 for attorney's fees, as the prevailing party under the arbitration provision of the subcontract. The superior court accepted Hedley's argument that it was the prevailing party because Creative recovered less than the full amount of the stipulated subcontract balance.

The court of appeal reversed, declaring that (1) under the authority of *Moncharsch,* the courts lack power to vacate an award, regardless of whether the arbitrator's decision was right or wrong on the facts or the law; (2) the only grounds for vacating an arbitration award are those set forth in Civil Procedure Code § 1286.2; and (3) this arbitrator had power under the agreement to award attorney's fees. Although the court of appeal agreed that the arbitrator's decision was correct (because the developer had sought to offset the entire contract balance by its delay claim), it held that even if his decision was wrong, so long as it was within his authority to decide the issue, it was binding upon the parties, and the courts would not interfere.

[D] Consolidation of Arbitration Proceedings

The question of whether separate arbitration proceedings, arising from three separate contracts (with unique and conflicting arbitration provisions) among several parties, could be consolidated was answered in an earlier case, *Garden Grove Community Church v. Pittsburgh Des Moines Steel Co.*[20] Pittsburgh Des Moines Steel Company (PDM) had a contract with the Garden Grove Community Church (Church) to construct the "Crystal Cathedral." The church had separately contracted with a construction manager and a project architect. The architect, in turn, hired a structural engineer.

The Church's contract with PDM contained an arbitration clause that stated that all disputes would be settled by arbitration in accordance with the Construction Industry Rules of the American Arbitration Association. It also stated that the owner shall not be obligated to arbitrate any such dispute if the owner, in order to fully protect its interests, desires in good faith to bring in, or make a party to the dispute, the construction manager, the architect, or any third party, who had not agreed to participate in and be bound by the same arbitration proceeding. The court referred to the foregoing clause as the "escape clause."

The Church had separate contracts with the construction manager and the architect, each of which contained arbitration clauses. The arbitration clause in the architect's contract, however, provided that no arbitration arising out of or relating to that agreement should include, by consolidation, joinder, or in any

---

[20] 140 Cal. App. 3d 251 (1983).

other manner, any additional party not a party to that agreement, except by written consent containing a specific reference to the agreement and signed by all the parties.

During the course of the job, numerous disputes arose between PDM and the Church concerning cost overruns as well as the balance due under the contract. The construction manager for the Church assured PDM that its claims would be reviewed in good faith and, if necessary, would be arbitrated as provided in PDM's contract. Ultimately, PDM's claims were rejected, and it instituted arbitration proceedings against the Church. The arbitrator rejected the Church's contention that the "escape clause" would preclude arbitration because the Church desired to resolve its potential claims for indemnity against the construction manager, architect, and engineer at the same time, in a single proceeding.

After the arbitration proceeding with PDM had commenced, the Church repeatedly asked the construction manager, architect, and engineer to join, but they refused to do so. The Church then initiated a lawsuit to stay the arbitration proceedings with PDM or, in the alternative, for an injunction requiring the other parties to join the PDM arbitration under Civil Procedure Code § 1281.3. That section provides that a party to an arbitration agreement may petition the court to consolidate separate arbitration proceedings when (1) separate arbitration agreements or proceedings exist between the parties or when one party is a party to a separate arbitration agreement or proceeding with a third party; (2) the dispute arises from the same transaction or series of transactions; and (3) there are common issues of law or fact creating the possibility of conflicting rulings by more than one arbitrator or panel of arbitrators. The statute provides that the court can order the matters to be heard before one arbitrator and/or may order consolidation of the separate proceedings. The statute further provides that if the arbitration agreements contain inconsistent provisions, the court shall resolve such conflicts and determine the rights and duties of the parties to achieve substantial justice under all circumstances.

The trial court stayed the arbitration with PDM and found that the Church's "escape clause" prevented arbitration because the church could not include third parties it believed in good faith were necessary to protect its interest.

PDM appealed, and the court of appeal reversed the stay of the pending PDM arbitration and remanded the matter to the trial court with directions to order the Church to institute separate proceedings against the project manager and the architect and to consolidate the separate arbitration proceedings. The court noted that California law controlled and that whether or not the "escape clause" should apply was a question for judicial decision because it pertained to whether or not the parties had agreed to arbitrate the dispute.

The *Garden Grove* court recognized that the Church's concern was understandable, as the bulk of PDM's claims related to alleged wrongdoing by the construction manager, architect, and engineer. Assuming that PDM's claims were correct, the Church would then be entitled to seek indemnity from the culpable parties. If there were separate arbitrations, the Church faced the danger of being trapped by conflicting findings in separate proceedings. For example, the arbitrator in the PDM arbitration could find the Church liable, while in separate proceedings with either the construction manager or the architect, those parties could prevail, thus resulting in substantial prejudice to the Church. The court stated that the trial court correctly found that the Church could properly exercise its contractual right to avoid a separate arbitration with PDM.

Finally, the *Garden Grove* opinion states that by adopting Civil Procedure Code § 1281.3, the California Legislature has manifested a strong policy favoring consolidation of arbitration proceedings involving common issues of fact and law in order to achieve efficient settling of private disputes, judicial economy, and the avoidance of contrary results. The court held that the three requirements for application of section 1281.3 were factually met in that (1) separate arbitration agreements existed between one party and each of the other parties to the coordinated arbitrations; (2) the disputes arose from the same transaction or series of transactions; and (3) there were common issues of law or fact, thereby creating the possibility of conflicting rulings by more than one arbitrator. Therefore, the court concluded that "[t]here is no reason to deny consolidation unless it would impair a substantial right or obligation of a party under the contract."[21]

The court further stated that the language in the contract between the Church and the architect, providing that arbitration with the architect could not be consolidated with any other arbitration without its consent, did not limit the right of the *court* to order consolidation. Under those circumstances, Civil Procedure Code § 1281.3 allows a court to order coordination of the separate arbitration proceedings. The *Garden Grove* case thus illustrates how strongly the courts favor arbitration and the liberality with which they will allow the consolidation process to join parties. Citing an earlier California Supreme Court holding, the *Garden Grove* court concluded: "Thus, a party may be forced into a coordinated arbitration proceeding in a dispute with a party with whom he has no agreement, before an arbitrator he has no voice in selecting and by a procedure he did not agree to."[22]

[E] Waiver of Right to Arbitrate

In spite of the judicial policy strongly favoring arbitration, a party may waive its right to arbitrate.[23] This occurs, most commonly, as a result of the party's failure to make a timely demand for arbitration or the party's appearance in a court action without asserting the arbitration rights.

The basic principle that a party may lose its right to arbitrate by failing to make a timely demand for arbitration is illustrated by *Platt Pacific, Inc. v. Andelson*,[24] in which the plaintiff contractor sued the defendant owners for nonpayment on a contract to construct a house and a garage. After nearly five years of litigation, the parties entered into an agreement to conduct settlement negotiations with a retired judge. The agreement provided that if settlement efforts were

---

[21] 140 Cal. App. 3d at 264.
[22] *Id.* at 263 (citing Keating v. Superior Court, 31 Cal. 3d 584, 603 (1982)).
[23] Cal. Civ. Proc. Code § 1281.2(a).
[24] 6 Cal. 4th 307 (1993).

unsuccessful, then the matter would be submitted to arbitration, "but in no event shall such a demand be filed later than August 31, 1989." The principal of the contracting company died, and therefore the plaintiff's attorney did not file the demand for arbitration until October 30, 1989. The American Arbitration Association determined that it was without authority to initiate arbitration without a court order, and the trial court denied the demand for arbitration because it was not filed within the time set forth in the agreement. The Supreme Court held that when the parties have agreed that a demand for arbitration must be made within a certain time, that time limitation is a condition precedent that must be performed before the contractual right to arbitration arises.

This case is thus a particular application of the general rule that when the parties contractually agree upon specific conditions precedent to the exercise of their rights to arbitrate, the courts will enforce those prerequisites. The most common example of this is failure of a party to abide by specific limits set forth in the arbitration provision with respect to the time and manner of initiating arbitration.[25] Even in the absence of a specific limitation as to time, a party who unduly delays in seeking arbitration may be deemed to have waived its rights.[26]

By far the most frequently recognized ground upon which California courts have sometimes concluded that a party has waived its right to arbitrate disputes that fall within a contractual arbitration provision is the filing of legal action without asserting the right to compel arbitration.[27] Some courts have ruled that the mere filing of a complaint is not sufficient to invoke the waiver doctrine—the waiving party must also engage in some litigation activities. However, it is not necessary for the litigation to have proceeded to judgment on the merits.[28]

Also, a defendant who raises the issue of arbitration promptly and in the form required by Civil Procedure Code §§ 1281.2 and 1281.4 will not be deemed to have waived its rights by having prepared its defense prior to the court's ruling on its petition to compel arbitration or by having continued to defend itself in the litigation after an adverse ruling thereon.[29]

The general rule with regard to the issue of waiver was simply stated in *Maddy v. Castle:* "The proper test as to whether the right to arbitrate has been waived is: do the circumstances of the case indicate an intent on the part of a party to waive his right to arbitrate by seeking the same type of relief in another forum."[30] The *Maddy* opinion also makes another basic and important point with respect to waiver; that is, while waiver of the right to arbitrate is generally nonrevocable, it can be revoked "only where the nonwaiving party has done nothing in reliance on that waiver."[31]

---

[25] *See also, e.g.,* Brunzell Constr. Co. of Nev. v. Harrah's Club, 228 Cal. App. 2d 764 (1967).
[26] *See, e.g.,* Allstate Ins. Co. v. Superior Court of S.F., 35 Cal. App. 3d 137 (1973).
[27] *See, e.g.,* Titan Enters., Inc. v. Armo Constr. Co., 32 Cal. App. 3d 828 (1973); Jones v. Pullock, 34 Cal. 2d 863 (1950).
[28] *See, e.g.,* McConnell v. Merrill Lynch, Pierce, Fenner & Smith, Inc., 105 Cal. App. 3d 946 (1980).
[29] Pacific Inv. Co. v. Townsend, 58 Cal. App. 3d 1 (1976).
[30] 58 Cal. App. 3d 716, 721 (1976).
[31] 58 Cal. App. 3d at 722.

Civil Procedure Code § 1281.5 creates a special application of this legal principle in the context of suits to enforce mechanic's lien rights. That statute specifically provides that a claimant who records a lien and then files an enforcement action will *not* be deemed to have waived the right to arbitrate "if, in filing an action to enforce the claim of lien, the claimant at the same time presents to the court an application that the action be stayed pending the arbitration of any issue, question or dispute which is claimed to be arbitrable under the agreement and which is relevant to the action to enforce the claim of lien." Subsection 1281.5(b) provides that a defendant in a lien enforcement action who fails to file a section 1281.2 petition to compel arbitration at or before the time it answers the complaint thereby waives its right to compel arbitration. Even before the enactment of this statute, in 1977, at least one court of appeal decision had reached the same result by ruling that a subcontractor had not waived its contractual right to demand arbitration by filing a mechanic's lien and an enforcement action thereon.[32]

Furthermore, as noted above, Rule 47 of the AAA Construction Industry Arbitration Rules specifically states that no judicial proceedings instituted by a party relating to the subject matter of the arbitration shall be deemed a waiver of the parties' right to arbitrate. Therefore, contractors who refer to the AAA's Construction Industry Arbitration Rules in their contractual arbitration provision agree in advance of any dispute that any action taken in court by the parties shall not be deemed a waiver of the party's right to arbitrate.

The courts are generally very reluctant to invoke the doctrine of waiver with respect to a highly favored procedure such as arbitration. Therefore, the courts are required to "closely scrutinize" any allegation that a party may have waived "such a favored right" as the right to arbitration of contract disputes.[33] Moreover, if determination of the issue of waiver requires consideration of the merits of the dispute submitted for arbitration, the issue of waiver may be determined by the arbitrator rather than the trial court.[34]

[F] Resisting Arbitration

Sometimes, even when there is an express arbitration provision in their contract, one of the parties may have good reason to resist arbitration of a particular dispute. Perhaps the most common example is when one of the parties cannot compel a third party to arbitrate at all and is concerned about inconsistent results between arbitration with one party and litigation with another. It is well established, under California law, that a party should not be compelled to participate in an arbitration if, at the same time, it faces another legal proceeding with a third party, arising out of the same transaction or series of related transactions, and

---

[32] Homestead Sav. & Loan Ass'n v. Superior Court of Marin County, 195 Cal. App. 2d 697 (1961).

[33] *See, e.g.,* Seidman & Seidman v. Wolfson, 50 Cal. App. 3d 826 (1975).

[34] *See, e.g.,* Butchers Union Local 532 v. Farmers Mkts., 67 Cal. App. 3d 905 (1977).

there is a possibility of conflicting rulings on common issues of law or fact.[35] Pursuant to Civil Procedure Code § 1281.2(c), even the *possibility* of such conflicting rulings is a ground for denying a petition to compel arbitration. Under that section, the court has four specific options. The court "(1) may refuse to enforce the arbitration agreement and may order intervention or joinder of all parties in a single action or special proceeding; (2) may order an intervention or joinder of all or only certain issues; (3) may order arbitration among the parties who have agreed to arbitration and stay the pending court action or special proceeding pending the outcome of the arbitration proceeding; or (4) may stay the arbitration pending the outcome of the court action or special proceeding." Upon being served with a suit and the petition to compel arbitration, the resisting party's opposition to the petition should set forth its grounds for asking the court to resort to one of the four options listed in section 1281.2(c). The statute is meant to give the court enough flexibility to structure the proceedings to avoid multiple proceedings and possibly conflicting outcomes.

This apparently covers a court's options when the party seeking arbitration petitions the court for an order under section 1281.2; but what happens if the party seeking arbitration proceeds with filing an arbitration directly with the American Arbitration Association or other such body without going to court? As there is neither a pending court action nor an opportunity to oppose a petition to compel arbitration, what is the other party to do if it believes (1) the particular dispute is not subject to arbitration under their agreement; or (2) that a party to the dispute who cannot be compelled to arbitrate (such as the project owner) may sue *later* on the same dispute?

One possible answer is the availability of injunctive relief to prevent a multiplicity of such judicial proceedings from going forward. Under California law, an injunction is proper where judicial restraint is necessary to prevent multiple legal proceedings on the same controversies.[36] Such a multiplicity of proceedings also constitutes grounds for a permanent injunction.[37] Moreover, where the injunction is sought to stay a proceeding already pending at the commencement of the action, prevention of a multiplicity of proceedings is still a statutory ground for issuance of an injunction.[38] Therefore, although there is no case law directly on point, a party should be able to commence an action in the superior court and then, by motion, seek an injunction to stop an arbitration proceeding that was previously initiated by the other party without first filing suit and petitioning to compel arbitration. As shown below, simply choosing to ignore the arbitration proceeding on such grounds is not a legally viable option.

---

[35] Cal. Civ. Proc. Code § 1281.2(c); Henry v. Alcove Inv., Inc., 233 Cal. App. 3d 93, 101 (1991); C.V. Starr & Co. v. Boston Reinsurance Corp., 190 Cal. App. 3d 1637, 1641 (1987).

[36] Cal. Civ. Proc. Code § 526(a)(6). *See also* Rynsburger v. Dairymen's Fertilizer Corp., 266 Cal. App. 2d 269, 279 (1968); Aldrich v. Transcontinental Land, etc., Co., 131 Cal. App. 2d 788, 796–797 (1955).

[37] Cal. Civ. Proc. Code § 526(a)(6); Cal. Civ. Code § 3422(3).

[38] Cal. Civ. Proc. Code § 526(b)(1).

[G] Binding Effect of Arbitration Awards

The danger being courted by any party who chooses to ignore an arbitration demand to which he or she even *might* be legally obligated to respond is dramatically demonstrated by the case of *Keller Construction Co. v. Kashani*,[39] which concerned the issue of who is bound by an award rendered after an arbitration. Although he had refused to participate in the arbitration proceedings, the court of appeal in *Keller* held that the general partner of a limited partnership was personally bound by an arbitration agreement and thus liable for an arbitration award rendered in excess of a million dollars.

In that case, a construction contract between Keller Construction (Keller) and Ramada of Simi Valley Ltd. (Ramada) was signed by Kazem Kashani (Kashani) under the words "Ramada Oi Simi Valley, Ltd." Ramada was a California limited partnership of which Kashani was the general partner. Keller demanded arbitration, serving the demand for arbitration on Ramada and on Kashani individually. At the arbitration proceedings, after Ramada had filed for bankruptcy, Kashani appeared specially, objected on the ground that he was not personally subject to the agreement, and then left the proceeding. An award of $1,417,456 was confirmed, and the court entered a judgment against Kashani that was then affirmed on appeal. The court of appeal held, first, that a general partner is liable for all debts and obligations of a limited partnership. The court further held that the general partner is an agent of the limited partnership and is therefore a beneficiary of any agreement entered into on behalf of the partnership. As a result, the court ruled that the general partner is individually bound by an agreement to arbitrate entered into by a limited partnership that he or she controls.

California courts have also held that arbitrated issues, once decided, are subject to the doctrine of *res judicata*. This is a Latin phrase meaning that "the thing has been judicially decided." When the merits of a legal dispute are determined by a trier of fact, the judgment rendered, once final, is deemed under this principle to be a conclusive determination of the rights of the parties with respect to all matters that were *or should have been* decided in that matter. The application of this principle to arbitration awards was confirmed by the court of appeal in *Thibodeaux v. Crum*,[40] a case in which homeowners engaged in an arbitration with the general contractor who had built a home for them, asserting numerous alleged construction defects. The arbitrator awarded the contractor the contract price plus some allowances and awarded the homeowners only some of their claimed damages. Significantly, the arbitrator also awarded the contractor attorney's fees as the prevailing party. The homeowners filed a petition to correct the award, contesting the attorney's fee issue. The contractor declared bankruptcy. Because of the automatic restraining order issued out of the bankruptcy court, the petition to correct the arbitration award was never ruled upon. In addition, neither the contractor nor the homeowners moved to confirm the arbitration award in court.

---

[39] 220 Cal. App. 3d 222 (1990).
[40] 4 Cal. App. 4th 749 (1992).

The homeowners then brought a court action against the subcontractor who constructed their driveway, seeking recovery for extensive cracks and other problems therewith. The trial judge entered judgment for the homeowners, rejecting the subcontractor's claim that the homeowners were estopped (i.e., legally precluded) from pursuing their claim because the entire subject had been litigated in the arbitration proceeding. The court of appeal reversed, holding that the action against the subcontractor was barred by the doctrine of *res judicata*. The court held that the doctrine of *res judicata* applies to arbitration proceedings as well as to judicial proceedings, and also to matters that could have been arbitrated but were not. The court noted that, in this particular case, if the problem of cracks in the driveway was not included in the arbitration proceedings, it should have been, because the cracks appeared immediately after construction and well before the arbitration took place. As a result, the homeowners knew and complained about the cracks before the arbitration and the driveway continued to deteriorate during the arbitration proceedings. The court held that the subcontractors had correctly raised *res judicata* as a defense and that the arbitration award still had *res judicata* effect on the rights of the homeowners, even though it was unconfirmed.

[H] Disclosure Requirements for Arbitrators

Finally, two provisions of the California Civil Procedure Code dealing with disclosures required of arbitrators directly relate to enforcement of arbitration agreements. Section 1281.9 provides that, when a person is proposed for nomination by all parties or by all party arbitrators, or is proposed for appointment by a court to serve as a neutral arbitrator, the proposed arbitrator shall disclose, within ten days: (1) any case in which the nominee served as a party arbitrator for any party to the current arbitration or for a lawyer for a party; or (2) any case in which the nominee served as a neutral arbitrator in any case involving any party or lawyer for any party; (3) the results of each case, including the amount of damages awarded; and (4) the names of the attorneys, the prevailing party, and the date of the award.

A proposed nominee may be disqualified by failure to comply with these disclosure requirements if any party serves a notice of disqualification within 15 days after the failure to disclose. Additionally, a proposed nominee who does comply with the disclosure requirements shall be disqualified if any party serves a notice of disqualification within 15 days after service of the disclosure statement. For purposes of disqualification, the term "lawyer for a party" includes any lawyer or law firm associated in the practice of law with the lawyer hired to represent a party. The disclosure requirements extend from three to five years prior to the date of the proposed nomination or appointment, depending on the date of the nomination or appointment.

This statute is problematic in several respects. First, it delays the appointment of arbitrators for up to 25 days to accommodate the disclosure and disqualification procedures. Second, the procedure provides for disqualification regardless of whether or not good cause exists. Finally, the new law defines "lawyer for a party" extremely broadly so as to include law firms and lawyers with whom the lawyer has been associated. This definition follows conflict-of-interest rules, which presume that lawyers regularly obtain confidential information about their clients and that lawyers in the same firm disclose such confidential information to one another. While the avoidance of conflict of interest due to exchange of confidential information would arguably apply where a proposed arbitrator previously served as a *party* arbitrator, the rationale does not transfer to *neutral* arbitrators.[41] In the judicial system, a judge is not disqualified to rule on a motion merely because it is made by a lawyer associated with another lawyer who appeared before the same judge previously in another, unrelated matter. The result of enforcement of this legislation may so narrow the field of potential qualified arbitrators—or cause such lengthy delays in the arbitration process—that parties may instead resort to litigation or other ADR procedures.

Section 1281.95 provides similar but much less onerous disclosure requirements relating to arbitration of any claim for more than $3,000 pursuant to a contract for the construction of improvement of a residential property consisting of four or fewer units.

§ 14.03 MEDIATION

Mediation is a nonbinding, structured process that may be used in conjunction with or as an alternative to either litigation or arbitration. A mediator does not render a decision but rather assists the participants in reaching a negotiated settlement of their differences. Like arbitrators, mediators are private individuals chosen by agreement between the parties themselves. Although experience and knowledge of the construction industry are significant factors in choosing a mediator, those chosen generally have professional legal and/or judicial experience. Thus, mediators are usually attorneys or retired judges and are seldom architects, engineers, contractors, or other construction industry professionals such as those who frequently serve as arbitrators.

Although mediation is generally voluntary, it may be required pursuant to statute or by local court rules (*see* discussion in § 14.04[B] below). Additionally, some construction contracts now require the parties to engage in non-binding mediation before submitting a dispute to arbitration or litigation (*see* discussion in § 14.04[A] below).

[A] The Mediation Process

While the parties and their chosen mediator are generally free to structure their process to their own liking, most mediations follow the same basic pattern. The first steps, of course, are agreeing to mediate, choosing a mediator, and setting up a schedule for the proceedings. This will sometimes include getting the

---

[41] A single arbitrator, having been chosen by the parties, is considered a "neutral" arbitrator. In cases in which multiple arbitrators are used, each party selects a "party arbitrator," and they, in turn, select a "neutral arbitrator," (who, for this reason, often has the "deciding vote" in the arbitrators' decisions).

trial court to agree to stay its own proceedings pending the outcome of the mediation. Within reason, courts are generally agreeable to doing this because of the strong public policy favoring negotiated settlements over trials.

The following is a brief discussion of what happens next: Typically, the parties will agree on the number, scope, and contents of a "mediation statement" that sets forth their basic positions to the mediator. These usually include a description of the dispute, the party's damages claims, and its defenses to the claims of its opponents. The parties have the option of exchanging their mediation statement with one another and/or submitting confidential mediation statements to be read only by the mediator. Confidential mediation statements usually also include some indication of the party's settlement position, which, naturally, the party does not want to share with the opposing party. On the other hand, it is useful for the parties to know one another's positions before the face-to-face mediation, and therefore, the parties usually choose to serve mediation statements on one another for that purpose. Sometimes, the parties may choose to exchange rebuttal statements in which they respond, in advance, to the points raised in one another's initial mediation statements. Otherwise, the parties may respond orally to one another's arguments at the mediation.

When there are disputed legal issues that are material to the parties' settlement decisions, they may also agree to "brief" those issues to the mediator in their mediation statements. As the mediator is not empowered to make legal decisions, this step is often abbreviated or even omitted altogether. When it is included, the mediator may be asked to render an informal opinion as to the relative strengths and weaknesses of the parties' legal positions.

In any event, most mediations begin with the mediator giving the parties an overview of how he or she intends to structure the mediation discussions, followed by each side's making a summary presentation of its position. Once the parties and the mediator feel that everyone is familiar with their adversaries' positions and agreed on how they will proceed, they move into the process of moving toward a mutually acceptable resolution of their disputes. This usually involves what is sometimes described as "shuttle diplomacy" wherein the mediator moves back and forth in private meetings with the parties, which are usually known as "caucuses."

In the separate caucuses with the mediator, a party and its counsel are expected to frankly discuss the issues and their settlement position. As quickly as possible, the mediator will turn the discussion of factual and legal issues to considerations of money. After all, the final resolution of most disputes is a determination of which party will pay how much to the other. The mediator is free to use whatever means of persuasion are at his or her disposal to convince one party to pay an amount that will be acceptable to the other party. Of course, in multiple-party disputes, this is a more complex process that involves determining how settlement obligations and proceeds are to be apportioned among the parties. This often involves consideration of available insurance coverage, and it is therefore common for sureties and insurers to be involved in mediations, even if they are not directly parties to the pending litigation or the underlying dispute.

The parties usually rely upon the mediator to control the sequence of separate and joint meetings and discussions between the parties; and there may be any combination of meetings with the mediator and one, all, or any number of the parties. The first and last meetings of the day are usually plenary sessions with everyone present: first, to set the ground rules and define the issues; and last, to either announce a settlement and "work out the details" or to decide that there is no point in continuing further. The latter may either be a decision that settlement is unlikely or impossible through mediation, in which case the parties initiate or return to the litigation process; or it may be a decision that the parties should reconvene at a future date to resume further negotiations. The parties are free at that time to decide, usually in conjunction with or at the request of the mediator, to exchange and/or submit further mediation statements on particular issues.

The success of many mediations turns on the skills, experience, and tenacity of the neutral mediator, who has great flexibility in choosing how to move the parties toward resolution. The best mediators usually have a "sixth sense" of where a case should settle, usually based upon the mediator's personal experience as an advocate and/or jurist in other cases. These same skills and experiences are used by the mediator to assess whether the parties are capable of reaching mutual agreement. In some cases, when the mediator has a strong opinion of where a case should settle but the parties themselves are unable to agree, the mediator may submit to the parties a "mediator's proposal" for settlement. This often happens near the end of the mediation, as a last resort, to break through an impasse rather than accepting failure. As with their back-and-forth negotiations, the parties must decide for themselves whether to reject or accept a mediator's proposal, as the mediator has no power to coerce a party to agree to any settlement.

If the parties, as is often the case, do agree to a settlement, it is vitally important that they commit their agreement to writing and sign it on the spot. It may be that further formalities are required that must be completed later, such as a comprehensive settlement and release document, dismissals of pending litigation, lien releases, or withdrawals of stop notices and the like. But it is absolutely vital that the essential "deal points" of the settlement be written and executed before concluding the mediation. The reasons for this are discussed in § 14.03[D] below. This process is vitally important and often involves unanticipated details and complications that require the mediator's participation. A good mediator will therefore not leave once a deal is struck but instead will stay to see the process through to completion. It is easy for an apparent settlement to fall apart if the mediator leaves the parties on their own to "work out the details."

[B] Advantages of Mediation

The primary advantages of mediation are the same as those of arbitration, only more so. The parties have great flexibility, along with their chosen mediator, to plan and schedule their process and to customize it for their particular needs. There is no discovery associated with the mediation process, though it is common for parties to have completed a certain level of discovery before they feel they are ready to proceed to mediation. There are even fewer formal legalities in a mediation than in an arbitration, as witnesses and evidence are usually not presented at all. Rather, the parties use their mediation statements to describe the support they believe is available to bolster their case and use oral presentations at the mediation to persuade one another of their positions. Therefore, the time, effort, and expense involved in mediation are even less than in arbitration—and are far less than in litigation.

Another advantage of mediation is that it is completely nonbinding. Thus, a party who may fear an adverse (and, for the most part, unappealable) outcome in arbitration need have no such concerns about voluntarily submitting a dispute to mediation. At any point that a party thinks that the mediation process is no longer in its interest, it is free to walk away without further obligation. However, the decision to abandon the mediation process should always be carefully weighed against the alternative of returning to arbitration and/or litigation.

Litigation often results because, during the arduous course of a construction project, the parties have become polarized, or even offended, by their interactions with one another. Many cases simply cannot settle until the parties have had an opportunity to express to one another how they feel about their treatment at one another's hands. Indeed, it is not uncommon for litigation to proceed all the way to trial because an offended party would otherwise have no forum for expressing real or perceived grievances against an adversary. Often, mediation can provide such a forum sooner, cheaper, and in a more private venue where mutual recriminations do not become part of the public record.

A final advantage of mediation, often overlooked but nonetheless important in many cases, is that there is no legal assessment of culpability even when a party agrees to pay substantial damages. While both mediation and arbitration usually result in "compromise" resolutions, an arbitration decision involves an assessment of fault, and a mediated settlement does not. Thus, a party that knows it is going to have to pay something but does not want to "admit wrongdoing" may well choose mediation over arbitration. This is, of course, a quality that is typical of any kind of settlement and is, in fact, why most cases settle rather than proceed to trial or other decision.

[C] Disadvantages of Mediation

Being nonbinding, the only discernible disadvantage of mediation is that it may prove to be a waste of time, effort, and resources. The parties bear the costs of their own efforts and must pay their attorneys and share the fees of the mediator. On the other hand, the mediator's fees are usually shared by two or more parties, and if the mediation is successful, it may bring to resolution in a day that which would require months of expensive litigation, including discovery and possible trial.

If there is no settlement, the parties return to their *status quo ante* positions and may well feel they have spent money and gained no results. Even this, however, may overlook the value of better understanding an adversary's case. Mediation statements and other information exchanged during the mediation process is not admissible evidence; however, a party may during an "unsuccessful" mediation learn much that is of value in assessing an adversary's strengths and weaknesses. Such information may be of great value in focusing discovery and shaping a party's case in preparation for trial. The efficiencies thus gained may be well worth the relatively low cost of the mediation itself.

[D] Enforcement of Mediation Agreements

Participants in mediation must be aware of the provisions of California Evidence Code § 1152, which provides that a settlement offer is inadmissible to prove liability for loss or damage. Therefore, any offer to compromise made during the course of mediation is inadmissible against the offering party. This remains the law today, even though former section 1152.5 (and its temporary successor, former section 1152.6) have recently been repealed and replaced by sections 1115 through 1128. These sections only became effective January 1, 1998, and have yet to generate appellate decisions. Therefore, in order to better understand these sections before specifically discussing them, a brief review of their predecessor is in order.

Former section 1152.5 provided that when persons agree to participate in a mediation, and except as otherwise provided in the statute, evidence of anything said or of any admission made in the course of the mediation is not admissible in evidence. Similarly, under section 1152.5, unless the document otherwise provides, no document prepared for the purpose of, or in the course of, a mediation was admissible in evidence.[42]

The effect of section 1152.5 on enforcement of an oral settlement reached at a mediation was interpreted by the court of appeal in *Ryan v. Garcia*,[43] in which, after commencing suit in superior court, the parties privately agreed to mediate their dispute. They then signed a confidentiality agreement acknowledging that the provisions of California Evidence Code § 1152.5 applied to the mediation. A settlement was reached, the terms of which were stated orally at the mediation session, and the attorney for the defendant was assigned to reduce the settlement to writing; however, the parties later disagreed about the terms of the settlement, and no written agreement was ever executed.

The plaintiffs amended their pending complaint by adding a cause of action to enforce the oral settlement agreement. At trial, the defendant objected to the introduction of anything said during the mediation. The trial court found that the parties had reached oral agreement, and judgment was rendered in the plaintiffs' favor. The appellate court, holding that admission of oral statements made at a mediation violates the prohibition of Evidence Code § 1152.5, reversed the trial court ruling.

---

[42] Pursuant to Evidence Code § 703.5, mediators and arbitrators are deemed not competent to testify as to any statement, conduct, decision, or ruling occurring at or in conjunction with the proceedings. Exceptions are limited to matters related to contempt, crimes, disqualification proceedings, and investigations by the state bar or the Judicial Council.

[43] 27 Cal. App. 4th 1006 (1994).

This case highlighted two important points concerning mediation. First, although oral settlements generally are enforced in the same manner as oral contracts, parties to a mediation may have been unable to prove the existence of an oral settlement agreement because of the evidentiary bar of section 1152.5. Second, even if a settlement agreement is reduced to writing at a mediation, it was unenforceable unless it included a waiver of section 1152.5.

In *Regents of University of California v. Sumner,*[44] the parties in a pending lawsuit participated in a voluntary mediation before a retired judge. After two days, the parties concluded their mediation sessions, then dictated the terms of the settlement into a tape recorder, and a transcript was prepared. It was further agreed that the terms of the dictated settlement agreement would be incorporated into a more formal release document. After the dictated settlement was concluded and before the typed release was prepared, the plaintiffs had second thoughts and refused to go through with the terms of the settlement. The University then filed suit to enforce the terms of the dictated settlement.

The trial court granted the motion to enforce the settlement agreement. On appeal, the plaintiffs asserted that the trial court had erred in considering the transcript of the dictated oral settlement, relying on Evidence Code § 1152.5 and *Ryan v. Garcia.*[45] The court of appeal held that *Ryan* was distinguishable because, in the instant case, the parties *concluded* the mediation session and then created a transcript of the settlement, and thus the transcript was not a part of the mediation session itself. Recognizing that certain language in *Ryan* was inconsistent with its ruling, the court of appeal criticized that decision, noting that "[t]he majority opinion in *Ryan* has never been cited or followed in a published opinion before today, and we respectfully decline to follow it."[46]

This then left a good deal of confusion over the issues of admissibility of statements made during the course of mediation as well as the enforceability of mediation settlement agreements. In response, the legislature temporarily replaced section 1152.5 with section 1152.6 and then enacted the more comprehensive scheme that is presently set forth in sections 1115 through 1128. The terms of the present statutory scheme are described at follows:

First, the terms "mediation," "mediator," and "mediation consultation" are defined in section 1115. Then, section 1116 provides that (1) the new statutory scheme does not impact a court's right to order parties to participate in ADR procedures; (2) it does not limit the enforceability of a contract clause in which parties agree to the use of mediation; and (3) it does not make admissible any evidence that is inadmissible under section 1152. Section 1117 makes clear the new scheme does not apply to mediations under the Family Code or settlement conferences under Rule 222 of the California Rules of Court.

Sections 1118 and 1124 provide criteria with respect to oral agreements reached during a mediation. Section 1118 provides that oral agreements are "in accordance with Section 1118" if they meet all four requirements: (a) they are

---

[44] 42 Cal. App. 4th 1209 (1996).
[45] 27 Cal. App. 4th 1006 (1994).
[46] *Regents of Univ. of Cal.*, 42 Cal. App. 4th at 1213.

taken down by a court reporter or otherwise reliably recorded; (b) the terms and the parties' agreement thereto are recited "on the record" in the presence of the mediator; (c) the parties state words to the effect that their agreement is to be binding and enforceable; and (d) the agreement is reduced to writing and signed by the parties within 72 hours of being recorded. Section 1124 provides that oral agreements made at or as a result of mediation are not made inadmissible or protected from disclosure if they meet any of three criteria: (a) they are made in accordance with section 1118; or (b) they are not in accord with the requirement of section 1118 as to being enforceable and binding, but the parties do expressly agree that they may be disclosed; or (c) subject to the same qualification, the agreement is "used to show fraud, duress or illegality that is relevant to an issue in dispute."

Section 1119 makes three provisions: (a) nothing said or admitted during the mediation is admissible or discoverable, nor can its disclosure be compelled; (b) no writing prepared during or for the mediation is admissible or discoverable, nor can its disclosure be compelled; and (c) all communications, discussions, and settlement negotiations by participants are to be confidential.

Section 1120 provides that evidence otherwise admissible or subject to discovery outside the mediation context will be made inadmissible or protected from disclosure by the mere fact of having been introduced or used during the mediation.

Section 1121 prohibits the use in court of "any statement, evaluation, recommendation or finding of any kind by the mediator"—other than a report mandated by law or court rule that states only whether an agreement was reached—unless the parties agree, either in writing or in accordance with section 1118.

Section 1122(a) provides that writings prepared during or for the mediation are not rendered inadmissible or protected from disclosure by the statutory scheme if either (1) all participants to the mediation agree the writing may be disclosed; or (2) all persons who participated in preparation of that writing agree to its disclosure and the writing does not disclose anything said, done, or admitted during the mediation. In both instances, such agreement to disclosure must either be in writing or in accordance with section 1118. Finally, section 1122(b) provides that if the "neutral person who conducts a mediation expressly agrees to disclosure, that agreement also binds any other [mediator]."

Section 1123 provides the following conditions for disclosure and admissibility of written settlement agreements prepared during or pursuant to a mediation: (a) the agreement itself provides that it is admissible or subject to disclosure; (b) the agreement itself provides that it is enforceable or binding; (c) all parties expressly agree (in writing or in accordance with section 1118) to its disclosure; and (d) "the agreement is used to show fraud, duress, or illegality that is relevant to an issue in dispute."

For confidentiality purposes, section 1125(a) defines when "a mediation ends" as being "when any one of the following conditions is satisfied": (1) the parties execute a written settlement agreement that fully resolves the dispute; (2) an oral agreement that fully resolves the dispute is reached in accordance with section 1118; (3) the mediator states in writing to the participants that the mediation is over; (4) a party provides a statement to the other parties that the mediation is terminated, in which event if there are two or more parties remaining, they may continue with the mediation; (5) there is no communication between the mediator and any of the parties for 10 calendar days (or any longer or shorter period to which the parties may agree). Next, section 1125(b) defines the end of mediation in circumstances where some, but not all, of the disputes are resolved in the mediation. This occurs when either (1) the parties execute a written settlement agreement that partially settles the dispute; or (2) such a partial agreement is reached orally in accordance with section 1118. Finally, section 1125(c) provides that the section neither precludes a party from ending a mediation without a settlement nor affects the extent to which a party may terminate a mediation.

Section 1126 provides that anything that is inadmissible, protected from disclosure, or confidential under the statutory scheme will remain so after the mediation ends. Section 1127 provides monetary sanctions against anyone who subpoenas or otherwise seeks to compel a mediator to testify or produce a writing that is made inadmissible or protected from disclosure.

Finally, section 1128 provides for vacating or modifying a decision, in whole or in part, or granting a new or further hearing on some or all issues, in a subsequent noncriminal proceeding in which reference is made to a mediation. It also provides that any such improper reference is an "irregularity in the proceedings of the trial" as that term is used in Civil Procedure Code § 657, which states the criteria for modifying or vacating decisions, in whole or in part, and for granting a new or further trial on all or part of the issues.

§ 14.04 OTHER AVAILABLE TYPES OF ADR

So long as they mutually agree, parties to a construction contract are free to enter into any form of ADR they may choose. In addition to arbitration and mediation, as discussed above, there is a variety of other available forms of ADR, the most common examples of which are discussed in the remaining subparts of this section. They include whatever form of contractual claims and disputes procedures may be set forth in the construction contract documents, discussed in § 14.04[A] below, including a growing trend in public contracts toward use of disputes review boards (DRBs) and "partnering" arrangements. The discussion in § 14.04[B] also covers ADR procedures (arbitration and mediation) that are required by California law or that may be imposed by court order. The use of private judges as an alternative to the regular court system is considered in the discussion of "private trials" and "mini-trials" in § 14.04[C]. Finally, § 14.04[D] sets forth legally required language that is necessary to include in arbitration provisions contained in residential construction contracts.

By far the most common and least structured form of ADR is direct settlement negotiations between the parties and their attorneys. At any time, either before or during the course of litigation or arbitration, the parties are free to come to an agreement resolving their disputes and enter into a binding settlement agreement. Such settlement agreements should be prepared by the attorneys, and if there is a court action on file, it would be dismissed. Settling claimants would also release any stop notices and/or mechanics' liens related to the settled claims. If the parties settle part but not all of their disputes, the litigation or arbitration would continue as to the remaining unresolved issues, and the claimant would only partially release any stop notices and/or mechanics' liens.

One item of particular note in California, which anyone entering into a settlement agreement should be aware of, is Civil Code § 1542, which provides:

> A general release does not extend to claims which the creditor does not know or suspect to exist in his favor at the time of executing the release, which if known by him must have materially affected his settlement with the debtor.

Therefore, any settlement agreement, whether reached informally or as a result of ADR, should include a contract provision acknowledging section 1542, quoting its provisions verbatim, and expressly releasing the parties' rights thereunder. To be completely safe from having all or part of a settlement agreement voided upon the discovery of "new or different" facts that later come to light, such a provision should also state that the parties have each consulted with legal counsel of their own choosing and have been advised as to the legal implications of waiving section 1542.

[A] Contractually Mandated ADR

Many construction contracts, especially those of public owners, now contain specific provisions for one form or another of ADR process. These may be as standard as an arbitration provision, such as the one contained in several of the AIA's standard form contracts (*see* Chapter 3 above); or they may be very detailed provisions especially customized to the needs of a particular owner. As construction contracts are usually prepared by project owners to suit their own interests and are rarely open to negotiation by the contractors, any contractor should carefully review all such provisions prior to executing the contract.

Many construction contracts contain, in addition to arbitration or mediation clauses, provisions requiring the contractor to first complete a specific claims and disputes process before resorting to such remedies, or to legal action of any kind to enforce a contract right (e.g., a time extension for owner-caused delays) or to recover on a claim for added compensation. These are not, strictly speaking, ADR processes, but they serve the same purpose of structuring negotiations between the owner and contractor to resolve their disputes.

Construction contracts often confer upon the project architect or engineer authority to resolve disputes and make decisions that involve interpretations of the plans and specifications. Not surprisingly, most of the form contracts prepared by the American Architects Association (another AAA) that are in widespread use throughout the industry contain such provisions. To the extent these processes may resolve disputes between the parties, they may in the broadest sense be considered as a form of ADR.

There is a growing trend, especially in public contracts, toward use of "partnering" programs whereby the parties to the construction contract meet to improve their communications, foster a sense of teamwork and cooperation, and discuss better ways of working together for the benefit of the project. Many contracts now have specific "partnering" provisions; however, even on projects that do not, the parties may mutually agree to try this method when they sense that the project is heading toward a major claim confrontation. Though not technically an ADR process, partnering may open channels of communication that can facilitate settlement negotiations.

Another trend in which public owners seem to be leading the way is the growing use of disputes review boards (DRBs). These are usually established, pursuant to specific contract provisions, to hear and resolve disputes that arise when one of the parties (usually the contractor) is not satisfied with the results of change order negotiations and/or the architect/engineer's decisions regarding interpretation of the plans and specifications. Most contract provisions make submission to the DRB a condition precedent to the contractor's right to pursue its claims into litigation. It is typical for DRBs to be constituted in a manner quite similar to the selection process for multiple arbitrators. That is, each side to the dispute chooses one or two members who, in turn, jointly select a third or fifth member. Then, depending upon the specific provisions of their contract, the parties present their disputes to the DRB for consideration and a decision. One major problem with DRBs is that it is an effort to constitute them and an even greater effort to keep them functioning well enough to keep up with the disputes on problem projects. Another is that they can become a drain on the parties' resources, especially if they involve preparation and submission of written claims and rebuttals. Usually, the only people with sufficient knowledge to prepare such submittals are the very project personnel who are already being overtaxed by having to deal with constructing the project. As a result, it is far too common that a DRB is unable to meet regularly, and it is not at all rare for the DRB to fall into disuse altogether. Perhaps the biggest problem with DRBs is that, having been established by the owner's contract, they tend to be biased in favor of the owner's interests. Whether or not this is actually true, that is how it is likely to be perceived by a contractor, especially one whose claims are denied by the DRB.

It is quite common for a construction contract to provide that such processes (i.e., DRB, review by architect/engineer, etc.) are the contractor's sole remedy for resolving disputes and/or that the decisions reached through such processes are "final and binding upon the parties." While some owners will assert that this means the contractor has no further right to contest adverse results by legal action, there is no case law in California to that effect. The authors are of the opinion that the legal effect of such provisions is to make the designated procedure the court of last resort within the context of the contract itself; however, in light of California Civil Code § 1670,[47] it is highly unlikely that the courts will

---

[47] Civil Code § 1670 specifically provides that "[a]ny dispute arising from a construction contract with a public agency, which contract contains a provision that one party to the contract or one party's agent or employee shall decide any disputes arising under that contract, shall be resolved by submitting the dispute to independent arbitration, if mutually agreeable, otherwise by litigation in a court of competent jurisdiction."

interpret them as foreclosing a party's right to seek redress through the legal system. Furthermore, a contractor who is disappointed with the functioning and/or decisions of a DRB and thus determined to sue may seek to invalidate the process on any one of several grounds that may be alleged in its complaint (e.g., bias, owner tampering, abandonment of contract, etc.).

In any event, if a construction contract provides for an ADR process, the parties who sign that contract are bound to follow it, just as they are bound to any other claims and disputes provisions of their agreement. Contracts usually provide that participation is a condition precedent to legal action, and the claimant who refuses to participate may find that such failure results in a waiver or forfeiture of further legal rights.

[B] Court-Ordered and Statutorily Required ADR

[1] Court-Ordered ADR

As stated above, the most common kind of ADR is the most simple—settlement negotiations between the parties. Accordingly, it is common practice for judges to encourage, and even conduct, voluntary settlement conferences (VSCs). A VSC can take place at any time, either at the instigation of the court or at the instigation of the parties themselves. Furthermore, it is standard procedure for a judge to set a "mandatory settlement conference" (MSC) for a date shortly before the date set for trial. Often the trial judge personally conducts the MSC, although some courts designate one or two judges to conduct MSCs on behalf of all of the judges in that particular courthouse. Usually such judges are chosen for having demonstrated an ability to get parties to settle their disputes without proceeding to a trial on the merits. In addition, there are a number of instances in which statutes require submission of certain matters to judicially sponsored arbitration and/or mediation.

[2] Statutorily Required ADR

Construction claims of less than $50,000, like any other such "small civil claims" that are sued upon in the courts, are subject to judicial arbitration, under Civil Procedure Code §§ 1141.10 *et seq.* and/or judicial mediation, under Civil Procedure Code §§ 1775 *et seq.*

[a] *Judicial Arbitration under Civil Procedure Code §§ 1141.10 et seq.*

If the superior court has 10 or more judges, or 18 or more judges and no municipal court, all such cases are required to be submitted to such arbitration; municipal courts and smaller courts are only subject to this requirement if adopted by local court rules. The court's decision to submit such matters to judicial arbitration are not appealable. Under section 1141.12(a), the parties to a dispute in excess of $50,000 may mutually stipulate to submit the matter to judicial arbitration. Under section 1141.2(b), municipal and smaller superior courts are empowered to adopt local rules to refer cases to judicial arbitration, regardless of the amount in controversy, if (i) the parties mutually stipulate; or (ii) the plaintiff elects to arbitrate and agrees that the arbitration award shall not exceed $50,000. Such election must be made after the case is at issue (i.e., all parties have been served and the complaint and all cross-complaints have been answered) and at least 90 days before trial. The rest of the statute is too lengthy and detailed to be completely summarized here, but the following points are of particular interest:

1. Actions seeking equitable relief are excluded (section 1141.13).

2. The Judicial Council is to establish practice and procedure rules (section 1141.14), which rules may provide for exceptions for cases not amenable to arbitration (section 1141.15).

3. Arbitrators are to be retired judges or commissioners, or attorneys admitted to practice; they shall sit individually; and their fees are limited to $150 per day (section 1141.18).

4. Discovery must be approved in advance by the court (section 1141.19.5).

5. Following arbitration, either party may ask for a trial *de novo,* in which case the court is to attempt to give the case the same priority it would have had for a trial date (section 1141.20).

6. In the event the party seeking trial *de novo* receives a judgment that is not more favorable than the arbitration award, that party may not recover its own costs, and the court shall order that party to reimburse the county and the other parties for their costs and fees (section 1141.21).

7. The arbitration award is to be in writing, signed by the arbitrator, and submitted to the court, where it has the same force and effect as a judgment and cannot be attacked or set aside except on the grounds stated in Civil Procedure Code §§ 473 and 1286.2 or Judicial Council rule.

8. An arbitrator may award an amount in excess of $50,000 (section 1141.26).

9. The statute does apply to actions in which a public entity is a party (section 1141.27).

10. With limited exceptions, administrative costs of the arbitration are paid by the county (section 1141.28).

[b] Judicial Mediation under Civil Procedure Code §§ 1775 et seq.

As a supplement to judicial arbitration, as discussed above, Civil Procedure Code §§ 1775 *et seq.* make very similar (and sometimes identical) provisions for judicial mediation. Section 1775 begins this statute with a recitation of the utility of mediation to the court and litigants alike and a strong policy statement in favor of "the use of court-annexed alternative dispute resolution methods in general, and mediation in particular." After defining the terms "Court" and "Mediation" in section 1775.1, the statute is made applicable to the courts of the county of Los Angeles and allows other counties to elect to apply it to their courts (section 1775.2). Section 1775.3 provides that all actions subject to arbitration under section 1141.10 (including actions to which a public entity is a party) may, as an alternative, be ordered to mediation, while section 1775.4 makes clear that an action can only be ordered into one program or the other, not both. Section 1175.5 parallels section 1141.16 and likewise limits the program to actions where the amount in controversy is less than $50,000.

Section 1775.6 provides that a mediator must be selected within 30 days of the matters being submitted to mediation. It allows the parties to decide for themselves the process for making this choice; but if they are unable to agree within 15 days of the submission, the court is empowered to select the mediator itself. Section 1775.7 provides for tolling certain time limits in the unusual event the mediation assignment period exceeds four and one-half years. Section 1775.8(a) provides that mediators are to be compensated the same as arbitrators are compensated under section 1141.18, and subsection (b) provides that administrative costs are to be paid in the same manner as for arbitration costs under section 1141.28.

Section 1775.9(a) provides that if the parties are unable to agree and any party wishes to terminate the mediation, the mediator must file with the court a "statement of nonagreement," and subsection (b) requires the court, upon receiving that statement to reset the matter for trial and attempt to put the case in the same position on the civil active list as it had before being sent to mediation. The confidential nature of the judicial mediation is preserved by making all statements made by the parties subject to the restrictions set forth in Evidence Code §§ 703.5 (establishing testimonial incompetence of arbitrators and mediators), 1152 (making offers to compromise inadmissible) and 1115 *et seq.* (the 1998 statute regarding mediation confidentiality, etc.).[48]

Section 1775.11 provides that normal discovery rights are not affected by a matter's being ordered into judicial mediation, while section 1775.12 makes it an irregularity (i.e., grounds for possible mistrial) if the judicial mediation or statement of nonagreement is referred to in a subsequent trial. A policy that judicial mediation shall not preempt other dispute resolution programs is stated in section 1775.14. Finally, section 1775.14 requires the Judicial Council to report on the success of the program by January 1, 1998, while § 1775.15 empowers the Judicial Council to provide rules and procedures for the statutory judicial mediation program.

[c] Arbitration Provisions in the Public Contract Code

There are two other statutory arbitration provisions that have particular significance to the construction industry, both having to do with public works contracts. California Public Contract Code §§ 22201 and 22202 provide that, unless

---

[48] All these provisions are discussed in earlier sections of this chapter.

otherwise prohibited by law, any public works project awarded by competitive bids or otherwise may include a provision for arbitration of any claim pursuant to Public Contract Code §§ 10240 *et seq.*, and arbitrations conducted thereunder are governed by Public Contract Code §§ 20104 *et seq.*

Public Contract Code §§ 10240 through 10240.13, collectively entitled "Resolution of Contract Claims," establish arbitration as the remedy for resolution of claims subject thereto[49] and require a claimant to initiate arbitration within 90 days of the final written decision by the department on the claim.[50] Section 10240.2 affirmatively bars a claimant, for 240 days following acceptance of the work, from proceeding to arbitration unless it diligently pursues and exhausts the required administrative procedures set forth in the contract under which the claim arose. Under section 10240.3, arbitration is to be held before a single arbitrator selected by the parties (or, if the parties cannot agree, by the court) from a list of arbitrators created by the Public Works Contract Arbitration Committee.[51] Section 10240.5 empowers the Departments of General Services, Transportation, and Water Resources to jointly adopt uniform regulations to implement this arbitration program. Under section 10240.11, waiver of the statute may only be by mutual agreement in writing to have the claim litigated in a court of competent jurisdiction. Section 10240.11 adopts the arbitration laws set forth in Civil Procedure Code §§ 1028 *et seq.*, as modified by the statute or the regulations adopted pursuant to section 10240.5.

Most importantly, the statute expressly changes the rule, generally applicable to other arbitration procedures, that arbitrators' decisions are not to be overturned by the courts based upon mistakes of law or fact. Thus, section 10240.8 provides that unless the parties agree otherwise, the arbitration decision "shall be decided under and in accordance with the laws of this state, supported by substantial evidence and, in writing, contain the basis for the decision, findings of fact, and conclusions of law." Section 10240.12 provides for judicial confirmation of awards and for correction or vacation of any "award, or part thereof, [that] is not supported by substantial evidence or . . . is not decided under or in accordance with the laws of this state." Therefore, one of the most significant disadvantages of arbitration, as discussed in § 14.02[B] above, does not apply to arbitration of claims on public works projects under this statute.[52]

Finally, section 10240.13 provides that the parties are to share equally the costs of conducting the arbitration, with each party separately responsible for its own filing fee, witness fees, costs of discovery, and other individually incurred costs. The arbitrator is given the same authority as a court in awarding interest. Reasonable attorney's fees may be awarded in two circumstances: (a) a party who

---

[49] Cal. Pub. Cont. Code § 10240.

[50] *Id.* § 10240.1. This section expressly excludes "any claim founded on any cost audit, latent defect, warranty, or guarantee under the contract."

[51] Public Contract Code §§ 10245 through 10245.4, in turn, provide for establishment and composition of this committee; their terms of office and appointment by the governor; methods of establishing procedures and regulations; standards, qualifications, certification, and removal of arbitrators; and administration of the program by the Office of Administrative Hearings.

[52] *See also* Cal. Civ. Code § 1296, discussed in § 14.02[B] above.

has made a settlement offer on more favorable terms than the arbitrator's award may be awarded reasonable attorney's fees incurred from and after that offer; and (b) when substantial evidence establishes that a party acted frivolously or in bad faith in its demand for or participation in the arbitration, reasonable attorney's fees may be awarded against that party.

Another arbitration statute that is of interest to construction contractors and public owners is Public Contract Code §§ 20104 *et seq.*, which applies to all public works claims of $375,000 or less that arise between a contractor and a local agency, other than a public agency that has elected to include in its contract a provision for resolving disputes under §§ 10240 *et seq.*, as discussed above. Claims are defined in section 20104(b)(2) to include demands for time extensions, monetary damages, and disputed payment amounts. The statute applies only to contracts entered into after January 1, 1991, that set forth the provisions of the statute, or a summary thereof, in the plans or specifications. Tort claims are expressly excluded from the statute in section 20104.2(f).

Section 20104.2 sets forth, in detail, procedures and time limits for the contractor and the public agency to submit and respond to claims and to engage in a "meet and confer" process to attempt to informally resolve their disputes. The time limit for the contractor to serve the public agency with a written claim under Government Code §§ 900 *et seq.* does not begin to run until this process is complete and all or part of the contractor's claim remains in dispute.[53]

After a Government Code claim is rejected and the contractor files a legal action, claims subject to the statute must proceed according to the provisions of section 20104.4. Within 30 to 60 days after the public agency responds to the contractor's complaint, the court must order the parties to a nonbinding mediation in accordance with the procedures and time limits set forth in section 20104.4(a). If the mediation does not resolve the dispute, section 20104.4(b)(1) requires the case to be submitted to judicial arbitration pursuant to Civil Procedure Code §§ 1141.10 *et seq.* (notwithstanding the provision of section 1141.11 restricting such judicial arbitrations to disputes regarding amounts of less than $50,000).

[C] "Private Trials" and "Mini-Trials"

In addition to mandatory and voluntary settlement conferences, California Civil Code § 638 sets up a "reference" process whereby, with the express consent of the parties, the court may refer all or some part of a matter to a private judge for determination. Section 638 contains two distinct subsections: Under section 638(1), the trial court makes a general reference to try all legal and factual issues in a proceeding and report a statement of decision to the trial court. The "referee" is vested with power to make definitive disposition of issues in the action, subject to the power of the court to affirm the referee's report.[54] Under

---

[53] Government Code claims are discussed above in Chapter 6.
[54] Murphy v. Padilla, 42 Cal. App. 4th 707, 713 (1996).

section 638(2), there is a special (i.e., specific) reference to ascertain some fact(s), but the referee's decision is not binding on the court.[55]

In addition, California Rules of Court provide specific requirements for general or special references,[56] which include the parties' presenting to the court a written agreement requesting an order of reference. The proposed order must include the name and address of the referee, the referee's signature, and the "scope of the reference" (i.e., general or specific reference).

Thus, the section 638 reference may involve a "private trial" in which a full trial on the merits is held before a retired judge who is paid by the parties themselves. Although such private trials are often conducted in a more informal setting, the same rules of evidence and procedure apply as if the trial were taking place in a regular courtroom. The advantages of this process are several. To begin with, such a trial can begin as soon as the parties are ready and an appropriate referee (i.e., "rent-a-judge") is available. In contrast, the parties might have to wait weeks or months for a courtroom to become available for a regular court trial. As with arbitration, the referee and the parties are free to set up a schedule of hearing dates that accommodates their convenience, so long as the process is completed within the time, if any, specified by the court in its referral order. Such trials may take place anywhere the parties and the referee agree, usually the offices of an ADR organization or public facilities such as a hotel conference room, which may be rented for that purpose.

This arrangement not only allows the parties to avoid having to meet at the courthouse, which many find to be an intimidating environment; it also makes it possible for greater efficiency than in a court. Most courts, even while conducting trials, must continue to deal with their caseloads, including hearing motions and other such time-consuming procedures. As a result, a typical court may convene an ongoing trial at 10:00 A.M. or later and conclude by 4:00 P.M. or sooner. When a lunch break is factored in, there may only be four to six hours per day when the trial is actually in session. At that pace, a court trial usually takes much longer than a private trial in which the parties and the judge are free to begin and end at their own convenience. Thus, it is not unusual for a private trial to commence at 8:00 A.M. and continue until 6:00 P.M., thus allowing eight or more hours per day of actual trial time. The savings in time and attorney's fees may very well make up for the fact that the litigants are paying for the referee's time and rental of the facility.

The term "mini-trial" is often used to describe a private trial of some, but not all, of the contested issues in a given dispute. Mini-trials are useful for some of the same, or similar, reasons that courts sometimes "bifurcate" their trials. Usually a mini-trial is conducted when a settlement is impossible due to an impasse as to one or more discrete legal or factual issues as to which all or some of the parties cannot agree without having their day in court. Such issues may include an affirmative defense, such as the running of a statute of limitations or

---

[55] Id.

[56] See California Rules of Court, Rule 244.1(a) for Superior Court, and Rule 532.1(a) for Municipal Court.

failure to comply with some other legal or contractual time limit. Often these are issues that might otherwise be presented to the court for summary adjudication under Civil Procedure Code § 437c, but the parties prefer to use live testimony and cross-examination rather than sworn declarations. Mini-trials may be entirely private in instances in which the parties do not want a formal legal determination and prefer to "test the waters" in a nonbinding circumstance. However, even if conducted under a special reference pursuant to Civil Code § 638(2), the results are not binding upon the court. Thus, only if the trial judge so chooses will the results of a special reference become part of the court's proceedings and have the same force and effect.

After conducting a mini-trial limited to those problematic issues, the parties may significantly change their attitudes toward settling the entire dispute. If so, they may then reenter ADR (usually mediation or settlement negotiations) and proceed to a settlement. If not, they may return to litigation and complete the rest of their trial. If the mini-trial is conducted pursuant to a section 638 reference, the parties would proceed in the same manner as in a bifurcated trial. If not, then a complete trial would be necessary.

[D] Arbitration Clauses in Residential Construction Contracts

One particular statutory requirement that merits special attention by contractors engaged in private residential construction is set forth in California Business and Professions Code § 7191, which requires that an arbitration clause in a contract for work on residential property with four or fewer units must be clearly titled "Arbitration of Disputes." Furthermore, the arbitration provision must be set out in at least 10-point Roman boldface type, or in contrasting red print in at least 8-point Roman boldface type. Also, if the provision is included in a typewritten contract, it shall be set out in capital letters.

The Code likewise provides that the contract shall contain a line or space for the parties to indicate their assent or non-assent to the arbitration provision. Furthermore, above the line for indicating assent or non-assent and immediately following the arbitration provision, the following language shall appear:

> NOTICE: BY INITIALING IN THE SPACE BELOW YOU ARE AGREEING TO HAVE ANY DISPUTE ARISING OUT OF THE MATTERS INCLUDED IN THE "ARBITRATION OF DISPUTES" PROVISION DECIDED BY NEUTRAL ARBITRATION AS PROVIDED BY CALIFORNIA LAW AND YOU ARE GIVING UP ANY RIGHTS YOU MIGHT POSSESS TO HAVE THE DISPUTE LITIGATED IN A COURT OR JURY TRIAL. BY INITIALING IN THE SPACE BELOW YOU ARE GIVING UP YOUR JUDICIAL RIGHTS TO DISCOVERY AND APPEAL, UNLESS THOSE RIGHTS ARE SPECIFICALLY INCLUDED IN THE "ARBITRATION OF DISPUTES" PROVISION. IF YOU REFUSE TO SUBMIT TO ARBITRATION AFTER AGREEING TO THIS PROVISION, YOU MAY BE COMPELLED TO ARBITRATE UNDER THE AUTHORITY OF THE BUSINESS AND PROFESSIONS CODE OR OTHER APPLICABLE LAWS. YOUR AGREEMENT TO THIS ARBITRATION PROVISION IS VOLUNTARY.

WE HAVE READ AND UNDERSTAND THE FOREGOING AND AGREE TO SUBMIT DISPUTES ARISING OUT OF THE MATTERS INCLUDED IN THE "ARBITRATION OF DISPUTES" PROVISION TO NEUTRAL ARBITRATION.

The provision quoted above must also be in 10-point Roman boldface type, or in contrasting red print in 8-point Roman boldface type, and, if in a typed contract, in capital letters. Finally, the statute provides that a provision for arbitration that does not comply with the foregoing requirements may not be enforced against the owner; however, the owner may still enforce the nonconforming arbitration provision against the licensed contractor.

§ 14.05 ENFORCEMENT OF ADR SETTLEMENT AGREEMENTS

Although it is not always necessary, parties who are successful in resolving their disputes through ADR may, for a variety of reasons, wish to have their settlement reduced to a judgment. This happens most often when all or part of the settlement involves continuing performance, such as extended payment terms or completion of such items as punch lists, repairs, or warranty work. The authority and procedure for translating a settlement into a legally enforceable court judgment is set forth in Civil Procedure Code § 664.6, entitled "Judgment by Stipulation." That statute provides that parties in a pending litigation may stipulate, in writing, to a full or partial settlement of their disputes and submit a motion to the court to enter judgment pursuant to the terms of the settlement. A judgment obtained under section 664.6 has the same full force and effect as that of a judgment following a normal trial and decision. The statute also provides that, if requested by the parties, the court may retain jurisdiction over the parties to enforce the judgment until the settlement terms are performed in full.

Section 664.6 also allows an unwritten settlement stipulation to be presented "orally before the court." However, the decision in *Murphy v. Padilla*[57] makes it clear that "the court" means a sitting superior or municipal court judge; a stipulation does not meet this requirement by being presented to a mediator who is not authorized by statute to act in a quasi-judicial manner. This case also illustrates the advantage of using the procedure for a formal reference under Civil Code § 638, as noted above in § 14.04[C]. The *Murphy* court, which refused to enforce the parties' oral agreement stated before a retired judge who served as a private mediator, stated that the agreement would have been enforceable under section 664.6 if the parties had fulfilled all of the requirements and obtained a formal court reference under Civil Code § 638. The court specifically noted that the reference process, if properly done, would have conferred "quasi-judicial" status upon the private judge, making statements before him the same as if made in court. The parties in *Murphy* did not complete the reference process due to failure to fulfill the requirement for written consent to the reference.

---

[57] 42 Cal. App. 4th 707 (1996).

If the parties do not avail themselves of these procedures, settlement agreements, including but not limited to those reached through ADR, may also be enforced through a separate motion for summary judgment, a suit for equity, or the amendment of existing pleadings.[58] The broad discretion vested in trial courts to enforce settlement agreements is illustrated by the case of *Malouf Bros. v. Dixon*,[59] in which the court granted (and the court of appeal affirmed) a simple motion, supported only by affidavits, to determine that a grading and excavation contractor had fulfilled its obligations in spite of the owner's contention he had not. Such enforcement is subject to normal principles applicable to contracts in general. Therefore, unless the subject matter is subject to the statute of frauds (e.g., transfers of real property, contracts that cannot be performed within a year, etc.), an oral settlement agreement is enforceable in the same manner as any other oral agreement.[60]

Another issue arising under Civil Procedure Code § 664.6 is whether the term "parties" includes an attorney representing a party. The first decision to interpret the meaning of the term "parties" was *Haldeman v. Boise Cascade*,[61] in which the court of appeal determined that the term meant both the litigants and their attorneys. In that case, the court of appeal affirmed a judgment that was entered based upon the stipulation of the plaintiff's attorney, in spite of the plaintiff's protest that she was not at the judicially supervised settlement conference and refused to personally sign a release that was an integral part of the settlement.

Thereafter, in a series of decisions, various districts of the court of appeal decided the issue both ways, creating a split in the legal authority. The issue was finally determined by the California Supreme Court, in *Levy v. Superior Court*,[62] which holds that due to the specific wording of section 664.6, an attorney representing a party may *not* sign the stipulation or present it "orally before the court," and in either situation, such a stipulation had to be signed and/or orally presented to the court *by the party*. In response, the legislature passed a bill in 1998, adding section 664.7, which expressly modifies the *Levy* decision with respect to such stipulations for entry of judgment in residential construction defect cases. Under section 664.7, in any action to recover monetary damages on a claim arising from residential construction defects, where a party's contribution is paid on its behalf pursuant to an insurance policy, the parties may stipulate to entry of judgment through their respective counsel. In all other cases, the rule remains as stated in *Levy*, that each party must personally sign the stipulation or present it orally to the court.

---

[58] *See, e.g.,* Robertson v. Chen, 44 Cal. App. 4th 1290 (1996); Murphy v. Padilla, 42 Cal. App. 4th 707 (1996).
[59] 230 Cal. App. 3d 280 (1991).
[60] *See, e.g.,* Nicholson v. Barab, 223 Cal. App. 3d 1671 (1991).
[61] 176 Cal. App. 3d 230 (1985).
[62] 10 Cal. 4th 578 (1995).

APPENDIXES

1. Conditional Waiver and Release Upon Progress Payment
2. Unconditional Waiver and Release Upon Progress Payment
3. Conditional Waiver and Release Upon Final Payment
4. Unconditional Waiver and Release Upon Final Payment
5. California Preliminary Notice, Together With Proof of Service
6. Mechanics' Lien
7. Stop Notice
8. Miller Act Form
9. Notice to Principal and Surety on Payment Bond on Private Work
10. Notice to Principal and Surety on Payment Bond on Public Work

Appendix 1
CONDITIONAL WAIVER AND RELEASE UPON PROGRESS PAYMENT

APPENDIX 1

Conditional Waiver and Release
Upon Progress Payment

CALIFORNIA CIVIL CODE SECTION 3262 (d)(1)

Upon receipt by the undersigned of a check from _____
(Maker of Check)

in the sum of $ _____
(Amount of Check)

payable to _____
(Payee or Payees of Check)

and when the check has been properly endorsed and has been paid by the bank upon which it is drawn, this document shall become effective to release any mechanics' lien, stop notice, or bond right the undersigned has on the job of

_____
(Owner)

located at _____
(Job Description)

to the following extent. This release covers a progress payment for labor, services, equipment, or material furnished

to _____
(Your Customer)

through _____
(Date)

only and does not cover any retentions retained before or after the release date; extras furnished before the release date for which payment has not been received; extras or items furnished after the release date. Rights based upon work performed or items furnished under a written change order which has been fully executed by the parties prior to the release date are covered by this release unless specifically reserved by the claimant in this release. This release of any mechanics' lien, stop notice, or bond right shall not otherwise affect the contract rights, including rights between parties to the contract based upon a rescission, abandonment, or breach of the contract, or the right of the undersigned to recover compensation for furnished labor, services, equipment, or material covered by this release if that furnished labor, services, equipment, or material was not compensated by the progress payment. Before any recipient of this document relies on it, said party should verify evidence of payment to the undersigned.

Dated: _____   _____
                                         (Company Name)

                                   By _____
                                         (Title)

NOTE: CIVIL CODE 3262 (d)(1) PROVIDES: *Where the claimant is required to execute a waiver and release in exchange for, or in order to induce the payment of, a progress payment and the claimant is not, in fact, paid in exchange for the waiver and release or a single payee check or joint payee check is given in exchange for the waiver and release, the waiver and release shall follow substantially the form set forth above.*

Form 6-A (Rev. 1/84) BICA (213) 251-1100 A Construction Credit Reporting Agency

Reprinted with permission from BICA. This form may be purchased from Building Industry Credit Association (BICA), 2351 West Third Street, Los Angeles, California, 90057; telephone (800) 722-2422.

Appendix 2
UNCONDITIONAL WAIVER AND RELEASE UPON PROGRESS PAYMENT

APPENDIX 2

Unconditional Waiver and Release
Upon Progress Payment

CALIFORNIA CIVIL CODE SECTION 3262 (d)(2)

The undersigned has been paid and has received a progress payment in the sum of $ _____

for labor, services, equipment, or material furnished to _____
(Your Customer)

on the job of _____
(Owner)

located at _____
(Job Description)

and does hereby release any mechanics' lien, stop notice, or bond right that the undersigned has on the above referenced job to the following extent. This release covers a progress payment for labor, services, equipment, or materials furnished to _____
(Your Customer)

through _____ only and does not cover any retentions retained
(Date)

before or after the release date; extras furnished before the release date for which payment has not been received; extras or items furnished after the release date. Rights based upon work performed or items furnished under a written change order which has been fully executed by the parties prior to the release date are covered by this release unless specifically reserved by the claimant in this release. This release of any mechanics' lien, stop notice, or bond right shall not otherwise affect the contract rights, including rights between parties to the contract based upon a rescission, abandonment or breach of the contract, or the right of the undersigned to recover compensation for furnished labor, services, equipment or material covered by this release if that furnished labor, services, equipment or material was not compensated by the progress payment.

Dated: _____   _____
(Company Name)

By _____
(Title)

"NOTICE TO PERSONS SIGNING THIS WAIVER: THIS DOCUMENT WAIVES RIGHTS UNCONDITIONALLY AND STATES THAT YOU HAVE BEEN PAID FOR GIVING UP THOSE RIGHTS. THIS DOCUMENT IS ENFORCEABLE AGAINST YOU IF YOU SIGN IT, EVEN IF YOU HAVE NOT BEEN PAID. IF YOU HAVE NOT BEEN PAID, USE A CONDITIONAL RELEASE FORM."

NOTE: CIVIL CODE 3262 (d)(2) PROVIDES: *Where the claimant is required to execute a waiver and release in exchange for, or in order to induce payment of, a progress payment and the claimant asserts in the waiver it has, in fact, been paid the progress payment, the waiver and release shall follow substantially the form set forth above.*

Form 6-C (Rev. 1/94) BICA (213) 251-1100 A Construction Credit Reporting Agency.

Reprinted with permission from BICA. This form may be purchased from Building Industry Credit Association (BICA), 2351 West Third Street, Los Angeles, California, 90057; telephone (800) 722-2422.

Appendix 3
CONDITIONAL WAIVER AND RELEASE UPON FINAL PAYMENT

APPENDIX 3

Conditional Waiver and Release
Upon Final Payment

CALIFORNIA CIVIL CODE SECTION 3262 (d)(3)

Upon receipt by the undersigned of a check from _____
(Maker of Check)
in the sum of $ _____ payable to
(Amount of Check)

_____ and when the check has been
(Payee or Payees of Check)
properly endorsed and has been paid by the bank upon which it is drawn, this document shall become effective to release any mechanics' lien, stop notice, or bond right the undersigned has on the job of
_____ located at
(Owner)
_____
(Job Description)

This release covers the final payment to the undersigned for all labor, services, equipment or material furnished on the job, except for disputed claims for additional work in the amount of $ _____

Before any recipient of this document relies on it, the party should verify evidence of payment to the undersigned.

Dated: _____   _____
                                        (Company Name)
                              By _____
                                        (Signature)
                                  _____
                                        (Title)

NOTE: CIVIL CODE 3262 (d)(3) PROVIDES: *Where the claimant is required to execute a waiver and release in exchange for, or in order to induce payment of, a final payment and the claimant is not, in fact, paid in exchange for the waiver and release or a single payee check or joint check is given in exchange for the waiver and release, the waiver and release shall follow substantially the form set forth above.*

Reprinted with permission from BICA. This form may be purchased from Building Industry Credit Association (BICA), 2351 West Third Street, Los Angeles, California, 90057; telephone (800) 722-2422.

Appendix 4
UNCONDITIONAL WAIVER AND RELEASE UPON FINAL PAYMENT

APPENDIX 4

Unconditional Waiver and Release Upon Final Payment

CALIFORNIA CIVIL CODE SECTION 3262 (d)(4)

The undersigned has been paid in full for all labor, services, equipment or material furnished to _____
(Your Customer)

on the job of _____
(Owner)

located at _____
(Job Description)

_____

and does hereby waive and release any right to a mechanics' lien, stop notice, or any right against a labor and material bond on the job, except for disputed claims for extra work in the amount of $ _____.

Dated: _____ _____
(Company Name)

By _____
(Signature)

_____
(Title)

NOTICE TO PERSONS SIGNING THIS WAIVER: THIS DOCUMENT WAIVES RIGHTS UNCONDITIONALLY AND STATES THAT YOU HAVE BEEN PAID FOR GIVING UP THOSE RIGHTS. THIS DOCUMENT IS ENFORCEABLE AGAINST YOU IF YOU SIGN IT, EVEN IF YOU HAVE NOT BEEN PAID. IF YOU HAVE NOT BEEN PAID, USE A CONDITIONAL RELEASE FORM.

NOTE: CIVIL CODE 3262 (d)(4) PROVIDES: *Where the claimant is required to execute a waiver and release in exchange for, or in order to induce payment of, a final payment and the claimant asserts in the waiver it has, in fact, been paid the final payment, the waiver and release shall follow substantially the form set forth above.*

BICA FORM 6-D (REV. 8/91)

Reprinted with permission from BICA. This form may be purchased from Building Industry Credit Association (BICA), 2351 West Third Street, Los Angeles, California, 90057; telephone (800) 722-2422.

Appendix 5
CALIFORNIA PRELIMINARY NOTICE, TOGETHER WITH PROOF OF SERVICE

APPENDIX 5

CALIFORNIA PRELIMINARY NOTICE

IN ACCORDANCE WITH SECTION 3097 AND 3098, CALIFORNIA CIVIL CODE

676863

THIS IS **NOT** A LIEN, THIS IS **NOT** A REFLECTION ON THE INTEGRITY OF ANY CONTRACTOR OR SUBCONTRACTOR

YOU ARE HEREBY NOTIFIED THAT . . .

LENDER'S COPY

CONSTRUCTION LENDER or
Reputed Construction Lender, if any

_____
(name of person or firm furnishing labor, services, equipment or material)

_____
(address of person or firm furnishing labor, services, equipment or materials)

has furnished or will furnish labor, services, equipment or materials of the following general description:

_____
(general description of the labor, services, equipment or material furnished or to be furnished)

for the building, structure or other work of improvement located at: _____

- - - - - - - - - - - - FOLD HERE - - - - - - - - - - - - -

(address or description of job site sufficient for identification)

OWNER or PUBLIC AGENCY
or Reputed Owner (on public work)
(on private work)

The name of the person or firm who contracted for the purchase of such labor, services, equipment or material is:

An estimate of the total price of the labor, services equipment or materials furnished or to be furnished is:

$ _____

ORIGINAL CONTRACTOR or
Reputed Contractor, if any

Trust Funds to which Supplemental Fringe Benefits are Payable. (Material men not required to furnish)

(name)	(address)
(name)	(address)
(name)	(address)
(name)	(address)

NOTICE TO PROPERTY OWNER
If bills are not paid in full for the labor, services, equipment, or materials furnished or to be furnished, a mechanic's lien leading to the loss, through court foreclosure proceedings, of all or part of your property being so improved may be placed against the property even though you have paid your contractor in full. You may wish to protect yourself against this consequence by (1) requiring your contractor to furnish a signed release by the person or firm giving you this notice before making payment to your contractor or (2) any other method or device that is appropriate under the circumstances.

SUBCONTRACTOR with whom claimant has contracted

Dated: _____

_____
signature   (title)

Telephone Number (_____) _____

This form (No. 594C Revised 1/1/95) ©Copyrighted by and distributed through BUILDING INDUSTRY CREDIT ASSOCIATION (213) 251-1100

Reprinted with permission from BICA. This form may be purchased from Building Industry Credit Association (BICA), 2351 West Third Street, Los Angeles, California, 90057; telephone (800) 722-2422.

CALIFORNIA CONSTRUCTION LAW

PRIVATE WORK
Proof of Service Affidavit
(Section 3097.1 (c) Calif. Civil Code)

IF BY MAIL

I, _____, declare:
That I served copies of this Preliminary Notice by first class registered / certified mail, postage prepaid, on the lender, owner, original contractor and subcontractor with whom I contracted at their respective addresses as shown on the reverse side, on _____
(date)

I declare, under penalty of perjury, that the foregoing is true and correct.

Executed on _____, at _____, California
                  (date)

_____
(Signature of person making service)

(Attach receipts and return receipts of registered or certified mail and/or photocopies of record of delivery and receipt maintained by the Post Office.)

IF BY PERSONAL SERVICE

I, _____, declare:
That I served copies of this Preliminary Notice on the parties named on the reverse side at the places and on the dates shown below:

Lender _____ Date _____
              (place of service)

Owner _____ Date _____
              (place of service)

Contractor _____ Date _____
              (place of service)

Subcontractor with whom I contracted _____ Date _____
              (place of service)

I declare, under penalty of perjury, that the foregoing is true and correct.

Executed on _____, at _____, California
                  (date)

_____
(Signature of person making service)

PUBLIC WORK
Proof of Service Affidavit
(Section 3098 Calif. Civil Code)

IF BY MAIL

I, _____, declare:
That I served copies of this Preliminary Notice by first class mail, registered mail or certified mail, postage prepaid addressed to the public agency, contractor and subcontractor with whom I contracted at the addresses shown on the reverse side, on _____
(date)

I declare, under penalty of perjury, that the foregoing is true and correct.

Executed on _____, at _____, California
                  (date)

_____
(Signature of person making service)

(Attach receipts of registered or certified mail when returned.)

IF BY PERSONAL SERVICE

I, _____, declare:
That I served copies of this Preliminary Notice on the parties named on the reverse side at the places and on the dates shown below:

Public Agency _____ Date _____
              (place of service)

Contractor _____ Date _____
              (place of service)

Subcontractor with whom I contracted _____ Date _____
              (place of service)

I declare, under penalty of perjury, that the foregoing is true and correct.

Executed on _____, at _____, California
                  (date)

_____
(Signature of person making service)

Reprinted with permission from BICA. This form may be purchased from Building Industry Credit Association (BICA), 2351 West Third Street, Los Angeles, California, 90057; telephone (800) 722-2422.

APPENDIX 6
MECHANICS' LIEN

APPENDIX 6

RECORDING REQUESTED BY

WHEN RECORDED MAIL TO

SPACE ABOVE THIS LINE FOR RECORDER'S USE ONLY

Mechanics' Lien
(Claim of Lien)
(To be recorded in the county recorder's office in the county in which the property is located.)

NOTICE IS HEREBY GIVEN: That _____
as claimant claims a lien for labor, service, equipment, or materials under Section 3082 et Seq. of the Civil Code of the State of California, upon the premises hereinafter described, and upon every estate or interest in such structures, improvements and premises held by any party holding any estate therein.

Said labor, service, equipment or materials, were furnished for the construction of those certain buildings, improvements, or structures, now upon that certain parcel of land situated in the County of _____,
State of California, said land described as follows:

STREET ADDRESS:

LEGAL DESCRIPTION:

Said lien is claimed for the following labor, services, equipment or materials: (describe labor, services, equipment, or materials in detail) _____

Amount due after deducting all just credits and offsets...$ _____

The name of the person or company by whom claimant was employed or to whom claimant furnished labor, services, equipment, or materials is _____

That _____
(Name)
_____
(Mailing Address)
is/are the reputed owner(s) of said building and/or premises, or some interest therein.

Date _____   Name of Claimant: _____
(Firm Name)

By: _____
(Signature)
_____
(Authorized Capacity)

VERIFICATION

I, the undersigned, state: I am the _____
("Agent of", "President of", "A Partner of", "Owner of", etc.)
the claimant named in the foregoing mechanics' lien; I have read said claim of mechanics' lien and know the contents thereof, and I certify that the same is true of my own knowledge.

I certify (or declare) under penalty of perjury under the laws of the State of California that the foregoing is true and correct.

Executed on _____, 19___, at _____
State of _____.   _____
(Signature of Claimant or Authorized Agent)

BICA Form 3 (Rev 5/94)

Reprinted with permission from BICA. This form may be purchased from Building Industry Credit Association (BICA), 2351 West Third Street, Los Angeles, California, 90057; telephone (800) 722-2422.

Appendix 7
STOP NOTICE

APPENDIX 7

Stop Notice
CALIFORNIA CIVIL CODE SECTION 3103

NOTICE TO _____
(Name of Construction Lender, Public Body or Owner)

_____
(Address)

(If Private Job — file with responsible officer or person at office or branch of construction lender administering the construction funds or with the owner — CIVIL CODE SECTIONS 3156 - 3175)

(If Public Job — file with office of controller, auditor, or other public disbursing officer whose duty it is to make payments under provisions of the contract — CIVIL CODE SECTIONS 3179 - 3214)

Prime Contractor: _____

Sub Contractor (If Any) _____

Owner or Public Body: _____

Improvement known as _____
(Name and address of project or work of improvement)

in the City of _____ , County of _____ ,
State of California.

_____ , Claimant, a _____
(Claimant)   (Corporation/Partnership/Sole Proprietorship)

furnished certain labor, service, equipment or materials used in the above described work of improvement. The name of the person or company by whom claimant was employed or to whom claimant furnished labor, service, equipment, or materials is _____
(Name of Subcontractor/Contractor/Owner-Builder)

The kind of labor, service, equipment, or materials furnished or agreed to be furnished by claimant was

_____
(Describe in detail)

Total value of labor, service, equipment, or materials agreed to be furnished......... $ _____

Total value of labor, service, equipment, or materials actually furnished is............ $ _____

Credit for materials returned, if any... $ _____

Amount paid on account, if any... $ _____

Amount due after deducting all just credits and offsets................................... $ _____

YOU ARE HEREBY NOTIFIED to withhold sufficient monies held by you on the above described project to satisfy claimant's demand in the amount of $ _____ and in addition thereto sums sufficient to cover interest, court costs and reasonable costs of litigation, as provided by law.

A bond (CIVIL CODE SECTION 3083) _____ attached. (Bond required with Stop
                                                     (is/is not)
Notice served on constructions lender on private jobs — bond not required on public jobs or on Stop Notice served on owner on private jobs).

Date _____ Name of Claimant _____
                                                              (Firm Name)

_____
(Mailing Address)

By _____
(Signature)

_____
(Official Capacity)

VERIFICATION

I, the undersigned, state: I am the _____
                                                        ("Agent of"; "President of"; "A Partner of"; "Owner of"; etc.)

the claimant named in the foregoing Stop Notice; I have read said claim of Stop Notice and know the contents thereof, and I certify that the same is true of my own knowledge.

I certify (or delcare) under penalty of perjury under the laws of the State of California that the foregoing is true and correct.

Executed on _____ , 19 _____ , at _____ ,

State of _____ . _____
                                                                    (Signature of Claimant or Authorized Agent)

REQUEST FOR NOTICE OF ELECTION
(Private Works Only)

If an election is made not to withhold funds pursuant to this stop notice by reason of a payment bond having been recorded in accordance with Sections 3235 or 3162, **please send notice** of such election and a copy of the bond within 30 days of such election in the enclosed preaddressed stamped envelope to the address of the claimant shown above. This information must be provided by you under Civil Code Sections 3159, 3161 or 3162.

Signed: _____
(Claimant must enclose self addressed stamped envelope)

BICA FORM 4 (FSNOI) 1/1/89, 5000

Reprinted with permission from BICA. This form may be purchased from Building Industry Credit Association (BICA), 2351 West Third Street, Los Angeles, California, 90057; telephone (800) 722-2422.

Appendix 8
MILLER ACT FORM

APPENDIX 8

Notice to Contractor

(For use on Federal Improvements)

AS PROVIDED FOR IN 40 U.S.C. 270a — 270e
(MILLER ACT).

TO: _____

_____
(Name and address of general contractor)

The undersigned is advised that you are the general or original contractor on that certain Federal Public Work of Improvement commonly known as the _____

_____ job,

located at _____ State of _____.

Please take notice that the undersigned has furnished _____

_____
(Describe labor, services, equipment, or material)

to _____
(Name of Subcontractor)

subcontractor engaged in performance of work under your general contract on the above described job; that the balance due and owing upon this account from said subcontractor amounts to the net sum of $ _____.

This Notice is notice to you, as the General Contractor on the project referred to above, that the undersigned claimant is looking to you and your payment bond surety for payment of the balance due set forth herein.

Date _____ Name of Claimant: _____
(Firm name)

By _____
(Signature)

_____
(Official Capacity)

BICA FORM 13 1000 (Rev. 4/91)

Reprinted with permission from BICA. This form may be purchased from Building Industry Credit Association (BICA), 2351 West Third Street, Los Angeles, California, 90057; telephone (800) 722-2422.

Appendix 9

NOTICE TO PRINCIPAL AND SURETY ON PAYMENT BOND ON PRIVATE WORK

APPENDIX 9

NOTICE TO PRINCIPAL AND SURETY ON PAYMENT BOND ON PRIVATE WORK
(CIVIL CODE §§ 3241 & 3242)

To: _____

(Name & Address of Principal)

To: _____

(Name & Address of Surety)

You, and each of you, are hereby notified that _____

(Name of Person or Firm Furnishing Labor, Services, Equipment, or Material)

has furnished labor, services, equipment, or material of the following general description:

_____

(General description of Labor, Services, Equipment or Material)

for the building, structure, or other work of improvement located at

_____

(Address or Description of the Job Site Sufficient for Identification)

The name of the person or firm who contracted for the purchase of such labor, services, equipment, or material is: _____

(Name of Person to or for Whom the Labor, Services, Equipment, or Material were Furnished)

The amount in value, as near as may be determined, of any labor, services, equipment, or materials already furnished, or to be furnished, is $ _____.

Date: _____

Name of Claimant: _____

(Firm Name)

By: _____

(Signature)

_____

(Official Capacity)

Form 11 (Rev.1/85) BICA (213) 251-1100 A Construction Credit Reporting Agency.

Reprinted with permission from BICA. This form may be purchased from Building Industry Credit Association (BICA), 2351 West Third Street, Los Angeles, California, 90057; telephone (800) 722-2422.

Appendix 10
NOTICE TO PRINCIPAL AND SURETY ON PAYMENT BOND ON PUBLIC WORK

APPENDIX 10

**NOTICE TO PRINCIPAL AND SURETY
ON PAYMENT BOND ON
PUBLIC WORK**
(CIVIL CODE §§ 3252 & 3243)

To: _____
_____
(Name & Address of Principal)

To: _____
_____
(Name & Address of Surety)

You, and each of you, are hereby notified that _____
(Name of Person or Firm Furnishing Labor, Services, Equipment, or Material)
has furnished labor, services, equipment, or material of the following general description:
_____
(General description of Labor, Services, Equipment or Material)
for the building, structure, or other work of improvement located at
_____
(Address or Description of the Job Site Sufficient for Identification)
The name of the person or firm who contracted for the purchase of such labor, services, equipment, or material is: _____
(Name of Person to or for Whom the Labor, Services, Equipment, or Material were Furnished)

The amount in value, as near as may be determined, of any labor, services, equipment, or materials already furnished, or to be furnished, is $ _____.

Date: _____

Name of Claimant: _____
By: _____
(Firm Name)
_____
(Signature)
_____
(Official Capacity)

Form 12 (Rev. 1/85) BICA (213) 251-1100 A Construction Credit Reporting Agency.

Reprinted with permission from BICA. This form may be purchased from Building Industry Credit Association (BICA), 2351 West Third Street, Los Angeles, California, 90057; telephone (800) 722-2422.

TABLE OF CASES

Case	Book §
A&A Elec., Inc. v. City of King, 54 Cal. App. 3d 457 (1976)	2.02[E]
AAS v. Superior Court of San Diego, 64 Cal. App. 4th 916 (1998)	5.08, 8.12, 8.13
Abbett Elec. Corp. v. California Fed. Sav. & Loan Ass'n, 230 Cal. App. 3d 355 (1991)	9.06[H]
Able Elec. Co. v. Vacanti & Randozzo Constr. Co., 212 Neb. 619, 324 N.W.2d 667 (1982)	7.03
Aced v. Hobbs-Sesack Plumbing Co., 55 Cal. 2d 573 (1961)	8.04
Acme Galvanizing Co. v. Firemans Fund Ins. Co., 221 Cal. App. 3d 170 (1990)	5.07[B]
A.D. Herman Constr., GSBCA No. 4823, 78-1 B.C.A. (CCH) ¶ 13,187	6.11
Adarand Constructors, Inc. v. Pena, 515 U.S. 200 (1995)	2.02[F][4][b]
Addison v. State, 21 Cal. App. 3d 313 (1978)	9.03[G][5]
Advanced Contractors, *In re,* 44 B.R. 239 (Bankr. M.D. Fla. 1984)	9.06[B]
Aesco Steel, Inc. v. J.A. Jones Constr. Co., 621 F. Supp. 1576 (E.D. La. 1985)	1.09[C]
Aetna Cas. & Sur. Co. v. Board of Trustees, 223 Cal. App. 2d 337 (1963)	6.02[B]
Aetna Cas. & Sur. Co. v. Butte-Meade Sanitary Water, 500 F. Supp. 193 (D.S.D. 1980)	6.02[B]
Aetna Cas. & Sur. Co. v. United States, 655 F.2d 1047 (9th Cir. 1981)	9.06[T]
Aircraft Gear Corp. v. Kaman Aerospace Corp., 875 F. Supp. 485 (N.D. Ill. 1995)	6.07[G]
A.J. Raisch Paving Co. v. Mountain View Sav. & Loan Ass'n, 28 Cal. App. 3d 832 (1972)	9.03[A]
A.J. Wolfe Co. v. Baltimore Contractors, Inc., 244 N.E.2d 717 (Mass. 1969)	1.09[C]
AKCON, Inc., ENGBCA No. 5593, 90-3 B.C.A. (CCH) ¶ 23,250 (1990)	6.14, 6.14[A]
Al Johnson Constr. Co. v. United States, 854 F.2d 467 (Fed. Cir. 1988)	7.02
Albaugh v. Moss Constr. Co., 125 Cal. App. 2d 126 (1954)	2.01[B]
Aldrich v. Transcontinental Land, etc., Co., 131 Cal. App. 2d 788 (1955)	14.02[F]
Ales-Peratis Foods v. American Can Co., 164 Cal. App. 3d 277 (1985)	3.04
All Bay Mill & Lumber Co. v. Surety Co., 208 Cal. App. 3d 11 (1989)	1.09[B], 1.09[J]
All W. Elecs., Inc. v. M-B-W, Inc., 64 Cal. App. 4th 717 (1998)	8.02
Allen v. Wilson, 178 Cal. 674 (1918)	9.03[C][2], 9.06[O]
Allied Materials & Equip. Co. v. United States, 569 F.2d 562, 215 Ct. Cl. 406 (1978)	6.10[C]
Allstate Ins. Co. v. Superior Court of S.F., 35 Cal. App. 3d 137 (1973)	14.02[E]
American Fidelity Fire Ins. Co. v. United States, 385 F. Supp. 1075 (N.D. Cal. 1974)	10.04[A]

Case	*Book §*
American Sheet Metal v. Em-Kay Eng'g, 478 F. Supp. 809 (E.D. Cal. 1979)	1.05, 1.06[B]
American Transit Mix Co. v. Weber, 106 Cal. App. 2d 74 (1951)	9.01[A][6], 9.03[C][1]
Amerson v. Christman, 261 Cal. App. 2d 811 (1968)	5.01[A], 5.01[B], 6.02[C]
Amp-Rite Elec. Co. v. Wheaton Sanitary Dist., 580 N.E.2d 622 (Ill. App. Ct. 1991)	4.14
Anderson Plumbing Co., *In re,* 71 B.R. 19 (Bankr. E.D. Cal. 1986)	9.06[B]
Angeles Elec. Co. v. Superior Court, 27 Cal. App. 4th 426, 32 Cal. Rptr. 2d 660 (2d Dist. 1994)	9.05[C], 9.06[B]
A-1 Door & Materials Co. v. Fresno Guar. Sav. & Loan Ass'n, 61 Cal. 2d 728 (1964)	9.01[B][2], 9.06[J], 10.04[A]
A.R. Moyer, Inc. v. Graham, 285 So. 2d 397 (Fla. 1973)	8.07
Arnold Braid Elec. Co., 908 F.2d 52 (6th Cir. 1990)	9.06[B]
Arthur v. New House Bldg. Corp., 217 Cal. App. 2d 526 (1963)	9.04[G]
Arvanis v. Noslo Eng'g Consultants, Inc., 739 F.2d 1287 (7th Cir. 1984), *cert. denied,* 469 U.S. 1911, 105 S. Ct. 964, 83 L. Ed. 2d 969 (1985)	11.06
Arvol D. Hays Constr. Co., ASBCA No. 25112, 84-3 B.C.A. (CCH) ¶ 17,661	6.08[B]
Asdourian v. Araj, 38 Cal. 3d 276 (1985)	1.06[B][2], 13.03
Associated Builders & Contractors v. Contra Costa Water Dist., 37 Cal. App. 4th 466 (1995)	2.02[B][2]
Associated Builders & Contractors, Inc. v. Metropolitan Water Dist. of S. Cal., 59 Cal. App. 4th 1503 (1997), *review granted,* 72 Cal. Rptr. 2d 215 (1998)	2.02[F][4][a]
Associated Gen. Contractors, Inc. v. City & County of S.F., 813 F.2d 922 (9th Cir. 1987)	2.02[F][4][a]
Associated Lathing Co. v. Louis C. Dunn, Inc., 135 Cal. App. 2d 40 (1955)	5.01[B]
Associated Wholesale Elec. v. S.H. Kress & Co., 11 Cal. App. 2d 592 (1936)	9.06[J]
AT&T Techs., Inc. v. Communication Workers of Am., 475 U.S. 643, 106 S. Ct. 1415, 89 L. Ed. 2d 648 (1986)	14.02[C]
Attorney's Office Management, Inc., *In re,* 29 B.R. 96 (Bankr. C.D. Cal. 1983)	12.05
Avner v. Longridge Estates, 272 Cal. App. 2d 607 (1969)	8.02
Azusa W., Inc. v. West Covina, 45 Cal. App. 3d 259 (1975)	10.04[A]
Bagwel Coatings, Inc. v. Middle S. Energy, Inc., 797 F.2d 1298 (5th Cir. 1986)	4.14
Bailey-Sperber, Inc. v. Yosemite Ins. Co., 64 Cal. App. 3d 725 (1976)	1.09[A], 1.09[G], 1.14
Baker v. Hubbard, 101 Cal. App. 3d 226 (1980)	9.01[A][6]
Baldwin-Lima-Hamilton Corp. v. Superior Court, 208 Cal. App. 2d 803 (1962)	2.02[C], 2.02[G][2], 2.02[G][3]
Ballenger Corp., DOTBCA Nos. 74-32, 74-32A, 74-32H, 84-1 B.C.A. (CCH) ¶ 16,973 (1983)	7.02

TABLE OF CASES

Case	Book §
Ballret Bros. Constr. v. Regents of the Univ. of Cal., 80 Cal. App. 3d 321 (1978)	2.02[E]
Bank of Italy v. MacGill, 93 Cal. App. 229 (1928)	9.04[A]
Barr Lumber Co. v. Old Ivy Home Builders, Inc., 34 Cal. App. 4th 1 (1995)	9.06[R]
Barr Lumber Co. v. Shaffer, 108 Cal. App. 2d 14 (1951)	9.04[A]
Barry v. Contractors State License Bd., 85 Cal. App. 2d 600 (1948)	1.09
Barth Elec. Co. v. Taylor Bros., Inc., 553 N.E.2d 504 (Ind. Ct. App. 1990)	4.14
Barton & Sons Co., ASBCA Nos. 9477, 9764, 65-2 B.C.A. (CCH) ¶ 4874 (1965)	4.13
Basic Modular Facilities, Inc. v. Ehsanipour, 83 Cal. Rptr. 2d 462 (Ct. App. 1999)	9.06[H], 9.06[O][2]
Baudier Marine Elecs. v. United States, 3 F.P.D. 41 (Cl. Ct. 1984)	11.06
Bay Lumber Co. v. Pickering, 120 Cal. App. 163 (1932)	9.06[F]
Bayuk v. Edson, 236 Cal. App. 2d 309 (1965)	5.01[A], 6.02[C]
B.C. Richter Contracting Co. v. Continental Cas. Co., 230 Cal. App. 2d 491 (1964)	3.03[E], 4.02, 4.07[E]
B.D. Collins Constr. Co., ASBCA No. 42662	7.05
Beach v. Contractors State License Bd., 151 Cal. App. 2d 117 (1957)	1.09[J]
Beckman Constr. Co., ASBCA No. 24725, 83-1 B.C.A. (CCH) ¶ 16,326 (1983)	7.05
Beco Corp., ASBCA No. 27090, 82-2 B.C.A. (CCH) ¶ 16,124 (1982)	4.13
Bel Marin Drywall v. Groves, 470 F.2d 934 (9th Cir. 1972)	9.06[B]
Beley v. Municipal Court, 100 Cal. App. 3d 5 (1979)	13.05
Bell Constr. Co., ASBCA No. 23376, 79-2 B.C.A. (CCH) ¶ 13,908	7.04
Bennett v. United States, 371 F.2d 859 (Ct. Cl. 1967)	4.07[D]
Benson Elec. Co. v. Hale Bros. Assocs., Inc., 246 Cal. App. 2d 686 (1966)	9.02[B][7]
Bentz Plumbing & Heating v. Favaloro, 128 Cal. App. 3d 145, 180 Cal. Rptr. 223 (1982)	9.05[B][1], 9.06[F]
Betancourt & Gonzales, S.E., DOT CAB Nos. 2785, 2789, 2799, 96-1 B.C.A. (CCH) ¶ 28,033	6.11
Bick-Com Corp., VACAB No. 1320, 80-1 B.C.A. (CCH) ¶ 14,285	6.11
Bigelow, Inc., ASBCA No. 24376, 81-2 B.C.A. (CCH) ¶ 15,300	7.04
Bird v. American Sur. Co., 175 Cal. 625 (1917)	3.02[B][10], 5.03, 6.02[A]
B&J Constr. Co. v. Spacious Homes, Inc., 204 Cal. App. 2d 216 (1962)	9.06[M]
Blackhawk Heating & Pumping Co., GSBCA No. 2432, 76-1 B.C.A. (CCH) ¶ 11,649	7.03
Blake Constr. Co., GSBCA No. 2477, 71-1 B.C.A. (CCH) ¶ 8870	6.10[B]
Blew v. Horner, 187 Cal. App. 3d 1380 (1986)	1.06[A]
Blinderman Constr. Co. v. United States, 39 Fed. Cl. 529 (1997)	7.02, 7.03, 7.04
Blinderman Constr. Co. v. United States, 695 F.2d 552 (Fed. Cir. 1982)	7.05
Bloom v. Bender, 48 Cal. 2d 793 (1957)	9.01[C][1]
Blount Bros. Constr. Co. v. United States, 346 F.2d 962 (Ct. Cl. 1965)	4.07[D]
Bohannan Bros. v. Lo Jean Dev. Co., 3 Cal. App. 3d 200 (1969)	9.05[F]
Bolivar v. Surety Co., 72 Cal. App. 3d Supp. 22 (1977)	1.14
Bolton Corp. v. T.A. Loving Co., 380 S.E.2d 796 (N.C. Ct. App. 1989)	4.14
Bonded Prods. Co. v. R.C. Gallyon Constr. Co., 228 Cal. App. 2d 186 (1964)	9.06[A], 10.04[A]

Case	*Book §*
Borello v. Eichler Homes, Inc., 221 Cal. App. 2d 487 (1963)	9.03[C][1], 9.06[O][1]
Boscus v. Waldmann, 31 Cal. App. 245 (1916)	9.03[E]
Bowden v. United States *ex rel.* Malloy, 239 F.2d 572 (9th Cir. 1956)	11.02
Boyajian v. United States, 423 F.2d 1231 (Ct. Cl. 1970)	6.14[A], 6.14[B]
Boyd & Lovesee Lumber Co. v. Western Pac. Fin. Corp., 44 Cal. App. 3d 460 (1975)	9.01[B][2][c]
Boydston v. Napa Sanitation Dist., 222 Cal. App. 3d 1362 (1990)	2.02[F][2]
Boys Club v. Fidelity & Deposit Co., 6 Cal. App. 4th 1266 (1992)	14.02[C]
Breda Construzioni Ferroviarie S.P.A. v. Los Angeles County Metro. Transit Auth., 56 Cal. App. 4th 1433 (1997)	10.02[C]
Broadway Maintenance Corp. v. Rutgers, 434 A.2d 1125 (N.J. Super. Ct. App. Div. 1981)	4.14
Brower Co. v. Garrison, 468 P.2d 469 (Wash. Ct. App. 1970)	6.06
Brown v. Aubrey, 22 Cal. 566 (1863)	9.05[B][1]
Brown Co. v. Superior Court, 148 Cal. App. 3d 891 (1983)	9.02[B][2], 9.02[B][6], 9.06[F]
Brunzell Constr. Co. of Nev. v. Harrah's Club, 228 Cal. App. 2d 764 (1967)	14.02[E]
Building & Constr. Trades Council of the Metro. Dist. v. Associated Builders & Contractors of Mass./R.I., Inc., 507 U.S. 218 (1993)	2.02[F][4][a]
Burke, *In re,* 76 F. Supp. 5 (S.D. Cal. 1948)	12.06
Burnett & Doty Dev. Co. v. Phillips, 84 Cal. App. 3d 384 (1978)	5.04, 6.02[A]
Burr v. Sherwin Williams Co., 42 Cal. 2d 682 (1954)	8.02
Burton v. Sosinsky, 203 Cal. App. 3d 569 (1988)	9.05[G]
Butchers Union Local 532 v. Farmers Mkts., 67 Cal. App. 3d 905 (1977)	14.02[E]
Butler v. Ng Chung, 160 Cal. 435 (1911)	9.06[A]
Buzgheia v. Leasco Sierra Grove, 60 Cal. App. 4th 374 (1997)	1.09[F], 1.12
Byler v. Great Am. Ins. Co., 395 F.2d 273 (10th Cir. 1968)	1.09[C]
C. Norman Peterson Co. v. Container Corp. of Am., 172 Cal. App. 3d 628 (1985)	4.09, 6.10[C]
C.A. Magistretti Co. v. Merced Irrigation Dist., 27 Cal. App. 3d 270 (1972)	10.04[A]
Cal-Air Conditioning, Inc. v. Auburn Union Sch. Dist., 21 Cal. App. 4th 65 (1993)	2.02[D][2]
Cal-Air Conditioning, Inc. v. M.P. Allen Gen. Contractors, 21 Cal. App. 4th 655 (1993)	2.02[D][2]
Cal-Pacific Materials Co. v. Redondo Beach City Sch. Dist., 94 Cal. App. 3d 652 (1979)	10.04[A]
Calhoun v. Huntington Park First Sav. & Loan Ass'n, 186 Cal. App. 2d 451 (1960)	9.01[B][2], 10.04[A]
California Elec. Supply Co. v. United Pac. Life Ins. Co., 227 Cal. App. 2d 138 (1964)	9.03[H][3], 10.04[A]
California Viking Sprinkler Co. v. Cheney, 182 Cal. App. 2d 564 (1960)	9.06[H]
Call v. Alcan Pac. Co., 251 Cal. App. 2d 442 (1967)	3.03[F]
Callahan v. Chatsworth Park, Inc., 204 Cal. App. 2d 597 (1962)	9.06[M]
Camper v. McDermott, 266 Cal. App. 2d 41 (1968)	5.03, 6.02[A]
Cannon Constr. Co., ASBCA No. 16142, 72-1 B.C.A. (CCH) ¶ 9404	7.04
Capitol Steel Fabricators, Inc. v. Mega Constr. Co., 58 Cal. App. 4th 1049 (1997)	1.09[C], 9.06[D], 9.06[T]

TABLE OF CASES

Case	*Book §*
Card Constr. Co. v. Ledbetter, 16 Cal. App. 3d 472 (1971)	9.06[A], 10.04[A]
Carpenters Health & Welfare Trust Fund v. Developers Ins. Co., 11 Cal. App. 4th 1539 (1992)	9.01[A][1]
Carpenters Health & Welfare Trust Fund v. Surety Co., 13 Cal. App. 4th 1406 (1993)	1.14, 9.01[A][1]
Carpenters Health & Welfare Trust Fund v. Tri Corp., 23 F.3d 849 (9th Cir. 1994)	9.01[A][1]
Carpenters S. Cal. Administrative Corp. v. El Capitan Dev. Co., 53 Cal. 3d 1041, 282 Cal. Rptr. 277 (1991)	9.01[A][1]
Carty v. American States Ins. Co., 7 Cal. App. 4th 399 (1992)	5.07[B]
Cates Constr., Inc. v. Talbot Partners, 21 Cal. 4th 28 (1999)	9.03[H][3]
Cates Constr., Inc. v. Talbot Partners, 53 Cal. App. 4th 1420 (1997)	9.03[H][3]
C.B.C. Enters., Inc. v. United States, 978 F.2d 669 (Fed. Cir. 1992)	6.07[G]
C.C. Cal. Plaza Assocs. v. Paller & Goldstein, 51 Cal. App. 4th 1042 (1996)	4.07[B]
Central Indus. Eng'g Co. v. Strauss Constr. Co., 98 Cal. App. 3d 460 (1979)	10.04[A]
Centric Corp. v. Morrison Knutsen Co., 731 P.2d 411 (Okla. 1986)	9.05[E]
C.G. Norton Co., ENGBCA No. 5182 (1988)	7.05
Chadwick v. Fire Ins. Exch., 17 Cal. App. 4th 1112 (1st Dist. 1993)	5.07[B]
Chameleon Eng'g Corp. v. Air Dynamics, Inc., 101 Cal. App. 3d 418 (1980)	8.06, 8.07
Chaney & James Constr., FAACAP No. 67-18, 66-2 B.C.A. (CCH) ¶ 6066 (1967)	7.02
Chaney & James Constr. Co. v. United States, 421 F.2d 728, 190 Ct. Cl. 699 (1970)	6.07[E]
Chang v. Redding Bank of Commerce, 29 Cal. App. 4th 672 (1994)	9.03[G][1]
Charles J. Rounds Co. v. Joint Council of Teamsters No. 42, 4 Cal. 3d 888 (1971)	14.02[C]
Charles L. Harney, Inc. v. Durkee, 107 Cal. App. 2d 803 (1962)	2.02[F]
Charles T. Parker Constr. Co. v. United States, 193 Ct. Cl. 320 (1970)	6.11
CIG Contractors, Inc. v. Mississippi State Bldg. Comm'n, 510 So. 2d 510 (Miss. 1987)	4.14
City Council of Beverly Hills v. Superior Court, 272 Cal. App. 2d 876 (1969)	2.02[F][3]
City St. Improvement Co. v. City of Marysville, 155 Cal. 419 (1909)	4.07[B]
C&K Eng'g Contractors v. Amber Steel Co., 23 Cal. 3d 1 (1978)	2.01[C]
C.L. Smith Co. v. Roger Ducharme, Inc., 65 Cal. App. 3d 735 (1977)	2.02[D][2]
Clifford F. MacEvoy Co. v. United States ex rel. Calvin Tomfine Co., 322 U.S. 102 (1944)	11.04
Clive Constr. Co., ASBCA No. 28600, 84-3 B.C.A. (CCH) ¶ 17,594	7.04, 7.05
C.O. Sparks, Inc. v. Pacific Constr. Paving Co., 159 Cal. App. 2d 513 (1959)	9.03[H][3]
COAC, Inc. v. Kennedy Eng'rs, 67 Cal. App. 3d 916 (1977)	8.05
Coalition for Economic Equity v. Wilson, 110 F.3d 1431 (9th Cir.), *cert. denied*, 118 S. Ct. 397 (1997)	2.02[F][4][b]
Coast Cent. Credit Union v. Superior Court, 209 Cal. App. 3d 703 (1989)	9.03[D]
Coast Pump Ass'n v. Stephen Tyler Corp., 62 Cal. App. 3d 421 (1976)	2.02[D][2]
Cocolat, Inc., *In re* (Cocolat, Inc. v. Fisher Dev., Inc.) 176 B.R. 540 (Bankr. N.D. Cal. 1995)	12.08
Coffey Constr. Co., Appeal of, VABCA No. 3361 (1993)	7.05
Coleman Eng'g Co. v. North Am. Aviation, 65 Cal. 2d 396 (1966)	4.07[E]
Coley Properties Corp., PODBCA No. 291, 76-1 B.C.A. (CCH) ¶ 11,701, and 75-2 B.C.A. (CCH) ¶ 11,514	6.08[B]

Case	*Book §*
Commercial Standard Ins. Co. v. Bank of Am., 57 Cal. App. 3d 241 (1976)	9.01[B][2][b]
Committee of Seven Thousand v. Superior Court, 45 Cal. 3d 491 (1988)	2.02[B][2]
Commonwealth of Pennsylvania State Pub. Sch. Bldg. Auth. v. Noble C. Quadel Co., 585 A.2d 1136 (Pa. Commw. Ct. 1991)	4.14
Community Lumber Co. v. Chute, 215 Cal. 268 (1932)	9.04[A]
Connolly Dev. Inc. v. Superior Court, 17 Cal. 3d 803 (1976)	9.01[A], 9.01[B], 9.01[B][2][f], 9.04[A], 9.06[B], 9.06[F]
Consolidated Elec. Distribs., Inc. v. Kirkham, Chaon & Kirkham, Inc., 18 Cal. App. 3d 54 (1971)	9.04[G]
Construction Fin. LLC v. Perlite Plastering Co., 53 Cal. App. 4th 170 (1997)	1.03, 1.05, 1.06[B][2]
Construction Indus. Force Account Council v. Delta Wetlands, 2 Cal. App. 4th 1589 (1992)	2.02[B], 2.02[B][2]
Continental Cas. Co. v. Hartford Accident & Indem. Co., 243 Cal. App. 2d 565 (1966)	9.03[H][3], 11.05
Continental Consolidated Corp., ENGBCA No. 2743, 67-2 B.C.A. (CCH) ¶ 6624 (1967)	7.02, 7.03
Contracting & Materials Co. v. City of Chicago, 314 N.E.2d 598 (Ill. App. Ct. 1974)	4.10
Contractors Dump Truck Serv., Inc. v. Gregg Constr. Co., 237 Cal. App. 2d 1 (1965)	1.01, 9.01[A][2], 9.01[A][4]
Contractors Labor Pool, Inc. v. Westway Contractors, Inc., 53 Cal. App. 4th 152 (1997)	1.01, 10.04[A]
Cooley v. Freeman, 204 Cal. 59 (1928)	10.01
Corbetta Constr. Co. v. United States, 461 F.2d 1330 (Ct. Cl. 1972)	4.07[D]
Corech v. Hornwood, 58 Cal. App. 4th 1412 (1997)	9.06[H]
Cornell v. Sennes, 18 Cal. App. 3d 126 (1971)	9.06[A]
Costello v. Campbell, 81 Cal. App. 2d 452 (1947)	1.06[B][1]
Creative Plastering, Inc. v. Hedley Builders, Inc., 19 Cal. App. 4th 1662, 24 Cal. Rptr. 2d 216 (1st Dist. 1994)	14.02[C]
Crestwood Lumber Co. v. Citizen's Sav. & Loan Ass'n, 83 Cal. App. 3d 819 (1978)	9.06[J]
Crofoot Lumber, Inc. v. Thompson, 163 Cal. App. 2d 324 (1958)	6.04
Crown Plastering v. Elite Assocs., 560 N.Y.S.2d 694 (App. Div. 1990)	1.09[C]
Culbertson v. Cizek, 225 Cal. App. 2d 451 (1964)	1.05, 1.06[B], 9.03[H][3]
Currie v. Stolowitz, 169 Cal. App. 2d 810 (1959)	1.03
C.V. Starr & Co. v. Boston Reinsurance Corp., 190 Cal. App. 3d 1637 (1987)	14.02[F]
CWC, Inc., Appeal of, ASBCA No. 26432, 82-2 B.C.A. (CCH) ¶ 15,907 (June 19, 1982)	4.13
Dahl-Beck Elec. Co. v. Rogge, 275 Cal. App. 2d 893 (1969)	1.05, 1.06[B], 9.01[A][4]
Darrell T. Stuart Contractor v. Bridges, 406 P.2d 143 (Ariz. Ct. App. 1965)	1.09[C]
Dart Equip. Corp. v. Mack Trucks, Inc., 9 Cal. App. 3d 837 (1970)	3.02[B][13]

TABLE OF CASES

Case	Book §
Daugherty Co. v. Kimberly Clark Corp., 14 Cal. App. 3d 151 (1971)	4.09
Davenport & Co. v. Spieker, 197 Cal. App. 3d 566 (1988) (depublished opinion)	13.03
Davis v. LaCrosse Hosp. Ass'n, 121 Wis. 579 (1904)	9.06[T]
Davis Co. v. Superior Court, 1 Cal. App. 3d 156 (1969)	1.05
Dawco Constr., Inc. v. United States, 18 Cl. Ct. 682 (1989), *aff'd in part and rev'd in part,* 930 F.2d 872 (Fed. Cir. 1991), *overruled in* Reflectone, Inc. v. Dalton, 60 F.3d 1572 (Fed. Cir. 1995)	6.14, 6.14[B]
Dawson v. Eldridge, 84 Idaho 331 (1962)	9.06[T]
Dawson Constr. Co., GSBCA No. 3820, 75-1 B.C.A. (CCH) ¶ 11,339	6.10[B]
Dawson Constr. Co., Appeal of, VABCA Nos. 2000, 2016, 86-3 B.C.A. (CCH) ¶ 19,322 (1986)	4.15
Dean Vincent, Inc. v. Redisco, Inc., 373 P.2d 995 (Or. 1962)	9.06[C][1]
Del E. Webb Corp. v. Structural Materials Co., 123 Cal. App. 3d 593 (1981)	8.05
Del Mar Beach Club Owner's Ass'n v. Imperial Contracting Co., 123 Cal. App. 3d 898 (1981)	8.02
Delco Elecs. Corp. v. United States, 17 Cl. Ct. 302 (1989), *aff'd,* 909 F.2d 1495 (Fed. Cir. 1990)	6.14[B]
Department of Indus. Relations v. Fidelity Roof Co., 57 Cal. App. 4th 445 (1997)	10.04[A]
Department of Indus. Relations v. Seaboard Sur., 50 Cal. App. 4th 1501 (1996)	10.04[A]
Department of the Army v. Blue Flocks, Inc., 119 S. Ct. 687 (1999)	11.07[E]
Design Assocs., Inc. v. Welch, 224 Cal. App. 2d 165 (1964)	9.01[A][2], 9.04[A]
Diamond Fruit Growers, Inc. v. Krack Corp., 794 F.2d 1440 (9th Cir. 1986)	3.04
Dick Henley, Inc., *In re,* 38 B.R. 210 (Bankr. M.D. La. 1984)	9.06[B]
Directors Guild of Am. v. Harmony Pictures, Inc., 32 F. Supp. 2d 1184 (C.D. Cal. 1998)	9.05[D]
Distefano v. Hall, 263 Cal. App. 2d 380 (1968)	9.06[H]
District Concrete Co. v. Bernstein Concrete Corp., 418 A.2d 1030 (D.C. 1980)	4.13
District of Columbia v. C.J. Langenfelder & Son, Inc., 558 A.2d 1155 (D.C. Ct. App. 1989)	4.15
Division of Labor Law Enforcement v. Egnew Inv., Inc., 247 Cal. App. 2d 863 (1966)	9.06[K]
Doherty v. Carruthers, 171 Cal. App. 2d 214 (1959)	9.03[A][1]
Domar Elec., Inc. v. City of L.A., 41 Cal. App. 4th 810 (1995)	2.02[F][4][b]
Domar Elec., Inc. v. City of L.A., 9 Cal. 4th 161 (1994)	2.02[F][4][b]
Douglas v. Donner Pines, 73 Cal. App. 3d 268 (1977)	9.06[K]
Downey, Inc. v. Bradley Ctr. Corp., 524 N.W.2d 915 (Wis. Ct. App. 1994)	4.11
Drennan v. Star Paving Co., 52 Cal. 2d 409 (1958)	2.01[C]
D.W. Sandau Dredging, ENGBCA No. 5812, 96-1 B.C.A. (CCH) ¶ 28,064	6.11
Dynamic Constr. Co. v. Barton Malow Co., 543 N.W.2d 31 (Mich. Ct. App. 1995)	4.14
E. Arthur Higgins, AGBCA No. 76-128, 79-2 B.C.A. (CCH) ¶ 14,050	6.11
E.A. Davis & Co. v. Richards, 120 Cal. App. 2d 237 (1953)	1.06[B][1]
Ealahan Elec. Co., Appeal of, DOTBCA No. 1959, 90-3 B.C.A. (CCH) ¶ 23,177	7.04
East Bay Garbage Co. v. Washington Township Sanitation Co., 52 Cal. 2d 708 (1959)	2.02[A]

Case	Book §
East River S.S. Corp. v. Transamerica Delaval, Inc., 476 U.S. 858 (1986)	8.07
E.C. Ernst, Inc. v. County of Contra Costa, 555 F. Supp. 122 (N.D. Cal. 1982)	1.05, 1.06[B]
E.C. Goldman, Inc. v. A/R/C Assocs., 543 So. 2d 1268 (Fla. Dist. Ct. App. 1989)	8.07
Ecker Bros. v. Jones, 186 Cal. App. 2d 775 (1960)	9.01[A][5], 9.01[A][6]
Ed Goetz Painting Co., Appeal of, DOTBCA No. 1168, 83-1 B.C.A. (CCH) ¶ 16,134 (1982)	4.15
E.D. McGillicuddy Constr. Co. v. Knoll Recreation Ass'n, Inc., 31 Cal. App. 3d 891 (1973)	9.04[E], 9.06[F]
Educational & Recreational Servs., Inc. v. Pasadena Unified Sch. Dist., 65 Cal. App. 3d 775 (1977)	2.02[F][1], 2.02[G][1]
E.F. Brady Co. v. M.H. Golden Co., 58 Cal. App. 4th 182 (1997)	2.02[D][2]
E.H. Morrill Co. v. State, 65 Cal. 2d 787 (1967)	3.02[B][12], 4.07[C], 6.11
Eichleay Corp., 60-2 B.C.A. (CCH) ¶ 2,688 (ASBCA 1960)	6.07[G]
E.J. Albrecht Co. v. New Amsterdam Cas. Co., 163 F.2d 16 (7th Cir. 1947)	4.13
Electric Supply Distrib. Co. v. Imperial Hot Mineral Spa, 122 Cal. App. 3d 131 (1981)	9.03[D]
Electrical Constr. & Maintenance Co. v. Maeda Pac. Corp., 764 F.2d 619 (9th Cir. 1985)	2.01[C]
Electronic & Missile Facilities, Inc. v. United States, 416 F.2d 1345 (Ct. Cl. 1969)	6.08[B]
Elsinore Union Sch. Dist. v. Kastorf, 54 Cal. 2d 380 (1960)	2.02[E]
Engineering Contractors Ass'n v. Metropolitan Dade County, 122 F.3d 895 (11th Cir. 1997)	2.02[F][4][b]
English v. Olympic Auditorium, Inc., 217 Cal. 631 (1933)	9.01[A][5], 9.01[A][6], 9.04[A], 9.06[F]
E.R. Fegert, Inc. v. Coral Constr., Inc. (*In re* E.R. Fegert, Inc.) 88 B.R. 258 (Bankr. 9th Cir. 1988), *aff'd sub nom.* O'Rourke v. Seaboard Sur. Co., 887 F.2d 955 (9th Cir. 1989)	9.06[B]
E.W. Eldridge, Inc., ENGBCA No. 5269, 90-3 B.C.A. (CCH) ¶ 23,080 (1990)	6.14[A]
Ewing Irrigation Prods. v. Rohnert Park Golf Course, 20 Cal. App. 3d 862 (1973)	9.05[A]
Fairlane Estates, Inc. v. Carrico Constr. Co., 228 Cal. App. 2d 65 (1964)	6.04[A]
FAJ, Inc. v. Surety Co. of Pac., 68 Cal. App. 3d Supp. 20 (1977)	1.09[E]
Familian Corp. v. Imperial Bank, 213 Cal. App. 3d 681 (1989)	9.01[B][2][a]
Fanderlik-Locke Co. v. United States *ex rel.* M.B. Morgan, 285 F.2d 939 (10th Cir. 1960)	11.07[B]
Farnsworth & Chambers, Inc. v. United States, 346 F.2d 577, 180 Ct. Cl. 992 (1965)	6.11
Federal Ins. Co. v. Superior Court, 60 Cal. App. 4th 1370 (1998)	10.04[A]
Feldman Constr. Co. v. Union Bank, 28 Cal. App. 3d 731 (1972)	9.06[L]
Fell v. Masseroff, 145 So. 2d 238 (Fla. Dist. Ct. App. 1962)	9.06[C][1]
Ferry v. Ohio Farmers Ins. Co., 211 Cal. App. 2d 651 (1963)	9.06[A]
Fidelity v. Parsons & Whittemore Constr., 397 N.E.2d 380, 421 N.Y.S.2d 869 (1979)	14.02[C]

TABLE OF CASES

Case	Book §
Fidelity Sound Sys., Inc. v. American Bonding Co., 85 Cal. App. 3d 13 (1978)	9.02[B][7], 10.04[A]
Finley Gordon Carpet Co. v. Bay Shore Homes, Inc., 247 Cal. App. 2d 131 (1966)	9.06[C][1]
Firestone Tire & Rubber Co. v. United States, 444 F.2d 547 (Ct. Cl. 1971)	4.07[D]
First Atl. Bldg. Corp. v. Neubar Constr. Co., 352 So. 2d 103 (Fla. Dist. Ct. App. 1977)	9.06[T]
First Options of Chicago, Inc. v. Kaplan, 514 U.S. 938, 115 S. Ct. 1920, 131 L. Ed. 2d 985 (1995)	14.02[C]
Fischbach & Moore Int'l Corp., ASBCA No. 18146, 77-1 B.C.A. (CCH) ¶ 12,300	7.04
Fishman Constr. Co. v. Hansen, 209 A.2d 605 (Md. 1965)	1.09[C]
Fleck v. Bollinger Home Corp., 54 Cal. App. 4th 926 (1997)	5.07[D]
Fleisher Eng'g & Constr. Co. v. United States *ex rel.* Hallenbeck, 311 U.S. 15 (1940)	11.02
Flickinger v. Swedlow Eng'g Co., 45 Cal. App. 2d 388 (1995)	10.04[C]
Flintkote Co. v. Lisa Constr. Co., 268 Cal. App. 2d 606 (1968)	9.04[D]
Flintkote Co. v. Presley of N. Cal., 154 Cal. App. 3d 458 (1984)	9.01[B][2][e], 9.03[G]
Flooring Concepts, Inc., *In re,* 37 B.R. 957 (Bankr. 9th Cir. 1984)	9.06[B], 12.09
Fluor, Utah, Inc., Appeal of, IBCA No. 1068-4-75, 81-1 B.C.A. (CCH) ¶ 14,876	6.08[B]
Fontana Paving, Inc. v. Hedley Bros., Inc., 38 Cal. App. 4th 146 (1995)	9.03[A][1]
Foo Catering & Hous., Inc., 971 F.2d 396 (9th Cir. 1992)	9.06[B]
Foremost Pro Color, Inc. v. Eastman Kodak Co., 703 F.2d 534 (9th Cir. 1983)	2.01[D]
Fortec Constructors v. United States, 8 Cl. Ct. 490 (1985)	7.02, 7.03
Foster Constr. C.A. & Williams Bros. Co. v. United States, 435 F.2d 873, 193 Ct. Cl. 587 (1970)	6.11
4-Star Constr. Corp. v. United States, 3 F.P.D. 46 (Cl. Ct. 1984)	11.06
Frank v. Kozlovsky, 13 Cal. App. 3d 120 (1970)	1.09[F], 9.01[A][4]
Frank Curran Lumber Co. v. Eleven Co., 271 Cal. App. 2d 175 (1969)	9.01[A], 9.06[S]
Frank M. Booth, Inc. v. Reynolds Metals Co., 754 F. Supp. 1441 (E.D. Cal. 1991)	3.04
Frank Pisano & Assocs. v. Taggart, 29 Cal. App. 3d 1 (1972)	9.02[B][7], 9.03[C][2], 9.06[N], 9.06[O]
Frank T. Hickey, Inc. v. L.A.J.C. Council, 128 Cal. App. 2d 676 (1954)	9.06[G]
Fraters Glass & Paint Co. v. Southwestern Constr. Co., 200 Cal. 688 (1927)	9.06[A]
Frazier-Fleming Co., ASBCA No. 34537, 91-1 B.C.A. (CCH) ¶ 23,378 (1990)	4.13
Fred A. Arnold, Inc., ASBCA No. 18915, 75-2 B.C.A. (CCH) ¶ 11,496	6.08[B]
Freeman v. State Farm Mut. Auto. Ins. Co., 14 Cal. 3d 473 (1975)	14.02[C]
Freeman & Mills, Inc. v. Belcher Oil Co., 11 Cal. 4th 85 (1995)	8.09
Freund v. United States, 260 U.S. 60 (1922)	6.10[A]
Fulghum Constr., *In re,* 23 B.R. 147 (Bankr. M.D. Tenn. 1982)	12.07
Fuller Co. v. Brown Minneapolis Tank & Fabricating Co., 678 F. Supp. 506 (E.D. Pa. 1987)	9.06[T]

Case	Book §
Fullerton Constr. Co., ASBCA No. 11500, 67-2 B.C.A. (CCH) ¶ 6394	6.08[B]
Gaines v. Eastern Pac., 136 Cal. App. 3d 679 (1982)	1.05, 1.06[B]
Garden Grove Community Church v. Pittsburgh Des Moines Steel Co., 140 Cal. App. 3d 251 (1983)	14.02[D]
Gardner Displays Co. v. United States, 346 F.2d 585, 171 Ct. Cl. 497 (1965)	4.13
Gartell Constr., Inc. v. Aubry, 940 F.2d 437 (9th Cir. 1991)	1.06[C]
Gary C. Tanko Well Drilling, Inc. v. Dodds, 117 Cal. App. 3d 588 (1981)	9.03[A][1]
Gateway Erectors v. Lutheran Gen. Hosp., 430 N.E.2d 20 (Ill. App. Ct. 1981)	4.14
General Elec. Co. v. Central Sur. & Ins. Corp., 232 Cal. App. 2d 590 (1965)	9.03[H][3], 10.04[A]
General Ins. Co. v. St. Paul Fire & Marine Ins. Co., 38 Cal. App. 3d 760 (1974)	1.24
General Ins. Co. of Am. v. Commerce Hyatt House, 5 Cal. App. 3d 460 (1970)	6.02[B], 6.07[E]
General Ins. Co. of Am. v. Mammoth Vista Owners Ass'n, Inc., 174 Cal. App. 3d 810 (1985)	9.03[H][3]
Gevyn Constr. Co., Appeal of, ENGBCA No. 3031, 83-1 B.C.A. (CCH) ¶ 16,428 (1983)	4.15
Gevyn Constr. Corp. v. United States, 827 F.2d 752 (Fed. Cir. 1987)	4.15
Ghilotti Constr. Co. v. City of Richmond, 45 Cal. App. 4th 897 (1996), *review denied*	2.02[D][2], 2.02[F][1], 2.02[G][3]
Gilbane Bldg. Co. v. Brish Waterproofing Co., 585 A.2d 248 (Md. 1991)	1.09[C]
Gilbert Fin. Corp. v. Steel Form Contracting Co., 82 Cal. App. 3d 65 (1978)	8.05, 8.07
Glendale Fed. Sav. & Loan Ass'n v. Marina View Heights Dev. Co., 66 Cal. App. 3d 101 (1977)	5.01[B]
Glenn Falls Indem. Co. v. United States *ex rel.* Westinghouse Elec. Supply Co., 229 F.2d 270 (9th Cir. 1955)	11.05
G.M. Shupe, Inc. v. United States, 5 Ct. Cl. 662, 3 F.P.D. ¶ 11 (1984)	4.13
Gogo v. Los Angeles County Flood Control Dist., 45 Cal. App. 2d 334 (1941)	2.01[A], 4.06, 6.02[B]
Golden Eagle Ins. Co. v. First Nationwide Fin. Corp., 26 Cal. App. 4th 160, 31 Cal. Rptr. 2d 815 (1994)	9.01[A][2]
Gonzales v. Covered Gardens Mobile Home Park, 90 Cal. App. 3d 871 (1979)	13.03
Gonzalez v. R.J. Novick Constr. Co., 20 Cal. 3d 798 (1978)	3.02[B][7]
Gordy v. United Pac. Ins. Group, 243 Cal. App. 2d 445 (1966)	9.03[H][3]
G.R. Smonagle & Sons, Inc. v. McKnight Constr. Co., 304 A.2d 339 (Del. 1973)	9.06[T]
Grade-Way Constr. Co. v. Golden Eagle Ins. Co., 13 Cal. App. 4th 826 (1993)	9.06[S]
Grand Chevrolet, Inc., *In re*, 25 F.3d 728 (9th Cir. 1994)	9.06[B]
Granite Constr. Co. v. American Motorists Ins. Co., 29 Cal. App. 4th 658 (1994)	10.04[B]
Gray v. Hickey, 94 Wash. 370 (1970)	9.06[T]
Graydon v. Pasadena Redevelopment Agency, 104 Cal. App. 3d 631 (1980), *cert. denied,* 449 U.S. 983, *reh'g denied,* 449 U.S. 1104 (1981)	2.02[A]
Green v. Clifford, 94 Cal. 49 (1892)	9.06[R]

TABLE OF CASES

Case	Book §
Green Builders, Inc., ASBCA No. 35518, 88-2 B.C.A. (CCH) ¶ 20,734 (1988)	4.13
Gribaldo, Jacobs, Jones & Assocs. v. Agrippina Versicherunges A.G., 3 Cal. 3d 434 (1970)	3.02[B][7], 3.03[C]
Grinell Fire Protection Sys. Co. v. American Sav. & Loan Ass'n, 183 Cal. App. 3d 352 (1986)	9.06[R]
Guardian Constr. Co. v. Tetra Tech Richardson, Inc., 583 A.2d 1378 (Del. Super. Ct. 1990)	8.07
Guinn v. Dotson, 23 Cal. App. 4th 262 (1994)	8.10
Gulf Oil Corp. v. Mobil Drilling Barge, 441 F. Supp. 1 (E.D. La. 1975)	3.02[B][13]
Haas & Haynie Corp., GSBCA Nos. 5530, 6224, 6638, 6919, 6920, 84-2 B.C.A. (CCH) ¶ 17,446	7.02
Halbert's Lumber v. Lucky Stores, Inc., 6 Cal. App. 4th 1233, 8 Cal. Rptr. 2d 298 (1992)	9.05[B], 9.05[B][2], 9.05[B][3], 9.05[B][4][a], 9.05[B][4][b], 9.05[B][5]
Halbert's Lumber, Inc. v. Burdett, 202 Cal. App. 3d 14 (1988)	9.06[K]
Haldeman v. Boise Cascade, 176 Cal. App. 3d 230 (1985)	14.05
Halspar, Inc. v. Barthe, 238 Cal. App. 2d 897 (1965)	9.02[B][7]
Hammond Lumber Co. v. Barth Inv. Corp., 202 Cal. 606 (1927)	9.03[E], 9.06[F]
Hammond Lumber Co. v. Moore, 104 Cal. App. 528 (1930)	9.06[F]
Hammond Lumber Co. v. Yeager, 185 Cal. 355 (1921)	9.03[A], 9.03[E]
Hanberry Corp. v. Mississippi State Bldg. Comm'n, 390 So. 2d 277 (Miss. 1980)	4.14
Hannah v. Steinman, 159 Cal. 142 (1911)	3.02[B][12]
Hardeman-Monier-Hutcheson, ASBCA No. 104444, 67-1 B.C.A. (CCH) ¶ 6158	7.04
Harold L. James, Inc. v. Five Points Ranch, Inc., 158 Cal. App. 3d 1 (1984)	9.02[B][4], 9.06[F]
Harry H. White Lumber Co. v. Crocker-Citizens Nat'l Bank, 253 Cal. App. 2d 368 (1967)	9.06[L]
Harsco Corp. v. Department of Pub. Works, 21 Cal. App. 3d 272 (1971)	10.04[A]
Harvis Constr., Inc. v. United States *ex rel.* Martin Constructors, Inc., 474 U.S. 817 (1985)	11.05
Hawley v. Orange County Flood Control Dist., 211 Cal. App. 2d 708 (1963)	3.02[B][11], 6.07[H], 6.08[A]
Hazard Gould & Co. v. Rosenberg, 177 Cal. 295 (1918)	9.06[R]
Healy v. Brewster, 59 Cal. 2d 455 (1963), *remanded,* 251 Cal. App. 2d 541 (1967)	9.06[G]
Hendrickson v. Bertelson, 1 Cal. 2d 430 (1934)	9.04[D], 9.06[F]
Heninger v. Dunn, 101 Cal. App. 3d 858 (1990)	5.01[A]
Henry v. Alcove Inv., Inc., 233 Cal. App. 3d 93 (1991)	14.02[F]
Henry Ericsson Co. v. United States, 62 F. Supp. 312 (Ct. Cl. 1945)	4.13
Hensler v. City of L.A., 124 Cal. App. 2d 71 (1954)	3.02[B][12]
Hercules, Inc. v. United States, 516 U.S. 417, 116 S. Ct. 981 (1996)	4.06

Case	Book §
Hernandez v. Badger Constr. Equip. Co., 28 Cal. App. 4th 1791 (1994)	3.03[C]
Herrick Corp. v. Canadian Ins. Co., 29 Cal. App. 4th 753 (1994)	3.02[B][7]
H.G. Fenton Material Co. v. Noble, 127 Cal. App. 338 (1932)	9.04[G]
Hiller v. City of L.A., 197 Cal. App. 2d 685 (1961)	2.02[B][2]
H&L Supply, Inc. v. Ewing, 253 Cal. App. 2d 283 (1967)	9.03[C][2]
H.O. Bragg Roofing, Inc. v. First Fed. Sav. & Loan Ass'n, 226 Cal. App. 2d 24 (1964)	9.01[B][2], 10.04[A]
Hobbs Lumber Co. v. Shidell, 326 N.E.2d 706 (Ohio C.P. 1974)	13.05
Hoffman v. United States, 340 F.2d 645, 166 Ct. Cl. 39 (1964)	6.11
Hogan Lumber Co. v. Boyk, 77 Cal. 477 (1918)	9.03[E]
Hol-Gar Mfg. Corp. v. United States, 351 F.2d 972 (Ct. Cl. 1965)	4.07[D]
Hollenbeck-Bush Planing Mill Co. v. Roman Catholic Bishop, 179 Cal. 229 (1918)	9.03[C][1]
Hollywood Wholesale Elec. Co. v. Jack Baskin, Inc., 146 Cal. App. 2d 399 (1956)	9.05[A]
Hollywood Wholesale Elec. Co. v. Jack Baskin, Inc., 121 Cal. App. 2d 415 (1953)	9.05[A], 9.05[F]
Home Depot USA, Inc. v. Contractors State License Bd., 41 Cal. App. 4th 1592 (1996)	1.06[B][2]
Homestead Sav. & Loan Ass'n v. Superior Court of Marin County, 195 Cal. App. 2d 697 (1961)	14.02[E]
Hood v. Gordy Homes, Inc., 267 F.2d 882 (4th Cir. 1959)	1.09[C]
Hood Plumbing, AGBCA No. 84-181-1 (1987)	7.05
Housing Auth. v. E.W. Johnson Constr. Co., 573 S.W.2d 316 (Ark. 1978)	4.13
Howard A. Deson & Co. v. Costa Tierra, Ltd., 2 Cal. App. 3d 742 (1969)	9.03[A], 9.03[C][1]
Howard Contracting, Inc. v. G.A. MacDonald Constr. Co., 71 Cal. App. 4th 38 (1999)	4.06, 4.07[A], 4.07[C], 4.13, 7.02
Howard-Green Elec. Co. v. Chaney & James Constr. Co., 182 S.E.2d 601 (N.C. Ct. App. 1971)	1.09[C]
Howell v. Gunderson, 250 Cal. App. 2d 961 (1967)	9.03[A][1]
Huang v. Garner, 157 Cal. App. 3d 404 (1984)	5.08
Hubbard v. Jurian, 47 Cal. App. 543 (1920)	9.03[E]
Huber, Hunt & Nichols, Inc. v. Moore, 67 Cal. App. 3d 278 (1977)	4.10, 6.14[A]
Humboldt Lumber Mill Co. v. Crisp, 146 Cal. 686 (1905)	9.06[A]
Hunters Run Ltd. Partnership, In re (Hand v. Hunters Run Ltd. Partnership) (9th Cir. 1989)	875 F.2d 1425 12.08
Hutnick v. United States Fidelity & Guar. Co., 199 Cal. App. 3d 49, rev'd, 47 Cal. 3d 456 (1988)	9.06[S]
H.W. Stanfield Constr. Corp. v. Robert McMullan & Sons, 14 Cal. App. 3d 848 (1971)	2.01[C]
Hydrotech Sys. Ltd. v. Oasis Water Park, 52 Cal. 3d 988 (1991)	1.01, 1.05, 1.06[B][2]
Idaco Lumber Co. v. Northwestern Sav. & Loan Ass'n, 265 Cal. App. 2d 490 (1968)	9.01[B][2], 10.04[A]
IGA Aluminum Prods., Inc. v. Manufacturers Bank, 130 Cal. App. 3d 699 (1982)	9.02[B][3]
Illinois Structural Steel Corp. v. Pathman Constr. Co., 23 Ill. App. 3d 1, 318 N.E.2d 232 (1974)	7.03
Indiana Plumbing-Supply Co. v. Bank of Am., 255 Cal. App. 2d 910 (1967)	9.06[L]

TABLE OF CASES

Case	*Book §*
Industrial Asphalt, Inc. v. Garrett Corp., 180 Cal. App. 3d 1001 (1986)	9.02[B][5], 9.06[F]
Inglewood-L.A. County Civic Ctr. Auth., City of, v. Superior Court of L.A. County, 7 Cal. 3d 861 (1972)	2.02[B][1], 2.02[C][2], 2.02[F][2], 2.02[G][1]
Inland Casino Corp. v. Superior Court, 8 Cal. App. 4th 770 (1992)	9.01[A][5]
Insul-Acoustics, Inc. v. Sung Soo Lee, 136 Cal. App. 3d 552 (1982)	9.03[A][1]
Integrated, Inc. v. Alec Fergusson Elec. Contractor, 250 Cal. App. 2d 287 (1967)	4.02
Interior Sys., Inc. v. Del. E. Webb Corp., 121 Cal. App. 3d 312 (1981)	2.02[D][2]
Interstate Gen. Gov't Contractors, Inc. v. United States, 12 F.3d 1053 (Fed. Cir. 1993)	6.07[G]
IRFM, Inc., *In re,* 52 F.3d 228 (9th Cir. 1995)	9.06[B]
J.A. Jones Constr. Co. v. Greenbriar Shopping Ctr., 332 F. Supp. 1336 (N.D. Ga. 1971), *aff'd,* 461 F.2d 1269 (5th Cir. 1972)	6.07[E], 7.05
J.A. Jones Constr. Co. v. Superior Court, 27 Cal. App. 4th 1568, 33 Cal. Rptr. 2d 206 (4th Dist. 1994)	9.05[B][3]
J.A. Thompson Corp. v. D.C. Contractors, 4 Cal. App. 4th 1355 (1992)	9.03[G][4]
Jackson v. Pancake, 266 Cal. App. 2d 307 (1968)	1.06[B][1], 9.01[A][4]
J'Aire Corp. v. Gregory, 24 Cal. 3d 799 (1979)	3.04, 8.06, 8.07
Jasper Constr., Inc. v. Foothill Junior College Dist., 91 Cal. App. 3d 1 (1979), *reh'g denied*	2.01[A], 4.06
Jay Bailey Constr. Co. v. Berry Hotel Corp., 221 Cal. App. 2d 135 (1963)	9.01[A][5], 9.01[A][6]
J.B.L. Constr. Co., Appeal of, VABCA No. 1799, 86-1 B.C.A. ¶ 18,529 (1985)	7.05
J.F., Inc. v. S.M. Wilson & Co., 504 N.E.2d 1266 (Ill. App. Ct. 1987)	4.14
Jimenez v. Pacific W. Constr. Co., 185 Cal. App. 3d 102 (1986)	3.02[B][7]
J.L. Simmons Co. v. United States, 412 F.2d 1360 (Ct. Cl. 1969)	4.07[A]
Joanaco Projects v. Nixon & Tierney Constr. Co., 248 Cal. App. 2d 821 (1967)	6.08
John A. Artukovich Sons, Inc. v. American Fidelity Fire Ins. Co., 72 Cal. App. 3d 940 (1977)	10.04[A]
John F. Burke Eng'g & Constr., Appeal of, ASBCA No. 8182, 1963 B.C.A. (CCH) ¶ 3,713 (Mar. 29, 1963)	4.13
John M. McShain, Inc. v. United States, 462 F.2d 489 (Ct. Cl. 1972)	4.07[D]
John Murphy Constr. Co., AGBCA No. 418, 79-1 B.C.A. (CCH) ¶ 73,836	7.04
John Reiner & Co. v. United States, 325 F.2d 438 (Ct. Cl. 1963), *cert. denied,* 377 U.S. 931 (1964)	2.02[B][2]
Johnson v. Mattox, 257 Cal. App. 2d 714 (1968)	1.06[B][1]
Johnson v. Smith, 97 Cal. App. 752 (1929)	9.06[A]
Johnston v. Groom, 99 Cal. App. 462 (1929)	9.05[A]
Jones v. Great S. Fireproof Hotel Co., 86 F. 370 (1898)	9.06[F]
Jones v. Kvistad, 19 Cal. App. 3d 836 (1971)	5.01[A], 6.02[C]
Jones v. Pullock, 34 Cal. 2d 863 (1950)	14.02[E]
Joseph Bell v. United States, 404 F.2d 975 (Ct. Cl. 1968)	4.15

Case	*Book §*
Joseph Pickard's Sons Co. v. United States, 532 F.2d 739, 209 Ct. Cl. 643 (1976)	6.14[B]
J.S. Schirm Co. v. Rollingwood Homes Co., 56 Cal. App. 2d 789 (1961)	9.05[F]
Judson Pacific-Murphy v. Durkee, 144 Cal. App. 2d 377 (1956)	2.02[G][1]
J.W. Bateson Co., ASBCA No. 27491, 84-3 B.C.A. (CCH) ¶ 17,566 (1984)	7.02
J.W. Bateson Co., Appeal of, VACAB No. 1148, 86-1 B.C.A. (CCH) ¶ 18,585 (1985)	4.15
Kajima/Ray Wilson v. Los Angeles County Metro. Transp. Auth., 69 Cal. App. 4th 1458 (1999)	2.02[G][2]
Kalfountzos v. Hartford Fire Ins. Co., 37 Cal. App. 4th 1655 (1995)	10.04[C]
Kalico, Inc. v. Dooley, 198 Cal. App. 2d 379 (1961)	10.04[A], 10.04[B]
KDR Bldg. Specialties, Inc., In re, 76 B.R. 778 (Bankr. S.D. Cal. 1987)	9.06[B]
Keating v. Superior Court, 31 Cal. 3d 584 (1982)	14.02[C], 14.02[D]
Keenan Pipe & Supply Co. v. Shields, 241 F.2d 486 (9th Cir. 1956)	9.06[B]
Keller Constr. Co. v. Kashani, 220 Cal. App. 3d 222 (1990)	14.02[G]
Keller Constr. Corp. v. George W. McCoy & Co., 119 So. 2d 450 (La. 1960)	4.07[A]
Kennedy v. Reece, 225 Cal. App. 2d 717 (1964)	3.02[B][12], 6.02[C]
Kenneth Reed Constr. Corp., ENGBCA Nos. 2748 et al., 72-1 B.C.A. (CCH) ¶ 9407	7.04
Kertscher & Co. v. Green, 205 N.Y. 522 (1911)	9.06[T]
K&F Constr. v. Los Angeles City Unified Sch. Dist., 123 Cal. App. 3d 1063 (1981)	3.02[B][11], 6.07[H]
Kiely Corp. v. H.C. Gibson, 231 Cal. App. 2d 39 (1964)	2.02[D][2]
Kim v. JF Enters., 42 Cal. App. 4th 849 (1996)	9.02[B][7]
King v. Hinderstein, 122 Cal. App. 3d 430 (1981)	1.06[B][1]
K&K Servs., Inc. v. City of Irwindale, 47 Cal. App. 4th 818 (1996)	1.05
Klan v. Hoffman, 176 Cal. 763 (1917)	9.03[E]
Klose v. Sequoia Union High Sch. Dist., 118 Cal. App. 2d 636 (1953)	2.01[B]
Knapp Dev. & Design v. Pal-Mal Properties, Ltd., 173 Cal. App. 3d 423 (1985)	1.06[B][2]
Kodiak Indus., Inc. v. Ellis, 185 Cal. App. 3d 75 (1986)	9.02[B][6], 9.02[B][7]
Kokomo F&W Traction Co. v. Kokomo Trust Co., 193 Ind. 219 (1923)	9.06[T]
Konica Bus. Machs. U.S.A., Inc. v. Regents of Univ. of Cal., 206 Cal. App. 3d 449 (1988)	2.02[F][3]
Korbel v. Sheung-Chi Chou, 26 Cal. App. 4th 1427 (1994)	8.10
Korherr v. Bumb, 262 F.2d 157 (9th Cir. 1958)	9.01[B][2]
Koudmani v. Ogle Enters., Inc., 47 Cal. App. 4th 1650, 55 Cal. Rptr. 2d 330 (1996)	9.03[D][1]
Kriegler v. Eichler Homes, Inc., 269 Cal. App. 2d 244 (1969)	8.02
Kritzer v. Tracy Eng'g Co., 16 Cal. App. 287 (1911)	9.04[D]
Kroeger v. Franchise Equities, Inc., 490 Neb. 731, 212 N.W.2d 348 (1973)	7.03
Kruger Bros. Builders, Inc. v. San Francisco Hous. Auth., 198 Cal. App. 3d 1 (1988)	10.04[A]
Kurland v. United Pac. Ins. Co., 251 Cal. App. 2d 112 (1967)	4.07[B], 6.02[C]

TABLE OF CASES

Case	Book §
L. Rosenman Corp. v. United States, 390 F.2d 711 (Ct. Cl. 1968)	4.07[D]
La Jolla Village Homeowners Ass'n v. Superior Court, 212 Cal. App. 3d 1131 (1989)	8.02
L.A.M. Constr., Inc. v. Kriz, 14 Cal. App. 4th Supp. 1 (1993)	9.06[K]
Lambert v. Superior Court, 228 Cal. App. 3d 383 (1991)	9.06[H]
Lambert Steel Co. v. Heller Fin. Inc., 16 Cal. App. 4th 1034 (1993)	9.04[A]
Larkin v. Cowert, 263 Cal. App. 2d 27 (1968)	9.06[C][1]
Latipac, Inc. v. Superior Court, 64 Cal. 2d 278 (1966)	1.06[B][2]
Law v. United States, 195 Ct. Cl. 370 (1971)	6.08[A]
Lawrence v. Walzer & Gabrielson, 207 Cal. App. 3d 1501 (1989)	14.02[C]
Leaf v. City of San Mateo, 104 Cal. App. 3d 398 (1980)	5.07[A]
Leatherby Ins. Co. v. Tustin, 76 Cal. App. 3d 678 (1977)	10.04[A]
Lee Hoffman v. United States, 340 F.2d 645, 166 Ct. Cl. 39 (1964)	6.11
Lemorge Elec. v. County of San Mateo, 46 Cal. 2d 659 (1956)	2.02[E]
Len Co. & Assocs. v. United States, 385 F.2d 438, 181 Ct. Cl. 29 (1967)	6.10[B]
Leslie Miller, Inc. v. Arkansas, 352 U.S. 187, 77 S. Ct. 257 (1956)	1.06[C]
Levy v. Superior Court, 10 Cal. 4th 578 (1995)	14.05
Lewis v. Hopper, 140 Cal. App. 2d 365 (1956)	9.03[E]
Lewis-Nicholson, Inc. v. United States, 550 F.2d 26, 213 Ct. Cl. 192 (1977)	6.08[B]
Lewis & Queen v. N.M. Ball Sons, 48 Cal. 2d 141 (1957)	1.05, 10.04[C]
Linda Jones Gen. Builder v. Contractors State License Bd., 194 Cal. App. 3d 1320 (1987)	1.09[D]
Link, Conservatorship of, 158 Cal. App. 3d 138 (1984)	3.05[A]
Litton Constr. Gen. Eng'g Contractor, Inc. v. United Pac. Ins., 16 Cal. App. 4th 557, 20 Cal. Rptr. 2d 200 (1993)	10.04[A], 10.04[B]
Lloyd H. Kessler, Inc., AGBCA No. 88-170-3, 91-2 B.C.A. (CCH) ¶ 23,802 (1991)	4.13
Loehr v. Great Republic Ins. Co., 226 Cal. App. 3d 727 (1990)	3.02[B][13]
Loral Terracom v. Valley Nat'l Bank, 49 F.3d 555 (9th Cir. 1995)	11.07[A]
Los Angeles Dredging Co. v. Long Beach, 210 Cal. 3d 348 (1930)	2.02[B][2]
Los Banos Gravel Co. v. Freeman, 58 Cal. App. 3d 785 (1976)	9.01[A][5], 9.01[A][6]
Louis Luskin & Sons, Inc. v. Samovitz, 166 Cal. App. 3d 533 (1985)	13.05
Lowy v. United Pac. Ins. Co., 67 Cal. 2d 87, 429 P.2d 577, 60 Cal. Rptr. 225 (1967)	5.01[B]
M. Arthur Gensler Jr. & Assocs., Inc. v. Larry Barrett, Inc., 7 Cal. 3d 695 (1972)	9.02[B][7]
M.A. Lombard & Son Co. v. Public Bldg. Comm'n of Chicago, 428 N.E.2d 889 (Ill. App. Ct. 1981)	4.14
MacIsaac & Menke Co. v. Cardon Corp., 193 Cal. App. 2d 661 (1961)	6.10[B]
Macomber v. State, 250 Cal. App. 2d 391 (1967)	4.07[A]
Maddy v. Castle, 58 Cal. App. 3d 716 (1976)	14.02[E]
Mai Steel Serv., Inc. v. Blake Constr. Co., 1992 U.S. App. LEXIS 14719, 92 Daily Journal D.A.R. 9116 (9th Cir. 1992)	11.05
Malouf Bros. v. Dixon, 230 Cal. App. 3d 280 (1991)	14.05
Manos v. Degen, 203 Cal. App. 3d 1237 (1988)	9.03[G]
Marin Mun. Water Dist. v. Peninsula Paving Co., 34 Cal. App. 2d 647 (1939)	4.07[B]
Maris Management Corp. v. Assured Drywall & Texture, 152 Cal. App. 3d 268 (1984)	9.03[D][1]

Case	Book §
Mark McDowell Corp. v. LSM, 214 Cal. App. 3d 1427 (1989)	9.06[J]
Markborough Cal., Inc. v. Superior Court, 227 Cal. App. 3d 705 (1991), *review denied*	3.05[A]
Markley v. Beagle, 66 Cal. 2d 951 (1967)	3.02[B][7]
Marshal v. Von Zumwalt, 120 Cal. App. 2d 807 (1953)	1.05, 1.06[B]
Martin v. Becker, 169 Cal. 301 (1915)	9.06[F]
Martin v. Mitchell Cement Contracting Co., 74 Cal. App. 3d 15 (1977)	9.01[A][4]
Martinez v. Traubner, 32 Cal. 3d 755 (1982)	5.07[A]
Maryland Dep't of Gen. Servs. v. Cherry Hill Sand & Gravel Co., 443 A.2d 628 (Md. 1982)	4.13
Mascioni v. I.B. Miller, Inc., 184 N.E. 473 (N.Y. 1933)	1.09[C]
Maurice L. Bein, Inc. v. Housing Auth., 157 Cal. App. 2d 670 (1958)	4.13
Maverick Diversified, Inc., ASBCA No. 19838, 76-2 B.C.A. (CCH) ¶ 12,104	6.11
Max Drill, Inc. v. United States, 427 F.2d 1233 (Ct. Cl. 1970)	4.07[D]
M&B Constr. v. Yuba County Water Agency, 68 Cal. App. 4th 1353 (1999)	1.04
McCarroll v. L.A. County, 49 Cal. 2d 45 (1957)	1.05
McCarty Corp. v. Pullman-Kellogg, 571 F. Supp. 1341 (M.D. La. 1983)	7.03
McConnell v. Corona City Water Co., 149 Cal. 60 (1906)	6.04[A]
McConnell v. Merrill Lynch, Pierce, Fenner & Smith, Inc., 105 Cal. App. 3d 946 (1980)	14.02[E]
McCutcheon-Peterson, IBCA No. 1392-9-80, 81-1 B.C.A. (CCH) ¶ 14,997	6.11
McDevitt & St. Co. v. Marriott Corp., 713 F. Supp. 906 (E.D. Va. 1989)	5.03
McDonald v. Filice, 252 Cal. App. 2d 613 (1967)	9.04[A]
McGuire & Hester v. City of S.F., 113 Cal. App. 186 (1952)	3.02[B][11]
MCM Constr., Inc. v. City & County of S.F., 66 Cal. App. 4th 359 (1998)	2.02[D][2], 2.02[F][1]
McMerit Constr. Co. v. Knightsbridge Dev. Co., 235 Va. 368, 367 S.E.2d 512 (1988)	9.06[T]
McMillan Bros. Constr. Inc., EBCA No. 328-10-84, 91-1 B.C.A. (CCH) ¶ 23,351	6.08[B]
McWilliams v. Holton, 248 Cal. App. 2d 447 (1967)	6.08
Mechanical Wholesale Corp. v. Fuji Bank, Ltd., 42 Cal. App. 4th 1647 (1996)	9.06[K]
Mega Constr. Co. v. United States, 29 Fed. Cl. 396 (1993)	7.02
Mendoyoma, Inc. v. County of Mendocino, 8 Cal. App. 3d 873 (1970)	6.02[A]
Menefee v. County of Fresno, 163 Cal. App. 3d 1175 (1985)	2.02[F][1]
Merchants & Mechanics Fed. Sav. & Loan v. Harold, 201 N.E.2d 237 (Ohio Ct. App. 1964)	9.06[C][1]
Metropolitan Paving Co. v. United States, 325 F.2d 241 (Ct. Cl. 1963)	4.13
Meyertech Corp., *In re*, 83 F.2d 410 (3d Cir. 1987)	6.14[A]
M.F. Kemper Constr. Co. v. City of L.A., 37 Cal. 2d 696 (1951)	2.02[E]
M.G.M. Constr. Corp. v. New Jersey Educ. Facilities Auth., 532 A.2d 764 (N.J. Super. Ct. Law Div. 1987)	4.14
Michel & Pfeffer v. Oceanside Properties, Inc., 61 Cal. App. 3d 433 (1976)	9.05[F]
Mickelson Concrete Co. v. Contractors State License Bd., 95 Cal. App. 3d 631 (1979)	1.09[D]
Mid Am. Homes, Inc. v. Eldon Horn, 377 N.E.2d 657 (Ind. Ct. App. 1978), *vacated,* 396 N.E.2d 879 (Ind. 1979)	13.05
Midstate Constructors, Inc., PSBCA No. 913, 81-1 B.C.A. (CCH) ¶ 14,898 (1991)	7.05

TABLE OF CASES

Case	*Book §*
Mignot v. Parkhill, 391 P.2d 755 (Or. 1964)	1.09[C]
Mike Moore's 24-Hour Towing v. City of San Diego, 145 Cal. App. 4th 1294 (1996)	2.02[G][2], 2.02[G][3]
Miller v. McKinnon, 20 Cal. 2d 83 (1942)	2.02[B][2], 2.02[G][2]
Miller v. Mountain View Sav. & Loan Ass'n, 238 Cal. App. 2d 644 (1965)	9.01[B][2], 9.01[B][2][b]
Miller Lee Corp., *In re,* 70 B.R. 780 (Bankr. S.D.N.Y. 1987)	12.08
Milovich v. City of L.A., 42 Cal. App. 2d 364 (1941)	3.02[B][11]
Milstein v. Security Pac. Nat'l Bank, 27 Cal. App. 3d 482 (1972)	6.08
Mineral Park Land Co. v. Howard, 172 Cal. 289 (1916)	3.02[B][12]
Minmar Builders, Inc., GSBCA No. 3430, 72-2 B.C.A. (CCH) ¶ 9599	7.02
Mobile Chem. Co. v. Blount Bros. Corp., 809 F.2d 1175 (5th Cir. 1987)	6.09[B]
Modesto Lumber Co. v. Wylde, 217 Cal. 421 (1933)	9.05[A]
Moncharsh v. Heily & Blase, 3 Cal. 4th 1 (1992)	14.02[B], 14.02[C]
Monte Vista Dev. Co. v. Superior Court (Wiley Tile Co.) 226 Cal. App. 3d 1681 (1991)	8.02
Monterey Mechanical Co. v. Sacramento Reg'l County Sanitation Dist., 44 Cal. App. 4th 1391 (1996)	2.02[F][1], 2.02[F][4][b]
Monterey Mechanical Co. v. Wilson, 125 F.3d 702 (9th Cir. 1997), *reh'g en banc denied,* 138 F.3d 1270 (9th Cir. 1998)	2.02[F][4][b]
Monterey S.P. Partnership v. W.L. Bangam, Inc., 49 Cal. 3d 454 (1989)	9.03[F]
Montgomery-Ross-Fisher, Inc., Appeal of, PSBCA Nos. 1033, 1096, 84-2 B.C.A. (CCH) ¶ 17,492 (1984)	4.13, 7.03
Moore Constr. Co. v. Clarksville, Dep't of Elec., 707 S.W.2d 1 (Tenn. Ct. App. 1986)	4.14
Morgan v. Stubblefield, 6 Cal. 3d 606 (1972)	3.02[B][7]
Morris v. Horton, 22 Cal. App. 4th 968 (1994)	5.08
Morrison-Knudsen Co. v. United States, 84 F. Supp. 282 (Ct. Cl. 1949)	4.13
Mortgages, Inc. v. U.S. Dist. Court, 934 F.2d 209 (9th Cir. 1991)	6.03
Moulding Brownell Corp. v. E.C. Defosse Constr. Co., 9 N.E.2d 459 (Ill. App. Ct. 1937)	9.06[T]
Mountain Home Contractors v. United States, 425 F.2d 1260 (Ct. Cl. 1970)	4.07[D]
Munger & Munger v. McBratney, 131 Cal. App. 2d Supp. 866 (1955)	9.03[A], 9.03[E]
Murphy v. A.A. Matthews, 841 S.W.2d 671 (Mo. 1992)	8.11
Murphy v. Padilla, 42 Cal. App. 4th 707 (1996)	14.04[C], 14.05
Mutual Constr., DOT CAB No. 1075, 80-2 B.C.A. (CCH) ¶ 14,630	6.11
Myers v. Alta Constr. Co., 37 Cal. 2d 739 (1951)	9.01[A][2]
Nathin & Co. v. George A. Fuller Co., 347 F. Supp. 17 (W.D. Mo. 1972)	7.03
National Identification Sys., Inc. v. State Bd. of Control, 11 Cal. App. 4th 1446 (1992)	2.02[F][1]
National Lumber & Supply, Inc. v. Orange Commercial Credit (*In re* National Lumber & Supply, Inc.) 184 B.R. 74 (Bankr. 9th Cir. 1995)	9.06[B]
Nelson v. Cohen, 160 Or. 336 (1938)	9.06[T]
Nelson Supply Co. v. Surety Co. of Pac., 161 Cal. App. 3d 490 (1984)	1.14
Nelson Valley Bldg. Co. v. Morrisey, 135 Cal. App. 2d 738 (1955)	1.09
Neptune Gunite Co. v. Monroe Enters., 229 Cal. App. 2d 439 (1964)	9.02[B][7]
Nevada County Lumber Co. v. Janiss, 25 Cal. App. 2d 579 (1938)	9.03[E]

Case	*Book §*
Newsom v. United States, 676 F.2d 647 (Ct. Cl. 1982)	4.07[D]
Nibbi Bros. v. Brannon St. Investors, 205 Cal. App. 3d 1415 (1988)	9.01[B][2][c], 9.01[B][2][d]
Nichols-Dynamics Co., ASBCA No. 17949, 75-2 B.C.A. (CCH) ¶ 11,556	6.08[B]
Nicholson v. Barab, 223 Cal. App. 3d 1671 (1991)	14.05
Nomellini Constr. Co. v. State *ex rel.* Department of Water Resources, 19 Cal. App. 3d 240 (1971)	6.02[B], 6.07[E]
Norcoast Constructors, Inc. v. United States, 488 F.2d 1400 (Ct. Cl. 1971)	4.07[D]
Norcross v. Winters, 209 Cal. App. 2d 207 (1962)	2.01[C]
Norfolk Dredging Co. v. United States, 360 F.2d 619, 175 Ct. Cl. 504, *cert. denied,* 385 U.S. 919 (1966)	6.11
North Coast Bus. Park v. Nielsen Constr. Co., 17 Cal. App. 4th 22 (4th Dist. 1993)	5.07[B]
Ogram v. Welchoff, 40 Cal. App. 298 (1919)	9.01[A][2]
Old Town Dev. Corp. v. Urban Renewal Agency, 249 Cal. App. 2d 313 (1967)	2.02[G][2], 2.02[G][3]
Operating Eng'rs Pension Trust v. Insurance Co. of the W., 35 Cal. App. 4th 59 (1995)	9.01[A][1]
Orndorff v. Christiana Community Builders, 217 Cal. App. 3d 683, 266 Cal. Rptr. 193 (1990)	5.01[A]
O'Rourke v. Seaboard Sur. Co., 887 F.2d 955 (9th Cir. 1989)	9.06[B]
Osborne v. Subaru of Am., Inc., 198 Cal. App. 3d 646 (1988)	8.02
Otis Elevator Co. v. Toda Constr., 27 Cal. App. 4th 559 (1994)	3.03[C]
Ott Hardware Co. v. Yost, 69 Cal. App. 2d 593 (1945)	9.01[A][5], 9.01[A][6]
Owen L. Schwam Constr. Co., ASBCA No. 22407, 79-2 B.C.A. (CCH) ¶ 13,919 (1979)	4.13
Pacific Employers Ins. Co. v. City of Berkeley, 158 Cal. App. 3d 145, 204 Cal. Rptr. 387 (1984)	5.02
Pacific Employers Ins. Co. v. State, 3 Cal. 3d 573 (1970)	10.04[A]
Pacific Inv. Co. v. Townsend, 58 Cal. App. 3d 1 (1976)	14.02[E]
Pacific Sash & Door Co. v. Bumillier, 162 Cal. 664 (1912)	9.06[A]
Pacific States Elec. Co. v. United States Fidelity & Guar. Co., 109 Cal. App. 691 (1930)	11.05
Packard Bell Elecs. Corp. v. Theseus, Inc., 244 Cal. App. 2d 355 (1966)	9.04[D], 9.06[R]
Palmer v. Lavigne, 104 Cal. 30 (1894)	9.01[A][2]
Pan Arctic Corp., ASBCA No. 20133, 77-1 B.C.A. (CCH) ¶ 12,514	6.08[B]
Panhandle Constr. Co., DOT CAB No. 77-31, 79-1 B.C.A. (CCH) ¶ 13,576	6.11
Parker v. Twentieth Century-Fox Film Corp., 118 Cal. App. 3d 895 (1981)	14.02[C]
Parsons Brinckerhoff Quade & Douglas, Inc. v. Kern County Employees Retirement Ass'n, 5 Cal. App. 4th 1264 (1992)	9.01[A][5]
Parvaneh Moradi Shalal v. Fireman's Fund Ins. Co., 46 Cal. 3d 287 (1988)	9.03[H][3]
Pasadena v. Charleville, 215 Cal. 384 (1932)	2.02[B][2]
Patrick J. Ruane, Inc. v. Parker, 185 Cal. App. 2d 488 (1960)	4.07[B]
Paul Potts Builders, Inc., *In re* (Grover v. County of Marin) (9th Cir. 1979)	608 F.2d 1279 9.03[D], 12.08

TABLE OF CASES

Case	Book §
Peacock Constr. Co. v. Modern Air Conditioning, Inc., 353 So. 2d 840 (Fla. 1977)	1.09[C]
People v. Custom Craft Carpets, Inc., 159 Cal. App. 2d 676 (1984)	9.06[C][1]
People v. Howard, 70 Cal. 2d 618 (1969)	9.06[E]
People v. Stark, 26 Cal. App. 4th 1179 (1994)	9.06[E]
Pepper Burns Insulation, Inc. v. Artco Corp., 970 F.2d 1340 (4th Cir. 1992)	11.02
Petaluma Bldg. Materials, Inc. v. Foremost Properties, Inc., 180 Cal. App. 2d 83 (1960)	9.05[F]
Peter & Burghard Stone v. Marion Nat'l Bank, 198 Ind. 581 (1926)	9.06[T]
Peter Kiewit & Sons Co. v. Iowa S. Util. Co., 355 F. Supp. 376 (S.D. Iowa 1973)	7.03
Petersen v. W.T. Grant Co., 41 Cal. App. 3d 217 (1974)	9.03[F], 9.06[I]
Phillips Constr. Co. v. Argonaut Ins. Co., 77 Cal. App. 3d 575 (1978)	9.05[F]
Phoenix Contractors, Inc. v. General Motors Corp., 355 N.W.2d 673 (Mich. Ct. App. 1984)	4.14
Pike v. Tuttle, 18 Cal. App. 3d 746 (1971)	9.04[A]
Piland Corp. v. League Constr. Co., 238 Va. 187, 38 S.E.2d 652 (1989)	5.06
Piledrivers Local Union No. 2375 v. City of Santa Monica, 151 Cal. App. 3d 509 (1984)	2.02[B][2], 2.02[F][3]
Pinney v. Phillips, 230 Cal. App. 3d 1570 (1991)	1.15
Pintail Plastering v. Mark Diversified Inc., 50 F.3d 15 (9th Cir. 1995)	1.06[B][2]
Piping Specialties Co. v. Kentile, Inc., 229 Cal. App. 2d 586 (1964)	9.01[A][4]
Platt Pac., Inc. v. Andelson, 6 Cal. 4th 307 (1993)	14.02[E]
Plough v. Petersen, 140 Cal. App. 2d 595 (1956)	9.06[C][1]
Plunkett v. Winchester, 98 Ark. 160 (1911)	9.06[T]
Pneucrete Co. v. U.S. Fidelity & Guar. Corp., 7 Cal. App. 2d 733 (1935)	10.01
Pollard v. Saxe & Yolles Dev. Co., 12 Cal. 3d 374 (1974)	8.04, 9.06[P]
Porier v. Desmond, 177 Mass. 201 (1900)	9.06[T]
Post Bros. Constr. Co. v. Yoder, 20 Cal. 3d 1 (1977)	9.05[F]
Powell Gen. Contracting Co., DOT CAB No. 1088, 80-2 B.C.A. (CCH) ¶ 14,680	6.11
Power City Constr. & Equip., Inc., IBCA No. 490-4-65, 68-2 B.C.A. (CCH) ¶ 7126	6.08[B]
Powers Regulator Co. v. Seaboard Sur. Co., 204 Cal. App. 2d 338 (1962)	9.06[A], 10.04[A]
Pozar v. Department of Transp., 145 Cal. App. 3d 269 (1983)	2.02[F][1]
Price v. Zim Israel Navigation Co., 616 F.2d 422 (9th Cir. 1980)	3.02[B][13]
Primo Team, Inc. v. Blake Constr. Co., 3 Cal. App. 4th 801 (1992)	10.04[A]
Pugh v. Moxley, 164 Cal. 374 (1912)	9.04[D]
R. Baker, Inc. v. Motel 6, Inc., 180 Cal. App. 3d 928 (1986)	9.03[D][2]
R.A. Glancy & Sons, Inc., VABCA No. 2327, 87-3 B.C.A. (CCH) ¶ 20,068	6.11
R.&A. Vending Serv., Inc. v. City of L.A., 172 Cal. App. 3d 1188 (1985)	2.02[F][2]
Ralph M. Parsons Co. v. Combustion Equip. Assocs., Inc., 172 Cal. App. 3d 211 (1985)	3.02[B][7]
Ranchwood Communities Ltd. Partnership v. Jim Beat Constr. Co., 49 Cal. App. 4th 1397 (1996)	1.05, 1.06[B]
Raymond v. Fresno City Unified Sch. Dist., 123 Cal. App. 2d 626 (1954)	2.02[F][2]
Raymond Constructors, Ltd. v. United States, 411 F.2d 1227, 188 Ct. Cl. 147 (1969)	6.07[E]

CALIFORNIA CONSTRUCTION LAW

Case	Book §
R.D. Reeder Lathing Co. v. Allen, 66 Cal. 2d 373 (1967)	9.06[Q]
Re-Bar Contractors, Inc. v. City of L.A., 219 Cal. App. 2d 134 (1963)	9.05[F]
R.E. Hazard, Jr. Enters., Inc. v. Insurance Co. of the W., 52 Cal. App. 4th 1088 (1997)	1.02
Reams v. Cooley, 171 Cal. 150 (1915)	2.02[B][2]
Red Alarm, Inc. v. Waycross, Inc., 47 F.3d 999 (9th Cir. 1995)	9.05[D]
Red Mountain Mach. Co. v. Grace Inv. Co., 29 F.3d 1408 (9th Cir. 1994)	9.01[A][5]
Reflectone, Inc. v. Dalton, 60 F.3d 1572 (Fed. Cir. 1995)	6.14[B]
Regan Roofing Co. v. Superior Court, 24 Cal. App. 4th 425 (1994), *review denied*	3.02[B][7], 3.03[C]
Regents of Univ. of Cal. v. Sumner, 42 Cal. App. 4th 1209 (1996)	14.03[D]
Regional Steel Corp. v. Superior Court, 25 Cal. App. 4th 525 (1994)	3.02[B][7], 3.03[C]
Reliance Enters., ASBCA No. 20808, 76-1 B.C.A. (CCH) ¶ 11,831	6.08[B]
Rexroth & Rexroth, Inc. v. General Cas. Co. of Am., 242 Cal. App. 2d 363 (1966)	9.03[H][3]
Rheem Mfg. Co. v. United States, 57 Cal. 2d 621 (1962)	9.04[A]
Rich-Lee Equip. Rentals, Inc. v. Intermountain Constr. Co., 79 Cal. App. 3d 581 (1978)	9.02[B][7]
Rich & Whillock, Inc. v. Ashton Dev., Inc., 157 Cal. App. 3d 1154 (1984)	9.05[E]
Richmond v. J.A. Croson Co., 488 U.S. 469 (1989)	2.02[F][4][b]
Rivera Contracting, Appeal of, ASBCA No. 28888 (1985)	7.05
Rob Glo, Inc., VABCA Nos. 2879, 2884, 91-1 B.C.A. (CCH) ¶ 23,357 (1990)	4.13
Robert E. McKee, Inc. v. City of Atlanta, 414 F. Supp. 957 (N.D. Ga. 1976)	4.06, 4.07[C]
Roberts v. Burleson, 284 S.W. 632 (Tex. Civ. App. 1926)	9.06[T]
Robertson v. Chen, 44 Cal. App. 4th 1290 (1996)	14.05
Robinson v. Diller, 274 Cal. App. 2d 813 (1969)	9.06[H]
Robinson v. S&S Dev., 256 Cal. App. 2d 13 (1967)	9.06[I]
Robinson v. United States, 261 U.S. 486 (1923)	6.02[B]
Rockwell v. Light, 6 Cal. App. 563 (1907)	9.03[E]
Rodeffer Indus., Inc. v. Chambers Estates, Inc., 263 Cal. App. 2d 116 (1968)	9.05[F]
Rodoni v. Harbor Eng'rs, 191 Cal. App. 2d 560 (1961)	1.01, 9.01[A][4]
Roen Salvage Co., ENGBCA No. 3670, 79-2 B.C.A. (CCH) ¶ 13,882	6.11
Romak Iron Works v. Prudential Ins. Co. of Am., 104 Cal. App. 3d 767 (1980)	9.02[B][1], 9.02[B][2], 9.02[B][6]
Ron Yates Constr. Co. v. Superior Court, 186 Cal. App. 3d 337 (1986)	1.03
Rossman Mill & Lumber Co. v. Fullerton Sav. & Loan Ass'n, 221 Cal. App. 2d 705 (1963)	9.01[B][2], 9.01[B][2][b], 10.04[A]
Rossmoor Sanitation, Inc. v. Pylon, Inc., 13 Cal. 3d 622 (1975)	3.02[B][7]
Roy Bros. Drilling Co. v. Jones, 123 Cal. App. 3d 175 (1981)	1.03
Royster Constr. Co. v. Urban W. Communities, 40 Cal. App. 4th 1158, 47 Cal. Rptr. 2d 684 (1995)	9.06[H]
Rubino v. Lolli, 10 Cal. App. 3d 1059 (1970)	2.02[G][2]
Rushing v. Powell, 61 Cal. App. 3d 597 (1976)	1.09[F], 9.01[A][4]
R.W. Contracting, Inc., ASBCA No. 24627, 84-2 B.C.A. (CCH) ¶ 17,302	7.02
Ryan v. Garcia, 27 Cal. App. 4th 1006 (1994)	14.03[D]
Rynsburger v. Dairymen's Fertilizer Corp., 266 Cal. App. 2d 269 (1968)	14.02[F]

TABLE OF CASES

Case	Book §
S. Leo Harmony, Inc. v. Binker Mfg. Co., 597 F. Supp. 1014 (S.D.N.Y. 1984)	7.03
Salem Eng'g & Constr. Corp. v. United States, 2 Cl. Ct. 808 (1983)	4.07[D]
Saliba-Kringlen Corp. v. Allen Eng'g Co., 15 Cal. App. 3d 95 (1971)	2.01[C]
Salinas, City of, v. Souza & McCue Constr. Co., 66 Cal. 2d 217 (1967)	4.06, 4.07[C]
San Diego, City of, v. United States Gypsum Co., 30 Cal. App. 4th 575 (1994)	5.07[B]
San Ore-Gardner v. Missouri Pac. R.R., 496 F. Supp. 1337 (E.D. Ark. 1980)	6.02[B]
San Pedro Lumber Co. v. Kreis, 111 Cal. App. 466 (1931)	9.04[G]
Sanchez v. Swinerton & Walberg Co., 47 Cal. App. 4th 1461 (1996)	5.07[C]
Santa Clara Land Title Co. v. Nowack Assocs., 226 Cal. App. 3d 1588 (1991)	9.03[D][1]
Santa Cruz Lumber Co. v. Bank of Am., 160 Cal. App. 3d 858 (1984)	9.04[A]
Santa Cruz P.C. Co. v. Snow Mountain Water & Power Co., 96 Cal. App. 615 (1929)	9.06[A]
Santa Fe, Inc., VABCA Nos. 1943 et al., 84-2 B.C.A. (CCH) ¶ 17,341	7.03
Sawyer Nurseries v. Galardi, 181 Cal. App. 3d 663 (1986)	9.03[F]
S.C. Anderson, Inc. v. Bank of Am. Nat'l Trust & Sav. Ass'n, 24 Cal. App. 4th 529 (1994)	4.12, 6.14[A]
Schaffer v. Debbas, 17 Cal. App. 4th 33, 21 Cal. Rptr. 2d 110 (1993)	5.01[A]
Schmid v. United States, 173 Ct. Cl. 302 (1965)	4.13
Schmitt v. Tri-Counties Bank, 70 Cal. App. 4th 1234 (1999)	9.04[A]
Schrader Iron Works, Inc. v. Lee, 26 Cal. App. 3d 621 (1972)	9.02[B][7]
Schuler-Haas Elec. v. Aetna Cas. & Sur., 371 N.Y.S.2d 207 (App. Div. 1975)	1.09[C]
Schwartz & Gottlieb, Inc. v. Marcuse, 175 Cal. 401 (1917)	9.03[D]
Schweigert Tire, Inc. v. United States, 388 F.2d 697 (Ct. Cl. 1967)	4.07[D]
Scott v. Continental Ins. Co., 44 Cal. App. 4th 24, 51 Cal. Rptr. 2d 566 (1996)	5.07[B]
Scott, Blake & Wynne v. Summit Ridge Estates, Inc., 251 Cal. App. 2d 347 (1967)	9.02[B][7]
Seal Tite Corp. v. Ehret, Inc., 589 F. Supp. 701 (D.N.J. 1984)	1.09[C]
Seaman's Direct Buying Serv., Inc. v. Standard Oil Co., 36 Cal. 3d 752 (1984)	8.09
Seattle W. Indus., Inc. v. David A. Mowat Co., 110 Wash. 2d 1, 750 P.2d 245 (1988)	6.14[A]
Seidman & Seidman v. Wolfson, 50 Cal. App. 3d 826 (1975)	14.02[E]
Service Employee Int'l Union, Local 347 v. City of L.A. Dep't of Transp., 24 Cal. App. 4th 136 (1994)	14.02[C]
Servidone Constr. Corp. v. United States, 19 Cl. Ct. 346 (1990), aff'd, 931 F.2d 860 (Fed. Cir. 1991)	4.15, 6.14[A]
Settimo Assocs. v. Environ Sys., Inc., 14 Cal. App. 4th 842 (1993)	1.05
Shell v. Schmidt, 164 Cal. App. 2d 350 (1958)	5.01[A], 5.01[B], 6.02[C]
Shell Oil Co. v. National Union Fire Ins. Co., 44 Cal. App. 4th 1633 (1996)	3.02[B][13]
Shelley v. Kofka, 107 Cal. App. 2d 827 (1951)	9.06[C][1]
Shoffner Indus., Inc. v. W.B. Lloyd Constr. Co., 257 S.E.2d 50 (N.C. 1979)	8.07
Shore Bridge Corp. v. State, 61 N.Y.S.2d 32 (Ct. Cl. 1946)	4.13
Shurpin v. Elmhirst, 148 Cal. App. 3d 193 (1983)	8.08
Sinnock v. Young, 61 Cal. App. 2d 130 (1943)	9.06[Q]

Case	*Book §*
Slaught v. Bencorno Roofing Co., 25 Cal. App. 4th 744 (1994)	14.02[C]
Slavin v. Borenstein, 25 Cal. App. 4th 713, 30 Cal. Rptr. 2d 745 (4th Dist. 1994)	4.02
Smith v. Riverside, 34 Cal. App. 3d 529 (1973)	2.02[B][2]
Smoketree-Lake Murray, Ltd. v. Mills Concrete Constr. Co., 234 Cal. App. 3d 1724 (1991)	3.02[B][7]
S.O.G. San Ore Gardner v. Missouri Pac. R.R., 658 F.2d 562 (8th Cir. 1981)	7.05
Solit v. The Tokai Bank, 68 Cal. App. 4th 35 (1999)	9.03[D]
Southern Cal. Acoustics Co. v. C.V. Holder, Inc., 71 Cal. 2d 719 (1969)	2.01[B], 2.02[D][2]
Southern Check Exch. v. San Diego, 5 Cal. App. 3d 81 (1970)	2.02[F][1]
Southern Fireproofing v. R.F. Ball Constr. Co., 334 F.2d 122 (8th Cir. 1964)	7.03
Southern Pac. Transp. Co. v. Sandy Land Protective Ass'n, 224 Cal. App. 3d 1494 (1990)	3.02[B][7]
Southwest Concrete Prods. v. Gosh Constr. Corp., 51 Cal. 3d 701 (1990)	9.06[J], 10.04[B]
Southwest Paving Co. v. Stone Hills, 206 Cal. App. 2d 548 (1962)	9.03[A]
Souza & McCue Constr. Co. v. Superior Court, 57 Cal. 2d 508 (1962)	2.01[A], 4.06, 4.07[A]
S&Q Constr. Co. v. Palma Seia Dev. Org., 179 Cal. App. 2d 364 (1960)	1.05, 1.06[B]
Stacy & Witbeck, Inc. v. City & County of S.F., 36 Cal. App. 4th 1074 (1995)	2.02[B][2]
Stark v. Shaw, 155 Cal. App. 2d 171 (1957)	4.08
State v. Guy F. Atkinson Co., 187 Cal. App. 3d 25 (1986)	6.07[E]
State ex rel. Department of Employment v. General Ins. Co. of Am., 13 Cal. App. 3d 853 (1970)	10.04[A]
State ex rel. Department of Transp. v. Guy F. Atkinson Co., 187 Cal. App. 3d 25 (1986) and 67 Cal. App. 3d 278 (1977)	6.14[A], 6.14[B]
Steelguard, Inc. v. Jannesen, 171 Cal. App. 3d 79 (1985)	2.02[A]
Steinbrenner v. Waterbury (J.A.) Constr. Co., 212 Cal. App. 2d 661 (1963)	1.06[B][1]
Steinwinter v. Maxwell, 183 Cal. App. 2d 34 (1960)	1.05, 1.06[B]
Steiny & Co. v. City Corp. Real Estate, Inc., 84 Cal. Rptr. 2d 38 (Ct. App. 1999)	9.01[B][2][a]
Stewart v. Cressview Mut. Water Co., 34 Cal. App. 3d 808 (1973)	8.02
Stimson v. Hanley, 151 Cal. 379 (1907)	2.02[F][1]
Structural Steel Fabricators v. City of Orange, 40 Cal. App. 4th 459 (1995)	9.03[G][5]
Structural Steel Fabricators v. City of Orange, 234 Cal. App. 3d 1206 (1991)	9.03[G][5]
Sturgis Sav. & Loan Ass'n v. Italian Village, Inc., 265 N.W.2d 755 (Mich. Ct. App. 1978)	9.06[T]
Sukut-Coulson, Inc. v. Allied Canon Co., 85 Cal. App. 3d 648 (1978)	9.01[C][2]
Sun Shipbuilding & Dry Dock Co. v. U.S. Lines, Inc., 439 F. Supp. 671 (E.D. Pa. 1977)	4.13
Sunbeam Constr. Co. v. Fisci, 2 Cal. App. 3d 181 (1969)	4.07[B], 6.02[C]
Sunlight Elec. Supply Co. v. McKee, 226 Cal. App. 2d 47 (1964)	9.03[G][4], 9.06[F], 10.04[A]
Sunlight Elec. Supply Co. v. Pacific Homes Corp., 226 Cal. App. 2d 110 (1964)	9.05[A]
Sunset Lumber Co. v. Smith, 91 Cal. App. 746 (1928)	9.03[H][3]

TABLE OF CASES

Case	Book §
Surety Co. of Pac. v. Piver, 149 Cal. App. 3d Supp. 29 (1983)	1.14
Sweeney v. McClaran, 58 Cal. App. 3d 824 (1976)	1.14
Sydney Constr. Co., ASBCA No. 21377, 77-2 B.C.A. (CCH) ¶ 12,719 (1977)	4.13
Systems Inv. Corp. v. National Auto. & Cas. Ins. Co., 25 Cal. App. 3d 1057 (1972)	9.01[B][2][e], 9.03[G]
Taylor Bus Serv., Inc. v. San Diego Bd. of Educ., 195 Cal. App. 3d 1331 (1987)	2.02[F][2]
Taylor Constr., Inc. v. Service Corp., 68 Cal. App. 4th 70 (1998)	11.07[D]
Terminix Co. v. Contractors State License Bd., 84 Cal. App. 2d 167 (1948)	1.09[L]
Theisen v. County of L.A., 54 Cal. 2d 170 (1960)	9.01[A][4]
Thibodeaux v. Crum, 4 Cal. App. 4th 749 (1992)	14.02[G]
Thomas Haverty Co. v. Pacific Indem. Co., 215 Cal. 555 (1932)	6.04[A]
Thos. J. Dyer Co. v. Bishop Int'l Eng'g Co., 303 F.2d 655 (6th Cir. 1962)	1.09[C]
Thomas Kelly & Sons v. City of L.A., 6 Cal. App. 2d 539 (1935)	3.02[B][12]
Titan Enters., Inc. v. Armo Constr. Co., 32 Cal. App. 3d 828 (1973)	14.02[E]
Tolstoy Constr. Co. v. Minter, 78 Cal. App. 3d 665 (1978)	5.01[B], 6.02[C]
Tom Shaw, Inc., DOTBCA Nos. 2106, 2108, 2109, 2110, 2131, 90-1 B.C.A. (CCH) ¶ 22,580 (1990)	6.14[B]
Tomko Woll Group Architects, Inc. v. Superior Court, 46 Cal. App. 4th 1326 (1996)	5.07[C]
Tonkin Constr. Co. v. County of Humboldt, 188 Cal. App. 3d 828 (1987)	2.01[A], 4.06
Tonn & Blank, Inc. v. Board of Comm'rs of LaPorte County, 554 N.E.2d 827 (Ind. Ct. App. 1990)	4.14
Toombs & Co. v. United States, 4 Cl. Ct. 535 (1984)	7.05
T&R Painting Constr., Inc. v. St. Paul Fire & Marine Ins. Co., 23 Cal. App. 4th 738 (1994)	9.03[H]
Tracy Price Assocs. v. Hebard, 266 Cal. App. 2d 778 (1968)	9.04[A]
Transamerica Ins. Co. v. Yonkers Contracting Co., 49 Misc. 2d 512, 267 N.Y.S.2d 669 (1966)	14.02[C]
Trinity Universal Ins. Co. v. Smithwick, 222 F.2d 16 (8th Cir.), cert. denied, 350 U.S. 837 (1955)	1.09[C]
Truestone, Inc. v. Simi W. Indus. Park II, 163 Cal. App. 3d 715 (1984)	9.02[B][7], 9.06[F]
Tucson House Constr. Co. v. Fulfor, 378 F.2d 734 (9th Cir. 1967)	12.09
Tutor-Saliba-Perini, PSBCA No. 1201, 87-2 B.C.A. (CCH) ¶ 19,775 (1987)	6.14[B]
Tzung v. State Farm Fire & Cas. Co., 873 F.2d 1338 (9th Cir. 1989)	5.07[B]
Ultramares Corp. v. Touche Ross & Co., 174 N.E. 441 (N.Y. 1931)	8.07
Union Asphalt, Inc. v. Planet Ins. Co., 18 Cal. App. 4th 633 (1993)	10.03[C]
Union Lumber Co. v. Simon, 150 Cal. 751 (1907)	9.03[C][1]
United Pub. Employees v. San Francisco, 53 Cal. App. 4th 1021 (1997)	14.02[C]
United States v. F.D. Rich Co., 439 F.2d 895 (8th Cir. 1971)	7.03
United States v. Gunnar I. Johnson & Son, Inc., 310 F.2d 899 (8th Cir. 1962)	10.04[B]
United States v. Peter Reiss Constr. Co., 273 F.2d 880 (2d Cir. 1959)	10.04[B]
United States v. Rice, 317 U.S. 61 (1942)	6.08[A]
United States v. Sovereign Constr. Co., 338 F. Supp. 659 (S.D.N.Y. 1972)	10.04[B]
United States v. Spearin, 248 U.S. 132, 39 S. Ct. 59 (1918)	2.01[A], 4.06, 4.07[B]
United States v. U.S. Fidelity & Guar. Co., 656 F.2d 993 (5th Cir. 1981)	10.04[B]

Case	Book §
United States Elevator Corp. v. Pacific Inv. Corp., 30 Cal. App. 4th 122 (1994) *review denied*	3.03[C]
United States *ex rel.* American Radiator & Standard Sanitary Corp. v. Northwestern Eng'g Co., 122 F.2d 600 (8th Cir. 1941)	11.02
United States *ex rel.* Austin v. Western Elec. Co., 337 F.2d 568 (9th Cir. 1964)	11.03
United States *ex rel.* B's Co. v. Cleveland Elec. Co., 373 F.2d 585 (4th Cir. 1967)	11.07[B]
United States *ex rel.* Carter-Schneider-Nelson, Inc. v. Campbell, 293 F.2d 816 (9th Cir. 1961)	11.02
United States *ex rel.* Consolidated Elec. Distribs., Inc. v. Altech, Inc., 929 F.2d 1089 (5th Cir. 1991)	11.02
United States *ex rel.* DDC Interiors, Inc. v. Dawson Constr. Co., 895 F. Supp. 270 (D. Colo. 1995)	11.07[B]
United States *ex rel.* Ehmcke Sheet Metal Works v. Wausau Ins. Co., 755 F. Supp. 906 (E.D. Cal. 1991)	9.03[H][3]
United States *ex rel.* Hawaiian Rock Prods. v. A.E. Lopez Enters., Ltd., 74 F.3d 972 (9th Cir. 1996)	11.05
United States *ex rel.* Heller Elec. Co. v. William F. Klingensmith, Inc., 670 F.2d 1227 (D.C. Cir. 1982)	7.03, 11.05
United States *ex rel.* Henderson v. Nucon Constr. Corp., 49 F.3d 1421 (9th Cir. 1995)	11.07[C]
United States *ex rel.* Hillsdale Rock Co. v. Cortelyou & Cole, Inc., 581 F.2d 239 (9th Cir. 1978)	11.05
United States *ex rel.* J.A. Edwards & Co. v. Thompson Constr. Corp., 273 F.2d 873 (2d Cir. 1959)	11.02
United States *ex rel.* Krupp Steel Prods., Inc. v. Aetna Ins. Co., 831 F.2d 978 (11th Cir. 1987)	11.05
United States *ex rel.* Lochridge-Priest, Inc. v. Kahn-Real Support Group, Inc., 950 F.2d 284 (5th Cir. 1992)	11.05
United States *ex rel.* Martin Steel Constructors, Inc. v. Avanti Constructors, Inc., 750 F.2d 759 (9th Cir. 1984)	11.05
United States *ex rel.* McGrath v. Travelers Indem. Co., 253 F. Supp. 330 (D. Ariz. 1966)	11.03
United States *ex rel.* Metal Mfg., Inc. v. Federal Ins. Co., 656 F. Supp. 1194 (D. Ariz. 1987)	11.04
United States *ex rel.* N.U., Inc. v. Gulf Ins. Co., 650 F. Supp. 557 (S.D. Fla. 1986)	11.03
United States *ex rel.* P.A. Bourquin & Co. v. Chester Constr. Co., 104 F.2d 648 (2d Cir. 1939)	10.04[B]
United States *ex rel.* Pertun Constr. Co. v. Harvesters Group, Inc., 918 F.2d 915 (11th Cir. 1990)	11.05
United States *ex rel.* Pippin v. J.R. Youngdale Constr. Co., 923 F.2d 146 (9th Cir. 1991)	11.03
United States *ex rel.* Superior Insulation Co. v. Robert E. McKee, Inc., 702 F. Supp. 1298 (N.D. Tex. 1988)	11.05
United States *ex rel.* T.M.S. Mechanical Contractors, Inc. v. Miller's Mut. Fire Ins. Co., 942 F.2d 946 (5th Cir. 1991)	11.05
United States *ex rel.* Westinghouse Elec. Supply Co. v. EndebrockWhite Co., 275 F.2d 57 (4th Cir. 1960)	11.05
United States *ex rel.* Whitmore Oxygen Co. v. Idaho Crane & Rigging Co., 193 F. Supp. 802 (D. Idaho 1961)	11.04

TABLE OF CASES

Case	Book §
United States Fidelity & Guar. Co. v. Orlando Utils. Comm'n, 564 F. Supp. 962 (M.D. Fla. 1983)	7.03
United States Indus., Inc. v. Vadnais, 270 Cal. App. 2d 520 (1969)	5.01[B]
United States Leasing Corp. v. Dupont, 69 Cal. App. 2d 275 (1968)	10.04[C]
United States Steel Corp. v. Missouri Pac. R.R., 668 F.2d 435 (8th Cir. 1982)	4.14
United Steelworkers of Am. v. Warrior & Gulf Navigation Co., 363 U.S. 574, 80 S. Ct. 1347, 4 L. Ed. 2d 1403 (1960)	14.02[C]
Universal By-Products v. Modesto, 43 Cal. App. 3d 145 (1974)	2.02[F][1]
University Casework Sys., Inc. v. Superior Court, 41 Cal. App. 3d 263 (1974)	5.01[B], 10.04[A]
Urban Masonry Corp. v. N&N Contractors, Inc., 676 A.2d 26 (D.C. Ct. App. 1996)	1.09[C]
U.S. Roofing, Inc. v. Credit Alliance Corp., 228 Cal. App. 3d 1431 (1991)	8.02
Vallejo Dev. Co. v. Beck Dev. Co., 24 Cal. App. 4th 929 (1994)	1.01, 1.05
Valley Constr. Co. v. City of Calistoga, 72 Cal. App. 2d 839 (1946)	6.10[A]
Valley Crest Landscape, Inc. v. City Council, 41 Cal. App. 4th 1432 (1996)	2.02[D][2]
Van Buren Plaza LLC, In re, 200 B.R. 384 (Bankr. C.D. Cal. 1996)	9.05[D]
Vanlar Constr., Inc. v. County of L.A., Case No. B005504 (Ct. App. 2d Dist., Div. 5 May 30, 1985)	8.07
Vaughn Materials Co. v. Security Pac. Nat'l Bank, 170 Cal. App. 3d 908 (1985)	9.03[F]
Vec, Inc., ASBCA No. 35988, 90-3 B.C.A. (CCH) ¶ 23,204 (1990)	4.13
Viking Pools, Inc. v. Maloney, 48 Cal. 3d 602 (1989)	1.09[A]
Villinger/Nicholls Dev. Co. v. Meleyco, 31 Cal. App. 4th 321 (1995)	9.01[B][3]
Vitek, Inc. v. Alvarado Ice Palace, Inc., 34 Cal. App. 3d 586 (1973)	9.01[A][4], 9.06[H]
Volpe-Head, Joint Venture, Appeal of, ENGBCA No. 4722, 89-3 B.C.A. (CCH) ¶ 22,105 (1989)	7.05
Vowels v. Witt, 149 Cal. App. 2d 257 (1957)	9.04[D], 9.06[J]
Walke v. Thornsbery, 97 Cal. App. 3d 842 (1979)	1.06[B][1]
Walker v. Lytton Sav. & Loan Ass'n, 2 Cal. 3d 152 (1970)	9.04[A]
Walker v. Signal Cos., 84 Cal. App. 3d 982 (1978)	5.05, 6.02[A], 8.03
Walter Dawgie Ski Corp. v. United States, 30 Fed. Cl. 115, 39 Cont. Cas. Fed. (CCH) ¶ 76,583	6.08[B]
Wand Corp. v. San Gabriel Valley Lumber Co., 236 Cal. App. 2d 855 (1965)	9.03[C][3]
Wang v. Division of Labor Standards Enforcement, 219 Cal. App. 3d 1152 (1990)	1.09[I]
Warner Constr. Corp. v. City of L.A., 2 Cal. 3d 285 (1970)	3.02[B][12], 4.07[A], 4.11, 4.12, 6.11
Warren Painting Co., ASBCA No. 18456, 74-2 B.C.A. (CCH) ¶ 10,834	6.11
Washington Int'l Ins. Co. v. Superior Court, 62 Cal. App. 4th 981 (1998)	1.09[C], 10.04[A]
Weaver-Bailey Contractors, Inc. v. United States, 19 Cl. Ct. 474 (1990)	4.13, 7.03

Case	*Book §*
Weeks v. Merritt Bldg. & Constr. Co., 39 Cal. App. 3d 520 (1974)	1.09[F], 9.01[A][4]
Weeshoff Constr. Co. v. Los Angeles County Flood Control Dist., 88 Cal. App. 3d 579 (1979)	9.06[G]
Wegner, *In re*, 839 F.2d 533 (9th Cir. 1988)	12.04
Welch v. State, 139 Cal. App. 3d 546 (1983)	3.02[B][12], 4.07[A]
West v. City of Oakland, 30 Cal. App. 566 (1916)	2.02[F][2], 2.02[F][3], 2.02[G][3]
West Coast Home Improvement Co. v. Contractors State License Bd., 72 Cal. App. 2d 287 (1945)	1.09, 1.09[D], 1.09[H]
West Coast Lumber Co. v. Newkirk, 80 Cal. 275 (1889)	9.03[C][2], 9.06[O]
Western Landscape Constr. v. Bank of Am., 58 Cal. App. 4th 57 (1997)	9.05[B][5]
Western Sierra, Inc. v. Ramos, 97 Cal. App. 3d 482 (1979)	9.06[G]
Westfour Corp. v. California First Bank, 3 Cal. App. 4th 1554 (1992)	9.02[B][7], 9.04[A]
Westinghouse Elec. Corp. v. County of L.A., 129 Cal. App. 3d 771 (1982)	10.04[A]
Westwood Bldg. Materials Co. v. Valdez, 158 Cal. App. 2d 107 (1958)	9.05[F]
W.F. Hayward Co. v. Transamerica Ins. Co., 16 Cal. App. 4th 1101 (1993)	9.03[A][1], 10.03[B]
Whittier v. Wilber, 48 Cal. 175 (1874)	9.05[B][1]
Wickham Contracting Co. v. Fischer, 12 F.3d 1574 (Fed. Cir. 1994)	6.07[G]
Wm. R. Clarke Corp. v. Safeco Ins. Co., 15 Cal. 4th 882 (1997)	1.09[C], 3.03[B], 4.03, 9.06[D], 9.06[T]
Wm. R. Clarke Corp. v. Safeco Ins. Co., 38 Cal. App. 4th 1655 (1995)	9.01[C][1]
Williams Enters. v. Strait Mfg. & Welding, Inc., 728 F. Supp. 12 (D.D.C. 1990)	7.05
Wilner v. United States, 23 Cl. Ct. 241 (1991)	7.02, 7.05
Winans v. State Farm Fire & Cas. Co., 968 F.2d 884 (9th Cir. 1992)	5.07[B]
Windowmaster Corp. v. B.G. Danis Co., 511 F. Supp. 157 (S.D. Ohio 1981)	14.02[C]
Windsor Mills v. Richard B. Smith, Inc., 272 Cal. App. 2d 336 (1969)	9.02[B][7]
Windsor Mills, Inc. v. Collins & Aikman Corp., 25 Cal. App. 3d 987 (1972)	3.05[A]
Winick Corp. v. Safeco Ins. Co., 187 Cal. App. 3d 1502 (1986)	10.04[A]
W.J. Lewis Corp. v. Harper Constr. Co., 116 Cal. App. 3d 27 (1981)	2.02[D][2]
Woodhead Lumber Co. v. E.G. Niemann Invs., 99 Cal. App. 456 (1929)	9.03[H][3]
W.P. Fuller & Co. v. McClure, 48 Cal. App. 185 (1920)	9.04[A]
W.P.C. Enters., Inc. v. United States, 323 F.2d 874 (Ct. Cl. 1963)	4.07[D]
WRB Corp. v. United States, 183 Ct. Cl. 409 (1968)	6.14[A], 6.14[B]
Wright v. United States Postal Serv., 29 F.3d 1426 (9th Cir. 1994)	11.04
Wunderlich v. State, 65 Cal. 2d 777 (1967)	2.01[A], 3.02[B][12], 4.06, 4.07[C], 6.11
Wunderlich Contracting Co. v. United States, 351 F.2d 956, 173 Ct. Cl. 180 (1965)	6.10[C]

TABLE OF CASES

Case	*Book* §
Yamanishi v. Bleily & Collishaw, Inc., 29 Cal. App. 3d 457 (1972)	1.09[C], 3.03[B], 6.07[H], 9.06[D]
Zamora v. Shell Oil Co., 55 Cal. App. 4th 204 (1997)	8.12

INDEX

References are to section numbers and appendixes.

A

Abandonment. *See also* Breach of contract by contractor; Breach of contract by owner
    by contractor, 5.01[B]
    by owner, 4.09
    use of, to satisfy lien claims, 12.08
Acceleration claims, 6.09
    actual acceleration, 6.09, 6.09[A]
    constructive acceleration, 6.09, 6.09[B]
        elements of claim, 6.09[B]
        examples, 6.09[B]
    damages recoverable, 4.10
    elements of, 4.10
Acceptance of bid, 2.01[B]
    silence, effect of, 2.01[B]
Acts of God, 6.07[C]
Actual changes, 6.10[A], 6.10[B]
    constructive changes compared, 6.10[B]
Actual cost method of proof
    quantifying and proving damages, 4.12, 6.14
Adequate assurance. *See* Bankruptcy
Advertising
    defined, 1.08, 1.17
    license number, inclusion of, 1.08
    unlicensed contractors, advertising by, 1.17
        penalty for, 1.17
Affirmative action programs, 2.02[F][4][b]
    forms to be submitted, 2.02[D][4]
    strict scrutiny, test of, 2.02[F][4][b]
    trend, current, 2.02[F][4][b]
AIA Document A201, General Conditions of the Contract for Construction, 3.02[B]
AIA Document A401, Standard Form Subcontract, 3.03
Alternative dispute resolution (ADR), 14.04
    arbitration. *See* Arbitration
    contractually mandated ADR, 14.04[A]
    court-ordered ADR, 14.04[B][1]
    direct settlement negotiations, 14.04
    disputes review boards. *See* Disputes review boards (DRBs)
    mandatory settlement conferences (MSCs), 14.04[B][1]
    mediation. *See* Mediation
    mini-trials, 14.04, 14.04[C]
    overview, 14.01
    partnering arrangements, 14.04
    private trials, 14.04, 14.04[C]
        advantages of process, 14.04[C]
    reference process, 14.04[C]
    settlement agreements, 14.05
        enforcement of, 14.04, 14.05
        general releases, 14.04
        judgment by stipulation, 14.05
        oral settlement agreements, 14.05
        settlement of part of but not all disputes, 14.04
    statutorily required ADR, 14.04[B][2]
        Civil Procedure Code, judicial arbitration under, 14.04[B][2], 14.04[B][2][a]
        Civil Procedure Code, judicial mediation under, 14.04[B][2], 14.04[B][2][b]
        Public Contract Code, arbitration provisions in, 14.04[B][2][c]
    voluntary settlement conferences (VSCs), 14.04[B][1]
Ambiguities
    contra proferentem, rule of, 4.07[D]
    in material supplier contracts, 3.04
    in plans and specifications. *See* Plans and specifications
    of unconditional release form, 9.05[B][4][c]
American Arbitration Association, 14.02
American Architects Association, 14.04[A]
Arbitration, 14.02
    advantages of, 14.02[A]
    American Arbitration Association, 14.02
    arbitrability of dispute, 14.02[C]

Arbitration (*continued*)
  arbitrators
    disclosure requirements, 14.02[H]
    disqualification of, 14.02[H]
    grounds for vacating arbitrator's award, 14.02[B]
  award, grounds for vacating, 14.02[B]
  binding effect of awards, 14.02[G]
  compelling arbitration, 14.02[C]
  conflicts between prime contract and subcontract, 14.02[B]
  consolidation of arbitration proceedings, 14.02[D]
  Construction Industry Arbitration Rules, 14.02[A], 14.02[B], 14.02[E]
  disadvantages of, 14.02[B]
  disclosure requirements for arbitrators, 14.02[H]
  flexibility of, 14.02[A]
  home improvement contracts, requirements for. *See* Home improvement contracts
  inconsistent provisions, 14.02[B], 14.02[D]
  incorporation of terms by reference, 14.02[C]
  judicial arbitration, 14.04[B][2], 14.04[B][2][a]
  under licensing laws, 1.10
  multiplicity of proceedings, prevention of, 14.02[F]
  of owner-contractor disputes, 3.02[B][9]
    consolidation/joinder of parties, limitation on, 3.02[B][9]
  popularity of, 14.02
  Public Contract Code provisions, 14.04[B][2][c]
    attorneys' fees, 14.04[B][2][c]
    costs of arbitration, 14.04[B][2][c]
    time limitations, 14.04[B][2][c]
    tort claims, exclusion of, 14.04[B][2][c]
    waiver of statute, 14.04[B][2][c]
  res judicata, doctrine of, 14.02[G]
  residential construction contracts, arbitration clauses in, 14.04[D]
  resisting arbitration, 14.02[F]
  statutorily required arbitration
    Civil Procedure Code, judicial arbitration under, 14.04[B][2][a]
    Public Contract Code, arbitration provisions in, 14.04[B][2][c]
  vacating award grounds for, 14.02[B]

waiver of right to arbitrate, 14.02[B], 14.02[E]
Architects
  liability
    limitation of liability, 3.05[A]
    for professional negligence, 8.07, 8.10
  owner-architect agreement, 3.05
Assignment for benefit of creditors
  effect of, on releases, 9.05[C], 9.06[B]
Assumption of contract. *See* Bankruptcy
Attorneys' fees, 3.03[C], 10.04[A]
  license bond, action on, 1.14
  mechanic's lien, includability in, 9.06[H]
  on payment bond on public works, 10.03[C], 10.04[B]
Automatic stay. *See* Bankruptcy

B

Backcharges
  offsetting backcharges, 5.06
Bankruptcy
  abandonment, use of, to satisfy lien claims, 12.08
  adequate assurance requirement, 12.06
    compensation for actual losses, 12.06
    continued performance, assurance of, 12.06
    ways to adequately assure future performance, 12.06
  assumption of contract, 12.04, 12.12
    time period for, 12.05
  automatic stay
    effect of, 12.02
    mechanics' liens, impact on, 12.03, 12.08
    purpose of, 12.02
    relief from, reasons to grant, 12.08
    violation of, 12.02, 12.07
  bonds, protection provided by, 12.11
  completing the work, 12.12
    substitution of completion contractor, 12.12
  cure of default, 12.06
  delay in project, recovery of damages for, 12.12
  double payment, protection against, 12.10
  evaluating potential claims, 12.09
  executory contracts, 12.04
    assumption or rejection of, 12.04, 12.12
    time periods for assumption or rejection, 12.05

INDEX

general contractor's vs. owner's
bankruptcy, 12.08
interpleader, use of, 12.10
of licensee, 1.09[E]
liens, release of, 12.10
losses resulting from default,
compensation for, 12.06
mechanic's lien issues, 12.08
automatic stay, impact of, 12.03, 12.08
foreclosure of lien, effect of bankruptcy
on, 9.03[F]
preservation of lien rights, 12.08
release of liens, 12.10
owner's vs. general contractor's
bankruptcy, 12.08
payment only if protected, 12.10
performance of project and, 12.03
preferences, 12.10
prepetition payment to non-insider
creditor, 12.10
process, 12.02
Chapter 7, 12.02
Chapter 11, 12.02
rejection of contract, 12.04, 12.05, 12.12
how to force rejection, 12.06
remedy for subcontractor, 12.05
time period for, 12.05
setoff, 12.07, 12.10
elements necessary to justify setoff,
12.07
waiver of right to claim setoff, 12.07
termination of contract before bankruptcy
filing, 12.06
validity of claim, establishing, 12.09
Bidding procedures for private works, 2.01
acceptance of bid, 2.01[B]
bidding documents, 2.01[A]
defective plans and specifications, risk of,
2.01[A]
defenses to contractor claim against
subcontractor, 2.01[C]
disclaimers, 2.01[A]
material supplier bids, 2.01[D]
mistakes in bid, 2.01[C]
open bidding procedures, 2.01
promissory estoppel, doctrine of, 2.01[C]
refusal to honor bid, 2.01[C]
short list, use of, 2.01
subcontractor bids, 2.01[C]
withdrawal of bid, 2.01[C]
Bidding procedures for public works, 2.02.
See also Public works
affirmative action documents, 2.02[D][4]
alternate bids, 2.02[F][3]

appellate review of award of contract,
2.02[G][3]
applicable rules, 2.02[B]
competitive bidding generally, 2.02[B][1]
benefits of, 2.02[B][3]
exceptions to competitive bidding
rules, 2.02[B], 2.02[B][2]
goals of competitive bidding, 2.02[A]
design-bid-build, use of, 2.02[B][1],
2.02[B][3]
design-build methods, use of. *See* Design-
build contracts; Design-build
methods
evaluation of bids, 2.02[F]
invitation to bid, contents of, 2.02[C]
late bids, 2.02[D]
material terms of bid, 2.02[F][1]
mistake in bid, 2.02[E]
negotiated proposals, 2.02[B][1]
non-collusion affidavit, 2.02[D][3]
nonresponsiveness of bid, 2.02[D][1],
2.02[D][4]
for failure to provide bid security,
2.02[D][1]
for failure to submit affirmative action
documents, 2.02[D][4]
notice requirements, noncompliance with,
2.02[E]
penalties for ignoring or evading
procedures, 2.02[B][2]
preferences, 2.02[F][4]
local firms and labor, 2.02[F][4][a]
minority-owned businesses,
2.02[F][4][b]
small businesses, 2.02[F][4][b]
prequalification of contractors,
2.02[C][2]
protests, 2.02[G]
damages, recoverability of, 2.02[G][2]
frivolous protests, 2.02[G][1]
injunctive relief, 2.02[G][2]
procedures, 2.02[G][1]
remedies, 2.02[G][2]
standard of review, 2.02[G][3]
standing, 2.02[G][1]
writ of mandate, 2.02[G][2]
public notice of projects, 2.02[B]
purpose, 2.02[A]
rescission, suits for, 2.02[E]
responsibility of bidder, 2.02[B], 2.02[F],
2.02[F][2]
factors considered, 2.02[F][2]
favoritism, minimizing exercise of,
2.02[F][3]

Bidding procedures for public works
protests, (*continued*)
  lowest bidders, determination of,
    2.02[F][3]
  relative superiority of bidders,
    2.02[F][2]
  responsiveness of bid, 2.02[B], 2.02[F],
    2.02[F][1]
    immaterial deviations, waiver of,
      2.02[F][1]
    nonresponsive bid. *See*
      nonresponsiveness of bid, this
      heading
    substantially conforming bids,
      2.02[F][1]
  security for bid, 2.02[D][1]
  sole-source specifications, 2.02[C]
  solicitation of bids, 2.02[C]
    or equal clause, 2.02[C][1]
  splitting of large projects into smaller
    pieces, 2.02[B]
  statutory framework, 2.02[A]
  subcontractor listing law, 2.02[D][2]
    failure to specify subcontractor,
      2.02[D][2]
    inadvertent clerical error, 2.02[D][2]
    substantial compliance with notice,
      2.02[D][2]
    substitutions, 2.02[D][2]
    violation of law, effect of, 2.02[D][2]
  submission of bids, 2.02[D]
  withdrawal of bid, 2.02[E]
Bona fide employee
  defined, 1.12
Bonding capacity, impairment of
  proof, methods of, 4.12
  recovery of lost profits, 4.12
Bonds, 12.11
  bid bonds, 2.02[D][1]
  capacity, bonding. *See* Bonding capacity,
    impairment of
  defenses
    surety's right to raise principal's
      defenses, 10.04[C]
  judgment bonds. *See* Judgment bonds
  labor and material bond. *See* Labor and
    material bond, public works;
    Payment (labor and material) bonds
    on private works
  license bonds. *See* License bonds
  payment bond. *See* Payment bonds on
    public works; Payment (labor and
    material) bonds on private works
  release bond. *See* Release bond

  requirements, bonding
    labor and materials bond, 3.03[D]
    performance bond, 3.03[D]
  use of, to avoid problems raised by
    bankruptcy, 12.11
Breach of contract by contractor. *See also*
  Construction claims; Damages
  abandonment of work, 5.01[B]
    measure of damages, 5.01[B]
  building code, violation of, 5.08
  defective construction. *See* Defective
    construction
  delay damages
    interest, 5.11
    liquidated damages, 5.02, 5.11
    profits, loss of, 5.04, 5.11
    rent, loss of, 5.03, 5.11
  failure to complete work, 5.01[B]
    measure of damages, 5.01[B]
    substantial performance defined,
      5.01[B]
  failure to perform work per plans and
    specifications, 5.01[A]
    measure of damages, 5.01[A]
  fraud, 5.05, 5.11
    punitive damages for, 5.05, 5.11
  offsetting backcharges, 5.06
  performance of contractor, measuring,
    5.01
  substantial completion of work and,
    5.01[B]
  termination of contractor, 5.09, 6.04
  terms of contract, impact of, 5.10
  violation of building code, damages for,
    5.08
Breach of contract by owner. *See also*
  Construction claims; Damages
  abandonment of contract, 4.09
  acceleration, 4.10
  change orders
    excessive numbers of, effect of, 4.09
    refusal to grant, 4.07[E]
  changed/extra work
    recoverability of interest on borrowed
      funds, 4.15
  damages
    consequential damages, recovery of,
      4.11
    delay damages on multi-prime projects,
      4.13
    for reduction of bonding capacity, 4.12
  defective plans and specifications. *See*
    Plans and specifications
  delay, 4.05

INDEX

in start of performance, 4.08
early completion date, delay or
    interference with, 4.13
material change in scope of contract, 4.09
nonpayment, 4.03
    contractor's right to stop private work
        of improvement, 4.03
    notice requirements, 4.03
    pay-if-paid clauses. See Pay-if-paid
        clauses
    stop work order remedy, 4.03
progress payments, failure to make, 4.02
    material breaches, 4.02
    options available to contractor, 4.02
types of breach, 4.01
Brooks Act
    Little Brooks Act, 2.02[B][1]
Building code
    damages for violation of, 5.08

C

California Records Act, 2.02[G][1]
Cardinal changes, 6.10[A], 6.10[C], 7.03
    effect of, 6.10[C]
Certificate of merit, 8.10
    privilege, 8.10
    purpose of, 8.10
    violation of statute, effect of, 8.10
Certification of claims
    risks of, 6.03
Cessation of work, 10.03[B]
    completion of work compared, 10.03[B]
Change orders
    excessive numbers of, effect of, 4.09
    refusal to grant, 4.07[E]
Changed conditions, 6.10
Changes in work
    actual changes, 6.10[A], 6.10[B]
        constructive changes compared,
            6.10[B]
    cardinal changes. See Cardinal changes
    changes clause, 6.10[A]
        limitations on, 6.10[A]
    constructive changes, 6.10[A], 6.10[B]
        actual changes compared, 6.10[B]
        examples, 6.10[B]
        notice of, 6.10[B]
        reasons for, 6.10[B]
    costs of changes, 6.10[A]
    in general scope of contract, 6.10[A]
Citation procedure, 1.11
    appeal of citation, 1.11[B]

Contractors State License Board rules,
    1.11[B]
corrective work
    extension of time to correct, 1.11[B]
    feasibility of, 1.11[B]
    time required to correct, 1.11[B]
criteria to evaluate gravity of violation,
    1.11[B]
disassociation from license, continuing
    responsibility after, 1.19
generally, 1.11[A]
penalties, recommended assessments of,
    1.11[B]
service of citation, 1.11[B]
Claims, construction-related. See
    Construction claims
Compensable delays, 6.07, 6.07[B], 6.07[D],
    7.05
    examples, 7.05
Completion of project
    cessation of work compared, 10.03[B]
    early finish
        additional compensation for, 4.13
        contractor's claim for right to, 4.13
        damages for wrongful interference with
            early completion date, 4.13
        interference, effect of owner's, 4.13
        notification to owner concerning, 4.13
    notice of completion. See Notice of
        completion
    what constitutes completion, 9.03[A],
        9.03[E]0
Concealed conditions, 3.02[B][12]. See also
    Differing site conditions
    misleading information from owner or
        agent, effect of, 3.02[B][12]
    time limitations, 3.02[B][12]
Concurrent delay, 6.07[B], 6.07[E], 7.05
    apportionment of fault/damages, 6.02[B],
        6.07[E], 7.05
    critical path delay analysis, 7.04. See also
        Critical path method (CPM)
        scheduling
    defense, use as, 7.05
    same delay, concurrently contributing to,
        7.05
    time extensions, 7.05
Condition precedent clauses. See Pay-if-paid
    clauses; Pay-when-paid clauses
Condominiums
    notice of completion, 9.04[E]
Consequential damages
    bonding capacity, lost profits based on
        impairment of, 4.12

Consequential damages (*continued*)
  limitation on, 5.10
  recovery of, 4.11
  types sought by contractors, 4.11
  waiver of, 3.02[B][8]
Construction claims
  causes of claims
    changed conditions, 6.10
    changes in work. *See* Changes in work
    differing site conditions. *See* Differing site conditions
    overview of leading causes, 6.10
  changes. *See* Changes in work
  contractor's claims
    acceleration claims. *See* Acceleration claims
    delay claims. *See* Delay claims
    disruption claims. *See* Disruption claims
    overview, 6.05
    payment-related claims, 6.05
    time-related claims, 6.05, 6.06–6.07
  damages. *See* Damages
  differing site conditions. *See* Differing site conditions
  early claims recognition, benefits of, 6.01
  false claims
    public works contractor's liabilities for, 6.03
  investigation of claims, 6.13
    claim documentation checklist, 6.13[A]
    claim outline and narrative, 6.13[B]
  owner's claims
    contractor's failure to perform per plans and specifications, 6.02[C]
    damages recoverable, 6.02[A]
    delay claims, 6.02[A]
    liquidated damages clauses, 6.02[B]
    overview, 6.02
  suspension of work, 6.10, 6.12
  termination of contractor. *See* Termination of contractor
  timely recognition of claims, importance of, 6.01
Construction contract. *See* Contract for construction
Construction defects. *See* Defects in construction
Construction Industry Arbitration Rules, 14.02[A], 14.02[B], 14.02[E]
Construction managers
  on multi-prime projects
    delay claims against managers, 4.14
    negligent direction of work, 4.14

Constructive acceleration, 3.02[A][3]
Constructive changes, 6.10[A], 6.10[B]
  actual changes compared, 6.10[B]
  examples, 6.10[B]
  notice of, 6.10[B]
  reasons for, 6.10[B]
Consultants
  liability of, 8.11
  witness immunity doctrine, 8.11
Continuing nuisance, 5.07[B]
Contra proferentem, rule of, 4.07[D]
Contract Disputes Act, 11.04
Contract documents, 3.02
Contract for construction
  breach of. *See* Breach of contract by contractor; Breach of contract by owner
  contract documents. *See* Contract documents
  contractor-subcontractor agreement. *See* Subcontracts
  design-bid-build contracts. *See* Design-bid-build process
  design-build contracts. *See* Design-build contracts
  disclaimers, protection against, 3.05[A]
  elements of, 3.02[A]
  exculpation clause, protection against, 3.05[A]
  home improvement contracts. *See* Home improvement contracts
  material supplier contracts. *See* Material supplier contracts
  modifications to standard forms, 3.01
  owner-architect agreement. *See* Owner-architect agreement
  owner-contractor agreement. *See* Owner-contractor agreement
  residential construction contracts. *See* Residential construction contracts
  rights and obligations of parties, 6.08
  standard forms most widely used, 3.01
  swimming pools. *See* Swimming pool construction contracts
  unambiguous, explicit disclaimers, protection against, 3.05[A]
Contractors
  business licenses, 1.20
  contractor-subcontractor agreement. *See* Subcontracts
  defined, 1.01, 1.02
  developers compared, 1.05
  disciplinary action, grounds for. *See* Disciplinary action

INDEX

general building contractor, 1.02
general engineering contractor, 1.03
home inspectors, 1.21
insurance adjuster, kickback by contractor to, 1.22
licensing requirement. *See* License law
owner-contractor agreement. *See* Owner-contractor agreement
specialty contractors, 1.02
unlicensed contractors, 1.05
Contractors State License Board
citation rules, 1.11[B]
Cooperation
implied duty of, 6.07[D], 6.08
Corporations
citations, continuing responsibility for, 1.19
disassociation of person from license, effect of, 1.19
license application by, 1.06[B][2], 1.12
Cost-plus agreements, 3.02[A][4]
CPM scheduling. *See* Critical path method (CPM) scheduling
Critical path method (CPM) scheduling
as-built schedule, 7.04
as-planned schedule, 7.04
burden of proof, 7.04
concurrent delays, effect of, 7.04
critical path defined, 7.02, 7.03
delay, use of CPM schedule to prove, 6.07[E], 7.04. *See also* Delay claims
entitlement schedule, 7.04
how CPM schedules are prepared, 7.03
inadequate scheduling, effect of, 7.02
incorporating delays as they occur, 7.02
ineffective use of bar chart, effect of, 7.02
management/administration of project, importance in, 7.02
preference for, in proof and defense of delay claims, 7.02
subcontractors, involvement of, in preparation of schedules, 7.03
up to date, importance of keeping, 7.02
use in the field, 7.02
use of, to prove delay, 7.04
what CPM schedule must show, 7.03

D

Damages. *See also* Breach of contract by contractor; Breach of contract by owner
for abandonment of work, 5.01[B]
acceleration claim, 4.10
for breach of contract, 5.05
for building code violation, 5.08
completion costs, 5.01[B], 5.11
consequential damages. *See* Consequential damages
cost of correction rule, 6.02[C]
for defective plans and specifications, 4.06
for defective work, 5.07, 5.11
for delay, 3.02[B][11], 5.02, 6.07[F], 6.07[G], 6.08[A]
delay in start of performance, 4.08
interest, 5.11
liquidated damages for delay, 5.02, 5.11
multi-prime projects, 4.14
profits, loss of, 5.04, 5.11
rent, loss of, 5.03, 5.11
diminution in market value, 5.01[A], 6.02[C]
for disruption, 6.08[A]
for extra or additional work performed by contractor, 4.12
actual cost method of proof, 4.12
jury verdict method of proof, 4.12
total cost method of proof, 4.12
for failure to complete work, 5.01[B]
for failure to perform work per plans and specifications
cost of remedying defect, 5.01[A]
diminution in market value, 5.01[A]
lesser measure rule, 5.01[A]
under false claims statutes, 6.03
for fraud, 6.02[A]
home office overhead, 3.02[B][8]
interest, recoverability of, for changed/extra work, 4.15
liquidated damages. *See* Liquidated damages
liquidated direct damages. *See* Liquidated direct damages
loss of use, 3.02[B][8]
lost profit damages, 4.12
methods available to prove damages, 4.12
measure of damages for breach, 6.02
loss of profits, 6.02[A]
punitive damages, 6.02[A]
multi-prime projects
delay damages on, 4.14
no damages for delay clause, 3.02[B][11], 6.08[A]
for partial performance, 6.02[C]

Damages (*continued*)
   punitive damages. *See* Punitive damages
   quantifying and proving damages, 6.14
      actual cost method, 4.12, 6.14
      competent but conflicting evidence, existence of, 6.14[B]
      jury verdict method, 4.12, 6.14, 6.14[B]
      modified total cost method, 6.14
      total cost method, 4.12, 6.14, 6.14[A]
   stigma damages, 8.13
   suspension of work and, 6.12
   terminated contractor, damages recoverable from, 6.04[A]
   terms of contract, impact of, 5.10
   for wrongful interference with early completion date, 4.13

Defective construction
   indemnity actions involving latent defects, 5.07[D]
   latent defects. *See* Latent defects
   patent defects. *See* Patent defects
   statute of repose, 5.07[A]
   time within which suit may be brought, 5.07[A]

Definitions. *See also* Ambiguities
   acceleration, 6.09
      actual acceleration, 6.09[A]
      constructive acceleration, 6.09[A]
   actual change, 6.10[B]
   advertising, 1.08, 1.17
   bona fide employee, 1.12
   California company, 2.02[F][4][a]
   cardinal change, 6.10[C]
   changes within scope of contract, 6.10[A]
   claim, 3.02[B][8], 6.03
   compensable delay, 6.07[D]
   completion, 4.04, 9.03[A], 9.03[E], 10.03[B]
   constructive change, 6.10[B]
   contractor, 1.01, 1.02
   critical path, 7.02, 7.03
   date of completion, 4.04
   economic loss, 3.04
   event-oriented investigation, 6.13
   excusable delay, 6.07[C]
   executory contract, 12.04
   express indemnity, 3.02[B][7]
   extras, 9.05[B][4][b]
   furnished, 9.05[B][2]
   general building contractor, 1.02, 1.03
   general engineering contractor, 1.03
   hard costs, 9.01[B][2][a]
   home improvement goods and services, 13.01
   home improvement inspector, 1.21
   home improvement salesperson, 13.02
   home improvements, 13.01
   home inspection, 1.21
   home inspection report, 1.21
   inexcusable delay, 6.07[C]
   latent act or omission, 1.18
   latent defect, 5.07[B]
   liquidated damages, 3.02[A][3]
   loss-oriented investigation, 6.13
   mediation, end of, 14.03[D]
   notice of nonresponsibility, 9.01[A][6]
   off-site work, 9.03[A]
   original source, 6.03
   overhead, 6.07[G]
   patent act or omission, 1.18
   patent defect, 5.07[A]
   plan-oriented investigation, 6.13
   soft costs, 9.01[B][2][a]
   subcontractor, 9.01[A][4]
   substantial performance, 5.01[B]

Delay claims, 3.02[B][10]
   apportionment of delay, 7.05
   burden of proof, 7.04, 7.06
   compensable delay. *See* Compensable delay
   concurrent delay. *See* Concurrent delay
   contractor's claims, 6.07
      causes of delay, 6.07[F]
      compensable delays, 6.07, 6.07[B], 6.07[D]
      concurrent delays, 6.07[B], 6.07[E]
      damages recoverable by contractor, 6.07[F]
      effects of delay, 6.07[F]
      Eichleay formula, 6.07[G]
      excusable delays, 6.07, 6.07[B], 6.07[C]
      extended home office overhead, 6.07[G]
      generally, 6.07[A]
      home office overhead, 6.07[G]
      inexcusable delays, 6.07, 6.07[B], 6.07[C]
      no damages for delay clauses, 6.07[H]
      noncompensable delays, 6.07, 6.07[B], 6.07[D], 6.07[E]
      time extension claims, 6.07[A]
      types of delay, 6.07[B]
   critical path network scheduling. *See also* Critical path method (CPM) scheduling
      use of, to prove delay, 6.07[E], 7.02
   damages suffered

INDEX

by contractor, 7.06
measure of, for breach, 3.02[B][10]
by owner, 7.06
disruption compared, 6.08[A]
excusable delay. *See* Excusable delay
inexcusable delay. *See* Inexcusable delay
no damages for delay clause, 3.02[B][11]
owner-caused delays, 4.05
  examples, 6.07[D]
owner's claims, 6.02[A]
prima facie case, establishing, 7.06
proof and defense of
  CPM scheduling techniques, preference for, 7.02
  delay to activity not on critical path, 7.04
  sequential delay, 7.05
  start of performance, delay in, 4.08
  type of delay, classification of, 7.05
  weather delays, 3.02[B][10]
Delegation of design, 3.02[B][6]
Design-bid-build process, 2.02[B][1], 2.02[B][3], 3.06
Design-build contracts, 3.06
  advantages of, 3.06
  considerations concerning appropriateness of, 3.06
  insurance considerations, 3.06
  project-specific provisions, importance of, 3.06
  risk allocation, 3.06
  risks to owner, 3.06
Design-build methods, 2.02[B][3]
  benefits of, 2.02[B][3]
  problems with, 2.02[B][3]
  state agencies, restrictions on, 2.02[B][3]
Design delegation, 3.02[B][6]
Design professionals
  liability of. *See* Liability
Design professionals lien, 9.01[A][3]. *See also* Mechanics' liens on private works
  conditions precedent to recording of lien, 9.01[A][3]
Developers, 1.05
  contractors compared, 1.05
Differing site conditions, 6.10, 6.11
  notice of changed conditions, 6.11
    failure to give notice, effect of, 6.11
  Type I conditions, 6.11
    examples of, 6.11
  Type II conditions, 6.11
    examples of, 6.11
  typical clause, 6.11

Disabled veteran–owned business enterprises (DVBEs), 2.02[F][4][b]
Disadvantaged business enterprises (DBEs), 2.02[F][4][b]
  Department of Transportation regulation, goals of, 2.02[F][4][b]
Disciplinary action
  abandonment of construction project, 1.09[A]
  avoidance of, 1.09[L]
  borrowing someone else's license, 1.09[F]
  defense to, contractor's, 1.09[L]
  disclosure of, 1.23
  diversion of funds, 1.09[B]
  evading license law, 1.09[F]
  failure to pay, 1.09[C]
  fraudulent act injuring another, 1.09[G]
  grounds for, 1.09
  misapplication of funds, 1.09[B]
  paying for use of someone else's license, 1.09[F]
  plans and specifications
    willful departure from, 1.09[D]
  preliminary notice, failure to give, 9.02[A][3], 10.02[A][8]
  purpose of, 1.09[H]
  records, failure to produce, 1.15
  settlement of obligations for less than full amount, 1.09[E]
  subcontractor listing law, violation of, 2.02[D][2]
  Terminix decision, effect of, 1.09[L]
  unlicensed individuals
    acting under unlicensed name or personnel, 1.09[H]
    entering into contract with unlicensed contractor/subcontractor, 1.09[I]
  using someone else's license, 1.09[F]
  willful act injuring another, 1.09[G]
  workers' compensation statement
    filing false statement, 1.09[K]
Disclaimers
  of coordination responsibilities, 4.14
  general vs. specific disclaimers, 4.07[C]
  of liability for delay damages, 4.14
  protection against, 3.05[A]
Dispute resolution. *See also* Alternative dispute resolution (ADR)
  owner-contractor disputes, 3.02[B][8]
  architect's role, 3.02[B][8]
Disputes review boards (DRBs), 14.04, 14.04[A]
  invalidation of process, 14.04[A]
  makeup of, 14.04[A]

Disputes review boards (DRBs) (*continued*)
  problem with, 14.04[A]
Disruption claims, 6.08
  delay compared, 6.08[A]
  examples on government projects, 6.08[B]
  proof, methods of, 6.08[C]
    historical comparison, 6.08[C]
    measured mile approach, 6.08[C]
Diversion of funds
  wrongful diversion of funds, 9.06[E]
Duress
  economic duress, 9.05[E]

E

Earthquakes
  insurance coverage, 3.02[B][13]
Economic duress, 9.05[E]
Economic loss
  defined, 3.04
    economic loss rule, 8.12
Eichleay formula, 6.07[G]
Employee
  bona fide employee defined, 1.12
Endorsements
  restrictive endorsement, 9.05[D]
Equitable liens
  non-enforceability of, 9.01[B][2][c]
Equitable tolling, doctrine of, 9.03[G][5]
Exculpatory clauses
  enforceability of, 4.07[C]
    general vs. specific disclaimers, 4.07[C]
  protection against, 3.05[A]
Excusable delays, 6.07, 6.07[B], 6.07[C], 7.05
  acts of God, 6.07[C]
  effect of, 7.05
  examples of, 6.07[C], 7.05
  force majeure, 6.07[C]
  inexcusable delays compared, 6.07[C]
  prevailing on claim of, 7.04
  strikes, 6.07[C]
  unavailability of materials, 6.07[C]
  unusually severe weather, 6.07[C]
Extra work. *See also* Changes in work
  borrowed funds, recoverability of interest on, 4.15
  extras defined, 9.05[B][4][b]
  oral agreements for, 9.06[G]
  release, exception to, 9.05[B][4][b]

F

False claims
  public works contractor's liabilities for, 6.03
False claims statutes, 6.03
  abuse, potential for, 6.03
  damages, 6.03
  excluded claims, 6.03
  original source defined, 6.03
  public works contractors and, 6.03
  specific acts prohibited, 6.03
  state and federal statutes compared, 6.03
  tips for contractors, 6.03
  whistle-blower provisions, 6.03
Federal construction projects. *See also* Public works
  statutory procedure. *See* Miller Act bond
Fixed-price contracts, 3.02[A][4]
Floods
  insurance coverage, 3.02[B][13]
Force majeure, 6.07[C]
Fraud
  liability for, 8.03
  punitive damages and, 6.02[A]

G

General building contractor, 1.02, 1.03
  defined, 1.02, 1.03
General engineering contractor, 1.03
  defined, 1.03

H

Hindrance. *See* Disruption claims
Home improvement contracts, 13.03. *See also* Residential construction contracts
  arbitration provisions, 13.06
    notice requirement, 13.06
    title requirement, 13.06
    typeface requirements, 13.06
  cancellation right of consumer. *See* rescission of home solicitation contracts, this heading
  mechanics' liens, 13.05
  notice regarding use of home as security, 13.03
  notice to owner, 1.07
  rescission of home solicitation contracts, 13.05

INDEX

emergency repairs or services, 13.05
notice of cancellation, form required, 13.05
time limitations, 13.05
statutory requirements, 13.01, 13.03
penalties for violations, 13.03
swimming pools, 13.04
written contract, requirements for, 13.03
Home improvement salespersons, 13.02
defined, 13.02
registration requirement, 13.02
denial of license, 13.02
failure to register, effect of, 13.02
Home inspections, 1.21
duty of care required, 1.21
prohibition against inspectors' offering/contracting to perform repairs, 1.21
time within which action against inspector must be filed, 1.21
Home solicitation contracts. *See* Home improvement contracts

I

Implied covenant of good faith and fair dealing, 9.03[H][3]
Implied warranties
implied warranty of design, 9.06[P]
liability for breach, 8.02, 8.04, 8.05
Impossibility of performance, 3.02[B][12]
rescission of contract, 3.02[B][12]
Indemnity obligations
contractor-subcontractor agreement, 3.03[C]
apportionment of fault, 3.03[C]
attorneys' fees, recoverability of, 3.03[C]
owner-contractor agreement, 3.02[B][7]
active negligence, 3.02[B][7]
duty to defend, 3.02[B][7]
express indemnity defined, 3.02[B][7]
general vs. specific indemnity clause, 3.02[B][7]
insurance policy compared, 3.02[B][7]
negligence of indemnitee, 3.02[B][7]
passive negligence, 3.02[B][7]
sole negligence, meaning of, 3.02[B][7]
Inexcusable delays, 6.07, 6.07[B], 6.07[C], 7.05
effect of, 7.05
examples, 7.05
excusable delays compared, 6.07[C]

Inspection of home. *See* Home inspections
Inspection of site, 3.02[B][3], 3.02[B][12].
See also Differing site conditions
concealed conditions, 3.02[B][12]
unknown conditions, 3.02[B][12]
Insurance, 3.02[B][13]
additional insured endorsements, 3.02[B][13]
certificate of insurance, 3.02[B][13]
design-build projects, 3.06
earthquake, coverage for, 3.02[B][13]
flood, coverage for, 3.02[B][13]
indemnity compared, 3.02[B][7]
Project Management Protective Liability Insurance, 3.02[B][7], 3.02[B][13]
property insurance, 3.02[B][13]
public liability insurance, 3.02[B][13]
self-insured retention, 3.02[B][13]
windstorm, coverage for, 3.02[B][13]
workers' compensation insurance, 3.02[B][13]
Insurance adjusters
kickback by contractor to adjuster, unlawfulness of, 1.22
Interest, recoverability of
for changed/extra work, 4.15
Interference. *See* Disruption claims
Investigation of claims, 6.13
claim documentation checklist, 6.13[A]
claim narrative, 6.13[B]
sections to include, 6.13[B]
claim outline, 6.13[B]
event-oriented investigation, 6.13
lawyer, importance of input from, 6.13[B]
loss-oriented investigation, 6.13
plan-oriented investigation, 6.13
records to maintain, 6.13[A]
steps to take, 6.13

J

Joint checks, 9.05[F]
exceptions to general rule, 9.05[F]
forged endorsement of, 9.06[L]
general rule, 9.05[F]
Judgment bonds, 1.14
Jury verdict method of proof
applications of, 6.14[B]
competent but conflicting evidence, existence of, 6.14[B]
conceptual touchstone for use of, 6.14[B]
constrained jury verdict method, 6.14[A]

Jury verdict method of proof (*continued*)
　quantifying and proving damages, 4.12,
　　6.14, 6.14[B]
　when appropriate, 6.14[B]

L

Labor and material bond, private works. *See*
　Payment (labor and material) bond
　on private works
Labor and material bond, public works
　completion vs. cessation of work, 10.03[B]
　court decisions, examples, 10.04[A]
　other requirements, 10.03[C]
　persons who can recover on bond,
　　10.03[C]
　preliminary 20-day notice, 9.03[H][1],
　　10.03[A]
　　contents of notice, 10.03[A][2]
　　methods of service, 10.03[A][2]
　　persons responsible for giving notice,
　　　10.03[A][1]
　　timing of notice, 10.03[A][2]
　procedural steps, 10.03[A]
　time for filing suit on bond, 10.03[B]
Last shot rule, 3.04
Latent defects, 5.07[A], 5.07[B], 5.07[D]
　defined, 5.07[B]
　limitations period, 5.07[A], 5.07[B]
Liability, 8.01
　of architect, for professional negligence,
　　8.07, 8.10
　certificate of merit. *See* Certificate of
　　merit
　of consultants, 8.11
　of contractor
　　for violation of industry standards,
　　　4.07[B]
　defective construction. *See* Defective
　　construction
　defective plans and specifications
　　negligent preparation of documents,
　　　8.07
　　owner's liability to subcontractor, 8.07
　defective products
　　failure to disclose defects, 8.02
　　product that has not yet caused damage,
　　　8.12
　of design professionals, for professional
　　negligence, 8.07, 8.10
　economic loss rule, 8.12
　of engineer, for professional negligence,
　　8.07, 8.10
　for fraud, 8.03
　home improvement salesperson, liability
　　for failure to register, 13.02
　implied warranty, breach of, 8.02, 8.04,
　　8.05
　interference with prospective economic
　　advantage, 8.06, 8.07
　of land surveyor, for professional
　　negligence, 8.07, 8.10
　limitation of liability, 3.05[A]
　of mass producers of homes, 8.02
　for negligence, 8.07, 8.10
　　certificate of merit, 8.10
　owner, personal liability of, 9.06[Q]
　privity of contract, lack of, 8.01, 8.02,
　　8.05, 8.06, 8.07, 8.09
　reasonable care, failure to exercise, 8.08
　stigma damages, 8.13
　strict liability, 8.02
　third-party beneficiary, 8.05
　tort recovery
　　for bad-faith denial of contract, 8.09
　　for contract breaches, 8.09
　　for negligence, 8.07
　trends, 8.09
　witness immunity doctrine, 8.11
License bonds, 1.14
　attorneys' fees, recoverability of, 1.14
　persons benefited by bond, 1.14
　priority of competing claims against
　　single bond, 1.14
　swimming pool contractor's bond, 1.14
　trust funds, action by, 1.14
　types of claims that may be made, 1.14
License law
　arbitration. *See* Arbitration proceedings
　bonds. *See* License bonds
　cases interpreting law, 9.01[A][4]
　citation procedure. *See* Citation procedure
　classifications, 1.02
　　incidental and supplemental work, 1.02
　　specialty contractors, 1.02
　contractor's business licenses, 1.20
　corporation, application by, 1.06[B][2],
　　1.12
　disassociation from license
　　continuing responsibility after, 1.19
　disciplinary action, grounds for. *See*
　　Disciplinary action
　dispute resolution. *See* Arbitration;
　　Citation procedure
　exemptions, 1.06
　　case law exemptions, 1.06[B]
　　laborer working for wages, 1.06[A]

INDEX

material supplier exemption, 1.06[B][1]
owner/builder exemption, 1.06[B][3]
substantial compliance doctrine,
  1.06[B][2]
federal projects, 1.06[B]
latent acts or omissions, 1.18
license number
  advertising, inclusion in, 1.08
  contracts, inclusion in, 1.08
material supplier exemption, 1.06[B][1]
mechanics' liens and, 9.01[A][4]
notice requirements
  contract, notice required by, 1.08
  license law, notice required by, 1.07
owner/builder exemption, 1.06[B][3]
partnership, application by, 1.12
patent acts or omissions, 1.18
penalties for not being properly licensed,
  1.05
public access to information, 1.12
public access to proceedings, 1.12
public works construction projects, 1.04
purpose of, 1.02, 1.05, 1.06[B][2]
requirement of, for contractors, 1.01
responsible managing employee,
  1.06[B][2], 1.12
responsible managing officer, 1.06[B][2],
  1.12
revocation of license, 1.10, 1.11[B], 1.23
  disclosure requirement, 1.23
  for failure to produce records, 1.15
services that can be performed, 1.03
substantial compliance doctrine,
  1.06[B][2]
sureties, license requirement for, 1.24
  exemption, availability of, 1.24
suspension of license, 1.10, 1.11[B], 1.23
  disclosure requirement, 1.23
time for filing of proceedings, 1.18
unlicensed contractors, 1.05
  advertising by, 1.17
Liens. *See* Design Professionals liens;
  Equitable liens; Mechanics' liens;
  Mechanics' liens on private works
Limitations periods
  construction defects
    latent defects, 5.07[A], 5.07[B]
    patent defects, 5.07[A]
Liquidated damages, 3.02[A][3]
  defined, 3.02[A][3]
  for delay, 5.02, 5.11
  liquidated damages clauses, 6.02[B]
  purpose of, 6.02[B]
  recoverability of, 5.10

Liquidated damages clauses, 6.02[B]
  factors against including clause, 6.02[B]
Liquidated direct damages, 3.02[B][8]
Lost profits, 6.02[A]
  methods available to prove damages,
  4.12

M

Mandatory settlement conferences (MSCs),
  14.04[B][1]
Market value
  diminution in, 5.01[A], 6.02[C]
Mass producers of homes
  strict liability of, 8.02
Material supplier contracts, 3.04
  ambiguous terms, 3.04
  credit application, 3.04
  economic damages, claim for, 3.04
  last shot rule, 3.04
  purchase orders, 3.04
Material suppliers
  bids, 2.01[D]
  Commercial Code requirements, 2.01[D]
  mechanics' liens, 3.03[D]
  quotations, rules for, 2.01[D]
  security on government projects. *See*
    Miller Act bond
  stop notices, 3.03[D]
Materials
  unavailability of materials, 6.07[C]
Mechanics' liens
  bankruptcy and. *See* Bankruptcy
  form, App. 8
  on private works. *See* Mechanics' liens on
    private works
  public works, no rights on, 10.01
Mechanics' liens on private works, 9.01[A]
  amount of lien, 9.06[H], 9.06[O][2]
    changes in contract, effect of,
      9.06[O][2]
    extended overhead, 9.06[H]
    sums over and above contract balance
      due, 9.06[H]
    what amounts are included, 9.06[H]
  assignment for benefit of creditors, effect
    of, 9.05[C], 9.06[B]
  attorneys' fees, includability of, 9.06[H]
  carpeting
    wall-to-wall carpeting, lienability of,
      9.06[C]
  classes of persons entitled to liens,
    9.01[A], 9.01[A][2]

Mechanics' liens on private works
(*continued*)
completion
  notice of completion. *See* notice of
    completion, this heading
  what constitutes completion, 9.03[A],
    9.03[E]
condominiums
  notice of completion, 9.04[E]
constitutionality of, 9.01[A]
contents of lien, 9.03[C], 9.06[O]
  amount of lien, 9.06[O][2]
  description of site, 9.03[C][1],
    9.06[O][1]
  errors in description, 9.03[C][1]
  name of owner/reputed owner,
    9.03[C][2]
  name of person by whom claimant was
    employed/to whom claimant
    furnished materials, 9.03[C][3]
  signature of claimant, 9.03[C][4]
core work vs. tenant work, 9.04[F]
defenses, 9.05
description of site, 9.03[C][1], 9.06[O][i]
  errors in, 9.03[C][1]
design professionals lien, 9.01[A][3]
  conditions precedent to recording of
    lien, 9.01[A][3]
  procedure, 9.01[A][3]
diversion of funds, wrongful, 9.06[E]
economic duress, 9.05[E]
extras, oral agreements for, 9.06[G]
fire, destruction by, 9.06[A]
foreclosure of lien
  arbitration, actions stayed pending,
    9.03[D][2]
  bankruptcy, effect of, 9.03[F]
  jurisdiction, 9.06[K]
  ninety-day foreclosure limit, 9.06[I]
  parties to lien foreclosure action,
    9.06[R]
  removal of lien not foreclosed,
    9.03[D][1]
  time for filing suit, 9.03[D], 9.03[D][2]
  waiver and, 9.03[F]
  where to bring action, 9.06[K]
forfeiture of lien, 9.06[M]
  willful overstatement, 9.06[M]
fraud by lien claimant, 9.05[G]
home improvement contracts, 13.05
implied warranty of design, 9.06[P]
interest, allowability of, 9.06[H], 9.06[J]
joint checks, 9.05[F]
  exceptions to general rule, 9.05[F]
  forged endorsement of, 9.06[L]
  general rule, 9.05[F]
  jurisdictional issues, 9.06[K]
  liberal construction of lien law, 9.06[F]
material suppliers, 3.03[D]
multiple contracts, 9.03[B]
multiple liens on same property for same
  work, 9.03[D], 9.03[D][1]
notice of completion, 9.03[A], 9.03[A][1]
  condominiums, 9.04[E]
  contents, 9.03[A][1]
  core work vs. tenant work, 9.04[F]
  errors in, 9.03[A][1]
  multiple contracts, 9.03[B]
  separate residential units, 9.04[C]
  separate works of improvement,
    9.04[B]
  substantial compliance, 9.03[A][1]
  tenant work vs. core work, 9.04[F]
  when recorded, 9.03[A][1]
  where recorded, 9.03[A][1]
notice of nonresponsibility, 9.01[A][6]
notice requirements. *See* preliminary
  notice, this heading
oral agreements for extras, 9.06[G]
other issues, 9.03[F]
overstatement, willful, 9.06[M]
  forfeiture of lien, 9.06[M]
owner, personal liability of, 9.06[Q]
payments
  pay-if-paid clauses, 9.06[D]
  payments, application of, 9.05[A]
  unreasonable delay in payment, 9.06[D]
persons entitled to file liens, 9.01[A],
  9.01[A][2]
persons not entitled to file liens,
  9.01[A][4]
preliminary notice, 9.02[A]
  cases interpreting notice, 9.02[B]
  contents of notice, 9.02[A][2]
  county recorder, filing with, 9.02[A][4]
  disciplinary action, 9.02[A][3]
  essential provisions, 9.02[A][1][f]
  exceptions to preliminary notice
    requirement, 9.02[B][7]
  failure to give notice, effect of,
    9.02[A][3]
  methods for giving notice,
    9.02[A][1][e]
  oral notice, 9.02[B][7]
  other requirements, 9.02[A][1][f]
  persons required to give notice,
    9.02[A][1][b]
  persons to whom given, 9.02[A][1][c]

INDEX

service of, 9.02[A][1][d], 9.02[A][1][e]
  when given, 9.02[A][1][a]
  where given, 9.02[A][1][d]
priority of claims, 9.04[A]
  commencement for priority purposes, 9.04[A]
  construction loan vs. mechanic's lien, 9.04[A]
  exceptions to general rule, 9.04[A]
  general rule, 9.04[A]
  off-site work under separate contract, 9.04[A]
privilege, absolute, 9.06[N]
problems with, 9.04
procedure for filing liens, 9.02
proof of use of labor and material in job, 9.04[G]
property to which liens may attach, 9.01[A][5]
  Native American–owned property, 9.01[A][5]
purpose of mechanic's lien statutes, 9.03[D][1]
recording the lien, 9.03
  completion, equivalents of, 9.03[A]
  notice of cessation of labor, 9.03[A]
  requirements, 9.03[A]
  timing requirements, 9.03[A]
releases. *See* Waivers and releases of mechanics' liens, stop notices, bond rights
removal of lien not foreclosed, 9.03[D][1]
removal of materials, 9.06[A]
  right to lien, 9.06[A]
restrictive endorsements, 9.05[D]
segregation of lien, 9.04[D]
separate residential units, 9.04[C]
separate works of improvement, 9.04[B]
slander of title, 9.06[N]
statutory implementation, 9.01[A][1]
tenant work vs. core work, 9.04[F]
time for filing liens, 9.03[A]
trust funds, mechanic's lien rights of, 9.01[A][1]
twenty-day notice. *See* preliminary notice, this heading
unclean hands, 9.05[G]
unlicensed contractors, 9.01[A][4]
voidable preferences, 9.06[B]
waivers. *See also* Waivers and releases of mechanics' liens, stop notices, bond rights
  contractual waivers of mechanic's lien remedies, 9.05[B], 9.06[T]

Mediation, 14.03
  admissibility of agreements, 14.03[D]
  advantages of, 14.03[B], 14.03[C]
  briefs, 14.03[A]
  caucuses, 14.03[A]
  disadvantages of, 14.03[B], 14.03[C]
  end of mediation defined, 14.03[D]
  enforceability of mediation settlement agreements, 14.03[D]
  oral agreements, 14.03[D]
  evidence, admissibility of, 14.03[D]
  judicial mediation, 14.04[B][2], 14.04[B][2][b]
  mediation statement, 14.03[A]
  mediators, factors in choosing, 14.03, 14.03[A]
  nonbinding nature, 14.03[B]
  offers to compromise, 14.03[D]
  of owner-contractor disputes, 3.02[B][9]
  process, 14.03[A]
  settlement offers, 14.03[D]
  statutorily required mediation
    Civil Procedure Code, judicial mediation under, 14.04[B][2][b]
Milestone dates, 6.06, 6.07[A]
  interim milestones, 6.02[B]
Miller Act bond
  attorneys' fees, recoverability of, 11.05
  delay damages, recoverability of, 11.05
  equitable lien not allowed, 11.07[E]
  failure of government to require bond, effect of, 11.06
  form, App. 8
  general contractor not indispensable party to suit, 11.07[C]
  illustrative cases, 11.07
  labor and material bond, 11.01
  limitations period, 11.03
  money due, recovery of, 11.07[D]
  ninety-day notice requirement, 11.02, 11.06
    substantial compliance, 11.02
  out-of-pocket costs, 11.05
  pay-if-paid clause, effect of, 11.07[B]
  payment bond, 11.05, 11.07[C]
  persons entitled to recover, 11.04
    claimants beyond third tier, 11.04
  practice notes, 11.06
  procedural steps to enforce claim on bond, 11.01
  prompt payment, surrender of rights to seek, 11.07[B]
  proof required, 11.05

Miller Act bond (*continued*)
   elements material suppliers are required to prove, 11.05
   purpose of, 11.07[B]
   savings clause
      recovery of money due under, 11.07[D]
   sovereign immunity, waiver of, 11.07[E]
   suit on the bond, 11.03, 11.06
      limitations period, 11.03
      where brought, 11.03
   surety, sufficiency of
      claims against person certifying sufficiency, 11.07[A]
   waiver of rights, 11.07[B]
Mini-trials, 14.04, 14.04[C]
Minority-owned business enterprises (MBEs), 2.02[F][4][b]
Mistake
   mutual mistake, rescission for, 3.02[B][12]
Modified total cost method of proof
   quantifying and proving damages, 6.14, 6.14[A]
Multi-prime contracts, 4.14
   benefits of, 4.14
   delay damages on, 4.14
   disclaimers
      of coordination responsibilities, 4.14
      of liability for delay damages, 4.14

N

Negligence
   active negligence, 3.02[B][7]
   architect's liability for, 8.07, 8.10
   construction manager's negligent direction of work, 4.14
   design professional's liability for, 8.07, 8.10
   in disbursing construction loan funds, 9.01[B][2][a], 9.01[B][2][b]
   engineer's liability for, 8.07, 8.10
   indemnitee's negligence, 3.02[B][7]
   land surveyor's liability for, 8.07, 8.10
   passive negligence, 3.02[B][7]
   in preparation of documents, 8.07
   sole negligence, meaning of, 3.02[B][7]
   tort recovery for, 8.07
No damages for delay clauses, 3.02[B][11], 4.13, 6.08[A]
   active interference exception, 4.14
   enforceability of, 6.07[H]
   on public projects, 6.07[H]

Noncompensable delays, 6.07, 6.07[B], 6.07[D], 6.07[E]
   concurrent delays, 6.07[E]
Nonpayment, 4.03
   contractor's right to stop private work of improvement, 4.03
   notice requirements, 4.03
   pay-if-paid clauses. *See* Pay-if-paid clauses
   stop work remedy (private works of improvement), 4.03
Nonresponsive bid. *See* Bidding procedures for public works
Notice of completion, 9.03[A][1]
   condominiums, 9.04[E]
   contents, 9.03[A][1]
   core work vs. tenant work, 9.04[F]
   errors in, 9.03[A][1]
   multiple contracts, 9.03[B]
   separate residential units, 9.04[C]
   separate works of improvement, 9.04[B]
   substantial compliance, 9.03[A][1]
   tenant work vs. core work, 9.04[F]
   when recorded, 9.03[A][1]
   where recorded, 9.03[A][1]
Notice of nonresponsibility, 9.01[A][6]
Notice requirements generally
   contract, notice required by, 1.08
   license law, notice required by, 1.07

O

Or equal clause, 2.02[C][1]
Overhead expenses
   delay damages for home office overhead expenses, 6.07[G]
Owner-architect agreement, 3.05
   limitation of liability, 3.05[A]
   project administration, 3.05
Owner-contractor agreement, 3.02[A]
   arbitration. *See* Arbitration; Dispute resolution
   claims, presentation of, 3.02[B][8]
      burden of proof, 3.02[B][8]
      consequential damages, waiver of, 3.02[B][8]
      time limit on claims, 3.02[B][8]
   concealed conditions, 3.02[B][12]
      time limitations, 3.02[B][12]
   contract price, 3.02[A][4]
   contractor's review of documents, field conditions, 3.02[B][3]
      purpose of review, 3.02[B][3]

INDEX

correction of work, 3.02[B][4]
damages. *See* Damages
delays, 3.02[B][10]
description of work, 3.02[A][2]
design delegation, 3.02[B][6]
dispute resolution. *See* Arbitration;
    Dispute resolution; Mediation
extensions of time, 3.02[A][3]
general conditions, 3.02[B]. *See also* AIA
    Document A201, General Conditions
    of the Contract for Construction
guarantees, 3.02[B][4]
identity of parties, 3.02[A][1]
indemnity. *See* Indemnity obligations
insurance. *See* Insurance
integration clause, 3.02[B][1]
interpretation of contract documents,
    3.02[B][2]
    reasonableness standard, 3.02[B][2]
    time allowed, 3.02[B][2]
mediation. *See* Dispute resolution;
    Mediation
no damages for delay clause, 3.02[B][11]
notice requirements, 3.02[B][14]
payment terms. *See* Payment terms
scope of work, 3.02[A][2]
subcontractor's adoption of contract
    documents, 3.02[B][5]
suspension of contract, 3.02[B][15]
termination of contract, 3.02[B][15]
    for convenience, 3.02[B][15]
time for performance, 3.02[A][3]
    time extensions, 3.02[B][10]
unknown conditions, 3.02[B][12]
    time limitations, 3.02[B][12]
warranties, 3.02[B][4]
    express warranties, 3.02[B][4]

P

Partnering arrangements, 14.04
Partnership
    citations, continuing responsibility for,
        1.19
    disassociation of person from license,
        effect of, 1.19
    license application by, 1.12
Patent defects, 5.07[A], 5.07[C]
    defined, 5.07[A]
    limitations period, 5.07[A]
Pay-if-paid clauses, 1.09[C], 3.03[B],
    9.06[T], 11.07[B]
    unenforceability of, 4.03, 9.06[D]

Pay-when-paid clauses, 1.09[C], 3.03[B]
Payment bond on public works
    amount of bond, calculating, 10.03[C]
    attorneys' fees, 10.04[A]
        on bond on public works, 10.03[C],
        10.04[B]
    notice to principal and surety
        form, App. 10
    preliminary notice, 9.03[H][1]
    proper claimants, 10.03[C], 10.04[A]
    surety's right to raise principal's defenses,
        10.04[C]
Payment (labor and material) bonds on
        private works, 9.01[C], 9.03[H]
    cases interpreting bond claims,
        9.03[H][3]
    notice to principal and surety
        form, App. 9
    off-site bonds, 9.01[C][2]
        release of bond, 9.01[C][2]
    perfecting payment bond claims,
        9.01[C][1]
    preliminary notice, 9.03[H][1]
    recovering on bond, 9.01[C][1]
    releases. *See* Waivers and releases of
        mechanics' liens, stop notices, bond
        rights
    stop notice, effect on, 9.01[B][2][f],
        9.03[G]
    time for filing suit, 9.03[H][2]
    waivers. *See* Waivers and releases of
        mechanics' liens, stop notices, bond
        rights
Payment terms
    contractor-subcontractor agreement,
        3.03[B]
        mechanic's lien releases, 3.03[B]
        pay-when-paid clause, 3.03[B]
    owner-contractor agreement, 3.02[A][5]
        final payment, 3.02[A][5]
        progress payments, 3.02[A][5]
        retention, 3.02[A][5]
Payments. *See also* Breach of contract by
        owner
    application of, 9.05[A]
    nonpayment. *See* Nonpayment
    pay-if-paid clauses. *See* Pay-if-paid
        clauses
    pay-when-paid clauses. *See* Pay-when-
        paid clauses
    progress payments, 1.09[C], 3.02[A][5],
        4.02, 4.04
    prompt payment legislation. *See* Prompt
        payment legislation

Payments (*continued*)
   retention payments, 1.09[C]
   retention, release of, 4.04
Performance bond, 3.03[D]
Performance of contract
   continued performance notwithstanding disputes, 4.02, 4.07[E]
   partial performance, damages for, 6.02[C]
Plans and specifications
   ambiguities in, 4.07[D]
      construction of, against drafter, 4.07[D]
      contra proferentem, rule of, 4.07[D]
      patent ambiguities, 4.07[D]
   contractor's failure to perform per, 6.02[C]
      cost of correction rule, 6.02[C]
      diminution in market value, 6.02[C]
   defective plans and specifications
      damages for, 4.06
      no liability of contractor for resulting defective product, 4.07[B]
   implied warranty of accuracy, 4.06, 4.07[A]
   interpretations of, by courts, 4.07
Preferences
   bankruptcy and, 12.10
   bidding procedures. *See* Bidding procedures for public works
   voidable preferences. *See* Voidable preferences
Preliminary notice
   form, with proof of service, App. 5
   private works. *See* Mechanics' liens on private works; Stop notices on private works
   public works. *See* Stop notices on public works
Private trials, 14.04, 14.04[C]
   advantages of process, 14.04[C]
Private works of improvement
   bidding procedures. *See* Bidding procedures for private works
   remedies available. *See* Remedies
Privity doctrine, 8.01
Profits, loss of
   for delay, 5.04
Progress payments, 1.09[C], 3.02[A][5], 4.02, 4.04
   release of, 4.04
Project delivery systems
   design-bid-build, 2.02[B][1], 2.02[B][3], 3.06
   design-build. *See* Design-build contracts
   fast-track projects, 4.14
   multi-prime contracts. *See* Multi-prime contracts
Project labor agreements
   nonunion contractors, challenges by, 2.02[F][4][a]
   use of, 2.02[F][4][a]
Project Management Protective Liability Insurance, 3.02[B][7], 3.02[B][13]
Promissory estoppel, doctrine of, 2.01[C]
   refusal to honor bid and, 2.01[C]
Prompt payment legislation, 4.04
   local governments, payments by, 4.04
   private works of improvement, 4.04
   public works of improvement, 4.04
   purpose of, 4.04
   state agencies, payments by, 4.04
   waiver of requirements, prohibition against, 4.04
Property insurance, 3.02[B][13]
Public liability insurance, 3.02[B][13]
Public records
   access to, 1.12, 2.02[G][1]
Public works
   bidding procedures. *See* Bidding procedures for public works
   bonds
      labor bonds. *See* Labor and material bond, public works
      payment bonds. *See* Payment bonds on public works
   license law. *See* License law
   mechanics' liens, no rights to, 10.01
   stop notices. *See* Stop notices on public works
Punitive damages
   for fraud, 5.05, 5.11, 6.02[A]
Purchase orders, 3.04

Q

Qui tam statutes, 6.03. *See also* False claims statutes

R

Rejection of contract. *See* Bankruptcy
Release bond, 9.06[5]
   enforcement of, 9.06[5]
   notice of recording, 9.06[5]
      failure to give notice, 9.06[5]

INDEX

Remedies
  for breach of contract. *See* Breach of contract by contractor; Breach of contract by owner
  damages. *See* Damages
  private works of improvement, 9.01
    bonds. *See* Payment (labor and material) bonds on private works
    mechanics' liens. *See* Mechanics' liens on private works
    stop notices. *See* Stop notices on private works
  public works of improvement
    bonds. *See* Labor and material bond, public works; Payment bonds on public works
    stop notices. *See* Stop notices on public works
Rents, loss of
  for delay, 5.03
Rescission of contract, 2.02[E], 3.02[B][12], 6.04, 6.04[B]
  for cause, 6.04
  extreme impracticability, 3.02[B][12]
  home solicitation contracts. *See* Home improvement contracts
  impossibility of performance, 3.02[B][12]
  mutual mistake, 3.02[B][12]
  termination compared, 6.04[B]
Residential construction contracts. *See also* Home improvement contracts
  arbitration clauses, 14.04[D]
    notice required, 14.04[D]
    typeface requirements, 14.04[D]
  single-family dwellings, 1.16
Responsible managing employee (RME), 1.06[B][2], 1.12
  citations, continuing responsibility for, 1.19
  disassociation of, from license, 1.19
Responsible managing officer (RMO), 1.06[B][2], 1.12
  citations, continuing responsibility for, 1.19
  disassociation of, from license, 1.19
Restrictive endorsements, 9.05[D]
Retail Sales Installment Act, 13.05
Retention payments, 1.09[C], 3.02[A][5]
  release of retention, 4.04, 4.05[B][4][b]
Revocation of license, 1.10, 1.11[B], 1.23
  disclosure requirement, 1.23
    penalties for violations, 1.23
  for failure to produce records, 1.15

S

Setoff, 12.07, 12.10
  elements necessary to justify setoff, 12.07
  waiver of right to claim setoff, 12.07
Settlement agreements, 14.03[D], 14.05
  enforcement of, 14.04, 14.05
  general releases, 14.04
  judgment by stipulation, 14.05
  oral settlement agreements, 14.05
  settlement of part of but not all disputes, 14.04
Single-family dwellings
  written contract required, 1.16
Slander of title, 9.06[N]
Sovereign immunity
  action seeking relief "other than money damages," 11.07[E]
  waiver of, 11.07[E]
Spearin doctrine, 2.01[A]
Specialty contractors, 1.02
Specifications. *See* Plans and specifications
Statutes of repose
  for construction defects suits, 5.07[A]
Stigma damages
  recoverability of, 8.13
Stop notices
  form, App. 7
  material suppliers, 3.03[D]
  private works. *See* Stop notices on private works
  public works. *See* Stop notices on public works
Stop notices on private works, 9.01[B]
  assignment for benefit of creditors, effect of, 9.05[C]
  attorneys' fees, entitlement to, 9.03[G][4]
    on bonded stop notice, 9.01[B][3]
  bonded stop notice, 9.01[B][3], 9.03[G][4]
  case law interpretation, 9.01[B][2]
  contents, 9.01[B][1], 9.03[G][3]
  defending claims, 9.01[B][2][e], 9.05
  economic duress, 9.05[E]
  enforcing stop notice, 9.03[G][5]
  equitable tolling, doctrine of, 9.03[G][5]
  elements of, 9.03[G][5]
  excessive disbursements, liability for, 9.01[B][2][b]
  general contractor's rights, 9.01[B][2][d]
  improper disbursement of loan funds, 9.01[B][2][b]
  interest, entitlement to, 9.03[G][4], 9.06[J]

Stop notices on private works (*continued*)
joint checks, 9.05[F]
exceptions to general rule, 9.05[F]
forged endorsement of, 9.06[L]
general rule, 9.05[F]
liberal construction of law, 9.03[G][4]
negligence in disbursing construction loan funds, 9.01[B][2][a], 9.01[B][2][b]
policy considerations regarding duty of care, 9.01[B][2][b]
payment bond, effect of, on stop notices, 9.01[B][2][f]
payments, application of, 9.05[A]
preliminary notice, 9.02[A]
cases interpreting notice, 9.02[B]
contents of notice, 9.02[A][2]
county recorder, filing with, 9.02[A][4]
disciplinary action, 9.02[A][3]
essential provisions, 9.02[A][1][f]
exceptions to preliminary notice requirement, 9.02[B][7]
failure to give notice, effect of, 9.02[A][3]
methods for giving notice, 9.02[A][1][e]
oral notice, 9.02[B][7]
other requirements, 9.02[A][1][f]
persons required to give notice, 9.02[A][1][b]
persons to whom given, 9.02[A][1][c]
service of, 9.02[A][1][d], 9.02[A][1][e], 9.03[G]
when given, 9.02[A][1][a]
where given, 9.02[A][1][d]
procedure for filing stop notices, 9.03[G]
releases. *See* Waivers and releases of mechanics' liens, stop notices, bond rights
restrictive endorsements, 9.05[D]
service of stop notice, 9.01[B][1], 9.03[G][2]
methods of service, 9.03[G][2]
persons upon whom notice must be served, 9.03[G][2]
when service is complete, 9.03[G][4]
where served, 9.03[G][2]
soft costs
types subject to stop notice, 9.01[B][2][a]
statute of limitations
equitable tolling doctrine, effect of, 9.03[G][5]
time for filing suit, 9.03[G][4], 9.03[G][5]
trust funds, construction funds as, 9.03[G][1]
waivers. *See* Waivers and releases of mechanics' liens, stop notices, bond rights
Stop notices on public works
completion vs. cessation of work, 10.03[B]
court decisions, examples, 10.04[A]
interest, recoverability of
interest earned on funds withheld, 10.02[C]
interest on unpaid balance due, 10.04[A]
interpleader, 10.04[A]
preliminary 20-day notice, 10.02[A]
contents of, 10.02[A][7]
disciplinary action, 10.02[A][8]
methods of giving notice, 10.02[A][4]
persons responsible for giving notice, 10.02[A][1]
persons to whom notice is given, 10.02[A][3]
service, completion of, 10.02[A][6]
timing of notice, 10.02[A][2]
where given, 10.02[A][5]
procedural steps, 10.02
release of funds, 10.02[C]
release of stop notice, 10.02[C]
by bond, 10.02[C]
burden of proof, 10.02[C]
by summary proceedings, 10.02[C]
requirements for stop notice, 10.02[B]
contents, 10.02[B][1]
methods for serving notice, 10.02[B][3]
persons upon whom notice is served, 10.02[B][4]
persons who may serve notice, 10.02[B][2]
service of, 10.02[B][2], 10.02[B][3], 10.02[B][4], 10.02[B][5]
timing of service, 10.02[B][5]
suit on stop notice, 10.02[D]
notice of, 10.02[D]
timing of, 10.02[D]
Stop work remedy, 4.03
Strikes, 6.07[C]
Subcontractors
AIA Document A401, Standard Form Subcontract, 3.03
contractor-subcontractor agreement. *See* Subcontracts
subcontractor listing law. *See* Bidding procedures on public works

INDEX

security on government projects. *See* Miller Act bond
termination of subcontractor, 3.03[F]
Subcontracts. *See also* AIA Document A401, Standard Form Subcontract
adoption of contract documents, 3.03[A]
bonding requirements, 3.03[D]
flow-down provisions, 3.03[A]
indemnity agreement. *See* Indemnity obligations
material breach, effect of, 3.03[F]
most widely used forms, 3.03
obligation to continue work if dispute arises, 3.03[E]
pay-when-paid clause, 3.03[B]
payment terms, 3.03[B]
termination of subcontractor, 3.03[F]
Subletting and Subcontracting Fair Practices Act, 2.02[B][3], 2.02[D][2]
purpose of Act, 2.02[D][2]
Substantial completion, 5.01[B]
Substantial compliance, doctrine of
licensing requirements and, 1.06[B][2]
Substantial performance, doctrine of, 5.01[B]
defined, 5.01[B]
Sureties
license requirement for, 1.24
exemption, availability of, 1.24
right of, to raise principal's defenses, 10.04[C]
Suspension of contract, 3.02[B][15]
Suspension of license, 1.10, 1.11[B], 1.23
disclosure requirement, 1.23
penalties for violations, 1.23
Suspension of work, 6.10, 6.12
constructive suspension of work, 6.12
examples of, 6.12
notice requirement, 6.12
Swimming pool construction contracts, 13.04
attorneys' fees, recoverability of, 13.04
bond, contractor's, 1.14
notice to owner, 1.07
penalties for failure to comply with law, 13.04
requirements of contractor, 13.04

T

Termination of contract, 3.02[B][15]
for convenience, 3.02[B][15]
Termination of contractor, 6.04
for cause, 6.04
completion of project, 6.04
for convenience, 6.04
damages recoverable from contractor, 6.04[A]
materiality of breach, 6.04
notice requirements, 6.04
public works contracts, 6.04
rescission of contract compared, 6.04[B]
Termination of subcontractor, 3.03[F]
Terminology. *See* Ambiguities; Definitions
Time extensions. *See also* Delays
bad weather, 3.02[B][10]
Tort recovery
for bad-faith denial of contract, 8.09
for contract breaches, 8.09
for negligence, 8.07
Total cost method of proof
criteria necessary to prevail, 6.14[B]
pros and cons of, 6.14[A]
quantifying and proving damages, 4.12, 6.14, 6.14[A]
Trial Delay Reduction Act, 14.01
Truth-in-Lending Act, 13.05

U

Unclean hands, doctrine of, 9.05[G]
Unfair Practices Act, 9.03[H][3]
improper conduct, types of, 9.03[H][3]
Unknown/unusual conditions, 3.02[B][12]. *See also* Differing site conditions
time limitations, 3.02[B][12]

V

Voidable preferences, 9.06[B]
assignment for benefit of creditors, 9.06[B]
new value, contemporaneous exchange for, 9.06[B]
ordinary course of business, transfers in, 9.06[B]
Voluntary settlement conferences (VSCs), 14.04[B][1]

W

Waivers
of consequential damages, 3.02[B][8]
foreclosure of lien and, 9.03[F]

Waivers (*continued*)
   of immaterial deviations in bid, 2.02[F][1]
   of Miller Act rights, 11.07[B]
   prompt payment requirements, 4.04
   of right to arbitrate, 14.02[B], 14.02[E], 14.04[B][2][c]
   of right to claim setoff, 12.07
   of sovereign immunity, 11.07[E]
Waivers and releases of mechanics' liens, stop notices, bond rights, 1.09[C]
   ambiguity of unconditional release form, 9.05[B][4][c]
   assignment for benefit of creditors, effect of, 9.05[C]
   contractual waivers of mechanic's lien remedies, 9.05[B], 9.06[T]
   definitions, 9.05[B][4][a]
   developments, 9.05[B][3]
   economic duress, 9.05[E]
   exceptions to release, 9.05[B][4][b]
      contract rights, 9.05[B][4][b]
      extras, 9.05[B][4][b]
      items, 9.05[B][4][b]
      retentions, 9.05[B][4][b]
      written change orders specifically reserved, 9.05[B][4][b]
   forms of waiver and release
      analysis of, 9.05[B][4]
      Conditional Waiver and Release upon Final Payment, 9.05[B], 9.05[B][1], App. 3
      Conditional Waiver and Release upon Progress Payment, 9.05[B], 9.05[B][1], 9.05[B][2], App. 1
      Unconditional Waiver and Release upon Final Payment, 9.05[B], 9.05[B][1], App. 4
      Unconditional Waiver and Release upon Progress Payment, 9.05[B], 9.05[B][1], App. 2
   Halbert's Lumber decision, 9.05[B][2]
   historical background, 9.05[B][1]
   lien waiver clauses, 9.06[T]
   pay-if-paid clauses, 9.06[T]
   recommendations, 9.05[B][5]
   release bond. *See* Release bond
   release forms, analysis of, 9.05[B][4]
   requirements and form of written waiver and release, 9.05[B]. *See also* forms of waiver and release, this heading
   rules, 9.05[B]
   unconditional release form, ambiguity of, 9.05[B][4][c]
Weather delays, 3.02[B][10]
   unusually severe weather, 6.07[C]
Whistle-blower provisions
   false claims statutes, 6.03
Windstorm
   insurance coverage, 3.02[B][13]
Witness immunity doctrine, 8.11
   absolute immunity, 8.11
Women-owned business enterprises (WBEs), 2.02[F][4][b]
Workers' compensation
   filing false statement, 1.09[K]
   insurance, 3.02[B][13]